GLOBAL POSITIONING SYSTEMS, INERTIAL NAVIGATION, AND INTEGRATION

The Wiley Bicentennial–Knowledge for Generations

Each generation has its unique needs and aspirations. When Charles Wiley first opened his small printing shop in lower Manhattan in 1807, it was a generation of boundless potential searching for an identity. And we were there, helping to define a new American literary tradition. Over half a century later, in the midst of the Second Industrial Revolution, it was a generation focused on building the future. Once again, we were there, supplying the critical scientific, technical, and engineering knowledge that helped frame the world. Throughout the 20th Century, and into the new millennium, nations began to reach out beyond their own borders and a new international community was born. Wiley was there, expanding its operations around the world to enable a global exchange of ideas, opinions, and know-how.

For 200 years, Wiley has been an integral part of each generation's journey, enabling the flow of information and understanding necessary to meet their needs and fulfill their aspirations. Today, bold new technologies are changing the way we live and learn. Wiley will be there, providing you the must-have knowledge you need to imagine new worlds, new possibilities, and new opportunities.

Generations come and go, but you can always count on Wiley to provide you the knowledge you need, when and where you need it!

William J. Pesce
President and Chief Executive Officer

Peter Booth Wiley
Chairman of the Board

GLOBAL POSITIONING SYSTEMS, INERTIAL NAVIGATION, AND INTEGRATION

SECOND EDITION

MOHINDER S. GREWAL
LAWRENCE R. WEILL
ANGUS P. ANDREWS

WILEY-INTERSCIENCE
A John Wiley & Sons, Inc., Publication

Published by John Wiley & Sons, Inc., Hoboken, New Jersey.
Published simultaneously in Canada.

Library of Congress Cataloging-in-Publication Data is available.

ISBN-13 978-0-470-04190-1
ISBN-10 0-470-04190-0

Printed in the United States of America.

10 9 8 7 6

M. S. G. dedicates this book to the memory of his parents, Livlin Kaur and Sardar Sahib Sardar Karam Singh Grewal.
L. R. W. dedicates his work to his late mother, Christine R. Weill, for her love and encouragement in pursuing his chosen profession.
A. P. A. dedicates his work to his wife Jeri, without whom it could not have been done.

CONTENTS

PREFACE TO THE SECOND EDITION

This book is intended for people who need to combine global navigation satellite systems (GNSSs), inertial navigation systems (INSs), and Kalman filters. Our objective is to give our readers a working familiarity with both the *theoretical* and *practical* aspects of these subjects. For that purpose we have included "real-world" problems from practice as illustrative examples. We also cover the more practical aspects of implementation: how to represent problems in a mathematical model, analyze performance as a function of model parameters, implement the mechanization equations in numerically stable algorithms, assess its computational requirements, test the validity of results, and monitor performance in operation with sensor data from GNSS and INS. These important attributes, often overlooked in theoretical treatments, are essential for effective application of theory to real-world problems.

The accompanying CD-ROM contains MATLAB m-files to demonstrate the workings of the Kalman filter algorithms with GNSS and INS data sets, so that the reader can better discover how the Kalman filter works by observing it in action with GNSS and INS. The implementation of GNSS, INS, and Kalman filtering on computers also illuminates some of the practical considerations of finite-wordlength arithmetic and the need for alternative algorithms to preserve the accuracy of the results. Students who wish to apply what they learn, must experience all the workings and failings of Kalman Filtering—and learn to recognize the differences.

The book is organized for use as a text for an introductory course in GNSS technology at the senior level or as a first-year graduate-level course in GNSS, INS, and Kalman filtering theory and application. It could also be used for self-instruction or review by practicing engineers and scientists in these fields.

This second edition includes some significant changes in GNSS/INS technology since 2001, and we have taken advantage of this opportunity to incorporate many of the improvements suggested by reviewers and readers. Changes in this second edition include the following:

1. New signal structures for GPS, GLONASS, and Galileo
2. New developments in augmentation systems for satellite navigation, including
   (a) Wide-area differential GPS (WADGPS)
   (b) Local-area differential GPS (LADGPS)
   (c) Space-based augmentation systems (SBASs)
   (d) Ground-based augmentation systems (GBASs)
3. Recent improvements in multipath mitigation techniques, and new clock steering algorithms
4. A new chapter on satellite system integrity monitoring
5. More thorough coverage of INS technology, including development of error models and simulations in MATLAB for demonstrating system performance
6. A new chapter on GNSS/INS integration, including MATLAB simulations of different levels of tight/loose coupling

The CD-ROM enclosed with the second edition has given us the opportunity to incorporate more background material as files. The chapters have been reorganized to incorporate the new material.

Chapter 1 informally introduces the general subject matter through its history of development and application. Chapters 2–7 cover the basic theory of GNSS and present material for a senior-level class in geomatics, electrical engineering, systems engineering, and computer science.

Chapters 8–10 cover GNSS and INS integration using Kalman filtering. These chapters could be covered in a graduate-level course in electrical, computer, and systems engineering. Chapter 8 gives the basics of Kalman filtering: linear optimal filters, predictors, nonlinear estimation by "extended" Kalman filters, and algorithms for MATLAB implementation. Applications of these techniques to the identification of unknown parameters of systems are given as examples. Chapter 9 is a presentation of the mathematical models necessary for INS implementation and error analysis. Chapter 10 deals with GNSS/INS integration methods, including MATLAB implementations of simulated trajectories to demonstrate performance.

MOHINDER S. GREWAL, PH.D., P.E.
CALIFORNIA STATE UNIVERSITY AT FULLERTON

LAWRENCE R. WEILL, PH.D.
CALIFORNIA STATE UNIVERSITY AT FULLERTON

ANGUS P. ANDREWS, PH.D.
ROCKWELL SCIENCE CENTER (RETIRED) THOUSAND OAKS, CALIFORNIA

ACKNOWLEDGMENTS

M. S. G. acknowledges the assistance of Mrs. Laura Cheung, graduate student at California State University at Fullerton, for her expert assistance with the MATLAB programs, and Dr. Jya-Syin Wu of the Boeing Company for her assistance in reviewing the earlier manuscript.

L. R. W. is indebted to the people of Magellan Navigation who so willingly shared their knowledge of the Global Positioning System during the development of the first handheld receiver for the consumer market.

A. P. A. thanks Captains James Black and Irwin Wenzel of American Airlines for their help in designing the simulated takeoff and landing trajectories for commercial jets, and Randall Corey from Northrop Grumman and Michael Ash from C. S. Draper Laboratory for access to the developing Draft IEEE Standard for Inertial Sensor Technology. He also thanks Dr. Michael Braasch at GPSoft, Inc. for providing evaluation copies of the GPSoft INS and GPS MATLAB Toolboxes.

ACRONYMS AND ABBREVIATIONS

A/D	Analog-to-digital (conversion)
ADC	Analog-to-digital converter
ADR	Accumulated delta range
ADS	Automatic dependent surveillance
AGC	Automatic gain control
AHRS	Attitude and heading reference system
AIC	Akaike information-theoretic criterion
AIRS	Advanced inertial reference sphere
ALF	Atmospheric loss factor
ALS	Autonomous landing system
altBOC	Alternate binary offset carrier
AODE	Age of data word, ephemeris
AOR-E	Atlantic Ocean Region East (WAAS)
AOR-W	Atlantic Ocean Region West (WAAS)
AR	Autoregressive
ARMA	Autoregressive moving average
ASD	Amplitude spectral density
ASIC	Application-specific integrated circuit
ASQF	Application-Specific Qualification Facility (EGNOS)
A-S	Antispoofing
ATC	Air traffic control
BOC	Binary offset carrier
BPSK	Binary phase-shift keying
C/A	Coarse acquisition (channel or code)
C&V	Correction and verification (WAAS)
CDM	Code-division multiplexing

CDMA	Code-division multiple access
CEP	Circle error probable
CNMP	Code noise and multipath
CONUS	Conterminous United States, also Continental United States
CORS	Continuously operating reference station
COSPAS	Acronym from transliterated Russian title "Cosmicheskaya Sistyema Poiska Avariynich Sudov," meaning "Space System for the Search of Vessels in Distress"
CPS	Chips per second
CRC	Cyclic redundancy check
CWAAS	Canadian WAAS
DGNSS	Differential GNSS
DGPS	Differential GPS
DME	Distance measurement equipment
DOD	Department of Defense (USA)
DOP	Dilution of precision
ECEF	Earth-centered, earth-fixed (coordinates)
ECI	Earth-centered inertial (coordinates)
EGNOS	European (also Geostationary) Navigation Overlay System
EIRP	Effective isotropic radiated power
EMA	Electromagnetic accelerator
EMA	Electromagnetic accelerometer
ENU	East–north–up (coordinates)
ESA	European Space Agency
ESG	Electrostatic gyroscope
ESGN	Electrically Supported Gyro Navigation (System; USA)
EU	European Union
EWAN	EGNOS Wide-Area (communication) Network (EGNOS)
FAA	Federal Aviation Administration (USA)
FEC	Forward error correction
FLL	Frequency-lock loop
FM	Frequency modulation
FOG	Fiberoptic gyroscope
FPE	Final prediction error (Akaike's)
FSLF	Free-space loss factor
FT	Feet
GAGAN	GPS & GEO Augmented Navigation (India)
GBAS	Ground-based augmentation system
GCCS	GEO communication and control segment
GDOP	Geometric dilution of precision

GEO	Geostationary earth orbit
GES	GPS Earth Station COMSAT
GIC	GPS Integrity Channel
GIPSY	GPS Infrared Positioning System
GIS	Geographic information system(s)
GIVE	Grid ionosphere vertical error
GLONASS	Global Orbiting Navigation Satellite System
GNSS	Global navigation satellite system
GOA	GIPSY/OASIS analysis
GPS	Global Positioning System
GUS	GEO uplink subsystem
GUST	GEO uplink subsystem type 1
HDOP	Horizontal dilution of precision
HMI	Hazardously misleading information
HOW	Handover word
HRG	Hemispheric resonator gyroscope
ICAO	International Civil Aviation Organization
ICC	Ionospheric correction computation
IDV	Independent Data Verification (of WAAS)
IF	Intermediate frequency
IFOG	Integrating or interferometric Fiberoptic gyroscope
IGP	Ionospheric grid point (for WAAS)
IGS	International GNSS Service
ILS	Instrument landing system
IMU	Inertial measurement unit
Inmarsat	International Mobile (originally "Maritime") Satellite Organization
INS	Inertial navigation system
IODC	Issue of data, clock
IODE	Issue of data, ephemeris
IONO	Ionosphere, Ionospheric
IOT	In-orbit test
IRU	Inertial reference unit
ISA	Inertial sensor assembly
ITRF	International Terrestrial Reference Frame
JPALS	Joint precision approach and landing system
JTIDS	Joint Tactical Information Distribution System
LAAS	Local-Area Augmentation System
LADGPS	Local-area differential GPS
LD	Location determination
LEM	Lunar Excursion module
LHCP	Left-hand circularly polarized
LORAN	Long-range navigation
LOS	Line of sight
LPV	Lateral positioning with vertical guidance

LSB	Least significant bit
LTP	Local tangent plane
M	Meter
MBOC	Modified BOC
MCC	Mission/Master Control Center (EGNOS)
MCPS	Million Chips Per Second
MEDLL	Multipath-estimating delay-lock loop
MEMS	Microelectromechanical system(s)
ML	Maximum likelihood
MLE	Maximum-likelihood estimate (or estimator)
MMSE	Minimum mean-squared error (estimator)
MMT	Multipath mitigation technology
MOPS	Minimum Operational Performance Standards
MSAS	MTSAT Satellite-based Augmentation System (Japan)
MTSAT	Multifunctional Transport Satellite (Japan)
MVUE	Minimum-variance unbiased estimator
MWG	Momentum wheel gyroscope
NAS	National Airspace System
NAVSTAR	Navigation system with time and ranging
NCO	Numerically controlled oscillator
NED	North–east–down (coordinates)
NGS	National Geodetic Survey (USA)
NLES	Navigation Land Earth Station(s) (EGNOS)
NPA	Nonprecision approach
NSRS	National Spatial Reference System
NSTB	National Satellite Test Bed
OASIS	Orbit analysis simulation software
OBAD	Old but active data
OD	Orbit determination
OPUS	Online Positioning User Service (of NGS)
OS	Open service (of Galileo)
PA	Precision approach
PACF	Performance Assessment and Checkout Facility (EGNOS)
P-code	Precision code
pdf	portable document format
PDOP	Position dilution of precision
PI	Proportional and integral (controller)
PID	Process Input Data (of WAAS); Proportional, integral, and differential (control)
PIGA	Pulse integrating gyroscopic accelerometer
PLL	Phase-lock loop
PLRS	Position Location and Reporting System (U.S. Army)
PN	Pseudorandom noise
POR	Pacific Ocean Region

PPS	Precise Positioning Service
PPS	Pulse(s) per second
PR	Pseudorange
PRN	Pseudorandom noise or pseudorandom number (=SVN for GPS)
PRS	Public Regulated service (of Galileo)
PSD	Power spectral density
RAG	Receiver antenna gain (relative to isotropic)
RAIM	Receiver autonomous integrity monitoring
RF	Radiofrequency
RHCP	Right-hand circularly polarized
RIMS	Ranging and Integrity Monitoring Station(s) (EGNOS)
RINEX	Receiver independent exchange format (for GPS data)
RLG	Ring laser gyroscope
RMA	Reliability, maintainability, availability
RMS	Root-mean-squared; reference monitoring station
RPY	Roll–pitch–yaw (coordinates)
RTCA	Radio Technical Commission for Aeronautics
RTCM	Radio Technical Commission for Maritime Service
RTOS	Real-time operating system
RVCG	Rotational vibratory coriolis gyroscope
s	second
SA	Selective availability (also abbreviated "S/A")
SAR	Search and Rescue (service; of Galileo)
SARP	Standards and Recommended Practices (Japan)
SARSAT	Search and rescue satellite–aided tracking
SAW	Surface acoustic wave
SBAS	Space-based augmentation system
SBIRLEO	Space-based infrared low earth orbit
SCOUT	Scripps coordinate update tool
SCP	Satellite Correction Processing (of WAAS)
SF	Scale Factor
SI	System International
SIS	Signal in space
SM	Solar magnetic
SNAS	Satellite Navigation Augmentation System (China)
SNR	Signal-to-noise ratio
SOL	Safety of Life Service (of Galileo)
SPS	Standard Positioning Service
STF	Signal Task Force (of Galileo)
SV	Space vehicle
SVN	Space vehicle number (= PRN for GPS)
TCS	Terrestrial communications subsystem (for WAAS)
TCXO	Temperature-compensated Xtal (crystal) oscillator
TDOA	Time difference of arrival

TDOP	Time dilution of precision
TEC	Total electron content
TECU	Total electron content units
TLM	Telemetry word
TOA	Time of arrival
TOW	Time of week
TTA	Time to alarm
TTFF	Time to first fix
UDRE	User differential range error
UERE	User-equivalent range error
URE	User range error
USAF	United States Air Force
USN	United States Navy
UTC	Universal Time, Coordinated (or Coordinated Universal Time)
UTM	Universal Transverse Mercator
VAL	Vertical alert limit
VCG	Vibratory coriolis gyroscope
VDOP	Vertical dilution of precision
VHF	Very high frequency (30–300 MHz)
VOR	VHF Omnirange (radionavigation aid)
VRW	Velocity Random Walk
WAAS	Wide-Area Augmentation System (U.S.)
WADGPS	Wide-area differential GPS
WGS	World Geodetic System
WMS	Wide-area Master Station
WN	Week number
WNT	WAAS network time
WRE	Wide-area reference equipment
WRS	Wide-area Reference Station
ZLG	Zero-Lock Gyroscope ("Zero Lock Gyro" and "ZLG" are trademarks of Northrop Grumman Corp.)

1

INTRODUCTION

There are five basic forms of navigation:

1. *Pilotage*, which essentially relies on recognizing landmarks to know where you are and how you are oriented. It is older than humankind.
2. *Dead reckoning*, which relies on knowing where you started from, plus some form of heading information and some estimate of speed.
3. *Celestial navigation*, using time and the angles between local vertical and known celestial objects (e.g., sun, moon, planets, stars) to estimate orientation, latitude, and longitude [186].
4. *Radio navigation*, which relies on radiofrequency sources with known locations (including global navigation satellite systems satellites).
5. *Inertial navigation*, which relies on knowing your initial position, velocity, and attitude and thereafter measuring your attitude rates and accelerations. It is the only form of navigation that does not rely on external references.

These forms of navigation can be used in combination as well [18, 26, 214]. The subject of this book is a combination of the fourth and fifth forms of navigation using Kalman filtering.

1.1 GNSS/INS INTEGRATION OVERVIEW

Kalman filtering exploits a powerful synergism between the *global navigation satellite systems* (GNSSs) and an *inertial navigation system* (INS). This synergism is possible, in part, because the INS and GNSS have very complementary

Global Positioning Systems, Inertial Navigation, and Integration, Second Edition, by M. S. Grewal, L. R. Weill, and A. P. Andrews

error characteristics. Short-term position errors from the INS are relatively small, but they degrade without bound over time. GNSS position errors, on the other hand, are not as good over the short term, but they do not degrade with time. The Kalman filter is able to take advantage of these characteristics to provide a common, integrated navigation implementation with performance superior to that of either subsystem (GNSS or INS). By using statistical information about the errors in both systems, it is able to combine a system with tens of meters position uncertainty (GNSS) with another system whose position uncertainty degrades at kilometers per hour (INS) and achieve bounded position uncertainties in the order of centimeters [with differential GNSS (DGNSS)] to meters.

A key function performed by the Kalman filter is the statistical combination of GNSS and INS information to track drifting parameters of the sensors in the INS. As a result, the INS can provide enhanced inertial navigation accuracy during periods when GNSS signals may be lost, and the improved position and velocity estimates from the INS can then be used to cause GNSS signal reacquisition to occur much sooner when the GNSS signal becomes available again.

This level of integration necessarily penetrates deeply into each of these subsystems, in that it makes use of partial results that are not ordinarily accessible to users. To take full advantage of the offered integration potential, we must delve into technical details of the designs of both types of systems.

1.2 GNSS OVERVIEW

There are currently three global navigation satellite systems (GNSSs) operating or being developed.

1.2.1 GPS

The Global Positioning System (GPS) is part of a satellite-based navigation system developed by the U.S. Department of Defense under its NAVSTAR satellite program [82, 84, 89–94, 151–153].

1.2.1.1 GPS Orbits The fully operational GPS includes 24 or more (28 in March 2006) active satellites approximately uniformly dispersed around six circular orbits with four or more satellites each. The orbits are inclined at an angle of 55° relative to the equator and are separated from each other by multiples of 60° right ascension. The orbits are nongeostationary and approximately circular, with radii of 26,560 km and orbital periods of one-half sidereal day ($\approx$11.967 h). Theoretically, three or more GPS satellites will always be visible from most points on the earth's surface, and four or more GPS satellites can be used to determine an observer's position anywhere on the earth's surface 24 h per day.

1.2.1.2 GPS Signals Each GPS satellite carries a cesium and/or rubidium atomic clock to provide timing information for the signals transmitted by the satellites. Internal clock correction is provided for each satellite clock. Each GPS satellite transmits two spread spectrum, L-band carrier signals—an L_1 signal with carrier frequency $f_1 = 1575.42$ MHz and an L_2 signal with carrier frequency $f_2 = 1227.6$

MHz. These two frequencies are integral multiples $f_1 = 1540 f_0$ and $f_2 = 1200 f_0$ of a base frequency $f_0 = 1.023$ MHz. The L_1 signal from each satellite uses *binary phase-shift keying* (BPSK), modulated by two *pseudorandom noise* (PRN) codes in phase quadrature, designated as the C/A-code and P-code. The L_2 signal from each satellite is BPSK modulated by only the P-code. A brief description of the nature of these PRN codes follows, with greater detail given in Chapter 3.

Compensating for Propagation Delays This is one motivation for use of two different carrier signals, L_1 and L_2. Because delay varies approximately as the inverse square of signal frequency f (delay $\propto f^{-2}$), the measurable differential delay between the two carrier frequencies can be used to compensate for the delay in each carrier (see Ref. 128 for details).

Code-Division Multiplexing Knowledge of the PRN codes allows users independent access to multiple GPS satellite signals on the same carrier frequency. The signal transmitted by a particular GPS signal can be selected by generating and matching, or correlating, the PRN code for that particular satellite. All PRN codes are known and are generated or stored in GPS satellite signal receivers carried by ground observers. A first PRN code for each GPS satellite, sometimes referred to as a *precision code* or *P-code*, is a relatively long, fine-grained code having an associated clock or chip rate of $10 f_0 = 10.23$ MHz. A second PRN code for each GPS satellite, sometimes referred to as a *clear* or *coarse acquisition code* or *C/A-code*, is intended to facilitate rapid satellite signal acquisition and handover to the P-code. It is a relatively short, coarser-grained code having an associated clock or chip rate $f_0 = 1.023$ MHz. The C/A-code for any GPS satellite has a length of 1023 chips or time increments before it repeats. The full P-code has a length of 259 days, during which each satellite transmits a unique portion of the full P-code. The portion of P-code used for a given GPS satellite has a length of precisely one week (7.000 days) before this code portion repeats. Accepted methods for generating the C/A-code and P-code were established by the satellite developer[1] in 1991 [61, 97].

Navigation Signal The GPS satellite bit stream includes navigational information on the ephemeris of the transmitting GPS satellite and an almanac for all GPS satellites, with parameters providing approximate corrections for ionospheric signal propagation delays suitable for single-frequency receivers and for an offset time between satellite clock time and true GPS time. The navigational information is transmitted at a rate of 50 baud. Further discussion of the GPS and techniques for obtaining position information from satellite signals can be found in Chapter 3 (below) and in Ref. 125, pp. 1–90.

1.2.1.3 Selective Availability Selective availability (SA) is a combination of methods available to the U.S. Department of Defense to deliberately derating the accuracy of GPS for "nonauthorized" (i.e., non-U.S. military) users during

[1] Satellite Systems Division of Rockwell International Corporation, now part of the Boeing Company.

periods of perceived threat. Measures may include pseudorandom time dithering and truncation of the transmitted ephemerides. The initial satellite configuration used SA with pseudorandom dithering of the onboard time reference [212] only, but this was discontinued on May 1, 2000.

Precise Positioning Service Formal, proprietary service Precise Positioning Service (PPS) is the full-accuracy, single-receiver GPS positioning service provided to the United States and its allied military organizations and other selected agencies. This service includes access to the unencrypted P-code and the removal of any SA effects.

Standard Positioning Service without SA Standard Positioning Service (SPS) provides GPS single-receiver (standalone) positioning service to any user on a continuous, worldwide basis. SPS is intended to provide access only to the C/A-code and the L_1 carrier.

Standard Positioning Service with SA The horizontal-position accuracy, as degraded by SA, currently is advertised as 100 m, the vertical-position accuracy as 156 m, and time accuracy as 334 ns—all at the 95% probability level. SPS also guarantees the user-specified levels of coverage, availability, and reliability.

1.2.2 GLONASS

A second configuration for global positioning is the Global Orbiting Navigation Satellite System (GLONASS), placed in orbit by the former Soviet Union, and now maintained by the Russian Republic [108, 123].

1.2.2.1 GLONASS Orbits GLONASS also uses 24 satellites, but these are distributed approximately uniformly in three orbital planes (as opposed to six for GPS) of eight satellites each (four for GPS). Each orbital plane has a nominal inclination of 64.8° relative to the equator, and the three orbital planes are separated from each other by multiples of 120° right ascension. GLONASS orbits have smaller radii than GPS orbits, about 25,510 km, and a satellite period of revolution of approximately $\frac{8}{17}$ of a sidereal day. A GLONASS satellite and a GPS satellite will complete 17 and 16 revolutions, respectively, around the earth every 8 days.

1.2.2.2 GLONASS Signals The GLONASS system uses frequency-division multiplexing of independent satellite signals. Its two carrier signals corresponding to L_1 and L_2 have frequencies $f_1 = (1.602 + 9k/16)$ GHz and $f_2 = (1.246 + 7k/16)$ GHz, where $k = 0, 1, 2, \ldots, 23$ is the satellite number. These frequencies lie in two bands at 1.597–1.617 GHz (L_1) and 1240–1260 GHz (L_2). The L_1 code is modulated by a C/A-code (chip rate = 0.511 MHz) and by a P-code (chip rate = 5.11 MHz). The L_2 code is presently modulated only by the P-code. The GLONASS satellites also transmit navigational data at a rate of 50 baud. Because the satellite frequencies are distinguishable from each other, the P-code and the C/A-code are the same for each satellite. The methods for receiving and analyzing GLONASS signals are similar to the methods used for GPS signals. Further details can be found in the patent by Janky [97].

GLONASS does not use any form of SA.

1.2.3 Galileo

The Galileo system is the third satellite-based navigation system currently under development. Its frequency structure and signal design is being developed by the European Commission's Galileo Signal Task Force (STF), which was established by the European Commission (EC) in March 2001. The STF consists of experts nominated by the European Union (EU) member states, official representatives of the national frequency authorities, and experts from the European Space Agency (ESA).

1.2.3.1 Galileo Navigation Services The EU intends the Galileo system to provide the following four navigation services plus one search and rescue (SAR) service.

Open Service (OS) The OS provides signals for positioning and timing, free of direct user charge, and is accessible to any user equipped with a suitable receiver, with no authorization required. In this respect it is similar to the current GPS L_1 C/A-code signal. However, the OS will be of higher quality, consisting of six different navigation signals on three carrier frequencies. OS performance will be at least equal to that of the modernized Block IIF GPS satellites, which began launching in 2005, and the future GPS III system architecture currently being investigated. OS applications will include the use of a combination of Galileo and GPS signals, thereby improving performance in severe environments such as urban canyons and heavy vegetation.

Safety of Life Service (SOL) The SOL service is intended to increase public safety by providing certified positioning performance, including the use of certified navigation receivers. Typical users of SOL will be airlines and transoceanic maritime companies. The EGNOS regional European enhancement of the GPS system will be optimally integrated with the Galileo SOL service to have independent and complementary integrity information (with no common mode of failure) on the GPS and GLONASS constellations. To benefit from the required level of protection, SOL operates in the L_1 and E_5 frequency bands reserved for the Aeronautical Radionavigation Services.

Commercial Service (CS) The CS service is intended for applications requiring performance higher than that offered by the OS. Users of this service pay a fee for the added value. CS is implemented by adding two additional signals to the OS signal suite. The additional signals are protected by commercial encryption and access protection keys are used in the receiver to decrypt the signals. Typical value-added services include service guarantees, precise timing, ionospheric delay models, local differential correction signals for very high-accuracy positioning applications, and other specialized requirements. These services will be developed by service providers, which will buy the right to use the two commercial signals from the Galileo operator.

Public Regulated Service (PRS) The PRS is an access-controlled service for government-authorized applications. It will be used by groups such as police, coast guards, and customs. The signals will be encrypted, and access by region or user group will follow the security policy rules applicable in Europe. The PRS will be operational at all times and in all circumstances, including periods of crisis. A major feature of PRS is the robustness of its signal, which protects it against jamming and spoofing.

Search and Rescue (SAR) The SAR service is Europe's contribution to the international cooperative effort on humanitarian search and rescue. It will feature near real-time reception of distress messages from anywhere on Earth, precise location of alerts (within a few meters), multiple satellite detection to overcome terrain blockage, and augmentation by the four low earth orbit (LEO) satellites and the three geostationary satellites in the current COSPAS-SARSAT system.

1.2.3.2 Galileo Signal Characteristics Galileo will provide 10 right-hand circularly polarized navigation signals in three frequency bands. The various signals fall into four categories: F/Nav, I/Nav, C/Nav, and G/Nav. The F/Nav and I/Nav signals are used by the Open Service (OS), Commercial Service (CS) and Safety of Life (SOL) service. The I/Nav signals contain integrity information, while the F/Nav signals do not. The C/Nav signals are used by the Commercial Service (CS), and the G/Nav signals are used by the Public Regulated Service (PRS). At the time of this writing not all of the signal characteristics described below have been finalized.

E_{5a}– E_{5b} Band This band, which spans the frequency range from 1164 to 1214 MHz, contains two signals, denoted E_{5a} and E_{5b}, which are respectively centered at 1176.45 and 1207.140 MHz. Each signal has an in-phase component and a quadrature component. Both components use spreading codes with chipping rate of 10 Mcps (million chips per second). However, the in-phase components are modulated by navigation data, while the quadrature components, called *pilot signals*, are data-free. The data-free pilot signals permit arbitrarily long coherent processing, thereby greatly improving detection and tracking sensitivity. A major feature of the E_{5a} and E_{5b} signals is that they can be treated as either separate signals or a single wide-band signal. Low-cost receivers can use either signal, but the E_{5a} signal might be preferred, since it is centered at the same frequency as the modernized GPS L_5 signal and would enable the simultaneous reception of E_{5a} and L_5 signals by a relatively simple receiver without the need for reception on two separate frequencies. Receivers with sufficient bandwidth to receive the combined E_{5a} and E_{5b} signals would have the advantage of greater ranging accuracy and better multipath performance.

Even though the E_{5a} and E_{5b} signals can be received separately, they actually are two spectral components produced by a single modulation called *alternate binary offset carrier* (altBOC) modulation. This form of modulation retains the simplicity of standard BOC modulation (used in the modernized GPS M-code military signals) and has a constant envelope while permitting receivers to differentiate the two spectral lobes. The current modulation choice is altBOC(15,10), but this may be subject to change.

The in-phase component of the E_{5a} signal is modulated with 50 symbols per second (sps) navigation data without integrity information, and the in-phase component of the E_{5b} signal is modulated with 250 sps (symbols per second) data with integrity information. Both the E_{5a} and E5b signals are available to the Open Service (OS), CS, and SOL services.

E_6 Band This band spans the frequency range from 1260 to 1300 MHz and contains a C/Nav signal and a G/Nav signal, each centered at 1278.75 MHz. The C/Nav signal is used by the CS service and has both an in-phase and quadrature pilot component using a BPSK spreading code modulation of 5 Mcps. The in-phase component contains 1000 sps data modulation, and the pilot component is data-free. The G/Nav signal is used by the PRS service and has only an in-phase component modulated by a BOC(10,5) spreading code and data modulation with a symbol rate that is to be determined.

E_2–L_1–E_1 Band The E_2–L_1–E_1 band (sometimes denoted as L_1 for convenience) spans the frequency range from 1559 to 1591 MHz and contains a G/Nav signal used by the PRS service and an I/Nav signal used by the OS, CS, and SOL services. The G/Nav signal has only an in-phase component with a BOC spreading code and data modulation; the characteristics of both are still being decided. The I/Nav signal has an in-phase and quadrature component. The in-phase component will contain 250 sps data modulation and will likely use BOC(1,1) spreading code, but this has not been finalized. The quadrature component is data-free.

1.3 DIFFERENTIAL AND AUGMENTED GPS

1.3.1 Differential GPS (DGPS)

Differential GPS (DGPS) is a technique for reducing the error in GPS-derived positions by using additional data from a reference GPS receiver at a known position. The most common form of DGPS involves determining the combined effects of navigation message ephemeris, conospheric and satellite clock errors (including the effects of SA) at a reference station and transmitting pseudorange corrections, in real time, to a user's receiver, which applies the corrections in the process of determining its position [94, 151, 153].

1.3.2 Local-Area Differential GPS

Local-area differential GPS (LAGPS) is a form of DGPS in which the user's GPS receiver also receives real-time pseudorange and, possibly, carrier phase corrections from a local reference receiver generally located within the line of sight. The corrections account for the combined effects of navigation message ephemeris and satellite clock errors (including the effects of SA) and, usually, propagation delay errors at the reference station. With the assumption that these errors are also common to the measurements made by the user's receiver, the application of the corrections will result in more accurate coordinates.

1.3.3 Wide-Area Differential GPS

Wide-area DGPS (WADGPS) is a form of DGPS in which the user's GPS receiver receives corrections determined from a network of reference stations distributed over a wide geographic area. Separate corrections are usually determined for specific error sources—such as satellite clock, ionospheric propagation delay, and ephemeris. The corrections are applied in the user's receiver or attached computer in computing the receiver's coordinates. The corrections are typically supplied in real time by way of a geostationary communications satellite or through a network of ground-based transmitters. Corrections may also be provided at a later date for postprocessing collected data [94].

1.3.4 Wide-Area Augmentation System

The WAAS enhances the GPS SPS over a wide geographic area. The U.S. Federal Aviation Administration (FAA), in cooperation with other agencies, is developing WAAS to provide WADGPS corrections, additional ranging signals from geostationary earth orbit (GEO) satellites, and integrity data on the GPS and GEO satellites.

1.4 SPACE-BASED AUGMENTATION SYSTEMS (SBASS)

Four space-based augmentation systems (SBASs) were under development at the beginning of the third millennium. These are the Wide-Area Augmentation System (WAAS), European Geostationary Navigation Overlay System (EGNOS), Multifunctional Transport Satellite (MTSAT)–based Augmentation System (MSAS), and GPS & GEO Augmented Navigation (GAGAN) by India.

1.4.1 Historical Background

Although GPS is inherently a very accurate system for positioning and time transfer, some applications require accuracies unobtainable without some form of performance augmentation, such as differential GPS (DGPS), in which position relative to a base (or reference) station can be established very accurately (in some cases within millimeters). A typical DGPS system employs an additional GPS receiver at the base station to measure the GPS signals. Because the coordinates of the base station are precisely known, errors in the received GPS signals can be calculated. These errors, which include satellite clock and position error, as well as tropospheric and ionospheric error, are very nearly the same for users at a sufficiently small distance from the base station. In DGPS the error values determined by the base station are transmitted to the user and applied as corrections to the user's measurements.

However, DGPS has a fundamental limitation in that the broadcast corrections are good only for users in a limited area surrounding the base station. Outside this area the errors tend to be decorrelated, rendering the corrections less accurate. An obvious technical solution to this problem would be to use a network of base stations, each with its own communication link to serve its geographic area. However, this would require a huge number of base stations and their associated communication links.

Early on it was recognized that a better solution would be to use a space-based augmentation system (SBAS) in which a few satellites can broadcast the correction data over a very large area. Such a system can also perform sophisticated computations to optimally interpolate the errors observed from relatively few ground stations so that they can be applied at greater distances from each station.

A major motivation for SBAS has been the need for precision aircraft landing approaches without requiring separate systems, such as the existing instrument landing systems (ILSs) at each airport. An increasing number of countries are currently developing their own versions of SBAS, including the United States (WAAS), Europe (EGNOS), Japan (NSAS), Canada (CWAAS), China (SNAS), and India (GAGAN).

1.4.2 Wide-Area Augmentation System (WAAS)

In 1995 the United States began development of the Wide Area Augmentation System (WAAS) under the auspices of the Federal Aviation Administration (FAA) and the Department of Transportation (DOT), to provide precision approach capability for aircraft. Without WAAS, ionospheric disturbances, satellite clock drift, and satellite orbit errors cause too much error in the GPS signal for aircraft to perform a precision landing approach. Additionally, signal integrity information as broadcast by the satellites is insufficient for the demanding needs of public safety in aviation. WAAS provides additional integrity messages to aircraft to meet these needs.

WAAS includes a core of approximately 25 wide-area ground reference stations (WRSs) positioned throughout the United States that have precisely surveyed coordinates. These stations compare the GPS signal measurements with the measurements that should be obtained at the known coordinates. The WRS send their findings to a WAAS master station (WMS) using a land-based communications network, and the WMS calculates correction algorithms and assesses the integrity of the system. The WMS then sends correction messages via a ground uplink system (GUS) to geostationary (GEO) WAAS satellites covering the United States. The satellites in turn broadcast the corrections on a per-GPS satellite basis at the same L_1 1575.42 MHz frequency as GPS. WAAS-enabled GPS receivers receive the corrections and use them to derive corrected GPS signals, which enable highly accurate positioning.

On July 10, 2003, Phase 1 of the WAAS system was activated for general aviation, covering 95% of the conterminous United States and portions of Alaska.

In September 2003, improvements enabled WAAS-enabled aircraft to approach runways to within 250 ft altitude before requiring visual control.

Currently there are two Inmarsat III GEO satellites serving the WAAS area: the Pacific Ocean Region (POR) satellite and the West Atlantic Ocean Region (AOR-W) satellite.

In March 2005 two additional WAAS GEO satellites were launched (PanAm-Sat Galaxy XV and Telesat Anik F1R), and are now operational. These satellites plus the two existing satellites will improve coverage of North America and all except the northwest part of Alaska. The four GEO satellites will be positioned at 54°, 107°, and 133° west longitude, and at 178° east longitude.

WAAS is currently available over 99% of the time, and its coverage will include the full continental United States and most of Alaska. Although primarily intended for aviation applications, WAAS will be useful for improving the accuracy of any WAAS-enabled GPS receiver. Such receivers are already available in low-cost handheld versions for consumer use.

Positioning accuracy using WAAS is currently quoted at less than 2 m of lateral error and less than 3 m of vertical error, which meets the aviation Category I precision approach requirement of 16 m lateral error and 4 m vertical error.

Further details of the WAAS system can be found in Chapter 6.

1.4.3 European Geostationary Navigation Overlay System (EGNOS)

The European Geostationary Navigation Overlay System (EGNOS) is Europe's first venture into satellite navigation. It is a joint project of the European Space Agency (ESA), the European Commission (EC), and Eurocontrol, the European organization for the safety of air navigation. Inasmuch as Europe does not yet have its own standalone satellite navigation system, initially EGNOS is intended to augment both the United States GPS and the Russian GLONASS systems, providing differential accuracy and integrity monitoring for safety-critical applications such as aircraft landing approaches and ship navigation through narrow channels.

EGNOS has functional similarity to WAAS, and consists of four segments: space, ground, user, and support facilities segments.

1.4.3.1 Space Segment The space segment consists of three geostationary (GEO) satellites, the Inmarsat-3 AOR-E, Inmarsat-3 AOR-W, and the ESA Artemis, which transmit wide-area differential corrections and integrity information throughout Europe. Unlike the GPS and GLONASS satellites, these satellites will not have signal generators aboard, but will be transponders relaying uplinked signals generated on the ground.

1.4.3.2 Ground Segment The EGNOS ground segment includes 34 Ranging and Integrity Monitoring Stations (RIMSs), four Mission/Master Control Centers

(MCCs), six Navigation Land Earth Stations (NLESs), and an EGNOS Wide-Area Network (EWAN).

The RIMS stations monitor the GPS and GLONASS signals. Each station contains a GPS/GLONASS/EGNOS receiver, an atomic clock, and network communications equipment. The RIMS tasks are to perform pseudorange measurements, demodulate navigation data, mitigate multipath and interference, verify signal integrity, and to packetize and transmit data to the MCC centers.

The MCC centers monitor and control the three EGNOS GEO satellites, as well as perform real-time software processing. The MCC tasks include integrity determination, calculation of pseudorange corrections for each satellite, determination of ionospheric delay, and generation of EGNOS satellite ephemeris data. The MCC then sends all the data to the NLES stations. Every MCC has a backup station that can take over in the event of failure.

The NLES stations receive the data from the MCC centers and generate the signals to be sent to the GEO satellites. These include a GPS-like signal, an integrity channel, and a wide-area differential (WAD) signal. The NLES send this data on an uplink to the GEO satellites.

The EWAN links all EGNOS ground-based components.

1.4.3.3 User Segment This segment consists of the user receivers. Although EGNOS has been designed primarily for aviation applications, it can also be used with land or marine EGNOS-compatible receivers, including low-cost handheld units.

1.4.3.4 Support Facilities Segment Support for development, operations, and verifications is provided by this segment.

The EGNOS system is currently operational. Positioning accuracy obtainable from use of EGNOS is approximately 5 m, as compared to 10–20 m with unaided GPS. There is the possibility that this can be improved with further technical development.

1.4.4 Japan's MTSAT Satellite-Based Augmentation System (MSAS)

The Japanese MSAS system, currently under development by Japan Space Agency and the Japan Civil Aviation Bureau, will improve the accuracy, integrity, continuity, and availability of GPS satellite signals throughout the Japanese Flight Information Region (FIR) by relaying augmentation information to user aircraft via Japan's Multifunctional Transport Satellite (MTSAT) geostationary satellites. The system consists of a network of Ground Monitoring Stations (GMS) in Japan, Monitoring and Ranging Stations (MRSs) outside of Japan, Master Control Stations (MCSs) in Japan with satellite uplinks, and two MTSAT geostationary satellites.

MSAS will serve the Asia–Pacific region with capabilities similar to the United States WAAS system. MSAS and WAAS will be interoperable and are compliant with the International Civil Aviation Organization (ICAO) Standards and Recommended Practices (SARP) for SBAS systems.

1.4.5 Canadian Wide-Area Augmentation System (CWAAS)

The Canadian CWAAS system is basically a plan to extend the U.S. WAAS coverage into Canada. Although the WAAS GEO satellites can be received in much of Canada, additional ground reference station sites are needed to achieve valid correctional data outside the United States. At least 11 such sites, spread over Canada, have been evaluated. The Canadian reference stations are to be linked to the U.S. WAAS system.

1.4.6 China's Satellite Navigation Augmentation System (SNAS)

China is moving forward with its own version of a SBAS. Although information on their system is incomplete, at least 11 reference sites have been installed in and around Beijing in Phase I of the program, and further expansion is anticipated. Receivers manufactured by Novatel, Inc. of Canada have been delivered for Phase II.

1.4.7 Indian GPS and GEO Augmented Navigation System (GAGAN)

In August 2001 the Airports Authority of India and the Indian Space Research Organization signed a memorandum of understanding for jointly establishing the GAGAN system. The system is not yet fully operational, but by 2007 a GSAT-4 satellite should be in orbit, carrying a transponder for broadcasting correction signals. On the ground, eight reference stations are planned for receiving signals from GPS and GLONASS satellites. A Mission Control Center, as well as an uplink station, will be located in Bangalore.

Once GAGAN is operational, it should materially improve air safety over India. There are 449 airports and airstrips in the country, but only 34 have instrument landing systems (ILSs) installed. With GAGAN, aircraft will be able to make precision approaches to any airport in the coverage area. There will undoubtedly be other uses for GAGAN, such as tracking of trains so that warnings can be issued if two trains appear likely to collide.

1.4.8 Ground-Based Augmentation Systems (GBASs)

Ground-based augmentation systems (GBASs) differ from the SBAS in that backup, aiding, and/or correction information is broadcast from ground stations instead of from satellites. Three major GBAS are LAAS, JPALS, and LORAN-C.

1.4.8.1 Local-Area Augmentation System (LAAS) LAAS is an augmentation to GPS that services airport areas approximately 20–30 mi in radius, and has been developed under the auspices of the Federal Aviation Administration (FAA). It broadcasts GPS correction data via a very high-frequency (VHF) radio data link from a ground-based transmitter, yielding extremely high accuracy, availability, and integrity deemed necessary for aviation Categories I, II, and III precision landing approaches. LAAS also provides the ability for flexible, curved aircraft approach trajectories. Its demonstrated accuracy is less than 1 m in both the horizontal and vertical directions.

A typical LAAS system, which is designed to support an aircraft's transition from en route airspace into and throughout terminal area airspace, consists of ground equipment and avionics. The ground equipment consists of four GPS reference receivers, a LAAS ground facility, and a VHF radio data transmitter. The avionics equipment includes a GPS receiver, a VHF radio data receiver, and computer hardware and software.

The GPS reference receivers and the LAAS ground facility work together to measure errors in GPS position that are common to the reference receiver and aircraft locations. The LAAS ground facility then produces a LAAS correction message based on the difference between the actual and GPS-calculated positions of the reference receivers. The correction message includes integrity parameters and approach-path information. The LAAS correction message is sent to a VHF data broadcast transmitter, which broadcasts a signal containing the correction/integrity data throughout the local LAAS coverage area, where it is received by incoming aircraft.

The LAAS equipment in the aircraft uses the corrections for position, velocity, and time to generate instrument landing system (ILS) lookalike guidance as low as 200 ft above touchdown. It is anticipated that further technical improvements will eventually result in vertical accuracy below 1 m, enabling ILS guidance all the way down to the runway surface, even in zero visibility (Category III landings).

A major advantage of LAAS is that a single installation at a major airport can be used for multiple precision approaches within its local service area. For example, if an airport has 12 runway ends, each with a separate ILS, all 12 ILS facilities can be replaced with a single LAAS installation. Furthermore, it is generally agreed that the Category III level of accuracy anticipated for LAAS cannot be supported by WAAS.

1.4.8.2 Joint Precision Approach and Landing System (JPALS) JPALS is basically a military version of LAAS that supports fixed-base, tactical, special-mission, and shipboard landing environments. It will allow the military to overcome problems of age and obsolescence of ILS equipment, and also will afford greater interoperability, both among systems used by the various services and between military and civilian systems.

The main distinction between LAAS and JPALS is that the latter can be quickly deployed almost anywhere and makes full use of military GPS functionality, which includes the use of the encrypted M-codes not available for civilian use. The requirement for deployment in a variety of locations not optimized for good GPS reception places great demands on the ability of JPALS equipment to handle poor signal environments and multipath. Such problems are not as severe for LAAS installations, where there is more freedom in site selection for best GPS performance of the reference receivers. Additionally, JPALS GPS receivers must be designed to foil frequent attempts by the enemy to jam the received GPS signals.

1.4.8.3 Long-Range Navigation (LORAN-C) LORAN-C is a low-frequency ground-based radionavigation and time reference system that uses stable 100 kHz transmissions to provide an accurate regional positioning service. Unlike LAAS and JPALS, LORAN-C is an independent, standalone system that does not provide corrections to GPS signals, but instead uses time difference of arrival (TDOA) to establish position.

LORAN-C transmitters are organized into chains of 3–5 stations. Within a chain one station is designated as the master (M) and the other secondary stations (slaves) are identified by the letters W, X, Y, and Z. The sequence of signal transmissions consists of a pulse group from the master station followed at precise time intervals by pulse groups from the secondary stations. All LORAN-C stations operate on the same frequency of 100 kHz, and all stations within a given chain use the same group repetition interval (GRI) to uniquely identify the chain. Within a chain, each of the slave stations transmits its pulse group with a different delay relative to the master station in such a way that the sequence of the pulse groups from the slaves is always received in the same order, independent of the location of the user. This permits identification of the individual slave station transmissions.

The basic measurements made by LORAN-C receivers are TDOAs between the master station signal pulses and the signal pulses from each of the secondary stations in a chain. Each time delay is measured to a precision of about 0.1 μs or better. LORAN-C stations maintain integrity by constantly monitoring their transmissions to detect signal abnormalities that would render the system unusable for navigation. If a signal abnormality is detected, the transmitted pulse groups "blink" on and off to notify the user that the transmitted signal does not comply with the system specifications.

LORAN-C, with an accuracy approaching approximately 30 m in regions with good geometry, is not as precise as GPS. However, it has good repeatability, and positioning errors tend to be stable over time. A major advantage of using LORAN-C as an augmentation to GPS is that it provides a backup system completely independent of GPS. A failure of GPS that would render LAAS or JPALS inoperable does not affect positioning using LORAN-C. On the other hand, LORAN-C is only a regional and not a truly global navigation system, covering significant portions, but not all, of North America, Canada, and Europe, as well as some other areas.

1.4.9 Inmarsat Civil Navigation

The Inmarsat overlay is an implementation of a wide-area differential service. Inmarsat is the International Mobile Satellite Organization (IMSO), an 80-nation international consortium, originally created in 1979 to provide maritime[2] mobile services on a global basis but now offering a much wider range of mobile satellite services. Inmarsat launched four geostationary satellites that provide complete

[2]The "mar" in the name originally stood for "maritime."

coverage of the globe from $\pm 70^\circ$ latitude. The data broadcast by the satellites are applicable to users in regions having a corresponding ground station network. The U.S. region is the continental U.S. (CONUS) and uses Atlantic Ocean Region West (AOR-W) and Pacific Ocean Region (POR) geostationary satellites. This is called the WAAS and is being developed by the FAA. The ground station network is operated by the service provider, that is, the FAA, whereas Inmarsat is responsible for operation of the space segment. Inmarsat affiliates operate the uplink Earth stations (e.g., COMSAT in the United States). WAAS is discussed further in Chapter 6.

1.4.10 Satellite Overlay

The Inmarsat Civil Navigation Geostationary Satellite Overlay extends and complements the GPS and GLONASS satellite systems. The overlay navigation signals are generated at ground-based facilities. For example, for WAAS, two signals are generated from Santa Paula, California—one for AOR-W and one for POR. The backup signal for POR is generated from Brewster, Washington. The backup signal for AOR-W is generated from Clarksburg, Maryland. Signals are uplinked to Inmarsat-3 satellites such as AOR-W and POR. These satellites contain special satellite repeater channels for rebroadcasting the navigation signals to users. The use of satellite repeater channels differs from the navigation signal broadcast techniques employed by GLONASS and GPS. GLONASS and GPS satellites carry their own navigation payloads that generate their respective navigation signals.

1.4.11 Future Satellite Systems

In Europe, activities supported by the European Tripartite Group [European Space Agency (ESA), European Commission (EC), EUROCONTROL] are underway to specify, install, and operate a future civil global navigation satellite system (GNSS) (GNSS-2 or Galileo).

Based on the expectation that GNSS-2 will be developed through an evolutionary process as well as long-term augmentations [e.g., EGNOS], short to midterm augmentation systems (e.g., differential systems) are being targeted.

The first steps toward GNSS-2 will be made by the Tripartite Group. The augmentations will be designed such that the individual elements will be suitable for inclusion in GNSS-2 at a later date. This design process will provide the user with maximum continuity in the upcoming transitions.

In Japan, the Japanese Commercial Aviation Board (JCAB) is currently developing the MSAS.

1.5 APPLICATIONS

Both GPS and GLONASS have evolved from dedicated military systems into true dual-use systems. Satellite navigation technology is utilized in numerous civil and military applications, ranging from golf and leisure hiking to spacecraft navigation. Further discussion on applications can be found in Chapters 6 and 7.

1.5.1 Aviation

The aviation community has propelled the use of GNSS and various augmentations (e.g., WAAS, EGNOS, GAGAN, MSAS). These systems provide guidance for en route through precision approach phases of flight. Incorporation of a data link with a GNSS receiver enables the transmission of aircraft location to other aircraft and/or to air traffic control (ATC). This function is called automatic dependent surveillance (ADS) and is in use in the POR. Key benefits are ATC monitoring for collision avoidance and optimized routing to reduce travel time and fuel consumption [153].

1.5.2 Spacecraft Guidance

The Space Shuttle utilizes GPS for guidance in all phases of its operation (e.g., ground launch, on-orbit and reentry, and landing). NASA's small satellite programs use and plan to use GPS, as does the military on SBIRLEO (space-based infrared low earth orbit) and GBI (ground-based interceptor) kill vehicles.

1.5.3 Maritime

GNSS has been used by both commercial and recreational maritime communities. Navigation is enhanced on all bodies of water, from oceanic travel to riverways, especially in inclement weather.

1.5.4 Land

The surveying community depends heavily on DGPS to achieve measurement accuracies in the millimeter range. Similar techniques are used in farming, surface mining, and grading for real-time control of vehicles and in the railroad community to obtain train locations with respect to adjacent tracks. GNSS is a key component in intelligent transport systems (ITSs). In vehicle applications, GNSS is used for route guidance, tracking, and fleet management. Combining a cellular phone or data link function with this system enables vehicle tracing and/or emergency messaging.

1.5.5 Geographic Information Systems (GISs), Mapping, and Agriculture

Applications include utility and asset mapping and automated airborne mapping, with remote sensing and photogrammetry. Recently, GIS, GPS, and remote sensing have matured enough to be used in agriculture. GIS companies such as the Environmental System Research Institute (Redlands, California) have developed software applications that enable growers to assess field conditions and their relationship to yield. Real time kinematic and differential GNSS applications for precision farming are being developed. This includes soil sampling, yield monitoring, chemical, and fertilizer applications. Some GPS analysts are predicting precision site-specific farming to become "the wave of the future."

PROBLEMS

1.1 How many satellites and orbit planes exist for GPS, GLONASS, and Galileo? What are the respective orbit plane inclinations?

1.2 List the differences in signal characteristics between GPS, GLONASS, and Galileo.

2

FUNDAMENTALS OF SATELLITE AND INERTIAL NAVIGATION

2.1 NAVIGATION SYSTEMS CONSIDERED

This book is about GNSS and INS and their integration. An inertial navigation system can be used anywhere on the globe, but it must be updated within hours of use by independent navigation sources such as GNSS or celestial navigation. Thousands of self-contained INS units are in continuous use on military vehicles, and an increasing number are being used in civilian applications.

2.1.1 Systems Other than GNSS

GNSS signals may be replaced by LORAN-C signals produced by three or more long-range navigation (LORAN) signal sources positioned at fixed, known locations for outside-the-building location determination. A LORAN-C system relies on a plurality of ground-based signal towers, preferably spaced 100–300 km apart, that transmit distinguishable electromagnetic signals that are received and processed by a LORAN signal antenna and LORAN signal receiver/processor that are analogous to the Satellite Positioning System signal antenna and receiver/processor. A representative LORAN-C system is discussed in the U.S. DOT *LORAN-C User Handbook* [127]. LORAN-C signals use carrier frequencies of the order of 100 kHz and have maximum reception distances of hundreds of kilometers. The combined use of FM signals for location determination inside a building or similar structure can also provide a satisfactory location determination (LD) system in most urban and suburban communities.

Global Positioning Systems, Inertial Navigation, and Integration, Second Edition, by M. S. Grewal, L. R. Weill, and A. P. Andrews

There are other ground-based radiowave signal systems suitable for use as part of an LD system. These include Omega, Decca, Tacan, JTIDS Relnav (U.S. Air Force Joint Tactical Information Distribution System Relative Navigation), and PLRS (U.S. Army Position Location and Reporting System) (see summaries in Ref. 125 pp. 6–7 and 35–60).

2.1.2 Comparison Criteria

The following criteria may be used in selecting navigation systems appropriate for a given application system:

1. Navigation method(s) used
2. Coordinates provided
3. Navigational accuracy
4. Region(s) of coverage
5. Required transmission frequencies
6. Navigation fix update rate
7. User set cost
8. Status of system development and readiness

2.2 FUNDAMENTALS OF INERTIAL NAVIGATION

The fundamental idea for inertial navigation (also called *Newtonian navigation*) comes from high-school physics:

The second integral of acceleration is position.

Given sensors that can measure the three components of acceleration over time, and initial values for position and velocity, the approach would appear to be relatively straightforward. As is often the case, however, "the devil is in the details."

This introductory section presents a descriptive overview of the fundamental concepts that have evolved in reducing this idea to practice. A more mathematical treatment is presented in Chapter 9, and additional application-specific details can be found in the literature [27, 36, 63, 88, 108, 120, 124, 139, 168, 169, 181, 189].

2.2.1 Basic Concepts

Inertia is the propensity of bodies to maintain constant translational and rotational velocity, unless disturbed by forces or torques, respectively (Newton's first law or motion).

An *inertial reference frame* is a coordinate frame in which Newton's laws of motion are valid. Inertial reference frames are neither rotating nor accelerating. They are not necessarily the same as the *navigation coordinates*, which are typically dictated by the navigation problem at hand. For example, "locally level" coordinates used for navigation near the surface of the earth are rotating (with the earth) and accelerating (to counter gravity). Such rotations and accelerations must be taken into account in the practical implementation of inertial navigation.

Inertial sensors measure inertial accelerations and rotations, both of which are vector-valued variables.

- *Accelerometers* are sensors for measuring inertial acceleration, also called *specific force* to distinguish it from what we call "*gravitational acceleration*." *Accelerometers do not measure gravitational acceleration*, which is perhaps more accurately modeled as a warping of the spacetime continuum in a gravitational field. An accelerometer in free fall (e.g., in orbit) in a gravitational field has no detectable input. What accelerometers measure is modeled by Newton's second law as $a = F/m$, where F is the physically applied force (not including gravity), m is the mass it is applied to, and specific force is the ratio F/m.
- *Gyroscopes* (usually shortened to *gyros*) are sensors for measuring rotation. *Rate gyros* measure rotation rate, and *displacement gyros* (also called *whole-angle gyros*) measure accumulated rotation angle. Inertial navigation depends on gyros for maintaining knowledge of how the accelerometers are oriented in inertial and navigational coordinates.
- The *input axis* of an inertial sensor defines which vector component of acceleration or rotation rate it measures. Multiaxis sensors measure more than one component.

An *inertial sensor assembly* (ISA) is an ensemble of inertial sensors rigidly mounted to a common base to maintain the same relative orientations, as illustrated in Fig. 2.1. Inertial sensor assemblies used in inertial navigation usually contain three accelerometers and three gyroscopes, as shown in the figure, or an equivalent configuration using multiaxis sensors. However, ISAs used for some other purposes (e.g., dynamic control applications such as autopilots or automotive steering augmentation) may not need as many sensors, and some designs provide more than three input axis directions for the accelerometers and gyroscopes. The term *inertial reference unit* (IRU) usually refers to an inertial sensor system for attitude information only (i.e., using only gyroscopes). Other terms used for the ISA are *instrument cluster* and (for gimbaled systems) *stable element* or *stable platform.*

An *inertial measurement unit* (IMU) includes an ISA and its associated support electronics for calibration and control of the ISA. The support electronics may also include thermal control or compensation, signal conditioning, and input/output control. The IMU may also include an IMU processor, and—for gimbaled systems—the gimbal control electronics.

An *inertial navigation system* (INS) consists of an IMU plus the following:

- *Navigation computers* (one or more) to calculate the gravitational acceleration (not measured by accelerometers) and process the outputs of the

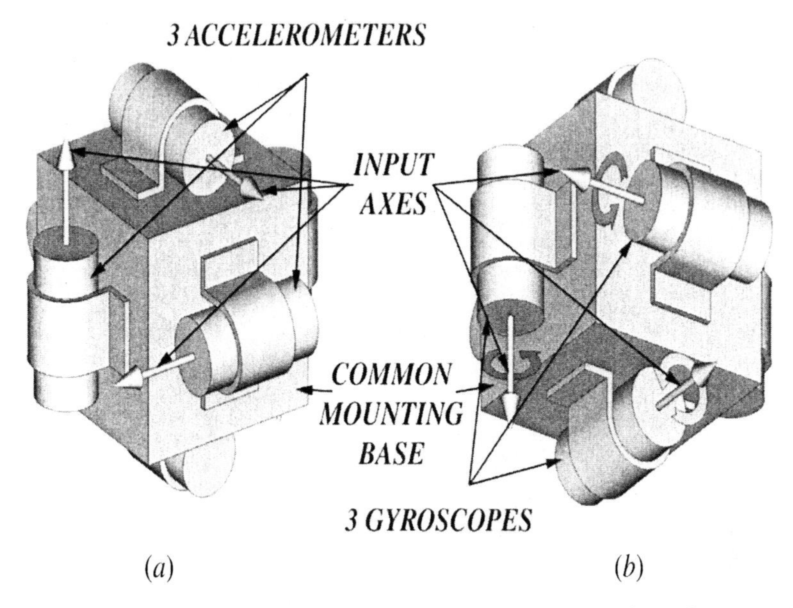

Fig. 2.1 Inertial sensor assembly (ISA) components: (*a*) top-Front view; (*b*) bottom-back view.

accelerometers and gyroscopes from the IMU to maintain an estimate of the position of the IMU. Intermediate results of the implementation method usually include estimates of velocity, attitude, and attitude rates of the IMU.

- *User interfaces*, such as display consoles for human operators and analog and/or digital data interfaces for vehicle guidance[1] and control functions.
- *Power supplies* and/or raw power conditioning for the complete INS.

2.2.1.1 Host Vehicles The term *host vehicle* is used to refer to the platform on or in which an INS is mounted. This could be a spacecraft, aircraft, surface ship, submarine, land vehicle, or pack animal (including humans).

2.2.1.2 What an INS Measures An INS estimates the *position of its ISA*, just as a GNSS receiver estimates the position of its antenna. The relative locations of the ISA and GNSS antenna on the host vehicle must be taken into account in GNSS/INS integration.

2.2.2 Inertial Navigation Systems

The first known inertial navigation systems are of the type you carry around in your head. There are two of them, and they are part of the *vestibular system* in

[1]*Guidance* generally includes the generation of command signals for controlling the motion and attitude of a vehicle to follow a specified trajectory or to arrive at a specified destination.

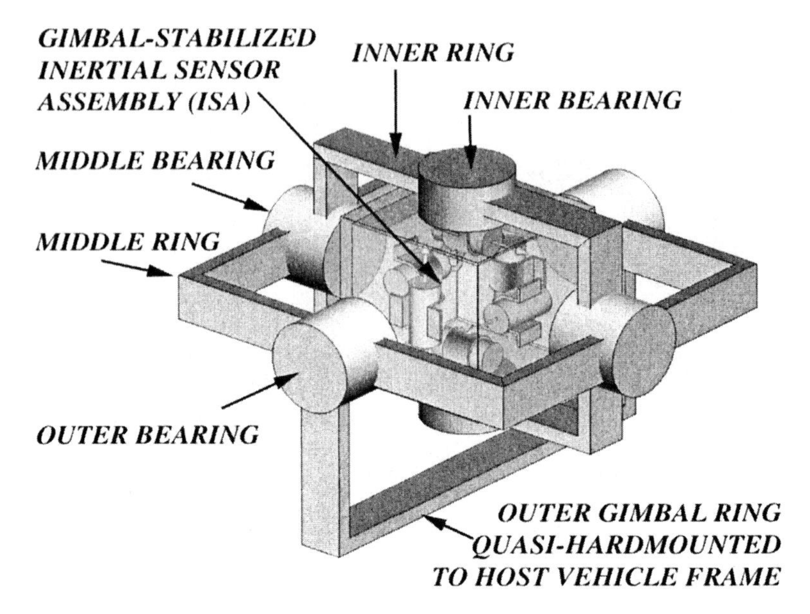

Fig. 2.2 Gimbaled IMU.

your inner ears. Each includes three rotational acceleration sensors (*semicircular canals*) and two dual-axis accelerometers (*otolith organs*). They are not accurate enough for long-distance navigation, but they do enable you to balance and walk in total darkness.

Engineers have designed many different inertial navigation systems with better long-distance performance characteristics. They generally fall into two categories:

- *Gimbaled* or *floated* systems, in which the inertial sensor assembly (ISA) is isolated from rotations of the host vehicle, as illustrated in Figs. 2.2 and 2.3. This rotation-isolated ISA is also called an *inertial platform*, *stable platform*, or *stable element*. In this case, the IMU includes the ISA, the gimbal/float structure and all associated electronics (e.g., gimbal wiring, rotary slip rings, gimbal bearing angle encoders, signal conditioning, gimbal bearing torque motors, and thermal control).
- *Strapdown* systems are illustrated in Fig. 2.4. In this case, the ISA is not isolated from rotations, but is "quasirigidly" mounted to the frame structure of the host vehicle.

2.2.2.1 Shock and Vibration Isolation We use the term *quasirigid* for IMU mountings that can provide some isolation of the IMU from shock and vibration transmitted through the host vehicle frame. Many host vehicles produce severe mechanical noise within their propulsion systems, or through vehicle contact with the environment. Both strapdown and gimbaled systems may require *shock and*

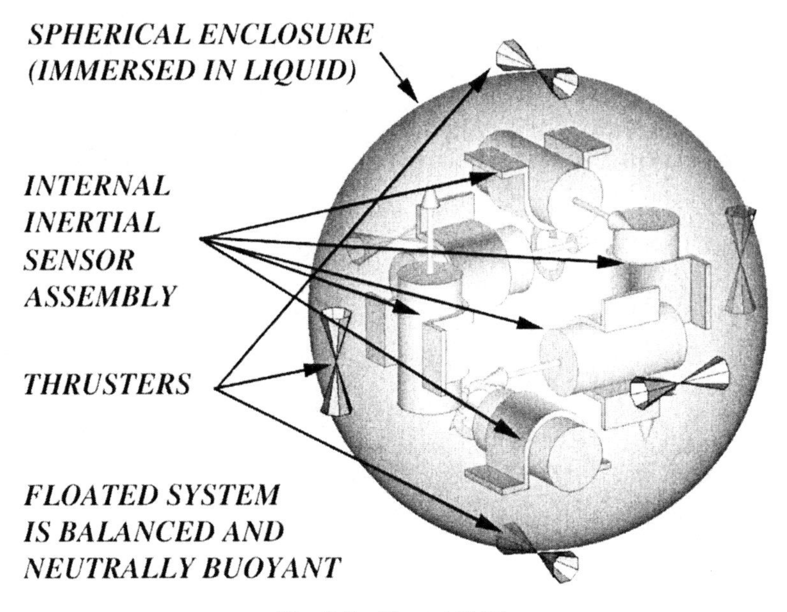

Fig. 2.3 Floated IMU.

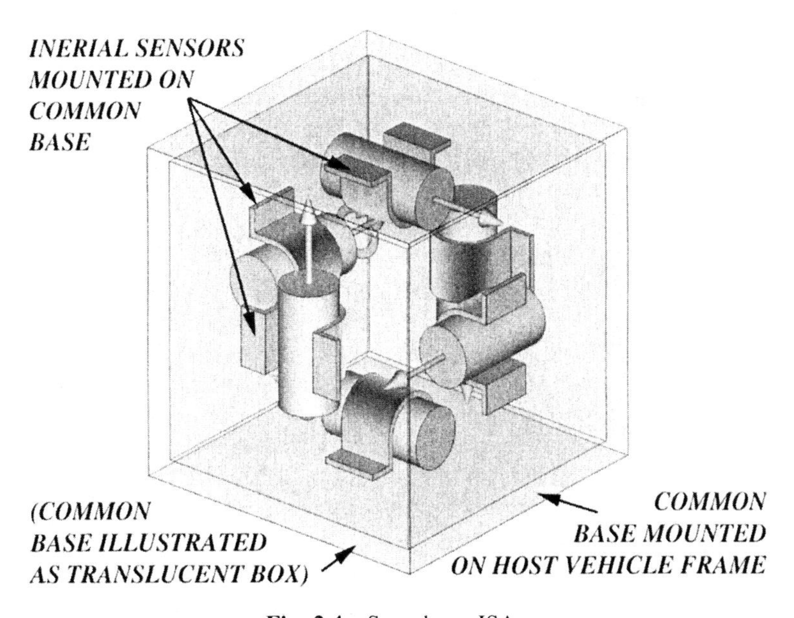

Fig. 2.4 Strapdown ISA.

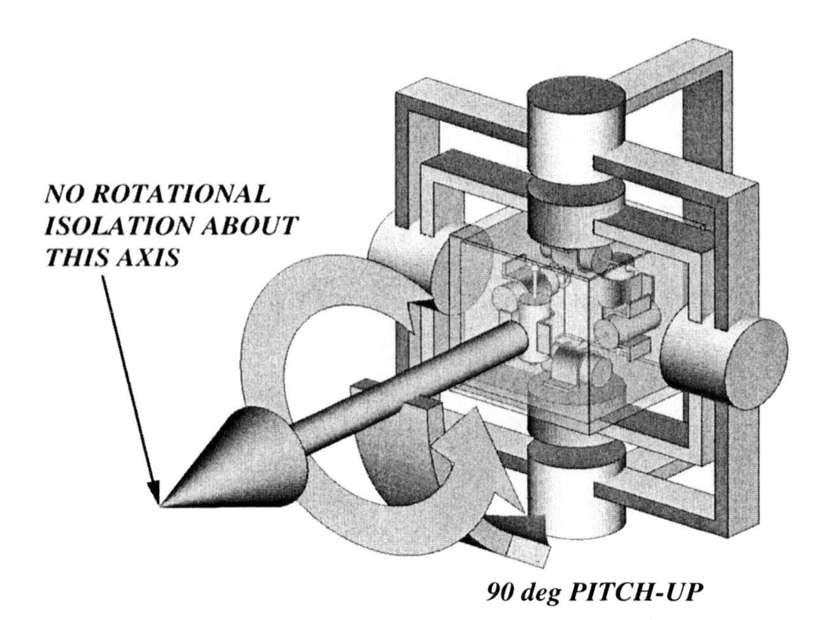

Fig. 2.5 Gimbal lock.

vibration isolators to dampen the vibrational torques and forces transmitted to the inertial sensors. These isolators are commonly made from "lossy" elastomers that provide some amount of damping, as well.

2.2.2.2 ***Gimbaled Systems*** Two-axis gimbals were in use in China (for carrying liquids with less sloshing) around the beginning of the Current Era. Gimbaled inertial navigation systems using feedback control technology were first developed around the middle of the twentieth century, when computers were too slow for strapdown calculations and too heavy for inflight applications.

Gimbals are nested ringlike structures with orthogonal rotation bearings (also called *gimbals*) that allow isolation of the inside from rotations of the outside. As illustrated in Fig. 2.2, three sets of gimbal bearings are sufficient for complete rotational isolation in applications with limited attitude mobility (e.g., surface ships), but applications in fully maneuverable hosts require an additional gimbal bearing to avoid the condition shown in Fig. 2.5, known as *gimbal lock*, in which the gimbal configuration no longer provides isolation from outside rotations about all three axes. The example shown in Fig. 2.5 cannot isolate the INS from rotations about the axis illustrated by the rotation vector.

Gyroscopes inside the gimbals can be used to detect any rotation of that frame due to torques from bearing friction or load imbalance, and torquing motors in the gimbal bearings can then be used to servo the rotation rates inside the gimbals to zero. For navigation with respect to the rotating earth, the gimbals can also be servoed to maintain the sensor axes fixed in locally level coordinates.

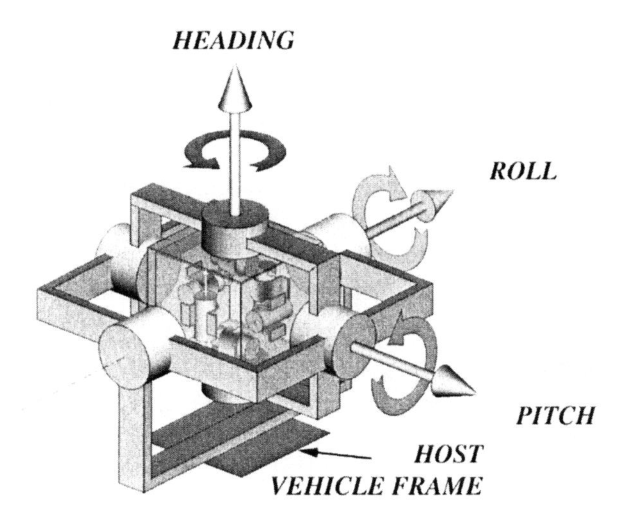

Fig. 2.6 Gimbals reading attitude Euler angles.

Design of gimbal torquing servos is complicated by the motions of the gimbals during operation, which determines how the torquing to correct for sensed rotation must be applied to the different gimbal bearings. This requires a bearing angle sensor for each gimbal axis.

The gimbal arrangement shown in Fig. 2.2 and 2.6, with the outer gimbal axis aligned to the roll (longitudinal) axis of the host vehicle and the inner gimbal axis maintained in the vertical direction, is a popular one. If the ISA is kept aligned with locally level east–north–up directions, the gimbal bearing angles shown in Fig. 2.6 will equal the heading, pitch and roll Euler angles defining the host vehicle attitude relative to east, north, and upward directions. These are the same Euler angles used to drive attitude and heading reference systems (AHRSs) (e.g., compass card and artificial horizon displays) in aircraft cockpits.

Advantages The principal advantage of both gimbaled and floated systems is the isolation of the inertial sensors from high angular rates, which eliminates many rate-dependent sensor errors (including gyro-scale factor sensitivity) and generally allows for higher accuracy sensors. Also, gimbaled systems can be self-calibrated by orienting the ISA with respect to gravity (for calibrating the accelerometers) and with respect to the earth rotation axis (for calibrating the gyros), and by using external optical autocollimators with mirrors on the ISA to independently measure its orientation with respect to its environment.

The most demanding INS applications for "cruise" applications (i.e., at ≈ 1 g) are probably for nuclear missile-carrying submarines, which must navigate submerged for about 3 months. The gimbaled Electrically Supported Gyro Navigation (ESGN, DOD designation AN/WSN-3 [34]) system developed in the 1970s at the Autonetics Division of Rockwell International for USN Trident-class submarines was probably the most accurate INS of that era [129].

Disadvantages The principal drawbacks of gimbals are cost, weight, volume, and gimbal flexure (in high-*g* environments). In traditional designs, electrical pathways are required through the gimbal structure to provide power to the IMU, and to carry power and the encoder, torquer, and sensor signals. These require slip rings (which can introduce noise) or cable wraps at the gimbal bearings. However, more recent designs have used wireless methods for signal and power transmission. The gimbal structure can interfere with air circulation used to maintain uniform temperatures within the IMU, and it can hamper access to the sensors during test and operation (e.g., for autocollimation off the ISA).

2.2.2.3 Floated Systems Gimbals and gimbal lock can be eliminated by floating the inertial sensor assembly in a liquid and operating it like a robotic submersible, using liquid thrusters to maintain its orientation and keep itself centered within the flotation cavity, as illustrated in Fig. 2.3. The floated assembly must also be neutrally buoyant and balanced to eliminate acceleration-dependent disturbances.

Advantages Floated systems have the advantage over gimbaled systems in that there are no gimbal structures to flex under heavy acceleration loading. Floated systems also have the same advantages as gimbaled systems over strapdown systems; isolation of the inertial sensors from high angular rates eliminates many rate-dependent error effects and generally allows for higher accuracy sensors. The ability to orient the sphere allows self-calibration capability, but the problem of autocollimation off the stable element is more difficult than for gimbaled systems. The floated *advanced inertial reference sphere* (AIRS) designed at the C. S. Draper Laboratory for MX/Peacekeeper and Minuteman III missiles is probably the most accurate (and most expensive) high-*g* INS ever developed [129].

Disadvantages A major disadvantage of floated systems is the difficulty of accessing the inertial sensor assembly for diagnostic testing, maintenance, or repair. The flotation system must be disassembled and the fluid drained for access, and then reassembled for operation. Floated systems also require means for determining the attitude of the floated assembly relative to the host vehicle, and wireless methods for providing power to the floated assembly and passing commands and sensor signals through the fluid.

2.2.2.4 Carouseling and Indexing

Carouseling A *carousel* is an amusement ride using continuous rotation of a circular platform about a vertical axis. The term "carouseling" has been applied to an implementation for gimbaled or floated systems in which the inertial sensor assembly revolves slowly around the local vertical axis—at rates in the order of a revolution per minute. The 3-gimbal configuration shown in Fig. 2.2 can implement carouseling using only the inner (vertical) gimbal axis. Carouseling significantly reduces long-term navigation errors due to some types of sensor errors (uncompensated biases of nominally level accelerometers and gyroscopes, in particular).

Indexing Alternative implementations called "indexing" or "gimbal flipping" use discrete rotations (usually by multiples of 90 degrees) to the same effect.

2.2.2.5 Strapdown Systems Strapdown systems use an inertial measurement unit that is not isolated from rotations of its host vehicle—except possibly by shock and vibration isolators. The gimbals are effectively replaced by software that uses the gyroscope outputs to calculate the equivalent accelerometer outputs in an attitude-stabilized coordinate frame, and integrates them to provide updates of velocity and position. This requires more computation (which is cheap) than does the gimbaled implementation, but it eliminates the gimbal system (which may not be cheap). It also exposes the accelerometers and gyroscopes to relatively high rotation rates, which can cause attitude-rate-dependent sensor errors.

Advantages The principal advantage of strapdown systems over gimbaled or floated systems is cost. The cost or replicating software is vanishingly small, compared to the cost of replicating a gimbal system for each IMU. For applications requiring attitude control of the host vehicle, strapdown gyroscopes generally provide more accurate rotation rate data than do the attitude readouts of gimbaled or floated systems.

Disadvantages Strapdown sensors must operate at much higher rotation rates, which can increase sensor cost. The dynamic ranges of the inputs to strapdown gyroscopes may be orders of magnitude greater than those for gyroscopes in gimbaled systems. To achieve comparable navigation performance, this generally requires orders of magnitude better scale factor stability for the strapdown gyroscopes. Strapdown systems generally require much shorter integration intervals—especially for integrating gyroscope outputs—and this increases the computer costs relative to gimbaled systems. Another disadvantage for strapdown is the cost of gyroscope calibration and testing, which requires a precision rate table. (Rate tables are not required for whole-angle gyroscopes—including electrostatic gyroscopes—or for the gyroscopes used in gimbaled systems.)

2.2.2.6 Strapdown Carouseling and Indexing For host vehicles that are nominally upright during operation (e.g., ships), a strapdown system can be rotated about the host vehicle yaw axis. So long as the vehicle yaw axis remains close to the local vehicle, slow rotation (carouseling) or indexing about this axis can significantly reduce the effects of uncompensated biases of the nominally level accelerometers and gyroscopes. The rotation is normally oscillatory, with reversal of direction after a full rotation, so that the connectors can be wrapped to avoid using slip rings (an option not generally available for gimbaled systems).

Disadvantages Carouseling or indexing of strapdown systems requires the addition of a rotation bearing and associated motor drive, wiring, and control electronics. The benefits are not without cost. For gimbaled and floated systems, the additional costs are relatively insignificant.

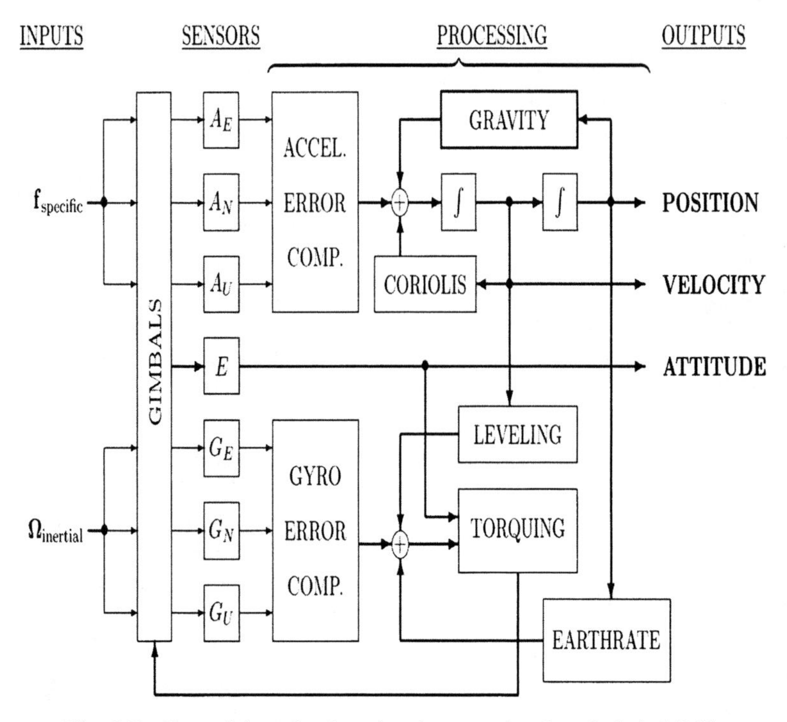

Fig. 2.7 Essential navigation signal processing for gimbaled INS.

2.2.3 Sensor Signal Processing

2.2.3.1 Gimbaled Implementations Figure 2.7 shows the essential navigation signal processing functions for a gimbaled INS with inertial sensor axes aligned to locally level coordinates, where

$\mathbf{f}_{specific}$ is the specific force (i.e., the sensible acceleration, exclusive of gravitational acceleration) applied to the host vehicle.

$\Omega_{inertial}$ is the instantaneous inertial rotation rate vector of the host vehicle.

A denotes a specific force sensor (accelerometer).

E denotes the ensemble of gimbal angle encoders, one for each gimbal angle (there are several possible formats for the gimbal angles, including digitized angles, 3-wire synchro signals, or sin / cos pairs).

G denotes an inertial rotation rate sensor (gyroscope).

POSITION is the estimated position of the host vehicle in navigation coordinates (e.g., longitude, latitude, and altitude relative to sea level).

VELOCITY is the estimated velocity of the host vehicle in navigation coordinates (e.g., east, north, and vertical).

ATTITUDE is the estimated attitude of the host vehicle relative to locally level coordinates. For some 3-gimbal systems, the gimbal angles are the Euler angles representing vehicle heading (with respect to north), pitch, and roll. Output attitude may also be used to drive cockpit displays such as compass cards or artificial horizon indicators.

ACCEL. ERROR COMP. and GYRO ERROR COMP. denote the calibrated corrections for sensor errors. These generally include corrections for scale factor variations, output biases and input axis misalignments for both types of sensors, and acceleration-dependent errors for gyroscopes.

GRAVITY denotes the gravity model used to compute the acceleration due to gravity as a function of position.

CORIOLIS denotes the acceleration correction for coriolis effect in rotating coordinates.

LEVELING denotes the rotation rate correction in locally level coordinates moving over the surface of the earth.

EARTHRATE denotes the model used to calculate the earth rotation rate in locally level INS coordinates.

TORQUING denotes the servo loop gain computations used in stabilizing the INS in locally level coordinates.

Not shown in the figure is the input altitude reference (e.g., barometric altimeter or GPS) required for vertical channel (altitude) stabilization.[2]

Initializing INS Alignment This signal processing schematic in Fig. 2.7 is for operation in the navigation mode. It does not show the implementation used for initial alignment of the sensor axes, which is done while the INS is essentially stationary. During initial alignment, the outputs of the east and north accelerometers (denoted by A_E and A_N) are used for leveling the INS, and the output of the east gyroscope (denoted by G_E) is used for aligning the INS in heading. When the INS is aligned, the east- and north-pointing accelerometer outputs will always be zero, and the east-pointing gyroscope output will also be zero.

A magnetic compass may be used to get an initial rough estimate of alignment, which can speed up the alignment process. Also—if the host vehicle has not been moved—the alignment information at system shutdown can be saved to initialize alignment at the next system turnon.

Initializing INS Position and Velocity The integrals shown in Fig. 2.7 require initial values for velocity and position. The INS normally remains stationary (i.e., with zero velocity) during INS alignment initialization, which solves the velocity initialization problem. The angle between the sensed acceleration vector and the sensed earthrate vector can be used to estimate latitude, but INS position

[2]Vertical channel instability of inertial navigation is caused by the decrease in modeled gravity with increasing altitude.

(including longitude and altitude) must ordinarily be initialized from external sources (such as GNSS). If the vehicle has not been moved too far during shutdown, the position from the last shutdown can be used to initialize position at turnon.

2.2.3.2 ***Strapdown Implementations*** The basic signal processing functions for a strapdown INS navigation are diagrammed in Fig. 2.8, where the common symbols used in Fig. 2.7 have the same meaning as before, and

G is the estimated gravitational acceleration, computed as a function of estimated position.

$\mathrm{POS}_{\mathrm{NAV}}$ is the estimated position of the host vehicle in navigation coordinates.

$\mathrm{VEL}_{\mathrm{NAV}}$ is the estimated velocity of the host vehicle in navigation coordinates.

$\mathrm{ACC}_{\mathrm{NAV}}$ is the estimated acceleration of the host vehicle in navigation coordinates, which may be used for trajectory control (i.e., vehicle guidance).

$\mathrm{ACC}_{\mathrm{SENSOR}}$ is the estimated acceleration of the host vehicle in sensor-fixed coordinates, which may be used for steering stabilization and control.

$\mathbf{C}_{\mathrm{NAV}}^{\mathrm{SENSOR}}$ is the 3×3 coordinate transformation matrix from sensor-fixed coordinates to navigation coordinates, representing the attitude of the sensors in navigation coordinates.

Ω_{SENSOR} is the estimated angular velocity of the host vehicle in sensor-fixed coordinates, which may be used for vehicle attitude stabilization and control.

Ω_{NAV} is the estimated angular velocity of the host vehicle in navigation coordinates, which may be used in a vehicle pointing and attitude control loop.

The essential processing functions include double integration (represented by boxes containing integration symbols) of acceleration to obtain position, and computation of (unsensed) gravitational acceleration as a function of position. The sensed angular rates also need to be integrated to maintain the knowledge of sensor attitudes. The initial values of all the integrals (i.e., position, velocity, and attitude) must also be known before integration can begin.

The position vector $\mathrm{POS}_{\mathrm{NAV}}$ is the essential navigation solution. The other outputs shown are not needed for all applications, but most of them (except Ω_{NAV}) are intermediate results that are available "for free" (i.e., without requiring further processing). The velocity vector $\mathrm{VEL}_{\mathrm{NAV}}$, for example, characterizes speed and heading, which are also useful for correcting the course of the host vehicle to bring it to a desired location. Most of the other outputs shown would be required for implementing control of an unmanned or autonomous host vehicle to follow a desired trajectory and/or to bring the host vehicle to a desired final position.

Navigation functions that are not shown in Fig. 2.8 include

1. How initialization of the integrals for position, velocity, and attitude is implemented. Initial position and velocity can be input from other sources (e.g., GNSS), and attitude can be inferred from some form of trajectory matching (using GNSS, e.g.) or by *gyrocompassing* (described below).

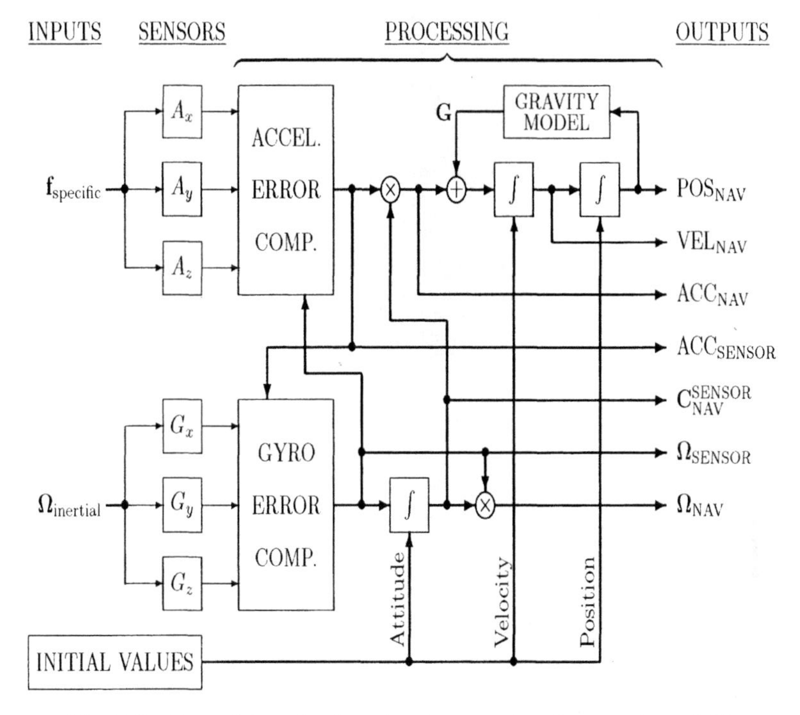

Fig. 2.8 Essential navigation signal processing for strapdown INS.

2. How attitude rates are integrated to obtain attitude. Because rotation operations are not commutative, attitude rate integration is not as straightforward as the integration of acceleration to obtain velocity and position. Special techniques required for attitude rate integration are described in Chapter 9.
3. For the case that navigation coordinates are earth-fixed, the computation of navigational coordinate rotation due to earthrate as a function of position, and its summation with sensed rates before integration.
4. For the case that navigation coordinates are locally level, the computation of the rotation rate of navigation coordinates due to vehicle horizontal velocity, and its summation with sensed rates before integration.
5. Calibration of the sensors for error compensation. If the errors are sufficiently stable, it needs to be done only once. Otherwise, it can be implemented using GNSS/INS integration techniques.

2.2.3.3 Gyrocompass Alignment

Alignment is the term used for a procedure to determine the orientation of the ISA relative to navigation coordinates.

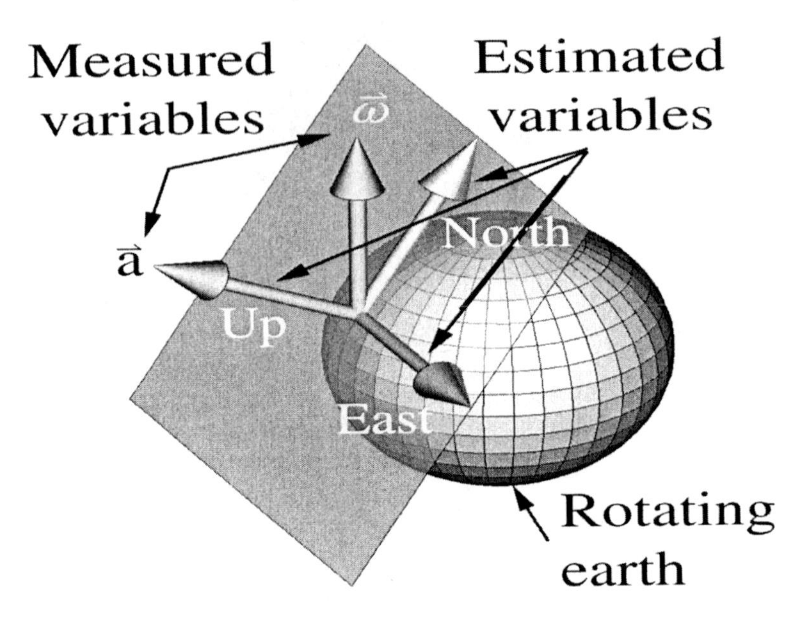

Fig. 2.9 Gyrocompassing determines sensor orientations with respect to east, north, and up.

A *gyrocompass* is an instrument that levels itself relative to true vertical, detects the rotation of the vertical due to the rotation of the earth, and uses the direction of rotation to align itself to true north.

Gyrocompassing of an INS is a procedure for self-contained alignment of its ISA.

Gyrocompassing must be performed when the host vehicle is essentially stationary. Under stationary conditions, the INS can use its accelerometers to determine the direction of the local vertical relative to the sensors, and its gyroscopes to determine the direction of the earth rotation axis relative to the sensors. The cross-product of local vertical vector and the earth rotation axis vector in sensor-fixed coordinates will point east, as illustrated in Fig. 2.9. The cross-product of local vertical (i.e., "Up") and "East," will be "North," as shown in the figure.

Performance Degrades Near the Poles Near the north or south pole of the earth, the direction of the earth rotation axis comes close to the local vertical direction, and their cross-product is not well determined. For that reason, gyrocompassing is not accurate near the poles.

2.2.4 Standalone INS Performance

2.2.4.1 Free Inertial Operation Operation of an INS without external aiding of any sort is called "free inertial" or "pure inertial." Because free inertial navigation in the near-earth gravitational environment is unstable[3] in the vertical direction, aiding by other sensors (e.g., barometric altimeters for aircraft or surface vehicles, radar altimeters for aircraft over water, or hydrostatic pressure for submersibles) is required to avoid vertical error instability. For that reason, performance of free-inertial navigation systems is usually specified for horizontal position errors only.

2.2.4.2 INS Performance Metrics

Simplified Error Model INS position is initialized by knowing where you are starting from at initial time t_0. The position error may start out very small, but it tends to increase with time due to the influence of sensor errors. Double integration of accelerometer output errors is a major source of this growth over time. Experience has shown that the variance and standard deviation of horizontal position error

$$\sigma^2_{\text{position}}(t) \overset{\propto}{\approx} (t - t_0)^2 , \tag{2.1}$$

$$\sigma_{\text{position}}(t) \approx C \times |t - t_0| , \tag{2.2}$$

with unknown positive constant C. This constant C would then characterize performance of an INS in terms of how fast its RMS position error grows.

A problem with this model is that actual horizontal INS position errors are two-dimensional, and we would need a 2×2 covariance matrix in place of C. That would not be very useful in practice. As an alternative, we replace C with something more intuitive and practical.

CEP The radius of a horizontal circle centered at the estimated position, and of sufficient radius such that it is equally probable that the true horizontal position is inside or outside the circle is called *circular error probable* (CEP). CEP is also used as an acronym for "circle of equal probability" (of being inside or outside).

CEP rate The time rate of change of circular error probable is CEP rate. Traditional units of CEP rate are nautical miles per hour or kilometers per hour. The nautical mile was originally intended to designate a surface distance equivalent to one arc minute of latitude change at sea level, but that depends on latitude. The SI-derived nautical mile is 1.852 km.

2.2.4.3 Performance Levels In the 1970s, before GPS became a reality, the U.S. Air Force had established the following levels of performance for INS:

High-accuracy systems have free inertial CEP rates in the order of 0.1 nautical miles per hour (nmi/h) ($\approx$ 185 m/h) or better. This is the order of

[3]This is due to the falloff of gravity with increasing altitude. The issue is covered in Section 9.5.2.2.

magnitude in accuracy required for intercontinental ballistic missiles and missile-carrying submarines, for example.

Medium-accuracy systems have free inertial CEP rates in the order of 1 nmi/h (nautical miles per hour) ($\approx$ 1.85 km/h). This is the level of accuracy deemed sufficient for most military and commercial aircraft [34].

Low-accuracy systems have free inertial CEP rates in the order of 10 nmi/h (nautical miles per hour) ($\approx$ 18.5 km/h) or worse. This range covered the requirements for many short-range standoff weapons such as guided artillery or tactical rockets.

However, after GPS became available, GPS/INS integration could make a low-accuracy INS behave more like a high-accuracy INS.

2.3 SATELLITE NAVIGATION

The GPS is widely used in navigation. Its augmentation with other space-based satellites is the future of near-earth navigation.

2.3.1 Satellite Orbits

GPS satellites occupy six orbital planes inclined 55° from the equatorial plane, as illustrated in Figs. 2.10 and 2.11. Each of the six orbit planes in Fig. 2.11 contains four or more satellites.

2.3.2 Navigation Solution (Two-Dimensional Example)

Antenna location in two dimensions can be calculated by using range measurements [65].

2.3.2.1 Symmetric Solution Using Two Transmitters on Land In this case, the receiver and two transmitters are located in the same plane, as shown in Fig. 2.12, with known positions x_1, y_1 and x_2, y_2. Ranges R_1 and R_2 of two transmitters from the user position are calculated as

$$R_1 = c\,T_1, \tag{2.3}$$

$$R_2 = c\,T_2, \tag{2.4}$$

where

c = speed of light (0.299792458 m/ns)

ΔT_1 = time taken for the radiowave to travel from transmitter 1 to the user

ΔT_2 = time taken for the radiowave to travel from transmitter 2 to the user

(X, Y) = user position

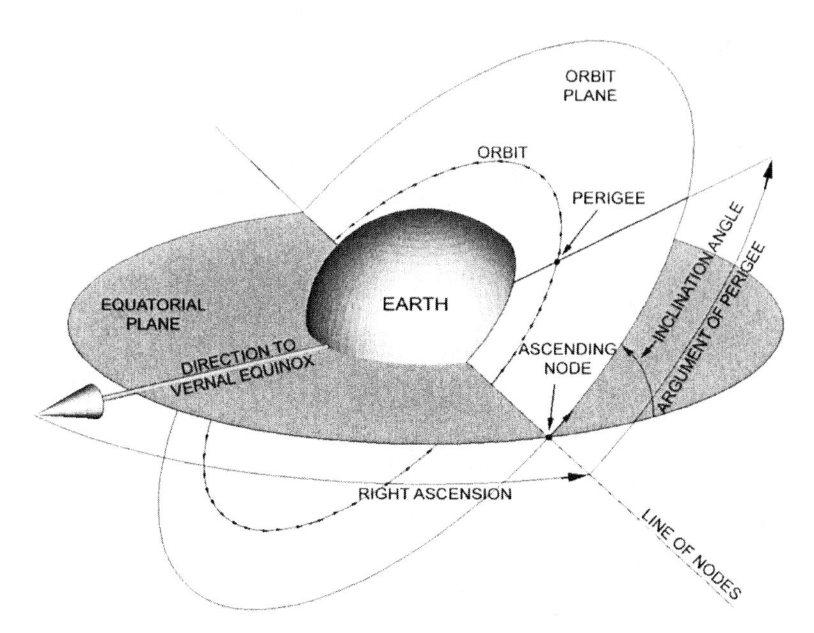

Fig. 2.10 Parameters defining satellite orbit geometry.

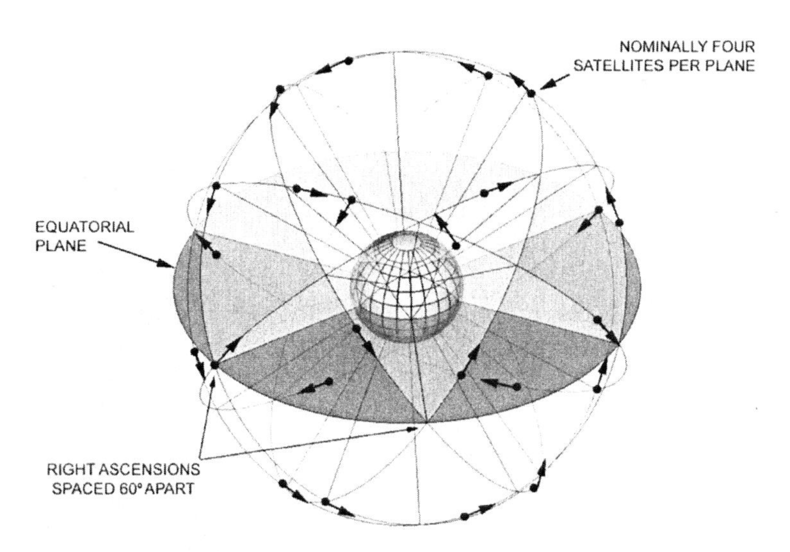

Fig. 2.11 GPS orbit planes.

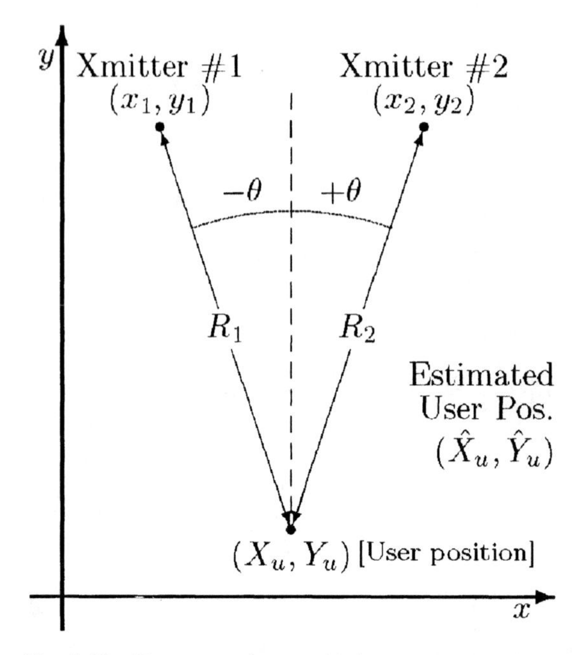

Fig. 2.12 Two transmitters with known 2D positions.

The range to each transmitter can be written as

$$R_1 = [(X - x_1)^2 + (Y - y_1)^2]^{1/2}, \tag{2.5}$$

$$R_2 = [(X - x_2)^2 + (Y - y_2)^2]^{1/2}. \tag{2.6}$$

Expanding R_1 and R_2 in Taylor series expansion with small perturbation in X by Δx and Y by Δy yields

$$\Delta R_1 = \frac{\partial R_1}{\partial X}\Delta x + \frac{\partial R_1}{\partial Y}\Delta y + u_1, \tag{2.7}$$

$$\Delta R_2 = \frac{\partial R_2}{\partial X}\Delta x + \frac{\partial R_2}{\partial Y}\Delta y + u_2, \tag{2.8}$$

where u_1 and u_2 are higher order terms. The derivatives of Eqs. 2.5 and 2.6 with respect to X, Y are substituted into Eqs. 2.7 and 2.8, respectively.

Thus, for the symmetric case, we obtain

$$\Delta R_1 = \frac{X - x_1}{[(X - x_1)^2 + (Y - y_1)^2]^{1/2}} \Delta x \tag{2.9}$$

$$+ \frac{Y - y_1}{[(X - x_1)^2 + (Y - y_1)^2]^{1/2}} \Delta y + u_1,$$

$$= \sin\theta\,\Delta x + \cos\theta\,\Delta y + u_1, \tag{2.10}$$

$$\Delta R_2 = -\sin\theta\,\Delta x + \cos\theta\,\Delta y + u_2. \tag{2.11}$$

To obtain the least-squares estimate of (X, Y), we need to minimize the quantity

$$J = u_1^2 + u_2^2, \tag{2.12}$$

which is

$$J = \Bigg(\underbrace{\Delta R_1 - \sin\ \theta\,\Delta x - \cos\ \theta\,\Delta y}_{u_1}\Bigg)^2 + \Bigg(\underbrace{\Delta R_2 + \sin\ \theta\,\Delta x - \cos\ \theta\,\Delta y}_{u_2}\Bigg)^2. \tag{2.13}$$

The solution for the minimum can be found by setting $\partial J/\partial\Delta x = 0 = \partial J/\partial\Delta y$, then solving for Δx and Δy:

$$0 = \frac{\partial J}{\partial \Delta x} \tag{2.14}$$

$$= 2(\Delta R_1 - \sin\theta\,\Delta x - \cos\theta\,\Delta y)\,(-\sin\theta) + 2(\Delta R_2 \tag{2.15}$$

$$+ \sin\theta\,\Delta x - \cos\theta\,\Delta y)\,(\sin\theta) \tag{2.16}$$

$$= \Delta R_2 - \Delta R_1 + 2\sin\theta\,\Delta x, \tag{2.17}$$

with solution

$$\Delta x = \frac{\Delta R_1 - \Delta R_2}{2\sin\theta}. \tag{2.18}$$

The solution for Δy may be found in similar fashion as

$$\Delta y = \frac{\Delta R_1 + \Delta R_2}{2\cos\theta}. \tag{2.19}$$

Navigation Solution Procedure Transmitter positions x_1, y_1, x_2, y_2 are given. Signal travel times ΔT_1, ΔT_2 are given. Estimated user position $\hat{X}_u, \hat{Y}_u$ are assumed.

Set position coordinates X, Y equal to their initial estimates:

$$X = \hat{X}_u, \qquad Y = \hat{Y}_u,$$

Compute the range errors:

$$\Delta R_1 = \overbrace{[(\hat{X}_u - x_1)^2 + (\hat{Y}_u - y_1)^2]^{1/2}}^{\text{Geometric ranges}} - \overbrace{C\Delta T_1}^{\text{Measured pseudoranges}}, \tag{2.20}$$

$$\Delta R_2 = [(\hat{X}_u - x_2)^2 + (\hat{Y}_u - y_2)^2]^{1/2} - C\Delta T_2. \tag{2.21}$$

Compute the theta angle:

$$\theta = \tan^{-1} \frac{\hat{X}_u - x_1}{\hat{Y}_u - y_1} \tag{2.22}$$

$$= \sin^{-1} \frac{\hat{X}_u - x_1}{\sqrt{\left(\hat{X}_u - x_1\right)^2 + \left(\hat{Y}_u - y_1\right)^2}}. \tag{2.23}$$

Compute user position corrections:

$$\Delta x = \frac{1}{2 \sin \theta}(\Delta R_1 - \Delta R_2), \tag{2.24}$$

$$\Delta y = \frac{1}{2 \cos \theta}(\Delta R_1 + \Delta R_2). \tag{2.25}$$

Compute a new estimate of position:

$$X = \hat{X}_u + \Delta x, \quad Y = \hat{Y}_u + \Delta y. \tag{2.26}$$

Continue to compute θ, ΔR_1 and ΔR_2 from these equations with new values of x and y.

Iterate Eqs. 2.20–2.26:

Correction equations	Iteration equations
$\Delta X_{\text{best}} = \frac{1}{2 \sin \theta}(\Delta R_1 - \Delta R_2)$,	$X_{\text{new}} = X_{\text{old}} + \Delta X_{\text{best}}$,
$\Delta Y_{\text{best}} = \frac{1}{2 \cos \theta}(\Delta R_1 + \Delta R_2)$,	$Y_{\text{new}} = Y_{\text{old}} + \Delta Y_{\text{best}}$.

2.3.3 Satellite Selection and Dilution of Precision

Just as in a land-based system, better accuracy is obtained by using reference points well separated in space. For example, the range measurements made to four reference points clustered together will yield nearly equal values. Position calculations involve range differences, and where the ranges are nearly equal, small relative errors are greatly magnified in the difference. This effect, brought about as a result of satellite geometry, is known as *dilution of precision* (DOP). This means that range errors that occur from other causes such as clock errors are also magnified by the geometric effect.

To find the best locations of the satellites to be used in the calculations of the user position and velocity, DOP calculations are needed.

The observation equations in three dimensions for each satellite with known coordinates (x_i, y_i, z_i) and unknown user coordinates (X, Y, Z) are given by

$$Z_{\rho i} = \rho^i = \sqrt{(x_i - X)^2 + (y_i - Y)^2 + (z_i - Z)^2} + C_b. \tag{2.27}$$

These are nonlinear equations that can be linearized using Taylor series (see, e.g., Chapter 5 of Ref. 66).

Let the vector of ranges be $Z_\rho = \mathbf{h}(\mathbf{x})$, a nonlinear function $\mathbf{h}(\mathbf{x})$ of the four-dimensional vector $\mathbf{x}$ representing user position and receiver clock bias, and expand the left-hand side of this equation in a Taylor series about some nominal solution $\mathbf{x}^{\text{nom}}$ for the unknown vector

$$\mathbf{x} = [X, Y, Z, C_b]^T \tag{2.28}$$

of variables

$$X \stackrel{\text{def}}{=} \text{east component of the user's antenna location}$$
$$Y \stackrel{\text{def}}{=} \text{north component of the user's antenna location}$$
$$Z \stackrel{\text{def}}{=} \text{upward vertical component of the user's antenna location}$$
$$C_b \stackrel{\text{def}}{=} \text{receiver clock bias}$$

for which

$$\begin{aligned} Z_\rho &= \mathbf{h}(\mathbf{x}) = \mathbf{h}(\mathbf{x}^{\text{nom}}) + \left.\frac{\partial \mathbf{h}(\mathbf{x})}{\partial \mathbf{x}}\right|_{\mathbf{x}=\mathbf{x}^{\text{nom}}} \delta\mathbf{x} + \text{H.O.T}\ , \\ \delta\mathbf{x} &= \mathbf{x} - \mathbf{x}^{\text{nom}}, \qquad \delta Z_\rho = \mathbf{h}(\mathbf{x}) - \mathbf{h}(\mathbf{x}^{\text{nom}}), \end{aligned} \tag{2.29}$$

where H.O.T stands for "higher-order terms."

These equations become

$$\delta Z_\rho = \left.\frac{\partial \mathbf{h}(\mathbf{x})}{\partial \mathbf{x}}\right|_{x=x^{\text{nom}}} \delta\mathbf{x} \quad = H^{[1]}\delta\mathbf{x},$$
$$\delta x = X - X_{\text{nom}},\ \delta y = Y - Y_{\text{nom}},\ \delta z = Z - Z_{\text{nom}}, \tag{2.30}$$

where $H^{[1]}$ is the first-order term in the Taylor series expansion

$$\delta Z_\rho = \rho(X,\ Y,\ Z) - \rho_r(X_{\text{nom}},\ Y_{\text{nom}},\ Z_{\text{nom}}) \tag{2.31}$$

$$\approx \underbrace{\left.\frac{\partial \rho_r}{\partial X}\right|_{X_{\text{nom}},\ Y_{\text{nom}},\ Z_{\text{nom}}}}_{H^{[1]}} \delta\mathbf{x} + v_\rho \tag{2.32}$$

for v_ρ = noise in receiver measurements. This vector equation can be written in scalar form where i = satellite number as

$$\left.\begin{aligned}
\frac{\partial \rho_r^i}{\partial X} &= \left.\frac{-(x_i - X)}{\sqrt{(x_i - X)^2 + (y_i - Y)^2 + (z_i - Z)^2}}\right|_{X=X_{\text{nom}},\ Y_{\text{nom}},\ Z_{\text{nom}}} \\
&= \frac{-(x_i - X_{\text{nom}})}{\sqrt{(x_i - X_{\text{nom}})^2 + (y_i - Y_{\text{nom}})^2 + (z_i - Z_{\text{nom}})^2}} \\
\frac{\partial \rho_r^i}{\partial Y} &= \frac{-(y_i - Y_{\text{nom}})}{\sqrt{(x_i - X_{\text{nom}})^2 + (y_i - Y_{\text{nom}})^2 + (z_i - Z_{\text{nom}})^2}} \\
\frac{\partial \rho_r^i}{\partial Z} &= \frac{-(z_i - Z_{\text{nom}})}{\sqrt{(x_i - X_{\text{nom}})^2 + (y_i - Y_{\text{nom}})^2 + (z_i - Z_{\text{nom}})^2}}
\end{aligned}\right\} \tag{2.33}$$

for

$$i = 1,\ 2,\ 3,\ 4 \text{ (i.e., four satellites)} \tag{2.34}$$

We can combine Eqs. 2.32 and 2.33 into the matrix equation

$$\underbrace{\begin{bmatrix} \delta z_\rho^1 \\ \delta z_\rho^2 \\ \delta z_\rho^3 \\ \delta z_\rho^4 \end{bmatrix}}_{4\times 1} = \underbrace{\begin{bmatrix} \frac{\partial\rho_r^1}{\partial x} & \frac{\partial\rho_r^1}{\partial y} & \frac{\partial\rho_r^1}{\partial z} & 1 \\ \frac{\partial\rho_r^2}{\partial x} & \frac{\partial\rho_r^2}{\partial y} & \frac{\partial\rho_r^2}{\partial z} & 1 \\ \frac{\partial\rho_r^3}{\partial x} & \frac{\partial\rho_r^3}{\partial y} & \frac{\partial\rho_r^3}{\partial z} & 1 \\ \frac{\partial\rho_r^4}{\partial x} & \frac{\partial\rho_r^4}{\partial y} & \frac{\partial\rho_r^4}{\partial z} & 1 \end{bmatrix}}_{4\times 4} \underbrace{\begin{bmatrix} \delta x \\ \delta y \\ \delta z \\ C_b \end{bmatrix}}_{4\times 1} + \underbrace{\begin{bmatrix} v_\rho^1 \\ v_\rho^2 \\ v_\rho^3 \\ v_\rho^4 \end{bmatrix}}_{4\times 1},$$

which we can write in symbolic form as

$$\overbrace{\delta Z_\rho}^{4\times 1} = \overbrace{H^{[1]}}^{4\times 4}\ \overbrace{\delta\mathbf{x}}^{4\times 1} + \overbrace{v_k}^{4\times 1}$$

(see Table 5.3 in Ref. 66).

To calculate $H^{[1]}$, one needs satellite positions and the nominal value of the user's position.

To calculate the geometric dilution of precision (GDOP) (approximately), we obtain

$$\overbrace{\delta Z_\rho}^{4\times 1} = \overbrace{H^{[1]}}^{4\times 1} \overbrace{\delta \mathbf{x}}^{4\times 1}. \tag{2.35}$$

Known are δZ_ρ and $H^{[1]}$ from the pseudorange, satellite position, and nominal value of the user's position. The correction δx is the unknown vector.

If we premultiply both sides of Eq. 2.35 by $H^{[1]T}$, the result will be

$$H^{[1]T}\,\delta Z_\rho = \underbrace{\overbrace{H^{[1]T}}^{4\times 4}\,\overbrace{H^{[1]}}^{4\times 4}}_{4\times 4}\,\delta \mathbf{x}. \tag{2.36}$$

Then we premultiply Eq. 2.36 by $\left(H^{[1]T}\,H^{[1]}\right)^{-1}$:

$$\delta \mathbf{x} = \left(H^{[1]T} H^{[1]}\right)^{-1} H^{[1]T}\delta Z_\rho. \tag{2.37}$$

If $\delta \mathbf{x}$ and δZ_ρ are assumed random with zero mean, the error covariance

$$E\langle(\delta \mathbf{x})\,(\delta \mathbf{x})^T\rangle$$

$$= E\langle\left(H^{[1]T} H^{[1]}\right)^{-1} H^{[1]T}\delta Z_\rho \left[\left(H^{[1]T} H^{[1]}\right)^{-1} H^{[1]T}\delta Z_\rho\right]^T\rangle \tag{2.38}$$

$$= \left(H^{[1]T} H^{[1]}\right)^{-1} H^{[1]T}\ \underbrace{E\langle\delta Z_\rho \delta Z_\rho^T\rangle}\left(H^{[1]T} H^{[1]}\right)^{-1}. \tag{2.39}$$

The pseudorange measurement covariance is assumed uncorrelated satellite-to-satellite with variance σ^2:

$$E\langle\delta Z_\rho \delta Z_\rho^T\rangle = \sigma^2 \mathbf{I}_4, \tag{2.40}$$

a 4×4 matrix.

Substituting Eq. 2.40 into Eq. 2.39 gives

$$E\langle\delta \mathbf{x}(\delta \mathbf{x})^T\rangle = \sigma^2 (H^{[1]T}\ H^{[1]})^{-1}\,\underbrace{(H^{[1]T}\ H^{[1]})(H^{[1]T}\ H^{[1]})^{-1}}_{\mathbf{I}} \tag{2.41}$$

$$= \sigma^2 (H^{[1]T}\ H^{[1]})^{-1}, \tag{2.42}$$

for

$$\overbrace{\delta \mathbf{x}}^{4\times 1} = \begin{bmatrix} \Delta E \\ \Delta N \\ \Delta U \\ C_b \end{bmatrix},$$

and

$$\begin{array}{l} \Delta E = \text{east error} \\ \Delta N = \text{north error} \\ \Delta U = \text{up error} \end{array} \qquad \left(\begin{array}{c} \text{locally} \\ \text{level} \\ \text{coordinate} \\ \text{frame} \end{array} \right),$$

and the covariance matrix becomes

$$\underbrace{E\left\langle \delta \mathbf{x}(\delta \mathbf{x})^{\mathrm{T}} \right\rangle}_{4\times 4} = \begin{bmatrix} \mathrm{E}\left\langle \Delta E^2 \right\rangle & \mathrm{E}\left\langle \Delta E \Delta N \right\rangle & \mathrm{E}\left\langle \Delta E \Delta U \right\rangle & \mathrm{E}\left\langle \Delta E \Delta Cb \right\rangle \\ \mathrm{E}\left\langle \Delta N \Delta E \right\rangle & \mathrm{E}\left\langle \Delta N^2 \right\rangle & \mathrm{E}\left\langle \Delta N \Delta U \right\rangle & \mathrm{E}\left\langle \Delta N \Delta Cb \right\rangle \\ \mathrm{E}\left\langle \Delta U \Delta E \right\rangle & \mathrm{E}\left\langle \Delta U \Delta N \right\rangle & \mathrm{E}\left\langle \Delta U^2 \right\rangle & \mathrm{E}\left\langle \Delta U \Delta Cb \right\rangle \\ \mathrm{E}\left\langle \Delta Cb \Delta E \right\rangle & \mathrm{E}\left\langle \Delta Cb \Delta N \right\rangle & \mathrm{E}\left\langle \Delta Cb \Delta U \right\rangle & \mathrm{E}\left\langle C_b^2 \right\rangle \end{bmatrix}. \tag{2.43}$$

We are principally interested in the diagonal elements of

$$(H^{[1]\mathrm{T}} H^{[1]})^{-1} = \begin{bmatrix} A_{11} & A_{12} & A_{13} & A_{14} \\ A_{21} & A_{22} & A_{23} & A_{24} \\ A_{31} & A_{32} & A_{33} & A_{34} \\ A_{41} & A_{42} & A_{43} & A_{44} \end{bmatrix}, \tag{2.44}$$

with $\sigma^2 = 1\ \mathrm{m}^2$ in the following combinations (see Fig. 2.13):

$$\begin{array}{lcll} \text{GDOP} & = & \sqrt{A_{11} + A_{22} + A_{33} + A_{44}} & \text{(geometric DOP)}, \\ \text{PDOP} & = & \sqrt{A_{11} + A_{22} + A_{33}} & \text{(position DOP)}, \\ \text{HDOP} & = & \sqrt{A_{11} + A_{22}} & \text{(horizontal DOP)}, \\ \text{VDOP} & = & \sqrt{A_{33}} & \text{(vertical DOP)}, \\ \text{TDOP} & = & \sqrt{A_{44}} & \text{(time DOP)}. \end{array}$$

Hereafter, all DOPs represent the sensitivities to pseudorange errors.

2.3.4 Example Calculation of DOPs

2.3.4.1 Four Satellites The best accuracy is found with three satellites equally spaced on the horizon, at minimum elevation angle, with the fourth satellite directly overhead, as listed in Table 2.1.

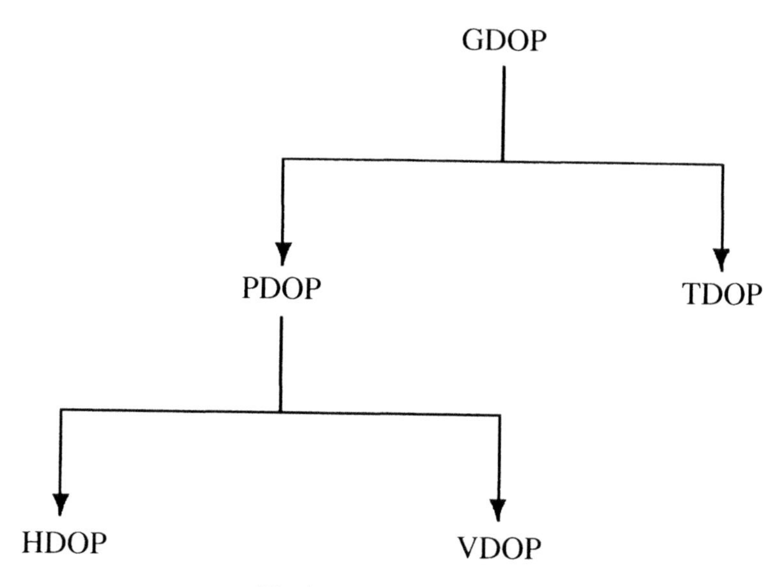

Fig. 2.13 DOP hierarchy.

The diagonal of the unscaled covariance matrix $H^{[1]\mathrm{T}} H^{[1]}$ then has the terms

$$\begin{bmatrix} (\text{east DOP})^2 & & & \\ & (\text{north DOP})^2 & & \\ & & (\text{vertical DOP})^2 & \\ & & & (\text{time DOP})^2 \end{bmatrix},$$

where

$$\mathrm{GDOP} = \sqrt{\operatorname{trace}(H^{[1]\mathrm{T}} H^{[1]})^{-1}},\ H^{[1]} = \left.\frac{\partial \rho}{\partial x}\right|_{X_{\mathrm{nom}}\ Y_{\mathrm{nom}},\ Z_{\mathrm{nom}}}.$$

TABLE 2.1. Example with Four Satellites

	Satellite location			
	1	2	3	4
Elevation (deg)	5	5	5	90
Azimuth (deg)	0	120	240	0

Typical example values of $H^{[1]}$ for this geometry are

$$H^{[1]} = \begin{bmatrix} 0.000 & 0.996 & 0.087 & 1.000 \\ 0.863 & -0.498 & 0.087 & 1.000 \\ -0.863 & -0.498 & 0.087 & 1.000 \\ 0.000 & 0.000 & 1.000 & 1.000 \end{bmatrix}.$$

The GDOP calculations for this example are

$$(H^{[1]T} H^{[1]})^{-1} = \begin{bmatrix} 0.672 & 0.000 & 0.000 & 0.000 \\ 0.000 & 0.672 & 0.000 & 0.000 \\ 0.000 & 0.000 & 1.600 & -0.505 \\ 0.000 & 0.000 & -0.505 & 0.409 \end{bmatrix},$$

$$\begin{aligned} \text{GDOP} &= \sqrt{0.672 + 0.672 + 1.6 + 0.409} \\ &= 1.83, \end{aligned}$$

$$\begin{aligned} \text{HDOP} &= 1.16, \\ \text{VDOP} &= 1.26, \\ \text{PDOP} &= 1.72, \\ \text{TDOP} &= 0.64. \end{aligned}$$

2.4 TIME AND GPS

2.4.1 Coordinated Universal Time Generation

Coordinated Universal Time (UTC) is the timescale based on the atomic second, but occasionally corrected by the insertion of leap seconds, so as to keep it approximately synchronized with the earth's rotation. The leap second adjustments keep UTC within 0.9 s of UT1, which is a timescale based on the earth's axial spin. UT1 is a measure of the true angular orientation of the earth in space. Because the earth does not spin at exactly a constant rate, UT1 is not a uniform timescale [3].

2.4.2 GPS System Time

The timescale to which GPS signals are referenced is referred to as *GPS time*. GPS time is derived from a composite or "paper" clock that consists of all operational monitor station and satellite atomic clocks. Over the long run, it is steered to keep it within about 1 μs of UTC, as maintained by the master clock at the U.S. Naval Observatory, ignoring the UTC leap seconds. At the integer second level, GPS time equaled UTC in 1980. However, due to the leap seconds that have been inserted into UTC, GPS time was ahead of UTC by 14 s in February 2006.

2.4.3 Receiver Computation of UTC

The parameters needed to calculate UTC from GPS time are found in subframe 4 of the navigation data message. These data include a notice to the user regarding the scheduled future or recent past (relative to the navigation message upload) value of the delta time due to leap seconds Δt_{LSF}, together with the week number WN_{LSF} and the day number DN at the end of which the leap second becomes effective. The latter two quantities are known as the *effectivity time* of the leap second. "Day 1" is defined as the first day relative to the end/start of a week and the WN_{LSF} value consists of the eight least significant bits (LSBs) of the full week number.

Three different UTC/GPS time relationships exist, depending on the relationship of the effectivity time to the user's current GPS time:

1. *First Case.* Whenever the effectivity time indicated by the WN_{LSF} and WN values is not in the past relative to the user's present GPS time, *and* the user's present time does not fall in the timespan starting at $DN + \frac{3}{4}$ and ending at $DN + \frac{5}{4}$, the UTC time is calculated as:

$$t_{UTC} = (t_E - \Delta t_{UTC}) \quad (\text{modulo } 86400)s,$$

where t_{UTC} is in seconds, 86400 is the number of seconds per day, and

$$\Delta t_{UTC} = \Delta t_{LS} + A_0 + A_1[t_E - t_{0t} + 604800(\text{WN} - \text{WN}_t)]\text{s},$$

where 604800 is the number of seconds per week, and

t_E = user GPS time from start of week (s)
Δt_{LS} = delta time due to leap seconds
A_0 = a constant polynomial term from the ephemeris message
A_1 = a first-order polynomial term from the ephemeris message
t_{0t} = reference time for UTC data
WN = current week number derived from subframe 1
WN_t = UTC reference week number

The user GPS time t_E is in seconds relative to the end/start of the week, and the reference time t_{0t} for UTC data is referenced to the start of that week, whose number WN_t is given in word 8 of page 18 in subframe 4. The WN_t value consists of the eight LSBs of the full week number. Thus, the user must account for the truncated nature of this parameter as well as truncation of WN, WN_t, and WN_{LSF} due to rollover of the full week number. These parameters are managed by the GPS control segment so that the absolute value of the difference between the untruncated WN and WN_t values does not exceed 127.

2. *Second Case.* Whenever the user's current GPS time falls within the time-span from DN + $\frac{3}{4}$ to DN + $\frac{5}{4}$, proper accommodation of the leap second event with a possible week number transition is provided by the following expression for UTC:

$$t_{\mathrm{UTC}} = W\,[\mathrm{modulo}(86400 + \Delta t_{\mathrm{LSF}} - \Delta t_{\mathrm{LS}})]\ \text{seconds},$$

where

$$W = (t_E - \Delta t_{\mathrm{UTC}} - 43200)\ (\text{modulo } 86400) + 43200\ \text{seconds},$$

and the definition of Δt_{UTC} given previously applies throughout the transition period.

3. *Third Case.* Whenever the effectivity time of the leap second event, as indicated by the $\mathrm{WN_{LSF}}$ and DN values, is in the past relative to the user's current GPS time, the expression given for t_{UTC} in the first case above is valid except that the value of Δt_{LSF} is used instead of Δt_{LS}. The GPS control segment coordinates the update of UTC parameters at a future upload in order to maintain a proper continuity of the t_{UTC} timescale.

2.5 EXAMPLE: USER POSITION CALCULATIONS WITH NO ERRORS

2.5.1 User Position Calculations

This section demonstrates how to go about calculating the user position, given ranges (pseudoranges) to satellites, the known positions of the satellites, and ignoring the effects of clock errors, receiver errors, propagation errors, and so on.

Then, the pseudoranges will be used to calculate the user's antenna location.

2.5.1.1 Position Calculations Neglecting clock errors, let us first determine position calculation with no errors:

ρ_r = pseudorange (known),
x, y, z = satellite position coordinates (known),
X, Y, Z = user position coordinates (unknown),

where x, y, z, X, Y, Z are in the earth-centered, earth-fixed (ECEF) coordinate system.

Position calculation with no errors is

$$\rho_r = \sqrt{(x - X)^2 + (y - Y)^2 + (z - Z)^2}. \tag{2.45}$$

Squaring both sides yields

$$\rho_r^2 = (x - X)^2 + (y - Y)^2 + (z - Z)^2 \quad (2.46)$$

$$= \underbrace{X^2 + Y^2 + Z^2}_{r^2 + C_{rr}} + x^2 + y^2 + z^2$$

$$-2Xx - 2Yy - 2Zz, \quad (2.47)$$

$$\rho_r^2 - (x^2 + y^2 + z^2) - r^2 = C_{rr} - 2Xx - 2Yy - 2Zz, \quad (2.48)$$

where r equals the radius of earth and C_{rr} is the clock bias correction. The four unknowns are (X, Y, Z, C_{rr}). Satellite position (x, y, z) is calculated from ephemeris data. For four satellites, Eq. 2.48 becomes

$$\begin{aligned} \rho_{r_1}^2 - (x_1^2 + y_1^2 + z_1^2) - r^2 &= \text{Crr} - 2Xx_1 - 2Yy_1 - 2Zz_1, \\ \rho_{r_2}^2 - (x_2^2 + y_2^2 + z_2^2) - r^2 &= \text{Crr} - 2Xx_2 - 2Yy_2 - 2Zz_2, \\ \rho_{r_3}^2 - (x_3^2 + y_3^2 + z_3^2) - r^2 &= \text{Crr} - 2Xx_3 - 2Yy_3 - 2Zz_3, \\ \rho_{r_4}^2 - (x_4^2 + y_4^2 + z_4^2) - r^2 &= \text{Crr} - 2Xx_4 - 2Yy_4 - 2Zz_4, \end{aligned} \quad (2.49)$$

with unknown 4×1 state vector

$$\begin{bmatrix} X \\ Y \\ Z \\ C_{rr} \end{bmatrix}.$$

We can rewrite the four equations in matrix form as

$$\begin{bmatrix} \rho_{r_1}^2 - (x_1^2 + y_1^2 + z_1^2) - r^2 \\ \rho_{r_2}^2 - (x_2^2 + y_2^2 + z_2^2) - r^2 \\ \rho_{r_3}^2 - (x_3^2 + y_3^2 + z_3^2) - r^2 \\ \rho_{r_4}^2 - (x_4^2 + y_4^2 + z_4^2) - r^2 \end{bmatrix} = \begin{bmatrix} -2x_1 - 2y_1 - 2z_1 \ 1 \\ -2x_2 - 2y_2 - 2z_2 \ 1 \\ -2x_3 - 2y_3 - 2z_3 \ 1 \\ -2x_4 - 2y_4 - 2z_4 \ 1 \end{bmatrix} \begin{bmatrix} X \\ Y \\ Z \\ C_{rr} \end{bmatrix}$$

or

$$\overbrace{Y}^{4\times 1} = \overbrace{M}^{4\times 4} \overbrace{\chi_\rho}^{4\times 1}, \quad (2.50)$$

where

$$Y = \text{vector (known)},$$

$$M = \text{matrix (known)},$$

$$\chi_\rho = \text{vector (unknown)}.$$

Then we premultiply both sides of Eq. 2.50 by M^{-1}:

$$\begin{aligned} M^{-1}Y &= M^{-1}M\chi_\rho \\ &= \chi_\rho \\ &= \begin{bmatrix} X \\ Y \\ Z \\ C_{rr} \end{bmatrix}. \end{aligned}$$

If the rank of M (defined in Section B.5.2), the number of linearly independent columns of the matrix M, is less than 4, then M will not be invertible. In that case, its determinant (defined in Section B.6.1) is given as

$$\det M = |M| = 0.$$

2.5.2 User Velocity Calculations

The governing equation in this case is

$$\dot{\rho}_r = \frac{(x - X)(\dot{x} - \dot{X}) + (y - Y)(\dot{y} - \dot{Y}) + (z - Z)(\dot{z} - \dot{Z})}{\rho_r}, \tag{2.51}$$

where

$$\begin{aligned} \dot{\rho}_r &= \text{range rate (known)} \\ \rho_r &= \text{range (known)} \\ (x, y, z) &= \text{satellite positions (known)} \\ (\dot{x},\ \dot{y},\ \dot{z}) &= \text{satellite rates (known)} \\ (X, Y, Z) &= \text{user position (known from position calculations)} \\ (\dot{X},\ \dot{Y},\ \dot{Z}) &= \text{user velocity (unknown)} \end{aligned}$$

and from Eq. 2.51,

$$\begin{aligned} -\dot{\rho}_r + \tfrac{1}{\rho_r}[\dot{x}(x - X) + \dot{y}(y - Y) + \dot{z}(z - Z)] \\ = \left(\tfrac{x-X}{\rho_r}\dot{X} + \tfrac{y-Y}{\rho_r}\dot{Y} + \tfrac{z-Z}{\rho_r}\dot{Z}\right). \end{aligned} \tag{2.52}$$

For three satellites, Eq. 2.52 becomes

$$\begin{bmatrix} -\dot{\rho}_{r_1} + \frac{1}{\rho_{r_1}}[\dot{x}_1(x_1 - X) + \dot{y}_1(y_1 - Y) + \dot{z}_1(z_1 - Z)] \\ -\dot{\rho}_{r_2} + \frac{1}{\rho_{r_2}}[\dot{x}_2(x_2 - X) + \dot{y}_2(y_2 - Y) + \dot{z}_2(z_2 - Z)] \\ -\dot{\rho}_{r_3} + \frac{1}{\rho_{r_3}}[\dot{x}_3(x_3 - X) + \dot{y}_3(y_3 - Y) + \dot{z}_3(z_3 - Z)] \end{bmatrix}$$

$$= \begin{bmatrix} \frac{(x_1-X)}{\rho_{r_1}} & \frac{(y_1-Y)}{\rho_{r_1}} & \frac{(z_1-Z)}{\rho_{r_1}} \\ \frac{(x_2-X)}{\rho_{r_2}} & \frac{(y_2-Y)}{\rho_{r_2}} & \frac{(z_2-Z)}{\rho_{r_2}} \\ \frac{(x_3-X)}{\rho_{r_2}} & \frac{(y_3-Y)}{\rho_{r_3}} & \frac{(z_3-Z)}{\rho_{r_3}} \end{bmatrix} \begin{bmatrix} \dot{X} \\ \dot{Y} \\ \dot{Z} \end{bmatrix}. \tag{2.53}$$

$$\overbrace{D_\xi}^{3\times1} = \overbrace{N}^{3\times3}\overbrace{U_\vee}^{3\times1}, \tag{2.54}$$

$$\overbrace{U_\vee}^{3\times1} = N^{-1} D_\xi. \tag{2.55}$$

However, if the rank of N (defined in Section B.5) is <3, N will not be invertible.

PROBLEMS

Refer to Appendix C for coordinate system definitions, and to Eqs. C.103 and C.104 for satellite orbit equations.

2.1 Which of the following coordinate systems is not rotating?

(a) North–east–down (NED)

(b) East–north–up (ENU)

(c) Earth-centered, earth-fixed (ECEF)

(d) Earth-centered inertial (ECI)

(e) Moon-centered, moon-fixed

2.2 What is the minimum number of two-axis gyroscopes (i.e., gyroscopes with two, independent, orthogonal input axes) required for inertial navigation?

(a) 1

(b) 2

(c) 3

(d) Not determined

2.3 What is the minimum number of gimbal axes required for gimbaled inertial navigators in fully maneuverable host vehicles? Explain your answer.

(a) 1

(b) 2

(c) 3

(d) 4

2.4 Define *specific force.*

2.5 An inertial sensor assembly (ISA) operating at a fixed location on the surface of the earth would measure

(a) No acceleration

(b) 1 g acceleration downward

(c) 1 g acceleration upward

2.6 Explain why an inertial navigation system is not a good altimeter.

2.7 The *inertial* rotation rate of the earth is

(a) 1 revolution per day

(b) 15 degrees per hour

(c) 15 arc-seconds per second

(d) $\approx$ 15.0411 arc-seconds per second

2.8 Define CEP and CEP rate for an INS.

2.9 The CEP rate for a *medium accuracy* INS is in the order of

(a) 2 meters per second (m/s)

(b) 200 meters per hour (m/h)

(c) 2000 m/h

(d) 20 km/h

2.10 For the following GPS satellites, find the satellite position in ECEF coordinates at $t = 3$ s. (*Hint*: see Appendix C.) Ω_0 and θ_0 are given below at time $t_0 = 0$:

	$\Omega 0$ (deg)	θ_0 (deg)
(a)	326	68
(b)	26	34

2.11 Using the results of the previous problem, find the satellite positions in the local reference frame. Reference should be to the COMSAT facility in Santa Paula, California, located at 32.4° latitude, −119.2° longitude. Use coordinate shift matrix $S = 0$. (Refer to Appendix C, Section C.3.9.)

2.12 Given the following GPS satellite coordinates and pseudoranges:

	Ω_0 (deg)	θ_0 (deg)	ρ (m)
Satellite 1	326	68	2.324×10^7
Satellite 2	26	340	2.0755×10^7
Satellite 3	146	198	2.1103×10^7
Satellite 4	86	271	2.3491×10^7

(a) Find the user's antenna position in ECEF coordinates.

(b) Find the user's antenna position in locally level coordinates referenced to 0° latitude, 0° longitude. Coordinate shift matrix $S = 0$.

(c) Find the various DOPs.

2.13 Given two satellites in north and east coordinates

$$\begin{aligned} x(1) &= 6.1464 \times 10^6,\ y(1) &= 2.0172 \times 10^7 \text{in meters,} \\ x(2) &= 6.2579 \times 10^6,\ y(2) &= -7.4412 \times 10^6 \text{in meters,} \end{aligned}$$

with pseudoranges

$$c\,\Delta t(1) = \rho_r(1) = 2.324 \times 10^7 \text{in meters,}$$

$$c\,\Delta t(2) = \rho_r(2) = 2.0755 \times 10^7 \text{in meters,}$$

and starting with an initial guess of $(x_{\text{est}},\ y_{\text{est}})$, find the user's antenna position.

2.14 Rank VDOP, HDOP and PDOP from smallest (best) to largest (worst) under normal conditions:

(a) VDOP $\leq$ HDOP$\leq$PDOP

(b) VDOP$\leq$PDOP$\leq$HDOP

(c) HDOP$\leq$VDOP$\leq$PDOP

(d) HDOP$\leq$PDOP$\leq$VDOP

(e) PDOP$\leq$HDOP$\leq$VDOP

(f) PDOP$\leq$VDOP$\leq$HDOP

2.15 UTC time and the GPS time are offset by an integer number of seconds (e.g., 14 s as of January 1, 2006) as well as a fraction of a second. The fractional part is approximately:

(a) 0.1–0.5 s

(b) 1–2 ms

(c) 100–200 ns

(d) 10–20 ns

2.16 Show that $C_{\text{ENU}}^{\text{ECEF}} \times C_{\text{ECEF}}^{\text{ENU}} = I$, the 3×3 identity matrix. (*Hint*: $C_{\text{ENU}}^{\text{ECEF}} = \left[C_{\text{ECEF}}^{\text{ENU}}\right]^T$.)

2.17 A satellite position at time $t = 0$ is specified by its orbital parameters as $\Omega_0 = 92.847°$, $\theta_0 = 135.226°$, $\alpha = 55°$, $R = 26,560,000$ m.

(a) Find the satellite position at $t = 1$ s, in ECEF coordinates.

(b) Convert the satellite position from (a) with user at

$$\begin{bmatrix} X_u \\ Y_u \\ Z_u \end{bmatrix}_{\text{ECEF}} = \begin{bmatrix} -2.430601 \\ -4.702442 \\ 3.546587 \end{bmatrix} \times 10^6 \text{ meter}$$

to WGS84 east–north–up (ENU) coordinates with origin at

$$\theta = \text{local reference longitude} = 32.4^\circ$$

$$\phi = \text{local reference latitude} = -119.2^\circ$$

3

SIGNAL CHARACTERISTICS AND INFORMATION EXTRACTION

Why is the GPS signal so complex? GPS was designed to be readily accessible to millions of military and civilian users. Therefore, it is a receive-only passive system for a user, and the number of users that can simultaneously use the system is unlimited. Because there are many functions that must be performed, the GPS signal has a rather complex structure. As a consequence, there is a correspondingly complex sequence of operations that a GPS receiver must carry out in order to extract desired information from the signal. In this chapter we characterize the signal mathematically, describe the purposes and properties of the important signal components, and discuss generic methods for extracting information from these components.

3.1 MATHEMATICAL SIGNAL WAVEFORM MODELS

Each GPS satellite simultaneously transmits on two L-band frequencies denoted by L_1 and L_2, which are 1575.42 and 1227.60 MHz, respectively. The carrier of the L_1 signal consists of an in-phase and a quadrature-phase component. The in-phase component is biphase modulated by a 50-bps (bits per second) data stream and a pseudorandom code called the *C/A-code* consisting of a 1023-chip sequence that has a period of 1 ms and a chipping rate of 1.023 MHz. The quadrature-phase component is also biphase modulated by the same 50-bps (bits per second) data stream but with a different pseudorandom code called the *P-code*, which

Global Positioning Systems, Inertial Navigation, and Integration, Second Edition, by M. S. Grewal, L. R. Weill, and A. P. Andrews

has a 10.23-MHz chipping rate and a one-week period. The mathematical model of the L_1 waveform is

$$s(t) = \sqrt{2P_I}d(t)c(t)\cos(\omega t + \theta) + \sqrt{2P_Q}d(t)p(t)\sin(\omega t + \theta), \qquad (3.1)$$

where P_I and P_Q are the respective carrier powers for the in-phase and quadrature-phase carrier components, $d(t)$ is the 50-bps (bits per second) data modulation, $c(t)$ and $p(t)$ are the respective C/A and P pseudorandom code waveforms, ω is the L_1 carrier frequency in radians per second, and θ is a common phase shift in radians. The quadrature carrier power P_Q is approximately 3 dB less than P_I.

In contrast to the L_1 signal, the L_2 signal is modulated with only the 50-bps (bits per second) data and the P-code, although there is the option of not transmitting the 50-bps (bits per second) data stream. The mathematical model of the L_2 waveform is

$$s(t) = \sqrt{2P_Q}d(t)p(t)\sin(\omega t + \theta). \qquad (3.2)$$

Figures 3.1 and 3.2 show the structure of the in-phase and quadrature-phase components, respectively, of the L_1 signal. The 50-bps (bits per second) data bit boundaries always occur at an epoch of the C/A-code. The C/A-code epochs mark the beginning of each period of the C/A-code, and there are precisely 20 code epochs per data bit, or 20,460 C/A-code chips. Within each C/A-code chip there are precisely 1540 L_1 carrier cycles. In the quadrature-phase component of the L_1 signal there are precisely 204,600 P-code chips within each 50-bps (bits per second) data bit, and the data bit boundaries always coincide with the beginning of a P-code chip [61, 84].

3.2 GPS SIGNAL COMPONENTS, PURPOSES, AND PROPERTIES

3.2.1 50-bps (bits per second) Data Stream

The 50-bps (bits per second) data stream conveys the *navigation message,* which includes, but is not limited to, the following information:

1. *Satellite Almanac Data.* Each satellite transmits orbital data called the *almanac*, which enables the user to calculate the approximate location of every satellite in the GPS constellation at any given time. Almanac data are not accurate enough for determining position but can be stored in a receiver where they remain valid for many months. They are used primarily to determine which satellites are visible at a given location so that the receiver can search for those satellites when it is first turned on. They can also be used to determine the approximate expected signal Doppler shift to aid in rapid acquisition of the satellite signals.

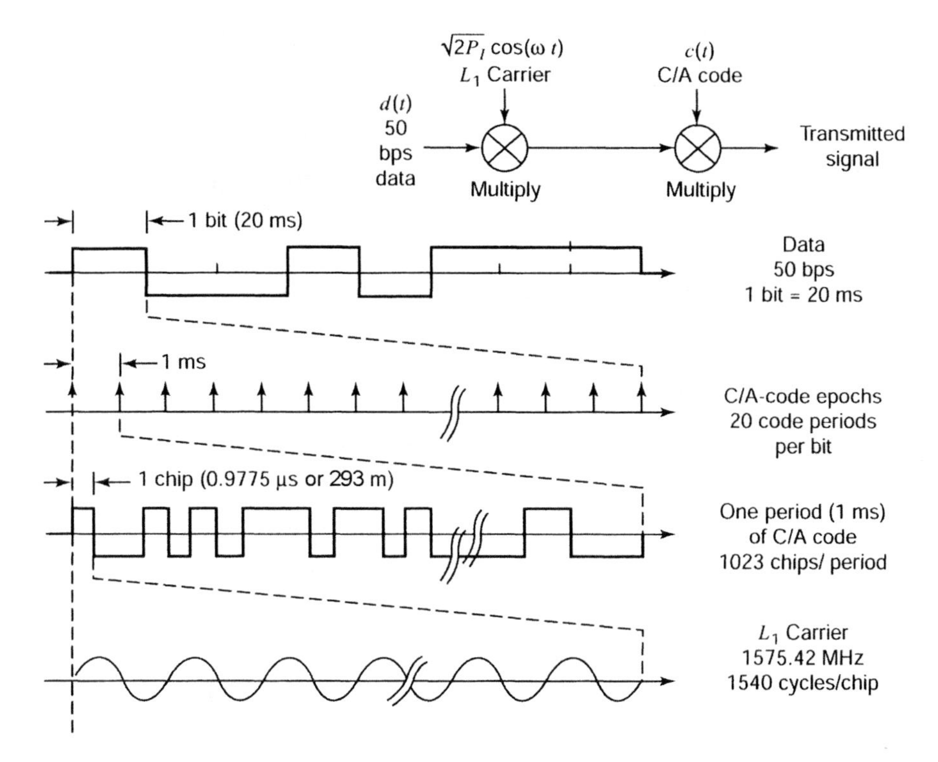

Fig. 3.1 Structure of in-phase component of the L_1 signal.

2. *Satellite Ephemeris Data.* Ephemeris data are similar to almanac data but enable a much more accurate determination of satellite position needed to convert signal propagation delay into an estimate of user position. In contrast to almanac data, ephemeris data for a particular satellite are broadcast only by that satellite, and the data are valid for only several hours.
3. *Signal Timing Data.* The 50-bps (bits per second) data stream includes time tagging, which is used to establish the transmission time of specific points on the GPS signal. This information is needed to determine the satellite-to-user propagation delay used for ranging.
4. *Ionospheric Delay Data.* Ranging errors due to ionospheric effects can be partially canceled by using estimates of ionospheric delay that are broadcast in the data stream.
5. *Satellite Health Message.* The data stream also contains information regarding the current health of the satellite, so that the receiver can ignore that satellite if it is not operating properly.

3.2.1.1 Structure of the Navigation Message The information in the navigation message has the basic frame structure shown in Fig. 3.3. A complete message

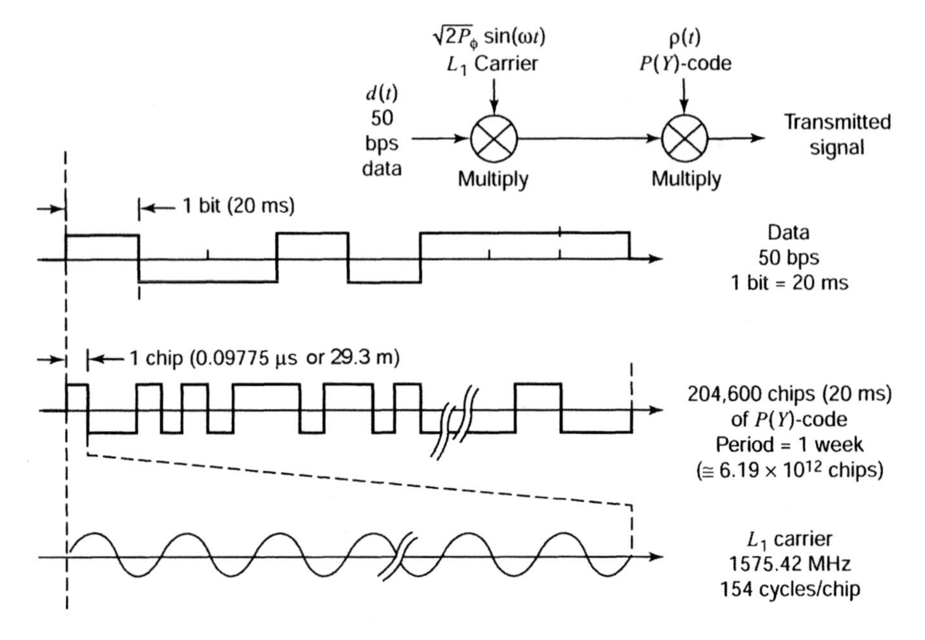

Fig. 3.2 Structure of quadrature-phase component of the L_1 signal.

consists of 25 frames, each containing 1500 bits. Each frame is subdivided into five 300-bit subframes, and each subframe consists of 10 words of 30 bits each, with the most significant bit (MSB) of the word transmitted first. Thus, at the 50-bps (bits per second) rate it takes 6 s to transmit a subframe and 30 s to complete one frame. Transmission of the complete 25-frame navigation message requires 750 s, or 12.5 min. Except for occasional updating, subframes 1, 2, and 3 are constant (i.e., repeat) with each frame at the 30-s frame repetition rate. On the other hand, subframes 4 and 5 are each subcommutated 25 times. The 25 versions of subframes 4 and 5 are referred to as pages 1–25. Hence, except for occasional updating, each of these pages repeats every 750 s, or 12.5 min.

A detailed description of all information contained in the navigation message is beyond the scope of this text. Therefore, we give only an overview of the fundamental elements. Each subframe begins with a *telemetry word* (TLM). The first 8 bits of the TLM is a preamble that enables the receiver to determine when a subframe begins. The remainder of the TLM contains parity bits and a telemetry message that is available only to authorized users and is not a fundamental item. The second word of each subframe is called the *handover word* (HOW).

3.2.1.2 Z-Count Information contained in the HOW is derived from a 29-bit quantity called the *Z-count*. The Z-count is not transmitted as a single word, but part of it is transmitted within the HOW. The Z-count counts *epochs* generated by the X_1 register of the P-code generator in the satellite, which occur every 1.5 s.

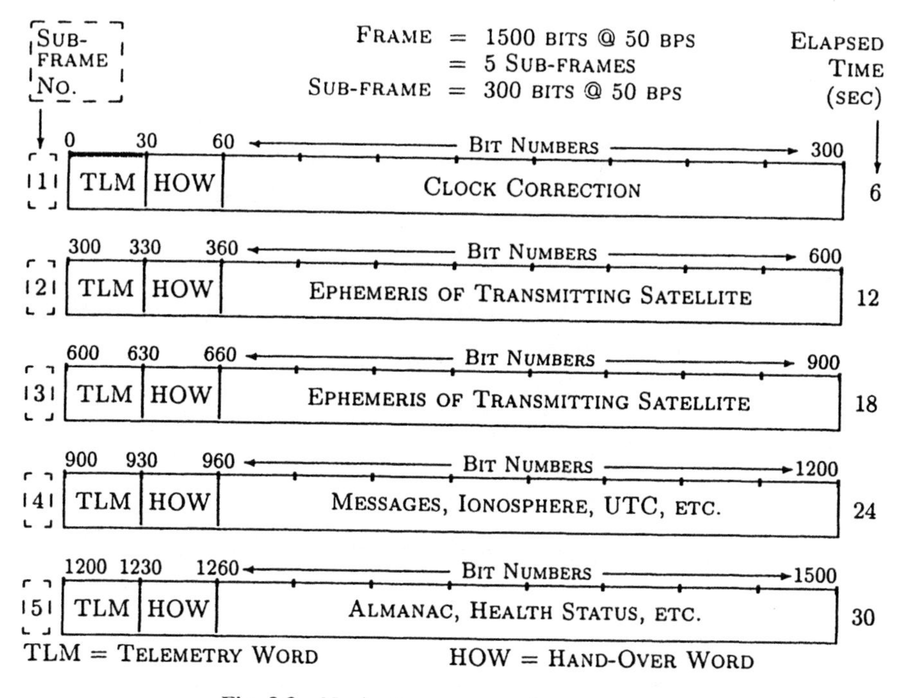

Fig. 3.3 Navigation message frame structure.

The 19 LSBs of the Z-count, called the *time-of-week* (TOW) count, indicate the number of X_1 epochs that have occurred since the start of the current week. The start of the current week occurs at the X_1 epoch, which occurs at approximately midnight of Saturday night/Sunday morning. The TOW count increases from zero at the start of the week to 403199 and then rolls over to zero again at the start of the following week. A TOW count of zero always occurs at the beginning of subframe 1 of the first frame (the frame containing page 1 of subcommutated subframes 4 and 5). A truncated version of the TOW count, containing its 17 MSBs, constitutes the first 17 bits of the HOW. Multiplication of this truncated count by 4 gives the TOW count at the start of the following subframe. Since the receiver can use the TLM preamble to determine precisely the time at which each subframe begins, a method for determining the time of transmission of any part of the GPS signal is thereby established. The relationship between the HOW counts and TOW counts is shown in Fig. 3.4.

3.2.1.3 GPS Week Number The 10 MSBs of the Z-count contain the GPS *week number* (WN), which is a modulo-1024 week count. The *zero state* is defined to be that week that started with the X_1 epoch occurring at approximately midnight on the night of January 5, 1980/morning of January 6, 1980. Because WN is a modulo-1024 count, an event called the *week rollover* occurs every 1024

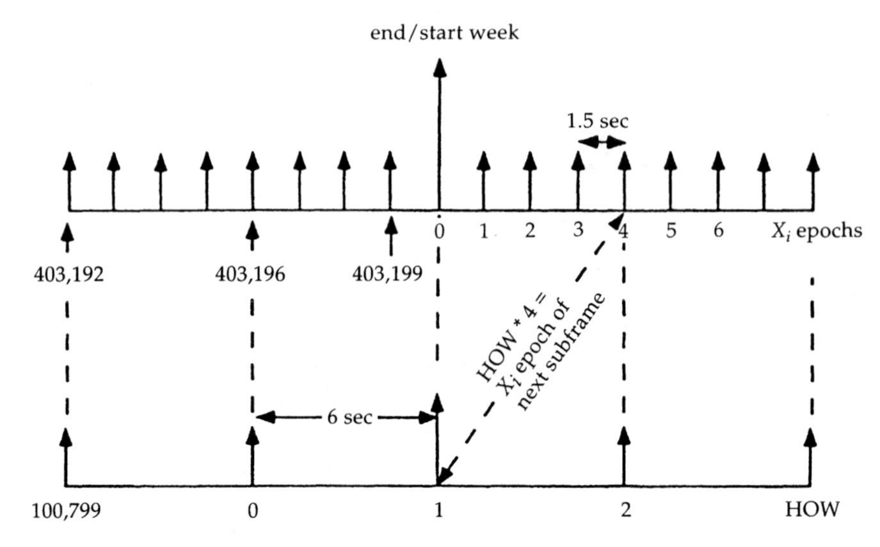

Fig. 3.4 Relationship between HOW counts and TOW counts.

weeks (a few months short of 20 years), and GPS receivers must be designed to accommodate it.[1]. The WN is not part of the HOW but instead appears as the first 10 bits of the third word in subframe 1.

Frame and Subframe Identification Three bits of the HOW are used to identify which of the five subframes is being transmitted. The frame being transmitted (corresponding to a page number from 1 to 25) can readily be identified from the TOW count computed from the HOW of subframe 5. This TOW count is the TOW at the start of the next frame. Since there are 20 TOW counts per frame, the frame number of that frame is simply (TOW/20) (mod 25).

3.2.1.4 Information by Subframe In addition to the TLM and HOW, which occur in every subframe, the following information is contained within the remaining eight words of subframes 1–5 (only fundamental information is described):

1. *Subframe 1.* The WN portion of the *Z*-count is part of word 3 in this subframe. Subframe 1 also contains GPS clock correction data for the satellite in the form of polynomial coefficients defining how the correction varies with time. Time defined by the clocks in the satellite is commonly called *SV time* (space vehicle time); the time after corrections have been applied is called *GPS time*. Thus, even though individual satellites may not have perfectly synchronized SV times, they do share a common GPS time.

[1]The most recent rollover occurred at GPS time zero on August 22, 1999, with little difficulty

Additional information in subframe 1 includes the quantities t_{0c}, T_{GD}, and IODC. The clock reference time t_{0c} is used as a time origin to calculate satellite clock error, the ionospheric group delay T_{GD} is used to correct for ionospheric propagation delay errors, and IODC (issue of date, clock) indicates the issue number of the clock data set to alert users to changes in clock parameters.

2. *Subframes 2 and 3.* These subframes contain the ephemeris data, which are used to determine the precise satellite position and velocity required by the navigation solution. Unlike the almanac data, these data are very precise, are valid over a relatively short period of time (several hours), and apply only to the satellite transmitting it. The components of the ephemeris data are listed in Table 3.1, and the algorithm that should be used to compute satellite position in WGS84 coordinates is given in Table 3.2. The satellite position computation using these data is implemented in the MATLAB m-file `ephemeris.m` on the accompanying CD. The IODE (issue of date, ephemeris) informs users when changes in ephemeris parameters have occurred. Each time new parameters are uploaded from the GPS control segment, the IODE number changes.
3. *Subframe 4.* The 25 pages of this subframe contain the almanac for satellites with PRN (pseudorandom code) numbers 25 and higher, as well as special messages, ionospheric correction terms, and coefficients to convert GPS time to UTC time. There are also spare words for possible future applications. The components of an almanac are very similar to those of the ephemeris, and the calculation of satellite position is performed in essentially the same way.
4. *Subframe 5.* The 25 pages of this subframe includes the almanac for satellites with PRN numbers from 1 to 24.

It should be noted that since each satellite transmits all 25 pages, almanac data for all satellites are transmitted by every satellite. Unlike ephemeris data, almanac data remain valid for long periods (months) but are much less precise. Additional data contained in the navigation message are user range error (URE), which estimate the range error due to errors in satellite ephemeris, timing errors, and selective availability (SA) and flags to indicate the health status of the satellites.

3.2.2 GPS Satellite Position Calculations

3.2.2.1 Transmission of Satellite Ephemerides The interface between the GPS space and user segments consists of two radiofrequency (RF) links, L_1 and L_2. The carriers of the L-band links are modulated by up to two bit trains, each of which normally is a composite generated by the modulo-2 addition of a PRN ranging code and the downlink system data. Utilizing these links, the space vehicles of the GPS space segment should provide continuous earth coverage for signals that provide to the user segment the ranging codes and system data needed to accomplish the GPS navigation mission. These signals are available to a suitably equipped user with RF visibility to a space vehicle. Therefore, the GPS users continuously receive navigation information from the space vehicles in the form of modulated data bits.

The received information is computed and controlled by the control segment and includes the satellite's time, its clock correction and ephemeris parameters, almanacs and health for all GPS space vehicles, and text messages. The precise position and clock offset of the space vehicle antenna phase center in the ECEF coordinates can be computed by receiving this information.

The ephemeris parameters describe the orbit during the interval of time (at least 1 h) for which the parameters are transmitted. This representation model is characterized by a set of parameters that is an extension (including drag) to the Keplerian orbital parameters. They also describe the ephemeris for an additional interval of time (at least 0.5 h) to allow time for the user to receive the parameters for the new interval of time. The definitions of the parameters are given in Table 3.1.

The *age of data word* (AODE) provides a confidence level in the ephemeris representation parameters. The AODE represents the time difference (age) between

TABLE 3.1. Components of Ephemeris Data

Term	Description	Units[a]
M_0	Mean anomaly at reference time	Semicircle
Δn	Mean motion difference from computed value	Semicircle/s
E	Eccentricity	Dimensionless
$\sqrt{a}$	Square root of semimajor axis	$m^{1/2}$
Ω_0	Longitude of ascending node of orbit plane at weekly epoch	Semicircle
i_0	Inclination angle at reference time	Semicircle
ω	Argument of perigee	Semicircle
$\dot{\Omega}$	Rate of right ascension	Semicircle/s
IDOT	Rate of inclination angle	Semicircle/s
C_{uc}	Amplitude of cosine harmonic correction term to the argument of latitude	rad
C_{us}	Amplitude of sine harmonic correction term to the argument of latitude	rad
C_{rc}	Amplitude of cosine harmonic correction term to the orbit radius	m
C_{rs}	Amplitude of sine harmonic correction term to the orbit radius	m
C_{ic}	Amplitude of cosine harmonic correction term to the angle of inclination	rad
C_{is}	Amplitude of sine harmonic correction term to the angle of inclination	rad
t_{0e}	Ephemeris reference time	s
IODE	Issue of data, ephemeris	Dimensionless

[a]Units used in MATLAB m-file `ephemeris.m` are different

TABLE 3.2. Algorithm for Computing Satellite Position

$\mu = 3.986005 \times 10^{14}\ \text{m}^3/\text{s}^2$	WGS84 value of earth's universal gravitational parameter
$\dot{\Omega}_e = 7.292115167 \times 10^{-5}$ rad/s	WGS84 value of earth's rotation rate
$a = (\sqrt{a})^2$	Semimajor axis
$n_0 = \sqrt{\mu / a^3}$	Computed mean motion, rad/s
$t_k = t - t_{0e}^{a}$	Time from ephemeris reference epoch
$n = n_0 + \Delta_n$	Corrected mean motion
$M_k = M_0 + nt_k$	Mean anomaly
$M_k = E_k - e \sin E_k$	Kepler's equation for eccentric anomaly
$f_k = \cos^{-1}\left(\frac{\cos E_k - e}{1 - e\cos E_k}\right)$	True anomaly from cosine
$f_k = \sin^{-1}\left(\frac{\sqrt{1-e^2} \sin E_k}{1 - e\cos E_k}\right)$	True anomaly from sine
$E_k = \cos^{-1}\left(\frac{e + \cos f_k}{1 + e\cos f_k}\right)$	Eccentric anomaly from cosine
$\phi_k = f_k + \omega$	Argument of latitude
$\delta\mu_k = C_{\mu c} \cos 2\phi_k + C_{\mu s} \sin 2\phi_k$	Second-harmonic correction to argument of latitude
$\delta g_k = C_{rc} \cos 2\phi_k + C_{rs} \sin 2\phi_k$	Second-harmonic correction to radius
$\delta g_k = C_{ic} \cos 2\phi_k + C_{is} \sin 2\phi_k$	Second-harmonic correction to inclination
$\mu_k = \phi_k + \delta\mu_k$	Corrected argument of latitude
$r_k = a(1 - e\cos E_k) + \delta r_k$	Corrected radius
$i_k = i_0 + \delta i_k + (IDOT)t_k$	Corrected inclination
$x'_k = r_k \cos \mu_k$	X coordinate in orbit plane
$y'_k = r_k \sin \mu_k$	Y coordinate in orbit plane
$\Omega_k = \Omega_0 + (\Omega - \dot{\Omega}_e)t_k - \dot{\Omega}_e t_{0e}$	Corrected longitude of ascending node
$x_k = x'_k \cos \Omega_k - y'_k \cos i_k \sin \Omega_k$	ECEF X coordinate
$y_k = x'_k \sin \Omega_k + y'_k \cos i_k \cos \Omega_k$	ECEF Y coordinate
$z_k = y'_k \sin i_k$	ECEF Z coordinate

[a] t is in GPS system time at time of transmission, i.e., GPS time corrected for transit time (range/speed of light). Furthermore, t_k shall be the actual total time difference between the time t and the time epoch t_{0e} and must account for beginning or end of week crossovers. Thus, if t_k is greater than 302,400 s, subtract 604800 s from t_k; if t_k is less than −302400 s, add 604,800 s to t_k

the reference time (t_{0e}) and the time of the last measurement update (t_L) used to estimate the representation parameters.

The ECEF coordinates for the phase center of the satellite's antennas can be calculated using a variation of the equations shown in Table 3.2. In this table, time t is the GPS system time at the time of transmission, that is, GPS time corrected for transit time (range/speed of light). Further, t_k is the actual total time difference between time t and epoch time t_{0e} and must account for beginning- or end-of-week crossovers. Thus, if t_k is greater than 302400 s, subtract 604800 s from t_k; if t_k is less than −302400 ss, add 604800 s to t_k.

3.2.2.2 Ephemeris Data Transmitted The ephemeris parameters and algorithms used for computing satellite positions are given in Tables 3.1 and 3.2,

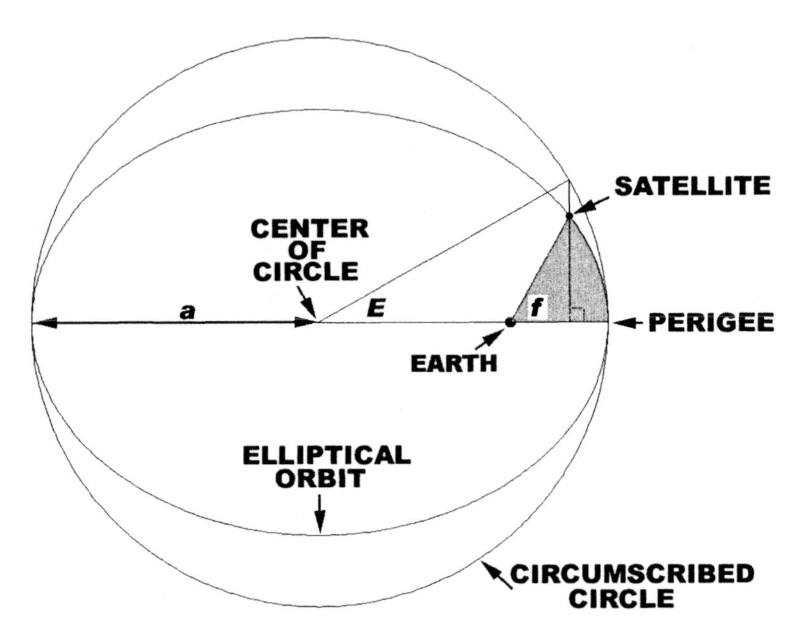

Fig. 3.5 Geometric relationship between true anomaly f and eccentric anomaly E.

respectively. The problem of determining satellite position from these data and equations is called the *Kepler problem*.

3.2.2.3 True, Eccentric, and Mean Anomaly Orbit phase variables used for determining the position of a satellite in its orbit are illustrated in Fig. 3.5. The variable f in the figure is called *true anomaly* in orbit mechanics. The hardest part of the Kepler problem, is the problem of determining true anomaly as a function of time. This problem was eventually solved by introducing two intermediate "anomaly" variables:

E, the *eccentric anomaly,* defined as a geometric function of true anomaly, as shown in Fig. 3.5. Eccentric anomaly E is defined by projecting the satellite position on the elliptical orbit out perpendicular to the semimajor axis a and onto the circumscribed circle. Eccentric anomaly is then defined as the central angle to this projection point on the circle, as shown in the figure. The shaded area represents the area swept out by the radius from the earth to the satellite as the satellite moves from its perigee. Kepler had determined that this area grows linearly with time, and he used this relationship to derive his equation for mean anomaly.

M, the *mean anomaly,* defined as a linear function of time:

$$M(t) = \frac{2\pi \left(t - t_{\text{perigee}}\right)}{T_{\text{period}}} \text{ (in radians)}, \tag{3.3}$$

where t is the time in seconds at which true anomaly is to be determined; t_{perigee} is the time at which the satellite was at its perigee, closest to the earth; and T_{period} is the orbit period in seconds.

From this equation, calculating mean anomaly as a function of time is relatively easy, but the solution for true anomaly as a function of eccentric anomaly is much more difficult.

For the low eccentricities of GPS orbits, the numerical values of the true, eccentric, and mean anomalies are quite close together. However, the precision required in calculating true anomaly will require that they be treated as separate variables.

3.2.2.4 Kepler's Equation The equation

$$M_k = E_k - e \sin E_k, \tag{3.4}$$

in Table 3.2 is called *Kepler's equation*. It relates the eccentric anomaly E_k of the kth satellite to its mean anomaly M_k and the orbit eccentricity e. This equation is the most difficult of all the equations in Table 3.2 to solve for E_k as a function of M_k.

3.2.2.5 Solution of Kepler's Equation Kepler's equation (Eq. 3.4) includes a transcendental function of eccentric anomaly E_k. It is impractical to solve for E_k in any way except by approximation. Standard practice is to solve the true anomaly equation iteratively for E_k, using the second-order Newton–Raphson method to solve

$$\varepsilon_k \stackrel{\text{def}}{=} M_k + E_k - e \sin E_k \tag{3.5}$$

$$= 0, \tag{3.6}$$

and then use the resulting value of E_k to calculate true anomaly. It starts by assigning an initial guess $E_k^{[0]}$ for E_k (the mean anomaly will do), and then forming successively better estimates $E_k^{[n+1]}$ by the second-order Newton–Raphson formula

$$E_k^{[n+1]} = E_k^{[n]} - \left\{ \frac{\varepsilon_k\left[E_k^{(n)}\right]}{\left.\frac{\partial \varepsilon_k}{\partial E_k}\right|_{E_k=E_k^{[n]}} - \left[\frac{\left.\frac{\partial^2 \varepsilon_k}{\partial E_k^2}\right|_{E_k=E_k^{[n]}} \varepsilon_k[E_k^{(n)}]}{2\left.\frac{\partial \varepsilon_k}{\partial E_k}\right|_{E_k=E_k^{[n]}}}\right]} \right\}. \tag{3.7}$$

The iteration of Eq. 3.7 can stop when the difference in the estimated E_k is sufficiently small, say

$$\left|E_k^{[n+1]} - E_k^{[n]}\right| < 10^{-6}.$$

3.2.2.6 Other Calculation Considerations The satellite's antenna phase center position is very sensitive to small perturbations in most ephemeris parameters. The sensitivity of position to the parameters $\sqrt{a}$, C_{rc}, and C_{rs} is about 1 m/m. This sensitivity to angular parameters is on the order of 10^8 m/semicircle and to the angular rate parameters on the order of 10^{12} m/semicircle/s. Because of this extreme sensitivity to angular perturbations, the required value of π (a mathematical constant, the ratio of a circle's circumference to its diameter) used in the curve fit is given as

$$\pi = 3.1415926535898.$$

The user must correct the time received from the space vehicle in seconds with the equation

$$t = t_{sv} - \Delta t_{sv}, \tag{3.8}$$

where t is GPS system time in seconds, t_{sv} is the effective SV PRN code phase time at message transmission time in seconds, and Δt_{sv} is the SV PRN code phase time offset in seconds. The SV PRN code phase offset is given by

$$\Delta t_{sv} = a_{f0} + a_{f1}(t - t_{0c}) + a_{f2}(t - t_{0c})^2 + \Delta t_r, \tag{3.9}$$

where a_{f0}, a_{f1}, a_{f2} are polynomial coefficients given in the ephemeris data file; t_{0c} is the clock data reference time in seconds; and Δt_r is the relativistic correction term in seconds, given by

$$\Delta t_r = Fe\sqrt{a}\,\sin E_k. \tag{3.10}$$

In Eq. 3.10, F is a constant whose value is given as

$$F = \frac{-2\sqrt{\mu}}{c^2} \tag{3.11}$$

$$= -4.442807633 \times 10^{-10}\ [\mathrm{s}/\sqrt{\mathrm{m}}], \tag{3.12}$$

where the speed of light $c = 2.99792458 \times 10^8$ m/s. Note that Eqs. 3.8 and 3.9 are coupled. While the coefficients a_{f0}, a_{f1}, and a_{f2} are generated by using GPS time as indicated in Eq. 3.9, sensitivity of t_{sv} to t is negligible. This negligible sensitivity will allow the user to approximate t by t_{sv} in Eq. 3.9. The value of t must account for beginning- or end-of-week crossovers. Thus, if the quantity $t - t_{0c}$ is greater than 302,400 s, subtract 604,800 s from t; if the quantity $t - t_{0c}$ is less than $-302,400$ s, add 604,800 s to t.

By using the value of the ephemeris parameters for satellite PRN 2 in the set of equations in Table 3.1 and Eqs. 3.8–3.12, we can calculate the space vehicle time offset and the ECEF coordinates of the satellite position [46]. The MATLAB m-file (`ephemeris.m`) on the accompanying CD calculates satellite position for one set of ephemeris data and one time. Other programs calculate satellite positions for a range of time. (See Appendix A.)

3.2.3 C/A-Code and Its Properties

The C/A-code has the following functions:

1. *To enable accurate range measurements and resistance to errors caused by multipath.* To establish the position of a user to within 10–100 m, accurate user-to-satellite range estimates are needed. The estimates are made from measurements of signal propagation delay from the satellite to the user. To achieve the required accuracy in measuring signal delay, the GPS carrier must be modulated by a waveform having a relatively large bandwidth. The needed bandwidth is provided by the C/A-code modulation, which also permits the receiver to use correlation processing to effectively combat measurement errors due to thermal noise. Because the C/A-code causes the bandwidth of the signal to be much greater than that needed to convey the 50-bps (bits per second) data stream, the resulting signal is called a *spread-spectrum* signal. Using the C/A-code to increase the signal bandwidth also reduces errors in measuring signal delay caused by multipath (the arrival of the signal via multiple paths such as reflections from objects near the receiver antenna) since the ability to separate the direct path signal from the reflected signal improves as the signal bandwidth is made larger.
2. *To permit simultaneous range measurement from several satellites.* The use of a distinct C/A-code for each satellite permits all satellites to use the same L_1 and L_2 frequencies without interfering with each other. This is possible because the signal from an individual satellite can be isolated by correlating it with a replica of its C/A-code in the receiver. This causes the C/A-code modulation from that satellite to be removed so that the signal contains only the 50-bps (bits per second) data and is therefore narrowband. This process is called *despreading* of the signal. However, the correlation process does not cause the signals from other satellites to become narrowband, because the codes from different satellites are orthogonal. Therefore the interfering signals can be rejected by passing the desired despread signal through a narrowband filter, a bandwidth-sharing process called *code-division multiplexing* (CDM) or *code-division multiple access* (CDMA).
3. *To provide protection from jamming.* The C/A-code also provides a measure of protection from intentional or unintentional jamming of the received signal by another man-made signal. The correlation process that despreads the desired signal has the property of spreading any other signal. Therefore, the signal power of any interfering signal, even if it is narrowband, will be spread over a large frequency band, and only that portion of the power lying in the narrowband filter will compete with the desired signal. The C/A-code provides about 20–30 dB of improvement in resistance to jamming from narrowband signals.

We next detail important properties of the C/A-code.

3.2.3.1 Temporal Structure Each satellite has a unique C/A-code, but all the codes consist of a repeating sequence of 1023 chips occurring at a rate of 1.023 MHz with a period of 1 ms, as previously shown in Fig. 3.1. The leading edge of a specific chip in the sequence, called the *C/A-code epoch,* defines the beginning of a new period. Each chip is either positive or negative with the same magnitude. The polarities of the 1023 chips appear to be randomly distributed but are in fact generated by a deterministic algorithm implemented by shift registers. The algorithm produces maximal-length *Gold codes*, which have the property of low cross-correlation between different codes (orthogonality) as well as reasonably small autocorrelation sidelobes.

3.2.3.2 Autocorrelation Function The autocovariance function of the C/A-code is

$$y(t) = \frac{1}{T} \int_0^T c(t)\, c(t - t)\; dt, \tag{3.13}$$

where $c(t)$ is the idealized C/A-code waveform (with chip values of ± 1), τ is the relative delay measured in seconds, and T is the code period (1 ms). The autocorrelation function is periodic in τ with a period of 1 ms. A single period is plotted in Fig. 3.6, which is basically a triangle two chips wide at its base with a peak located at $\tau = 0$ [in reality $\psi(\tau)$ contains small-sidelobe structures outside the triangular region, but these are of little consequence].

The C/A-code autocorrelation function plays a substantial role in GPS receivers, inasmuch as it forms the basis for code tracking and accurate user-to-satellite range measurement. In fact, the receiver continually computes values of this function in which $c(t)$ in the integral in Eq. 3.13 is the signal code waveform and $c(t - \tau)$ is an identical reference waveform (except for the relative delay τ)

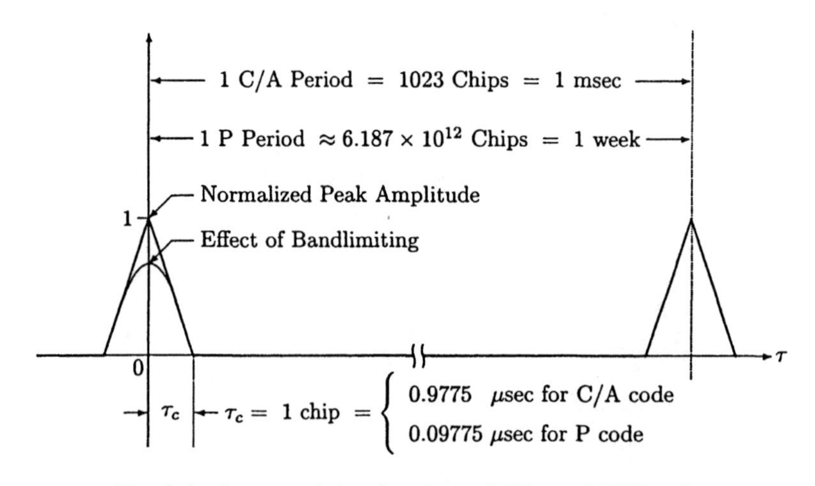

Fig. 3.6 Autocorrelation functions of C/A- and P(Y)-codes.

generated in the receiver. Special hardware and software enable the receiver to adjust the reference waveform delay so that the value of τ is zero, thus enabling determination of the time of arrival of the received signal.

3.2.3.3 Power Spectrum The power spectrum $\Psi(f)$ of the C/A-code describes how the power in the code is distributed in the frequency domain. It can be defined in terms of either a Fourier series expansion of the code waveform or, equivalently, the code autocorrelation function. Using the latter, we have

$$\Psi(f) = \lim_{T\to\infty} \frac{1}{2T} \int_{-T}^{T} \psi(\tau)\, e^{-j2\pi f\tau}\, d\tau. \tag{3.14}$$

A plot of $\Psi(f)$ is shown as a smooth curve in Fig. 3.7; however, in reality $\Psi(f)$ consists of spectral lines with 1-kHz spacing due to the 1-ms periodic structure of $\psi(\tau)$. The power spectrum $\Psi(f)$ has a characteristic $\sin^2(x)/x^2$ shape with first nulls located 1.023 MHz from the central peak. Approximately 90% of the signal power is located between these two nulls, but the smaller portion lying outside the nulls is very important for accurate ranging. Also shown in the figure for comparative purposes is a typical noise power spectral density found in a GPS receiver after frequency conversion of the signal to baseband (i.e., with carrier removed). It can be seen that the presence of the C/A-code causes the

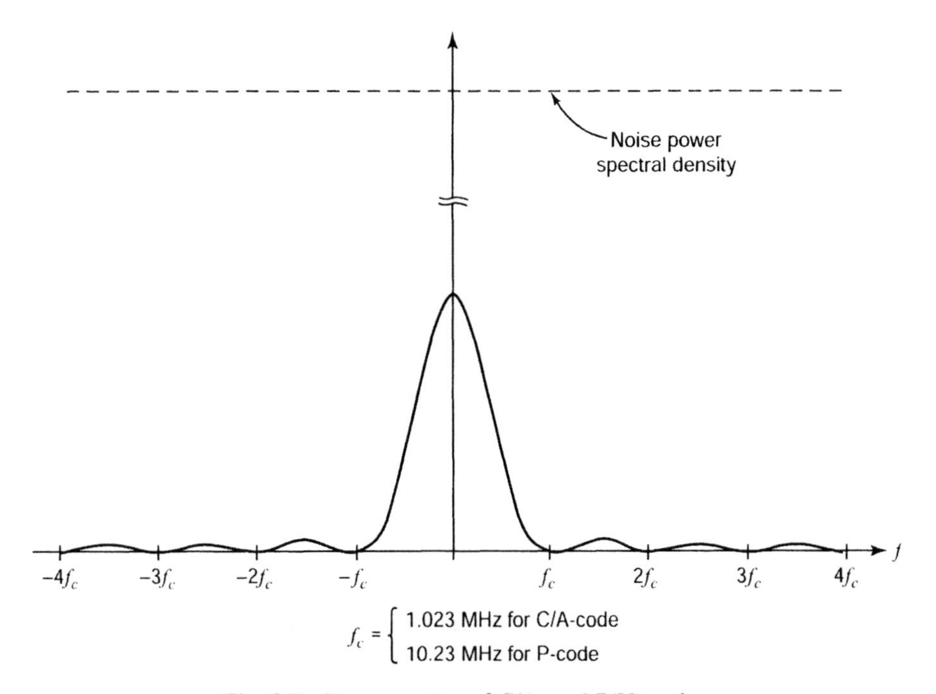

Fig. 3.7 Power spectra of C/A- and P(Y)-codes.

entire signal to lie well below the noise level, because the signal power has been spread over a wide frequency range (approximately ±1 MHz).

3.2.3.4 ***Despreading of the Signal Spectrum*** The mathematical model of the signal modulated by the C/A-code is

$$s(t) = \sqrt{2P_I}d(t)c(t)\cos(\omega t + \theta), \tag{3.15}$$

where P_I is the carrier power, $d(t)$ is the 50-bps (bits per second) data modulation, $c(t)$ is the C/A-code waveform, ω is the L_1 carrier frequency in radians per second, and θ is the carrier phase shift in radians. When this signal is frequency-shifted to baseband and tracked with a phase-lock loop, the carrier is removed and only the data modulation and the C/A-code modulation remain. The resulting signal, which in normalized form is

$$s(t) = d(t)c(t), \tag{3.16}$$

has a power spectrum similar to that of the C/A-code in Fig. 3.7. As previously mentioned, the signal in this form has a power spectrum lying below the receiver noise level, making it inaccessible. However, if the signal is multiplied by a replica of $c(t)$ in exact alignment with it, the result is

$$s(t)c(t) = d(t)c(t)c(t) = d(t)c^2(t) = d(t), \tag{3.17}$$

where the last equality arises from the fact that the values of the ideal C/A-code waveform are ±1 (in reality the received C/A-code waveform is not ideal, due to bandlimiting in the receiver; however, the effects are usually minor). This procedure, called *code despreading*, removes the C/A-code modulation from the signal. The resulting signal has a two-sided spectral width of approximately 100 Hz due to the 50-bps (bits per second) data modulation. From the above equation it can be seen that the total signal power has not been changed in this process, but it now is contained in a much narrower bandwidth. Thus the magnitude of the power spectrum is greatly increased, as indicated in Fig. 3.8. In fact, it now exceeds that of the noise, and the signal can be recovered by passing it through a small-bandwidth filter (signal recovery filter) to remove the wideband noise, as shown in the figure.

3.2.3.5 ***Role of Despreading in Interference Suppression*** At the same time that the spectrum of the desired GPS signal is narrowed by the despreading process, any interfering signal that is not modulated by the C/A-code will instead have its spectrum *spread* to a width of at least 2 MHz, so that only a small portion of the interfering power can pass through the signal recovery filter. The amount of interference suppression gained by using the C/A-code depends on the bandwidth of the recovery filter, the bandwidth of the interfering signal, and the bandwidth of the C/A-code. For narrowband interferors whose signal can be

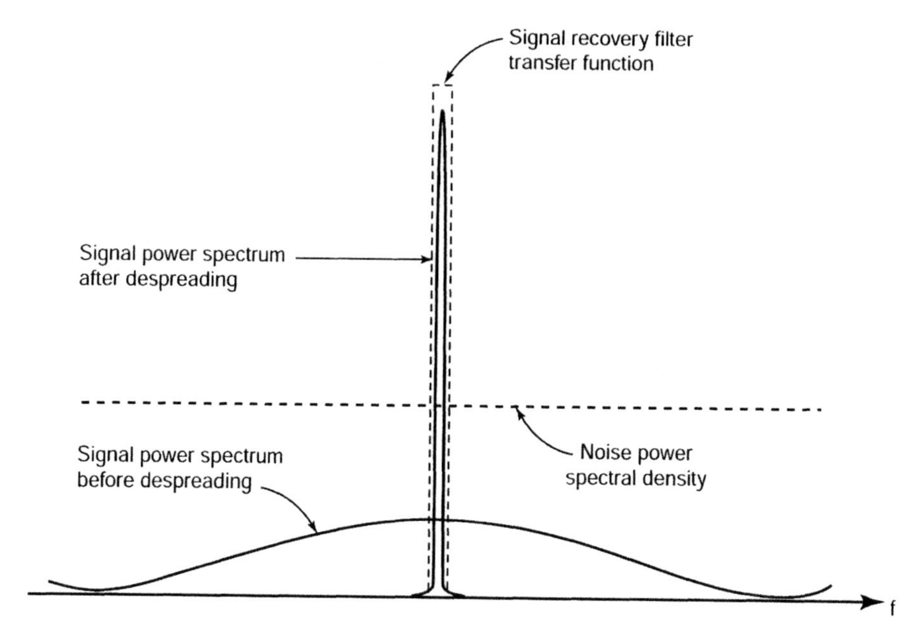

Fig. 3.8 Despreading of the C/A-code.

modeled by a nearly sinusoidal waveform and a signal recovery filter bandwidth of 1000 Hz or more, the amount of interference suppression in decibels is given approximately by

$$\eta = 10 \log \left(\frac{W_c}{W_f} \right) \qquad \text{dB}, \tag{3.18}$$

where W_c and W_f are respectively the bandwidths of the C/A-code (2.046 MHz) and the signal recovery filter. If $W_f = 2000$ Hz, about 30 dB of suppression can be obtained for narrowband interferors. When the signal recovery filter has a bandwidth smaller than 1000 Hz, the situation is more complicated, since the despread interfering sinusoid will have discrete spectral components with a 1000-Hz spacing. As the bandwidth of the interfering signal increases, the C/A-code despreading process provides a decreasing amount of interference suppression. For interferors having a bandwidth greater than that of the signal recovery filter, the amount of suppression in decibels provided by the C/A-code is approximately

$$\eta = 10 \log \left(\frac{W_I + W_c}{W_I} \right) \qquad \text{dB}, \tag{3.19}$$

where W_I is the bandwidth of the interferor. When $W_I >> W_c$, the C/A-code provides essentially no interference suppression at all compared to the use of an unspread carrier.

3.2.3.6 Code-Division Multiplexing Property The C/A-codes from different satellites are *orthogonal*, which means that for any two codes $c_1(t)$ and $c_2(t)$ from different satellites, the cross-covariance

$$\frac{1}{T}\int_0^T c_1(t)\, c_2(t-\tau)\ dt \cong 0 \quad \text{for all } \tau. \tag{3.20}$$

Thus, when a selected satellite signal is despread using a replica of its code, the signals from other satellites look like wideband interferors that are below the noise level. This permits a GPS receiver to extract a multiplicity of individual satellite signals and process them individually, even though all signals are transmitted at the same frequency. This process is called *code-division multiplexing* (CDM).

3.2.4 P-Code and Its Properties

The P-code, which is used primarily for military applications, has the following functions:

1. *Increased Jamming Protection.* Because the bandwidth of the P-code is 10 times greater than that of the C/A-code, it offers approximately 10 dB more protection from narrowband interference. In military applications the interference is likely to be a deliberate attempt to jam (render useless) the received GPS signal.
2. *Provision for Antispoofing.* In addition to jamming, another military tactic that an enemy can employ is to radiate a signal that appears to be a GPS signal (*spoofing*), but in reality is designed to confuse the GPS receiver. This is prevented by encrypting the P-code. The would-be spoofer cannot know the encryption process and cannot make the contending signal look like a properly encrypted signal. Thus the receiver can reject the false signal and decrypt the desired one.
3. *Denial of P-Code Use.* The structure of the P-code is published in the open literature, so than anyone may generate it as a reference code for despreading the signal and making range measurements. However, encryption of the P-code by the military will deny its use by unauthorized parties.
4. *Increased Code Range Measurement Accuracy.* All other parameters being equal, accuracy in range measurement improves as the signal bandwidth increases. Thus, the P-code provides improved range measurement accuracy as compared to the C/A-code. Simultaneous range measurements using both codes is even better. Because of its increased bandwidth, the P-code is also more resistant to range errors caused by multipath.

3.2.4.1 P-Code Characteristics Unlike the C/A-code, the P-code modulates both the L_1 and L_2 carriers. Its chipping rate is 10.23 MHz, which is precisely 10 times the C/A rate, and it has a period of one week. It is transmitted synchronously with the C/A-code in the sense that each chip transition of the C/A-code always corresponds to a chip transition in the P-code. Like the C/A-code, the P-code autocorrelation function has a triangular central peak centered at $\tau = 0$, but with one-tenth the base width, as shown in Fig. 3.6. The power spectrum also has a $\sin^2(x)/x^2$ characteristic, but with 10 times the bandwidth, as indicated in Fig. 3.6. Because the period of the P-code is so long, the power spectrum may be regarded as continuous for practical purposes. Each satellite broadcasts a unique P-code. The technique used to generate it is similar to that of the C/A-code, but somewhat more complicated, and will not be covered in this book.

3.2.4.2 Y-Code The encrypted form of the P-code used for antispoofing and denial of the P-code to unauthorized users is called the *Y-code*. The Y-code is formed by multiplying the P-code by an encrypting code called the *W-code*. The W-code is a random-looking sequence of chips that occur at a 511.5-kHz rate. Thus there are 20 P-code chips for every W-code chip. Since both the P-code and the W-code have chip values of ± 1, the resulting Y-code has the same appearance as the P-code; that is, it also has a 10.23-MHz chipping rate. However, the Y-code cannot be despread by a receiver replica P-code unless it is decrypted. Decryption consists of multiplying the Y-code by a receiver-generated replica of the W-code that is made available only to authorized users. Since the encrypting W-code is also not known by the creators of spoofing signals, it is easy to verify that such signals are not legitimate.

3.2.5 L_1 and L_2 Carriers

The L_1 (or L_2) carrier is used for the following purposes:

1. *To provide very accurate range measurements for precision applications using carrier phase.*
2. *To provide accurate Doppler measurements.* The phase rate of the received carrier can be used for accurate determination of user velocity. The integrated Doppler, which can be obtained by counting the cycles of the received carrier, is often used as a precise delta range observable that can materially aid the performance of code tracking loops. The integrated Doppler history is also used as part of the carrier phase ambiguity resolution process.

3.2.5.1 Dual-Frequency Operation The use of *both* the L_1 and L_2 frequencies provides the following benefits:

1. *Provides accurate measurement of ionospheric signal delay.* A major source of ranging error is caused by changes in both the phase velocity and group velocity of the signal as it passes through the ionosphere. Range errors of 10–20 m are commonplace and sometimes much larger. Because the delay induced by the ionosphere is known to be inversely proportional to the square of frequency, ionospheric range error can be estimated accurately

by comparing the times of arrival of the L_1 and L_2 signals. Details on the calculations appear in Chapter 5.

2. *Facilitates carrier phase ambiguity resolution.* In high-accuracy GPS differential positioning, the range estimates using carrier phase measurements are precise but highly ambiguous due to the periodic structure of the carrier. The ambiguity is more easily resolved (by various methods) as the carrier frequency decreases. By using L_1 and L_2 carrier frequencies, the ambiguity resolution can be based on their frequency difference (1575.42–1227.6 MHz), which is smaller than either carrier frequency alone, and hence will result in better ambiguity resolution performance.
3. *Provides system redundancy (primarily for the military user).*

3.3 SIGNAL POWER LEVELS

3.3.1 Transmitted Power Levels

The L_1 C/A-code signal is transmitted at a minimum level of 478.63 W (26.8 dBW) effective isotropic radiated power (EIRP), which means that the minimum received power is the same as that that would be obtained if the satellite radiated 478.63 W from an isotropic antenna. This effective power level is reached by radiating a smaller total power in a beam approximately 30° wide toward the earth. The radiated power level was chosen to provide a signal-to-noise ratio sufficient for tracking of the signal by a receiver on the earth with an unobstructed view of the satellite. However, the chosen power has been criticized as being inadequate in light of the need to operate GPS receivers under less desirable conditions, such as in heavy vegetation or in urban canyons where considerable signal attenuation often occurs. For this reason, future satellites may have higher transmitted power.

3.3.2 Free-Space Loss Factor

As the signal propagates toward the earth, it loses power density due to spherical spreading. The loss is accounted for by a quantity called the *free-space loss factor* (FSLF), given by

$$\text{FSLF} = \left(\frac{\lambda}{4\pi R}\right)^2. \tag{3.21}$$

The FSLF is the fractional power density at a distance R meters from the transmitting antenna compared to a value normalized to unity at the distance $\lambda/4\pi$ meters from the antenna phase center. Using $R = 2 \times 10^7$ and $\lambda = 0.19$ m at the L_1 frequency, the FSLF is about 5.7×10^{-19}, or -182.4 dB.

3.3.3 Atmospheric Loss Factor

An additional atmospheric loss factor (ALF) of about 2.0 dB occurs as the signal becomes attenuated by the atmosphere. If the receiving antenna is assumed to be

TABLE 3.3. Calculation of Minimum Received Signal Power

Minimum transmitted signal power (EIRP)	26.8[a]	dBW
Free-space loss factor (FSLF)	−182.4	dB
Atmospheric loss factor (ALF)	−2.0	dB
Receiver antenna gain relative to isotropic (RAG)	3.0	dB
Minimum received signal power (EIRP − FSLF − ALF + RAG)	−154.6	dBW

[a]Including antenna gain

isotropic, the received signal power is EIRP − FSLF − ALF = 26.8 − 182.4 − 2.0 = −157.6dBW.

3.3.4 Antenna Gain and Minimum Received Signal Power

Since a typical GPS antenna with right-hand circular polarization and a hemispherical pattern has about 3.0 dB of gain relative to an isotropic antenna, the minimum received signal power for such an antenna is about 3.0 dB larger. These results are summarized in Table 3.3.

3.4 SIGNAL ACQUISITION AND TRACKING

When a GPS receiver is turned on, a sequence of operations must ensue before information in a GPS signal can be accessed and used to provide a navigation solution. In the order of execution, these operations are as follows:

1. Determine which satellites are visible to the antenna.
2. Determine the approximate Doppler of each visible satellite.
3. Search for the signal both in frequency and C/A-code phase.
4. Detect the presence of a signal and confirm detection.
5. Lock onto and track the C/A-code.
6. Lock onto and track the carrier.
7. Perform data bit synchronization.
8. Demodulate the 50-bps (bits per second) navigation data.

3.4.1 Determination of Visible Satellites

In many GPS receiver applications it is desirable to minimize the time from receiver turnon until the first navigation solution is obtained. This time interval is commonly called *time to first fix* (TTFF). Depending on receiver characteristics, the TTFF might range from 30 s to several minutes. An important consideration in minimizing the TTFF is to avoid a fruitless search for those satellite signals that are blocked by the earth, that is, below the horizon. A receiver can restrict its search to only those satellites that are visible if it knows its approximate location (within several hundred miles) and approximate time (within approximately 10 min) and has satellite almanac data obtained within the last several months. The approximate location can be manually entered by the user or it can be the position obtained by GPS when the receiver was last in operation. The approximate time can also be entered manually, but most receivers have a sufficiently accurate real-time clock that operates continuously, even when the receiver is off.

Using the approximate time, approximate position, and almanac data, the receiver calculates the elevation angle of each satellite and identifies the visible satellites as those whose elevation angle is greater than a specified value, called the *mask angle*, which has typical values of 5° to 15°. At elevation angles below the mask angle, tropospheric attenuation and delays tend to make the signals unreliable.

Most receivers automatically update the almanac data when in use, but if the receiver is just "out of the box" or has not been used for many months, it will need to search "blind" for a satellite signal to collect the needed almanac. In this case the receiver will not know which satellites are visible, so it simply must work its way down a predetermined list of satellites until a signal is found. Although such a "blind" search may take an appreciable length of time, it is infrequently needed.

3.4.2 Signal Doppler Estimation

The TTFF can be further reduced if the approximate Doppler shifts of the visible satellite signals are known. This permits the receiver to establish a frequency search pattern in which the most likely frequencies of reception are searched first. The expected Doppler shifts can be calculated from knowledge of approximate position, approximate time, and valid almanac data. The greatest benefit is obtained if the receiver has a reasonably accurate clock reference oscillator.

However, once the first satellite signal is found, a fairly good estimate of receiver clock frequency error can be determined by comparing the predicted Doppler shift with the measured Doppler shift. This error can then be subtracted out while searching in frequency for the remaining satellites, thus significantly reducing the range of frequencies that need to be searched.

3.4.3 Search for Signal in Frequency and C/A-Code Phase

Why is a signal search necessary? Since GPS signals are radio signals, one might assume that they could be received simply by setting a dial to a particular frequency, as is done with AM and FM broadcast band receivers. Unfortunately, this is not the case.

1. GPS signals are *spread-spectrum* signals in which the C/A- or P-codes spread the total signal power over a wide bandwidth. The signals are therefore virtually undetectable unless they are *despread* with a replica code in the

receiver that is precisely aligned with the received code. Since the signal cannot be detected until alignment has been achieved, a search over the possible alignment positions (code search) is required.

2. A relatively narrow postdespreading bandwidth (perhaps 100–1000 Hz) is required to raise the signal-to-noise ratio to detectable and/or usable levels. However, because of the high carrier frequencies and large satellite velocities used by GPS, the received signals can have large Doppler shifts (as much as ± 5 kHz), which may vary rapidly (by as much as 1 Hz/s). The observed Doppler shift also varies with location on earth, so that the received frequency will generally be unknown a priori. Furthermore, the frequency error in typical receiver reference oscillators will typically cause several kilohertz or more of frequency uncertainty at L-band. Thus, in addition to the code search, there is also the need for a search in frequency.

Therefore, a GPS receiver must conduct a two-dimensional search in order to find each satellite signal, where the dimensions are C/A-code delay and carrier frequency. A search must be conducted across the full delay range of the C/A-code for each frequency searched. A generic method for conducting the search is illustrated in Fig. 3.9, in which the received waveform is multiplied by delayed replicas of the C/A-code, translated by various frequencies, and then passed through a baseband correlator containing a lowpass filter which has a relatively small bandwidth (perhaps 100–1000 Hz). The output energy of the detection filter serves as a signal detection statistic and will be significant only if both the selected code delay and frequency translation match that of the signal. When the

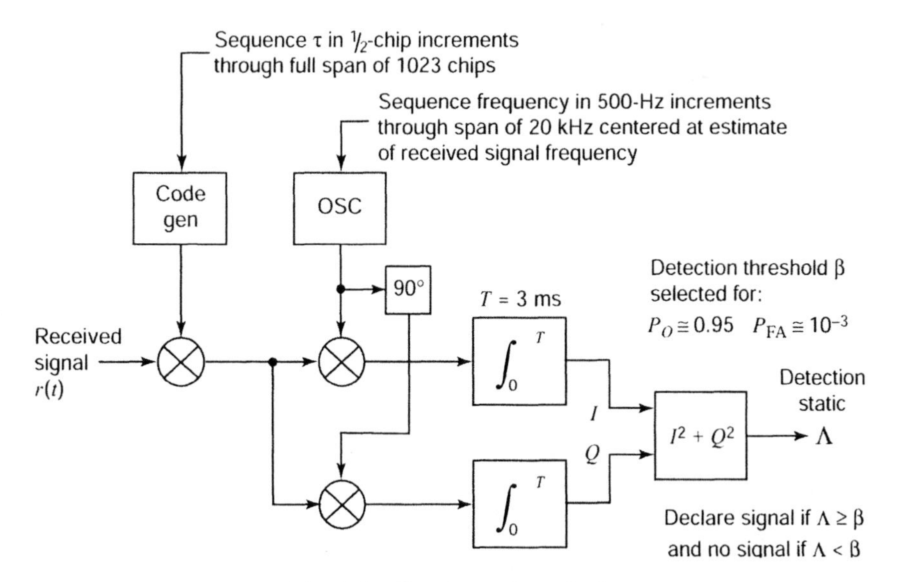

Fig. 3.9 Signal search method.

energy exceeds a predetermined threshold β, a tentative decision is made that a signal is being received, subject to later confirmation. The value chosen for the threshold β is a compromise between the conflicting goals of maximizing the probability P_D of detecting the signal when it is actually present at a given Doppler and code delay and minimizing the probability P_{FA} of false alarm when it is not.

3.4.3.1 Searching in Code Delay For each frequency searched, the receiver generates the same PRN code as that of the satellite and moves the delay of this code in discrete steps (typically 0.5 chip) until approximate alignment with the received code (and also a match in Doppler) is indicated when the correlator output energy exceeds threshold β. A step size of 0.5 code chip, which is used by many GPS receivers, is an acceptable compromise between the conflicting requirements of search speed (enhanced by a larger step size) and guaranteeing a code delay that will be located near the peak value of the code correlation function (enhanced by a smaller step size). For a search conducted in 1-chip increments, the best situation occurs when one of the delay positions is at the correlation function peak, and the worst one occurs when there are two delay positions straddling the peak, as indicated in Fig. 3.10. In the latter case, the effective SNR is reduced by as much as 6 dB. However, the effect is ameliorated because, instead of only one delay position with substantial correlation, there are two that can be tested for the presence of signal.

An important parameter in the code search is the dwell time used for each code delay position, since it influences both the search speed and the detection/falsealarm performance. The dwell time should be an integral multiple of

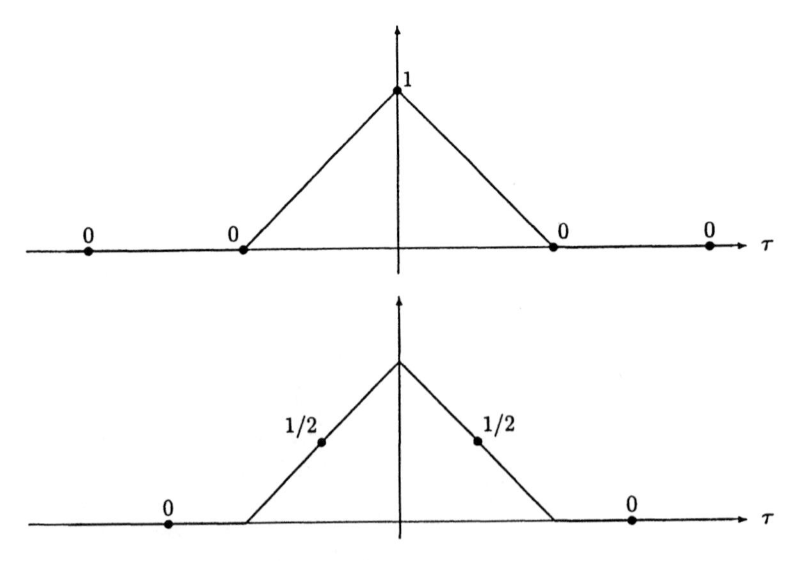

Fig. 3.10 Effect of 1-chip step size in code search.

1 ms to assure that the correct correlation function, using the full range of 1023 code states, is obtained. Satisfactory performance is obtained with dwell times from 1 to 4 ms in most GPS receivers, but longer dwell times are sometimes used to increase detection capability in weak-signal environments. However, if the dwell time for the search is a substantial fraction of 20 ms (the duration of one data bit), it becomes increasingly probable that a bit transition of the 50-Hz data modulation will destroy the coherent processing of the correlator during the search and lead to a missed detection. This imposes a practical limit for a search using coherent detection.

The simplest type of code search uses a fixed dwell time, a single detection threshold value β, and a simple yes/no binary decision as to the presence of a signal. Many receivers achieve considerable improvement in search speed by using a sequential detection technique in which the overall dwell time is conditioned on a ternary decision involving an upper and a lower detection threshold. Details on this approach can be found in the treatise by Wald [199].

3.4.3.2 Searching in Frequency The range of frequency uncertainty that must be searched is a function of the accuracy of the receiver reference oscillator, how well the approximate user position is known, and the accuracy of the receiver's built-in real-time clock. The first step in the search is to use stored almanac data to obtain an estimate of the Doppler shift of the satellite signal. An interval $[f_1, f_2]$ of frequencies to be searched is then established. The center of the interval is located at $f_c + f_d$, where f_c is the L_1 (or L_2) carrier frequency and f_d is the estimated carrier Doppler shift. The width of the search interval is made large enough to account for worst-case errors in the receiver reference oscillator, in the estimate of user position, and in the real-time clock. A typical range for the frequency search interval is $f_c + f_d \pm 5$ kHz.

The frequency search is conducted in N discrete frequency steps that cover the entire search interval. The value of N is $(f_2 - f_1)/\Delta f$, where Δf is the spacing between adjacent frequencies (bin width). The bin width is determined by the effective bandwidth of the correlator. For the coherent processing used in many GPS receivers, the frequency bin width is approximately the reciprocal of the search dwell time. Thus, typical values of Δf are 250–1000 Hz. Assuming a $\pm$5-kHz frequency search range, the number N of frequency steps to cover the entire search interval would typically be 10–40.

3.4.3.3 Frequency Search Strategy Because the received signal frequency is more likely to be near to—rather than far from—the Doppler estimate, the expected time to detect the signal can be minimized by starting the search at the estimated frequency and expanding in an outward direction by alternately selecting frequencies above and below the estimate, as indicated in Fig. 3.11. On the other hand, the unknown code delay of the signal can be considered to be uniformly distributed over its range so that each delay value is equally likely. Thus, the delays used in the code search can simply sequence from 0 to 1023.5 chips in 0.5-chip increments.

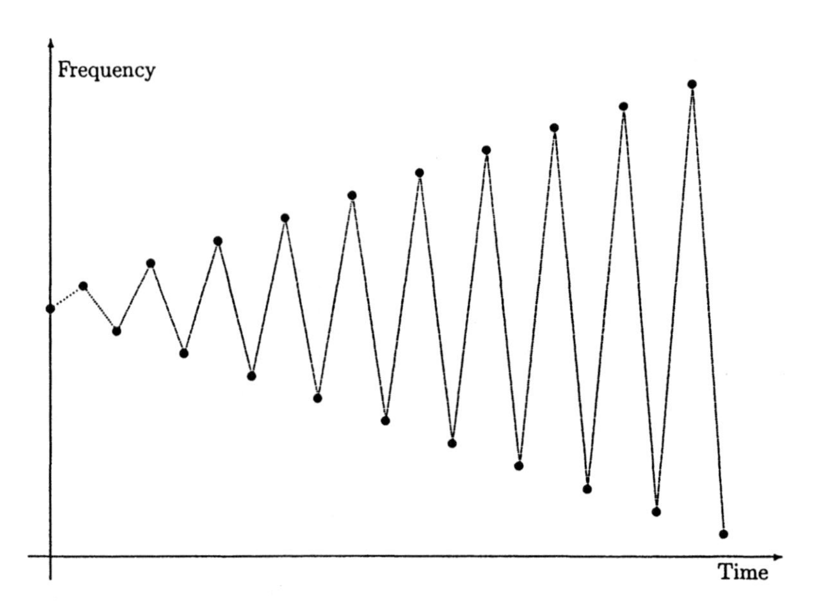

Fig. 3.11 Frequency search strategy.

3.4.3.4 Sequential versus Parallel Search Methods Almost all current GPS receivers are multichannel units in which each channel is assigned a satellite and processing in the channels is carried out simultaneously. Thus, simultaneous searches can be made for all usable satellites when the receiver is turned on. Because the search in each channel consists of sequencing through all possible frequency and code delay steps, it is called a *sequential search*. In this case, the expected time required to acquire as many as eight satellites is typically 30–100 s, depending on the specific search parameters used.

Certain applications (mostly military) demand that the satellites be acquired much more rapidly (perhaps within a few seconds). This can be accomplished by using a *parallel search* technique in which extra hardware permits many frequencies and code delays to be searched at the same time. However, this approach is seldom used in commercial receivers because of its high cost.

3.4.4 Signal Detection and Confirmation

As previously mentioned, there is a tradeoff between the probability of detection P_D and false alarm P_{FA}. As the detection threshold β is decreased, P_D increases but P_{FA} also increases, as illustrated in Fig. 3.12. Thus, the challenge in receiver design is to achieve a sufficiently large P_D so that a signal will not be missed but at the same time keep P_{FA} small enough to avoid difficulties with false detections. When a false detection occurs, the receiver will try to lock onto and track a nonexistent signal. By the time the failure to track becomes evident, the receiver will have to initiate a completely new search for the signal. On the other

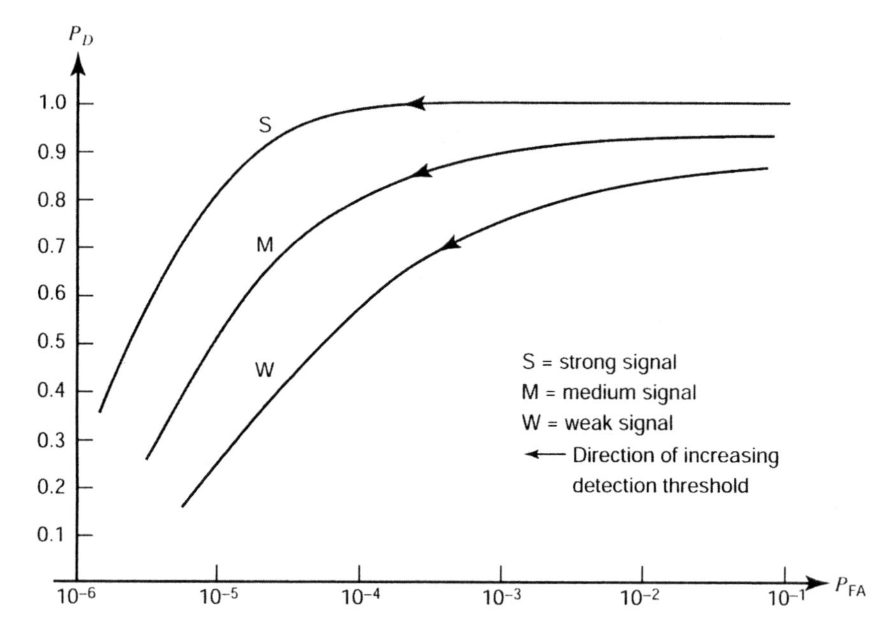

Fig. 3.12 Illustration of tradeoff between P_D and P_{FA}.

hand, when a detection failure occurs, the receiver will waste time continuing to search remaining search cells that contain no signal, after which a new search must be initiated.

3.4.4.1 Detection Confirmation One way to achieve both a large P_D and a small P_{FA} is to increase the dwell time so that the relative noise component of the detection statistic is reduced. However, to reliably acquire weak GPS signals, the required dwell time may result in unacceptably slow search speed. An effective way around this problem is to use some form of *detection confirmation*.

To illustrate the detection confirmation concept, suppose that to obtain the detection probability $P_D = 0.95$ with a typical medium-strength GPS signal, we obtain the false-alarm probability $P_{FA} = 10^{-3}$. (These are typical values for a fixed search dwell time of 3 ms.) This means that on the average, there will be one false detection in every 1000 frequency/code cells searched. A typical two-dimensional GPS search region might contain as many as 40 frequency bins and 2046 code delay positions, for a total of $40 \times 2046 = 81,840$ such cells. Thus we could expect about 82 false detections in the full search region. Given the implications of a false detection discussed previously, this is clearly unacceptable.

However, suppose that we change the rules for what happens when a detection (false or otherwise) occurs by performing a confirmation of detection before turning the signal over to the tracking loops. Because a false detection takes place only once in 1000 search cells, it is possible to use a much longer dwell (or a sequence of repeated dwells) for purposes of confirmation without markedly

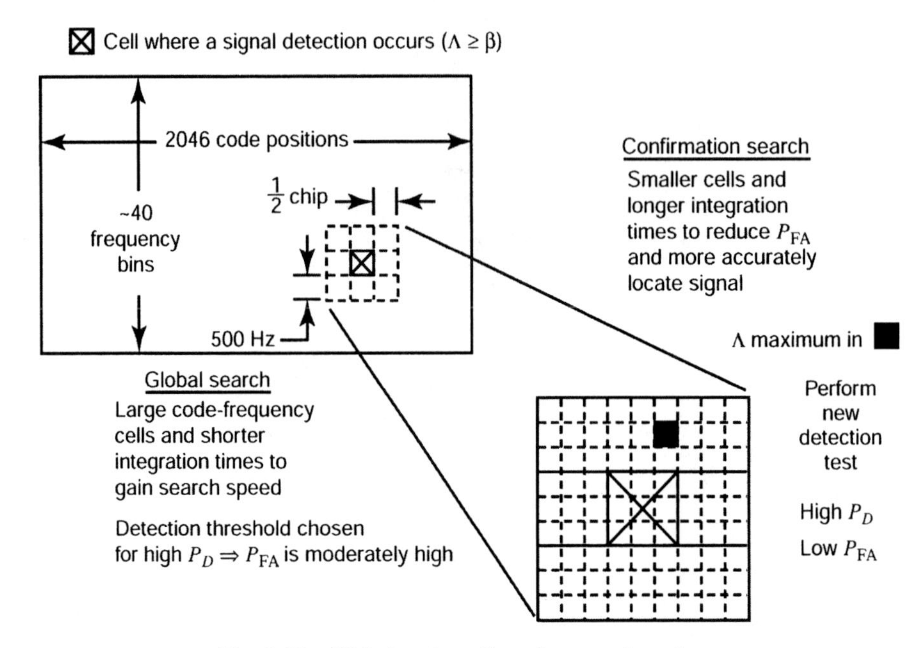

Fig. 3.13 Global and confirmation search regions.

increasing the overall search speed, yet the confirmation process will have an extremely high probability of being correct. In the event that confirmation indicates no signal, the search can continue without interruption by the large time delay inherent in detecting the failure to track. In addition to using longer dwell times, the confirmation process can also perform a *local search* in which the frequency/code cell size is smaller than that of the main, or *global*, search, thus providing a more accurate estimate of signal frequency and code phase when a detection is confirmed. Figure 3.13 depicts this scheme. The global search uses a detection threshold β that provides a high P_D and a moderate value of P_{FA}. Whenever the detection statistic Λ exceeds β at a frequency/delay cell, a confirmation search is performed in a local region surrounding that cell. The local region is subdivided into smaller cells to obtain better frequency delay resolution, and a longer dwell time is used in forming the detection statistic Λ. The longer dwell time makes it possible to use a value of β that provides both a high P_D and a low P_{FA}.

3.4.4.2 Adaptive Signal Searches Some GPS receivers use a simple adaptive search in which shorter dwell times are first used to permit rapid acquisition of moderate to strong signals. Whenever a search for a particular satellite is unsuccessful, it is likely that the signal from that satellite is relatively weak, so the receiver increases the dwell time and starts a new search that is slower but has better performance in acquiring weak signals.

3.4.4.3 Coordination of Frequency Tuning and Code Chipping Rate As the receiver is tuned in frequency during search, it is advantageous to precess the chipping rate of the receiver generated code so that it is in accordance with the Doppler shift under consideration. The relationship between Doppler shift and the precession rate of the C/A-code is given by $p(t) = f_d/1540$, where $p(t)$ is the code precession rate in chips per second, f_d is the Doppler shift in hertz, and a positive precession rate is interpreted as an increase in the chipping rate. Precession is not required while searching because the dwell times are so short. However, when detection of the signal occurs, it is important to match the incoming and reference code rates during the longer time required for detection confirmation and/or initiation of code tracking to take place.

3.4.5 Code Tracking Loop

At the time of detection confirmation the receiver-generated reference C/A-code will be in approximate alignment with that of the signal (usually within 0.5 chip), and the reference code chipping rate will be approximately that of the signal. Additionally, the frequency of the signal will be known to within the frequency bin width Δf. However, unless further measures are taken, the residual Doppler on the signal will eventually cause the received and reference codes to drift out of alignment and the signal frequency to drift outside the frequency bit at which detection occurred. If the code alignment error exceeds one chip in magnitude, the incoming signal will no longer despread and will disappear below the noise level. The signal will also disappear if it drifts outside the detection frequency bin. Thus there is the need to continually adjust the timing of the reference code so that it maintains accurate alignment with the received code, a process called *code tracking*. The process of maintaining accurate tuning to the signal carrier, called *carrier tracking*, is also necessary and will be discussed in following sections.

Code tracking is initiated as soon as signal detection is confirmed, and the goal is to make the receiver-generated code line up with incoming code as precisely as possible. There are two objectives in maintaining alignment:

1. *Signal Despreading.* The first objective is to fully despread the signal so that it is no longer below the noise and so that information contained in the carrier and the 50-bps (bits per second) data modulation can be recovered.
2. *Range Measurements.* The second objective is to enable precise measurement of the time of arrival (TOA) of received code for purposes of measuring range. Such measurements cannot be made directly from the received signal, since it is below the noise level. Therefore, a code tracking loop, which has a large processing gain, is employed to generate a reference code precisely aligned with that of the received signal. This enables range measurements to be made using the reference code instead of the much noisier received signal code waveform.

Figure 3.14 illustrates the concept of a code tracking loop. It is assumed that a numerically controlled oscillator (NCO) has translated the signal to complex

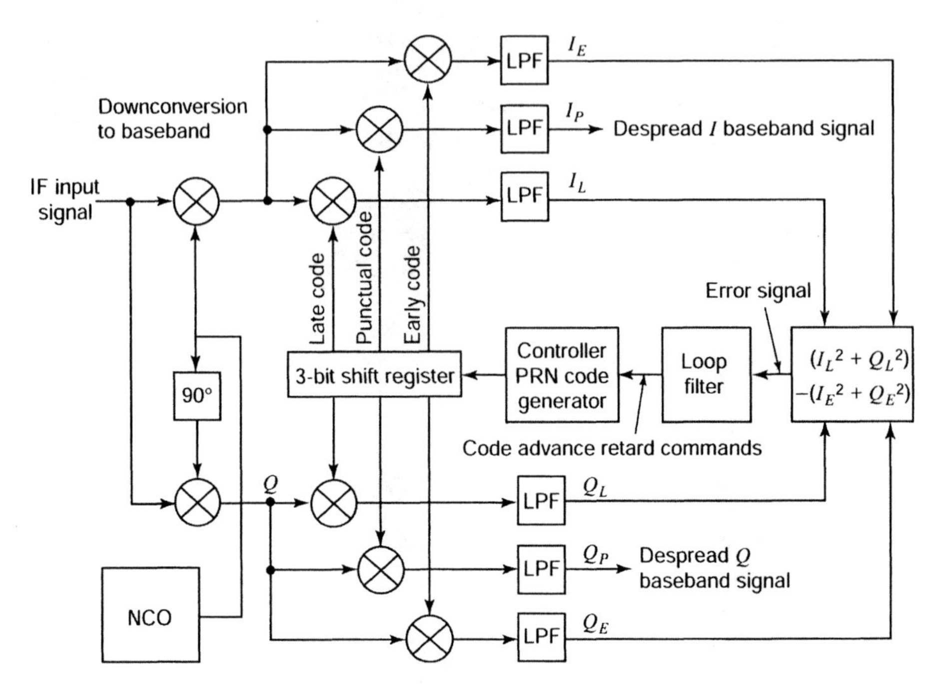

Fig. 3.14 Code tracking loop concept.

baseband form (i.e., zero frequency). Each component (I and Q) of the baseband signal is multiplied by three replicas of the C/A-code that are formed by delaying the output of a single code generator by three delay values called *early, punctual, and late*. In typical GPS receivers the early and late codes respectively lead and lag the punctual code by 0.05 to 0.5 code chips and always maintain these relative positions. Following each multiplier is a lowpass filter (LPF) or integrator that, together with its associated multiplier, forms a correlator. The output magnitude of each correlator is proportional to the cross-correlation of its received and reference codes, where the cross-correlation function has the triangular shape previously shown in Fig. 3.6. In normal operation the punctual code is aligned with the code of the incoming signal so that the squared magnitude $I_P^2 + Q_P^2$ of the punctual correlator output is at the peak of the cross-correlation function, and the output magnitudes of the early and late correlators have smaller but equal values on each side of the peak. To maintain this condition, a loop error signal

$$e_c(\tau) = I_L^2 + Q_L^2 - \left(I_E^2 + Q_E^2\right), \tag{3.22}$$

is formed, which is the difference between the squared magnitudes of the late and early correlators. The loop error signal as a function of received code delay is shown in Fig. 3.15. Near the tracking point the error is positive if the received code is delayed relative to the punctual code and negative if it is advanced.

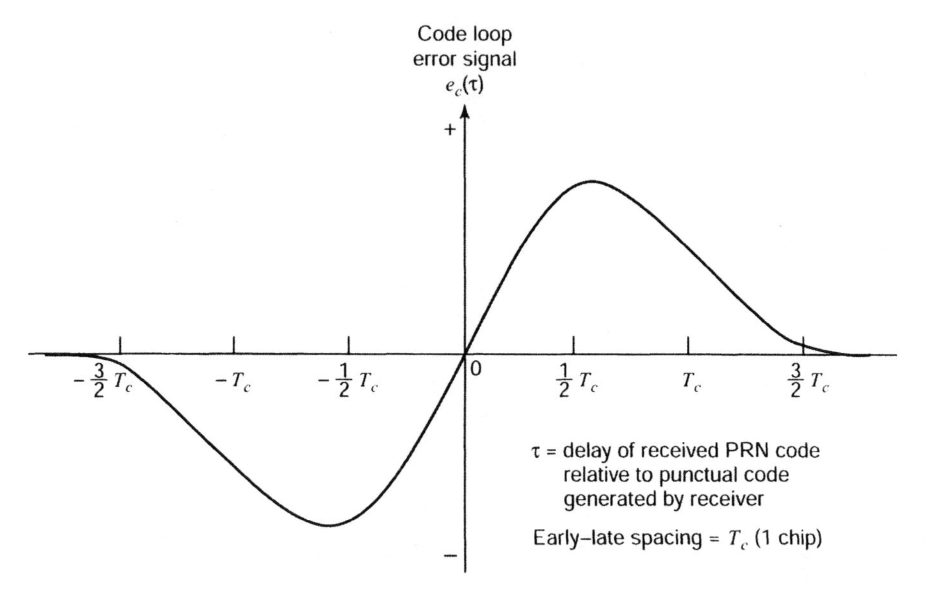

Fig. 3.15 Code tracking loop error signal.

Alignment of the punctual code with the received code is maintained by using the error signal to delay the reference code generator when the error signal is positive and to advance it when the error signal is negative. Since $e_c(\tau)$ is generally quite noisy, it is sent through a lowpass *loop filter* before it controls the timing of the reference code generator, as indicated in Fig. 3.14. The bandwidth of this filter is usually quite small, resulting in a closed-loop bandwidth typically less than 1 Hz. This is the source of the large processing gain that can be realized in extracting the C/A-code from the signal.

When the code tracking loop is first turned on, the integration time T for the correlators is usually no more than a few milliseconds, in order to minimize corruption of the correlation process by data bit transitions of the 50-bps (bits per second) data stream whose locations in time are not yet known. However, after bit synchronization has located the data bit boundaries, the integration interval can span a full data bit (20 ms) in order to achieve a maximum contribution to processing gain.

3.4.5.1 Coherent versus Noncoherent Code Tracking If the error signal is formed from only the squared magnitudes of the (complex) early and late correlator outputs as described above, the loop is called a *noncoherent* code tracking loop. A distinguishing feature of such a loop is its insensitivity to the phase of the received signal. Insensitivity to phase is desirable when the loop is first turned on, since at that time the signal phase is random and not yet under any control.

On the other hand, once the phase of the signal is being tracked, a *coherent* code tracker can be employed, in which the outputs of the early and late correlators are purely real. In this situation the loop error signal can be formed directly from the difference of the early and late squared magnitudes from only the I correlator. By avoiding the noise in the Q correlator outputs, a 3-dB SNR advantage is thereby gained in tracking the code. However, a price is paid in that the code loop error signal becomes sensitive to phase error in tracking the carrier. If phase tracking is ever lost, complete failure of the code tracking loop could occur. This is a major disadvantage, especially in mobile applications where the signal can vary rapidly in magnitude and phase. Since noncoherent operation is much more robust in this regard and is still needed when code tracking is initiated, most GPS receivers use only noncoherent code tracking.

3.4.5.2 Factors Affecting Code Tracking Performance The bandwidth of the code tracking loop is determined primarily by the loop filter and needs to be narrow for best ranging accuracy but wide enough to avoid loss of lock if the receiver is subject to large accelerations that can suddenly change the apparent chipping rate of the received code. Excessive accelerations cause loss of lock by moving the received and reference codes too far out of alignment before the loop can adequately respond. Once the alignment error exceeds approximately 1 code chip, the loop loses lock because it no longer has the ability to form the proper error signal.

In low-dynamics applications with lower-cost receivers, code tracking loop bandwidths on the order of 1 Hz permit acceptable performance in handheld units and in receivers with moderate dynamics (e.g., in automobiles). For high-dynamics applications, such as missile platforms, loop bandwidths might be on the order of 10 Hz or larger. In surveying applications, which have no appreciable dynamics, loop bandwidths can be as small as 0.01 Hz to obtain the required ranging accuracy. Both tracking accuracy and the ability to handle dynamics are greatly enhanced by means of *carrier aiding* from the receiver's carrier phase tracking loop, which will be be discussed subsequently.

3.4.6 Carrier Phase Tracking Loops

The purposes of tracking carrier phase are to

1. Obtain a phase reference for coherent detection of the GPS biphase modulated data
2. Provide precise velocity measurements (via phase rate)
3. Obtain integrated Doppler for rate aiding of the code tracking loop
4. Obtain precise carrier phase pseudorange measurements in high-accuracy receivers

Tracking of carrier phase is usually accomplished by a phase-lock loop (PLL). A Costas-type PLL or its equivalent must be used to prevent loss of phase coherence induced by the biphase data modulation on the GPS carrier. The origin of the Costas PLL is described in [40]. A typical Costas loop is shown in

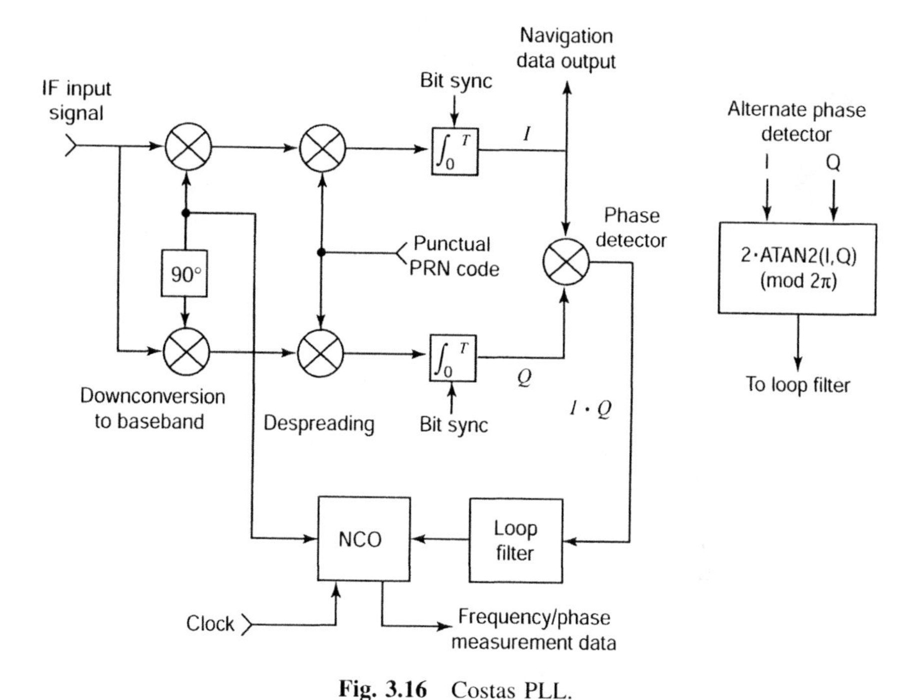

Fig. 3.16 Costas PLL.

Fig. 3.16. In this design the output of the receiver last intermediate-frequency (IF) amplifier is converted to a complex baseband signal by multiplying the signal by both the in-phase and quadrature-phase outputs of an NCO and integrating each product over each 20-ms data bit interval to form a sequence of phasors. The phase angle of each phasor is the phase difference between the signal carrier and the NCO output during the 20-ms integration. A loop phase error signal is formed by multiplying together the I and Q components of each phasor. This error signal is unaffected by the biphase data modulation because the modulation appears on both I and Q and is removed in forming the $I \times Q$ product. After passing through a lowpass loop filter the error signal controls the NCO phase to drive the loop error signal $I \times Q$ to zero (the phase-locked condition). In some receivers the error signal is generated by forming twice the four-quadrant arctangent of the I and Q phasor components, as indicated in the figure.

Because the Costas loop is unaffected by the data modulation, it will achieve phase lock at two stable points where the NCO output phase differs from that of the signal carrier by either 0° or 180°, respectively. This can be seen by considering $I = A \cos \theta$ and $Q = A \sin \theta$, where A is the phasor amplitude and θ is its phase. Then

$$I \times Q = A^2 \cos\theta \sin\theta = \tfrac{1}{2} A^2 \sin 2\theta. \tag{3.23}$$

There are four values of θ in $[0, 2\pi)$ where the error signal $I \times Q = 0$. Two of these are the stable points, namely, $\theta = 0$ and $\theta = 180^\circ$, toward which the loop tends to return if perturbed. Since $\sin 2\theta$ is unchanged by 180° changes in θ caused by the data bits, the data modulation will have no effect. At either of the two stable points the Q integrator output is nominally zero and the I integrator output contains the demodulated data stream, but with a polarity ambiguity that can be removed by observing frame preamble data. Thus the Costas loop has the additional feature of serving as a data demodulator.

In the Costas loop design shown the phase of the signal is measured by comparing the phase of the NCO output with a reference signal. Normally the reference signal frequency is a rational multiple of the same crystal-controlled oscillator that is used in frequency shifting the GPS signal down to the last IF. When the NCO is locked to the phase of the incoming signal, the measured phase rate will typically be in the range of ± 5 kHz due to signal Doppler shift. Two types of phase measurement are usually performed on a periodic basis (the period might be every 20 ms). The first is an accurate measurement of the phase modulo 2π, which is used in precision carrier phase ranging. The second is the number of cycles (including the fractional part) of phase change that have occurred from a defined point in time up to the present time. The latter measurement is often called integrated Doppler and is used for aiding the code tracking loop. By subtracting consecutive integrated Doppler measurements, extremely accurate average frequency measurements can be made, which can be used by the navigation filter to accurately determine user velocity.

Although the Costas loop is not disturbed by the presence of data modulation, at low SNR its performance degrades considerably from that of a loop designed for a pure carrier. The degradation is due to the noise $\times$ noise component of the $I \times Q$ error signal. Furthermore, the 20-ms duration of the I and Q integrations represents a limit to the amount of coherent processing that can be achieved. If it is assumed that the maximum acceptable bit error rate for the 50-bps (bits per second) data demodulation is 10^{-5}, GPS signals become unusable when C/N_0 falls below about 25 dB-Hz.

The design bandwidth of the PLL is determined by the SNR, desired tracking accuracy, signal dynamics, and ability to "pull in" when acquiring the signal or when lock is momentarily lost.

3.4.6.1 PLL Capture Range An important characteristic of the PLL is the ability to "pull in" to the frequency of a received signal. When the PLL is first turned on following code acquisition, the difference between the signal carrier frequency and the NCO frequency must be sufficiently small, or the PLL will not lock. In typical GPS applications, the PLL must have a relatively small bandwidth (1–10 Hz) to prevent loss of lock due to noise. However, this results in a small pullin (or capture) range (perhaps only 3–30 Hz), which would require small (hence many) frequency bins in the signal acquisition search algorithm.

3.4.6.2 Use of Frequency-Lock Loops for Carrier Capture Some receivers avoid the conflicting demands of the need for a small bandwidth and a large capture range in the PLL by using a frequency-lock loop (FLL). The capture range of a FLL is typically much larger than that of a PLL, but the FLL cannot lock to phase. Therefore, a FLL is often used to pull the NCO frequency into the capture range of the PLL, at which time the FLL is turned off and the PLL is turned on. A typical FLL design is shown in Fig. 3.17. The FLL generates a loop error signal e_{FLL} that is approximately proportional to the rotation rate of the baseband signal phasor and is derived from the vector cross-product of successive baseband phasors $[I(t-\tau), Q(t-\tau)]$ and $[I(t), Q(t)]$, where τ is a fixed delay, typically 1–5 ms. More precisely

$$e_{\mathrm{FLL}} = Q(t)I(t-\tau) - I(t)Q(t-\tau). \tag{3.24}$$

3.4.6.3 PLL Order The *order* of a PLL refers to its capability to track different types of signal dynamics. Most GPS receivers use second- or third-order PLLs. A second-order loop can track a constant rate of phase change (i.e., constant frequency) with zero average phase error and a constant rate of frequency change with a nonzero but constant phase error. A third-order loop can track a constant rate of frequency change with zero average phase error and a constant acceleration of frequency with nonzero but constant phase error. Low-cost receivers typically use a second-order PLL with fairly low bandwidth because the user dynamics are minimal and the rate of change of the signal frequency due to satellite motion is sufficiently low (<1 Hz/s) that phase tracking error is negligible. On the other hand, receivers designed for high dynamics (i.e., missiles) will sometimes use third-order or even higher-order PLLs to avoid loss of lock due to the large accelerations encountered.

The price paid for using higher-order PLLs is a somewhat lower robust performance in the presence of noise. If independent measurements of platform dynamics are available (such as accelerometer or INS outputs), they can be used to aid the PLL by reducing stress on the loop. This can be advantageous because it often renders the use of higher-order loops unnecessary.

3.4.7 Bit Synchronization

Before bit synchronization can occur, the PLL must be locked to the GPS signal. This is accomplished by running the Costas loop in a 1-ms integration mode where each interval of integration is over one period of the C/A-code, starting and ending at the code epoch. Since the 50-Hz biphase data bit transitions can occur only at code epochs, there can be no bit transitions while integration is taking place. When the PLL achieves lock, the output of the I integrator will be a sequence of values occurring once per millisecond or 20 times per data bit. With nominal signal levels the processing gain of the integrator is sufficient to guarantee with high probability that the polarity of the 20 integrator outputs will remain constant during each data bit interval and will change polarity when a data bit transition occurs.

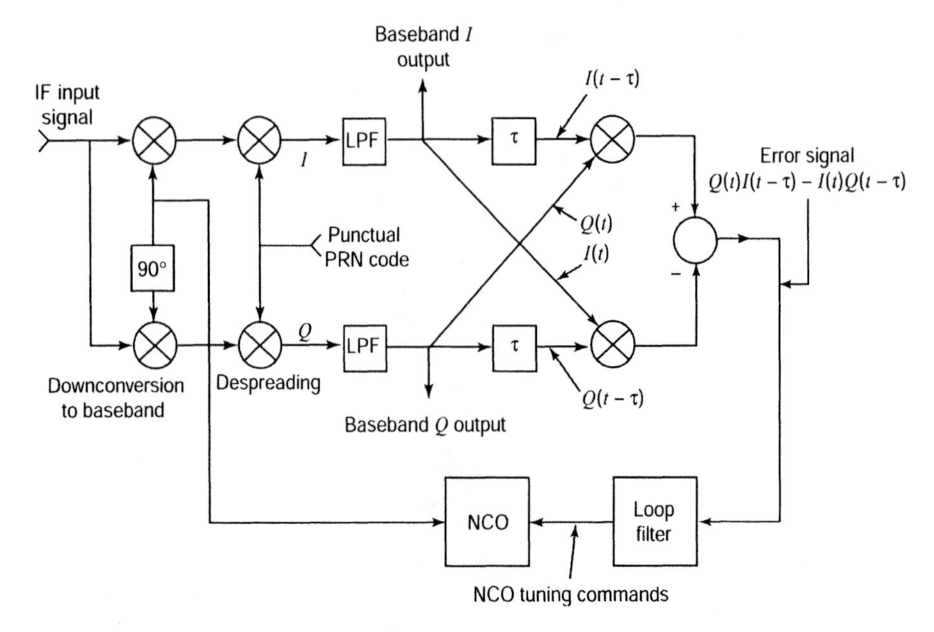

Fig. 3.17 Frequency-lock loop.

A simple method of bit synchronization is to clock a modulo-20 counter with the epochs of the receiver-generated reference C/A-code and record the count each time the polarity of the I integrator output changes. A histogram of the frequency of each count is constructed, and the count having the highest frequency identifies the epochs that mark the data bit boundaries.

3.4.8 Data Bit Demodulation

Once bit synchronization has been achieved, demodulation of the data bits can occur. As previously described, many GPS receivers demodulate the data by integrating the in-phase (I) component of the baseband phasor generated by a Costas loop, which tracks the carrier phase. Each data bit is generated by integrating the I component over a 20-ms interval from one data bit boundary to the next. The Costas loop causes a polarity ambiguity of the data bits that can be resolved by observation of the subframe preamble in the navigation message data.

3.5 EXTRACTION OF INFORMATION FOR NAVIGATION SOLUTION

After data demodulation has been performed, the essential information in the signal needed for the navigation solution is at hand. This information can be classified into the following three categories:

1. The information needed to determine signal transmission time
2. The information needed to establish the position and velocity of each satellite
3. The various pseudorange and Doppler measurements made by the receiver

3.5.1 Signal Transmission Time Information

In our previous discussion of the Z-count, we saw that the receiver can establish the time of transmission of the beginning of each subframe of the signal and of the corresponding C/A-code epoch that coincides with it. Since the epochs are transmitted precisely 1 ms apart, the receiver labels subsequent C/A-code epochs merely by counting them. This enables the determination of the transmission time of *any* part of the signal by a process to be described later.

3.5.2 Ephemeris Data

The ephemeris data permit the position and velocity of each satellite to be computed at the signal transmission time. The calculations were outlined in Table 3.2.

3.5.3 Pseudorange Measurements Using C/A-Code

In its basic form, finding the three-dimensional position of a user would consist of determining the *range*, that is, the distance of the user from each of three or more satellites having known positions in space, and mathematically solving for a point in space where that set of ranges would occur. The range to each satellite can be determined by measuring how long it takes for the signal to propagate from the satellite to the receiver and multiplying the propagation time by the speed of light.

Unfortunately, however, this method of computing range would require very accurate synchronization of the satellite and receiver clocks used for the time measurements. GPS satellites use very accurate and stable atomic clocks, but it is economically infeasible to provide a comparable clock in a receiver. The problem of clock synchronization is circumvented in GPS by treating the receiver clock error as an additional unknown in the navigation equations and using measurements from an additional satellite to provide enough equations for a solution for time as well as for position. Thus the receiver can use an inexpensive clock for measuring time. Such an approach leads to perhaps the most fundamental measurement made by a GPS receiver: the *pseudorange* measurement, computed as

$$\rho = c(t_{\mathrm{rcve}} - t_{\mathrm{xmit}}), \tag{3.25}$$

where t_{rcve} is the time at which a specific, identifiable portion of the signal is received, t_{xmit} is the time at which that same portion of the signal is transmitted,

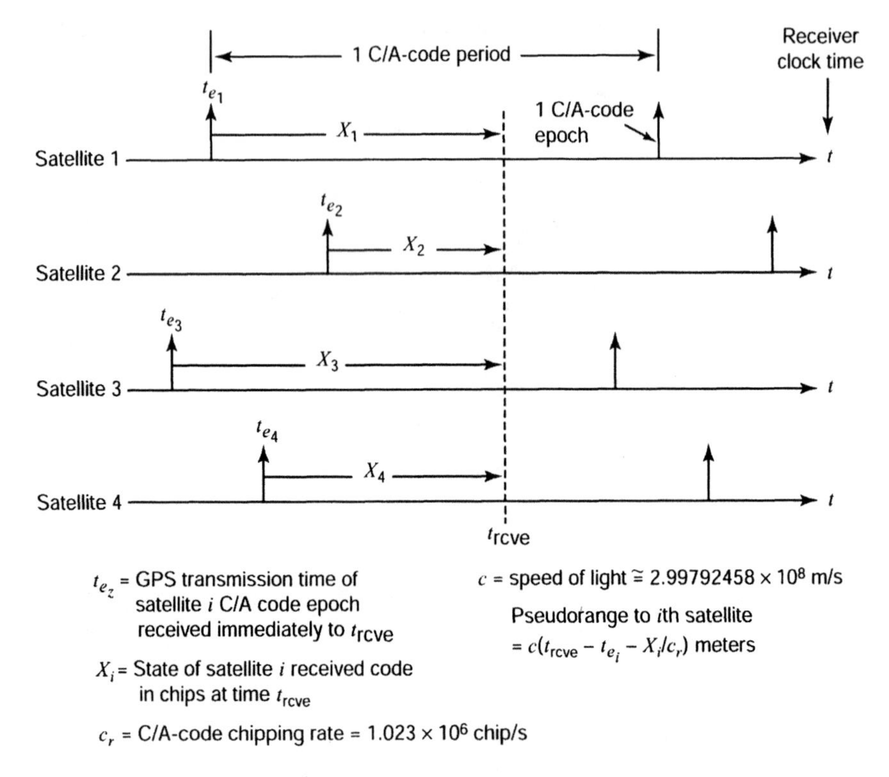

Fig. 3.18 Pseudorange measurement concept.

and c is the speed of light (2.99792458×10^8 m/s). It is important to note that t_{rcve} is measured according to the receiver clock, which may have a large time error, but t_{xmit} is in terms of GPS time, which in turn is SV (spacecraft vehicle) time plus a time correction transmitted by the satellite. If the receiver clock were synchronized to GPS time, then the pseudorange measurement would in fact be the range to the satellite.

Figure 3.18 shows the pseudorange measurement concept with four satellites, which is the minimum number needed for a three-dimensional position solution without synchronized clocks. The raw measurements are simultaneous snapshots at time t_{rcve} of the states of the received C/A-codes from all satellites. This is accomplished indirectly by observation of the receiver-generated code state from each code tracking loop. For purposes of simplicity we define the *state* of the C/A-code to be the number of chips (including the fractional part) that have occurred since the last code epoch. Thus the state is a real number in the interval [0, 1023).

As discussed earlier, the receiver has been able to tag each code epoch with its GPS transmission time. Thus, it is a relatively simple matter to compute the time of transmission of the code state that is received at time t_{rcve}. For a given satellite let t_e denote the GPS transmission time of the last code epoch received prior to t_{rcve}, let X denote the code state observed at t_{rcve}, and let c_r denote the C/A-code chipping rate (1.023×10^6 chips/s). Then the transmission time of that code state is

$$t_{\text{xmit}} = t_e + \frac{X}{c_r}. \tag{3.26}$$

3.5.3.1 Basic Positioning Equations If pseudorange measurements can be made from at least four satellites, enough information exists to solve for the unknown position (X, Y, Z) of the GPS user and for the receiver clock error C_b (often called the clock *bias*). The equations are set up by equating the measured pseudorange to each satellite with the corresponding unknown user-to-satellite distance plus the distance error due to receiver clock bias:

$$\begin{aligned}
\rho_1 &= \sqrt{(x_1 - X)^2 + (y_1 - Y)^2 + (z_1 - Z)^2} + C_b, \\
\rho_2 &= \sqrt{(x_2 - X)^2 + (y_2 - Y)^2 + (z_2 - Z)^2} + C_b, \\
&\vdots \\
\rho_n &= \sqrt{(x_n - X)^2 + (y_n - Y)^2 + (z_n - Z)^2} + C_b,
\end{aligned} \tag{3.27}$$

where ρ_i denotes the measured pseudorange of the ith satellite whose position in ECEF coordinates at t_{xmit} is (x_i, y_i, z_i) and $n \geq 4$ is the number of satellites observed. The unknowns in this nonlinear system of equations are the user position (X, Y, Z) in ECEF coordinates and the receiver clock bias C_b.

3.5.4 Pseudorange Measurements Using Carrier Phase

Although pseudorange measurements using the C/A-code are the most commonly employed, a much higher level of measurement precision can be obtained by measuring the received phase of the GPS L_1 or L_2 carrier. Because the carrier waveform has a very short period (6.35×10^{-10} s at the L_1 frequency), the noise-induced error in measuring signal delay by means of phase measurements is typically 10–100 times smaller than that encountered in code delay measurements.

However, carrier phase measurements are highly ambiguous because phase measurements are simply modulo 2π numbers. Without further information such measurements determine only the fractional part of the pseudorange when measured in carrier wavelengths. Additional measurements are required to effect *ambiguity resolution*, in which the integer number of wavelengths in the pseudorange measurement can be determined. The relation between the measured signal phases ϕ_i and the unambiguous pseudoranges ρ_i can be expressed as

$$
\begin{aligned}
\rho_1 &= \lambda\left(\frac{\phi_1}{2\pi} + k_1\right), \\
\rho_2 &= \lambda\left(\frac{\phi_2}{2\pi} + k_2\right), \\
&\vdots \\
\rho_n &= \lambda\left(\frac{\phi_n}{2\pi} + k_n\right),
\end{aligned}
\tag{3.28}
$$

where n is the number of satellites observed, λ is the carrier wavelength, and k_i is the unknown integral number of wavelengths contained in the pseudorange. The additional measurements required for determination of the k_i may include C/A- and/or P(Y)-code pseudorange measurements from the same satellites used for the phase measurements. Since the code measurements are unambiguous, they significantly narrow the range of admissible integer values for the k_i. Additionally, phase measurements made on both the L_1 and L_2 signals can be used to obtain a virtual carrier frequency equal to the difference of the two carrier frequencies $(1575.42 - 1227.60 = 347.82$ MHz). The 86.3-cm wavelength of this virtual carrier thins out the density of pseudorange ambiguities by a factor of about 4.5, making the ambiguity resolution process much easier. Redundant code and phase measurements from extra satellites can also be used to aid the process; the extra code measurements further narrow the range of admissible integer values for the k_i, and the extra phase measurements thin out the phase ambiguity density by virtue of satellite geometry.

Because of unpredictable variations in propagation delay of the code and carrier due to the ionosphere and other error sources, it is virtually impossible to obtain ambiguity resolution with single-receiver positioning. Therefore, carrier phase measurements are almost always relegated to high-accuracy applications in which such errors are canceled out by differential operation with an additional receiver (base station).

In GPS receivers, carrier phase is usually measured by sampling the phase of the reference oscillator of the carrier tracking loop. In most receivers this oscillator is an NCO that tracks the phase of the incoming signal at a relatively low intermediate frequency. The signal phase is preserved when the incoming signal is frequency-downconverted. The NCO is designed to provide a digital output of its instantaneous phase in response to a sampling signal. Phase-based pseudorange measurements are made by simultaneously sampling at time t_{rcve} the phases of the NCOs tracking the various satellites. As with all receiver measurements, the reference for the phase measurements is the receiver's clock reference oscillator.

3.5.5 Carrier Doppler Measurement

Measurement of the received carrier frequency provides information that can be used to determine the velocity vector of the user. Although this could be done by forming differences of code-based position estimates, frequency measurement is inherently much more accurate and has faster response time in the presence of user dynamics. The equations relating the measurements of Doppler shift to the user velocity are

$$\begin{aligned} f_{d1} &= \tfrac{1}{\lambda}(\boldsymbol{v} \cdot \boldsymbol{u}_1 - \boldsymbol{v}_1 \cdot \boldsymbol{u}_1) + f_b, \\ f_{d2} &= \tfrac{1}{\lambda}(\boldsymbol{v} \cdot \boldsymbol{u}_2 - \boldsymbol{v}_2 \cdot \boldsymbol{u}_2) + f_b, \\ &\vdots \\ f_{dn} &= \tfrac{1}{\lambda}(\boldsymbol{v} \cdot \boldsymbol{u}_n - \boldsymbol{v}_n \cdot \boldsymbol{u}_n) + f_b, \end{aligned} \tag{3.29}$$

where the unknowns are the user velocity vector $\mathbf{v} = (v_x, v_y, v_z)$ and the receiver reference clock frequency error f_b in hertz and the known quantities are the carrier wavelength λ and the measured Doppler shifts f_{di} in hertz, satellite velocity vectors $\mathbf{v}_i$, and unit satellite direction vectors $\mathbf{u}_i$ (pointing from the receiver antenna toward the satellite antenna) for each satellite index i. The unit vectors $\mathbf{u}_i$ are determined by computing the user-to-ith satellite displacement vectors ρ_i and normalizing them to unit length:

$$\begin{aligned} \rho_i &= [(x_i - X),\ (y_i - Y),\ (z_i - Z)]^T, \\ \boldsymbol{u}_i &= \frac{\rho_i}{|\rho_i|}. \end{aligned} \tag{3.30}$$

In these expressions the ith satellite position (x_i, y_i, z_i) at time t_{xmit} is computed from the ephemeris data and the user position (X, Y, Z) can be determined from solution of the basic positioning equations using the C/A- or P(Y)-codes.

In GPS receivers, the Doppler measurements f_{di} are usually derived by sampling the frequency setting of the NCO (Fig. 3.16) that tracks the phase of the incoming signal. An alternate method is to count the output cycles of the NCO over a relatively short time period, perhaps 1 s or less. However, in either case, the measured Doppler shift is not the raw measurement itself, but the deviation from what the raw NCO measurement would be without any signal Doppler shift, assuming that the receiver reference clock oscillator had no error.

3.5.6 Integrated Doppler Measurements

Integrated Doppler can be defined as the number of carrier cycles of Doppler shift that have occurred in a given interval $[t_0, t]$. For the ith satellite the relation between integrated Doppler F_{di} and Doppler shift f_{di} is given by

$$F_{di}(t) = \int_{t_0}^{t} f_{di}(t)\, dt. \tag{3.31}$$

However, accurate calculation of integrated Doppler according to this relation would require that the Doppler measurement be a continuous function of time. Instead, GPS receivers take advantage of the fact that by simply observing the output of the NCO in the carrier tracking loop (Fig. 3.16), the number of cycles that have occurred since initial time t_0 can be counted directly.

Integrated Doppler measurements have several uses:

1. *Accurate Measurement of Receiver Displacement over Time*. Motion of the receiver causes a change in the Doppler shift of the incoming signal. Thus, by counting carrier cycles to obtain integrated Doppler, precise estimates of the *change* in position (*delta position*) of the user over a given time interval can be obtained. The error in these estimates is much smaller than the error in establishing the absolute position using the C/A- or P(Y)-codes. The capability of accurately measuring changes in position is used extensively in *real-time kinematic* surveying with differential GPS. In such applications the user needs to determine the locations of many points in a given land area with great accuracy (perhaps to within a few centimeters). When the receiver is first turned on, it may take a relatively long time to acquire the satellites, to make both code and phase pseudorange measurements, and to resolve phase ambiguities so that the location of the first surveyed point can be determined. However, once this is done, the relative displacements of the remaining points can be found very rapidly and accurately by transporting the receiver from point to point while it continues to make integrated Doppler measurements.
2. *Positioning Based on Received Signal Phase Trajectories*. In another form of differential GPS, a fixed receiver is used to measure the integrated Doppler function, or *phase trajectory curve*, from each satellite over relatively long periods of time (perhaps 5–20 min). The position of the receiver can be determined by solving a system of equations relating the shape of the trajectories to the receiver location. The accuracy of this positioning technique, typically within a few decimeters, is not as good as that obtained with carrier phase pseudoranging but has the advantage that there is no phase ambiguity. Some handheld GPS receivers employ this technique to obtain relatively good positioning accuracy at low cost.
3. *Carrier Rate Aiding for the Code Tracking Loop*. In the code tracking loop, proper code alignment is achieved by using observations of the loop error signal to determine whether to advance or retard the state of the otherwise free-running receiver-generated code replica. Because the error signal is relatively noisy, a narrow loop bandwidth is desirable to maintain good pseudoranging accuracy. However, this degrades the ability of the loop to maintain accurate tracking in applications where the receiver is subject to substantial accelerations. The difficulty can be substantially mitigated with *carrier rate aiding*, in which the primary code advance/retard commands are not derived from the code discriminator (early–late correlator) error signal but instead are derived from the Doppler-induced accumulation of carrier cycles in the integrated Doppler function. Since there are 1540 carrier cycles per C/A-code chip, the code will therefore be advanced by precisely one chip for every 1540 cycles of accumulated count of integrated Doppler. The advantage of this approach is that, even in the presence of dynamics, the integrated Doppler can track the received code *rate* very

accurately. As a consequence, the error signal from the code discriminator is "decoupled" from the dynamics and can be used for very small and infrequent adjustments to the code generator.

3.6 THEORETICAL CONSIDERATIONS IN PSEUDORANGE AND FREQUENCY ESTIMATION

In a well-designed GPS receiver the major source of measurement error within the receiver is thermal noise, and it is useful to know the best performance that is theoretically possible in its presence. Theoretical bounds on errors in estimating code-based and carrier-based pseudorange, as well as in Doppler frequency estimates, have been developed within an interesting branch of mathematical statistics called *estimation theory*. There it is seen that a powerful estimation approach called the *method of maximum likelihood* (ML) can often approach theoretically optimum performance (see Section 7.2.4). ML estimates of pseudorange (using either the code or the carrier) and frequency are *unbiased*, which means that the expected value of the error due to random noise is zero.

An important lower bound on the error variance of any unbiased estimator is provided by the *Cramer–Rao bound*, and any estimator that reaches this lower limit is called a *minimum-variance unbiased estimator* (MVUE). It can be shown that at the typical SNRs encountered in GPS, ML estimates of code pseudorange, carrier pseudorange, and carrier frequency are all MVUEs. Thus, these estimators are optimal in the sense that no unbiased estimator has a smaller error variance [197].

3.6.1 Theoretical versus Realizable Code-Based Pseudoranging Performance

It can be shown that a ML estimate τ_{ML} of signal delay based on code measurements is obtained by maximizing the cross-correlation of the received code $c_r(t)$ with a reference code $c_{\text{ref}}(t)$ that is an identical replica (including bandlimiting) of the received code:

$$t_{\text{ML}} = \max_{\tau} \int_0^T c_r(t)\, c_{\text{ref}}(t-\tau)\, dt, \tag{3.32}$$

where $[0, T]$ is the signal observation interval. Here we assume coherent processing for purposes of simplicity. This estimator is a MVUE, and it can be shown that the error variance of τ_{ML} (which equals the Cramer–Rao bound) is

$$\sigma^2_{\tau_{\text{ML}}} = \frac{N_0}{2\int_0^T \left[c_r'(t)\right]^2 dt}. \tag{3.33}$$

This is a fundamental relation that in temporal terms states that the error variance is proportional to the power spectral density N_0 of the noise and inversely proportional to the integrated square of the derivative of the received code waveform. It is generally more convenient to use an expression for the standard deviation, rather than the variance, of delay error, in terms of the bandwidth of the C/A-code. The following formula is derived in Ref. 201:

$$\sigma_{\tau_{\mathrm{ML}}} = \frac{3.444 \times 10^{-4}}{\sqrt{(C/N_0)WT}}. \tag{3.34}$$

In this expression it is assumed that the received code waveform has been bandlimited by an ideal lowpass filter with one-sided bandwidth W. The signal observation time is still denoted by T, and C/N_0 is the ratio of power in the code waveform to the one-sided power spectral density of the noise. A similar expression is obtained for the error variance using the P(Y)-code, except the numerator is $\sqrt{10}$ times smaller.

Figure 3.19 shows the theoretically achievable pseudoranging error using the C/A-code as a function of signal observation time for various values of C/N_0. The error is surprisingly small if the code bandwidth is sufficiently large. As an example, for a moderately strong signal with $C/N_0 = 31,623$ (45 dB-Hz), a bandwidth $W = 10$ MHz, and a signal observation time of 1 s, the standard deviation of the ML delay estimate obtained from Eq. 3.34 is about 6.2×10^{-10} s, corresponding to 18.6 cm after multiplying by the speed of light.

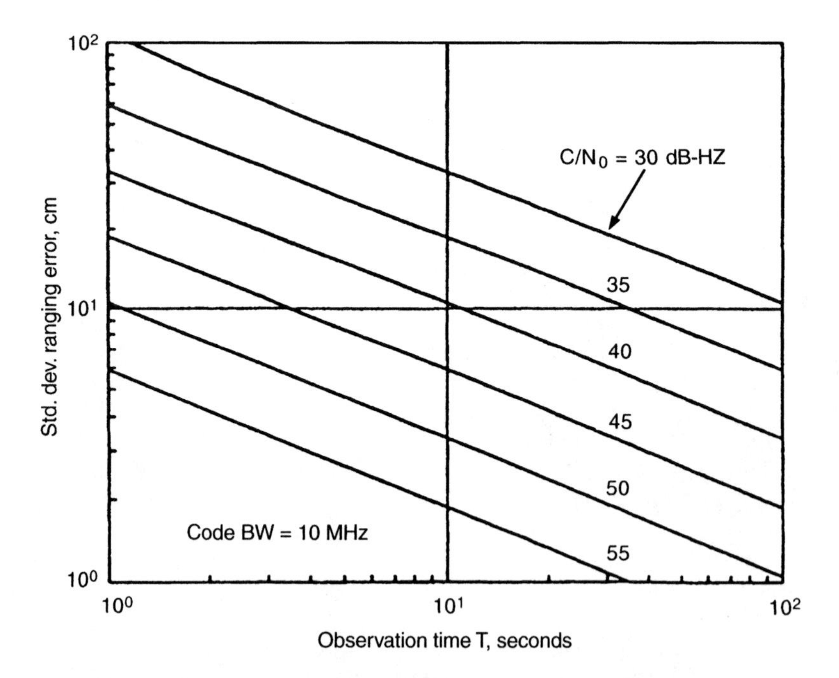

Fig. 3.19 Theoretically achievable C/A-code pseudoranging error.

3.6.1.1 Code Pseudoranging Performance of Typical Receivers Most GPS receivers approximate the ML estimator by correlating the incoming signal with an ideal code waveform that does not include bandlimiting effects and use early and late correlators in the code tracking loop that straddle the location of the correlation function peak rather than find its actual location. As a result, the code tracking error can be significantly larger than the theoretical minimum discussed above. One-chip early–late spacing of the tracking correlators was common practice for the several decades preceding the early 1990s. It is somewhat surprising that the substantial amount of performance degradation resulting from this approach went unnoticed for so long. Not until 1992 was it widely known that significant error reduction could be obtained by narrowing the spacing down to 0.1–0.2 C/A-code chips in combination with a large precorrelation bandwidth. Details of this approach, dubbed *narrow-correlator technology*, can be found in Ref. 195. With narrow early–late spacing the random noises on the early and late correlator outputs become highly correlated and therefore tend to cancel when the difference error signal is formed. A large precorrelation bandwidth sharpens the peak of the correlation function so that the closely spaced early and late correlators can still operate on the high-slope portion of the correlation function, thus preserving SNR in the loop.

It can be shown that the variance of the code tracking error continues to decrease as the early–late spacing approaches zero but approaches a limiting value. Some researches are aware that forming a difference signal with early and late correlators is mathematically equivalent to a single correlation with the difference of the early and late codes, which in the limit (as the early–late spacing goes to zero) becomes equivalent to polarity-modulated sampling of the received code at the punctual reference code transitions and summing the sample values to produce the loop error signal. Some GPS receivers already put this principle into practice.

Figure 3.20 [201] compares the performance of several correlator schemes, including the narrow correlator, with theoretical limits. It is seen that the narrow correlator approaches the theoretical performance limit given by the Cramer–Rao bound as the early–late spacing $2e$ approaches zero.

3.6.2 Theoretical Error Bounds for Carrier-Based Pseudoranging

At typical GPS signal-to-noise ratios the ML estimate τ_{ML} of signal delay using carrier phase is a MVUE, and it can be shown that the error standard deviation is

$$\sigma_{\tau_{\text{ML}}} = \frac{1}{2\pi f_c \sqrt{2(C/N_0)T}}, \tag{3.35}$$

where f_c is the GPS carrier frequency and C/N_0 and T have the same meaning as in Eq. 3.34. This result is also reasonably accurate for a carrier tracking loop if T is set equal to the reciprocal of the loop bandwidth. As an example of the much greater accuracy of carrier phase pseudoranging compared with code pseudoranging, a signal at $C/N_0 = 45$ dB-Hz observed for 1 s can theoretically

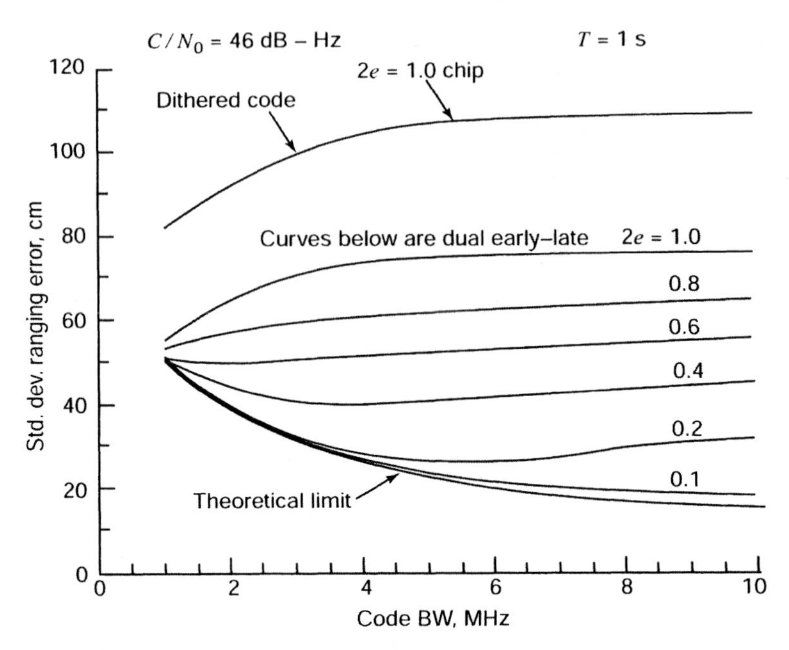

Fig. 3.20 Performance of various correlators. (Reproduced from Ref. 76, with permission.)

yield an error standard deviation of 4×10^{-13} s, which corresponds to only 0.12 mm. However, typical errors of 1–3 mm are experienced in most receivers as a result of random phase jitter in the reference oscillator.

3.6.3 Theoretical Error Bounds for Frequency Measurement

The ML estimate f_{ML} of the carrier frequency is also a MVUE, and the expression for its error standard deviation is

$$\sigma_{f_{\mathrm{ML}}} = \sqrt{\frac{3}{2\pi^2 (C/N_0) T^3}}. \tag{3.36}$$

A 1-s observation of a despread GPS carrier with $C/N_0 = 45$ dB-Hz yields a theoretical error standard deviation of about 0.002 Hz, which could also be obtained with a phase tracking loop having a bandwidth of 1 Hz. As in the case of phase estimation, however, phase jitter in the receiver reference oscillator yields frequency error standard deviations from 0.05 to 0.1 Hz.

3.7 MODERNIZATION OF GPS

Since it was declared fully operational in April 1995, the GPS has been operating continuously with 24 or more operational satellites, and user equipment has evolved rapidly, especially in the civil sector. As a result, radically improved levels of performance have been reached in positioning, navigation, and time transfer. However, the availability of GPS has also spawned new and demanding applications that reveal certain shortcomings of the present system. Therefore, since the mid-1990s numerous governmental and civilian committees have investigated the needs and deficiencies of the existing system in order to conceive a plan for GPS modernization.

The modernization of GPS is a difficult and complex task that requires tradeoffs in many areas. Major issues include spectrum needs and availability, military and civil performance, signal integrity and availability, financing and cost containment, and potential competition from Europe's Galileo system. However, after many years of hard work it now appears that critical issues have been resolved. Major decisions have been made for the incorporation of new civil frequencies, new civil and military signals, and higher transmitted power levels.

3.7.1 Deficiencies of the Current System

The changes that are planned for GPS address the following needs:

1. *Civil users need two-frequency ionospheric correction capability in autonomous operation.* Since only the encrypted P-code appears at the L_2 frequency, civil users are denied the benefit of dual-frequency operation to remove ionospheric range error in autonomous (i.e., nondifferential) operation. Although special techniques such as signal squaring can be used to recover the L_2 carrier, the P-code waveform is lost and the SNR is dramatically reduced. Consequently, such techniques are of little value to the civil user in reducing ionospheric range error.

2. *Signal blockage and attenuation are often encountered.* In some applications heavy foliage in wooded areas can attenuate the signal to an unusable level. In certain locations, such as in urban canyons, a satellite signal can be completely blocked by buildings or other features of the terrain. In such situations there will not be enough visible satellites to obtain a navigation solution. New applications, such as emergency 911 position location by GPS receivers embedded in cellular telephone handsets, will require reliable operation of GPS receivers inside buildings, despite heavy signal attenuation due to roof, floors, and walls. Weak signals are difficult to acquire and track.

3. *Ability to resolve ambiguities in phase measurements needs improvement.* High-accuracy differential positioning at the centimeter level by civil users requires rapid and reliable resolution of ambiguities in phase measurements. Ambiguity resolution with single-frequency (L_1) receivers generally requires a sufficient length of time for the satellite geometry to change significantly. Performance is improved with dual-frequency receivers. However, the effective SNR of the L_2 signal is dramatically reduced because the encrypted P-code cannot be despread by the civil user.

4. *Selective Availability is detrimental to performance in civil applications.* SA has been suspended as of 8 p.m. EDT on May 1, 2000. The degradation in autonomous positioning performance by SA (about 50 m RMS error) is of concern in many civil applications requiring the full accuracy of which GPS is capable. A prime example is vehicle tracking systems in which an accuracy of 5–10 m RMS is needed to establish the correct city street on which a vehicle is located. Moreover, many civil and military committees have found that a military adversary can easily mitigate errors due to SA by using differential positioning. In the civil sector, a large and costly infrastructure has developed to overcome its effects.

5. *Improvements in system integrity and robustness are needed.* In applications involving public safety the integrity of the current system is judged to be marginal. This is particularly true in aviation landing systems that demand the presence of an adequate number of healthy satellite signals and functional cross-checks during precision approaches. Additional satellites and higher transmitted power levels are desirable in this context.

6. *Improvement is needed in multipath mitigation capability.* Multipath remains a dominant source of GPS positioning error and cannot be removed by differential techniques. Although certain mitigation techniques, such as multipath mitigation technology (MMT), approach theoretical performance limits for in-receiver processing, the required processing adds to receiver costs. In contrast, effective multipath rejection could be made available to all receivers by using new GPS signal designs.

7. *The military needs improved acquisition capability and jamming immunity.* Because the P(Y)-code has an extremely long period (seven days), it is difficult to acquire unless some knowledge of the code timing is known. In the current system P(Y) timing information is supplied by the HOW. However, to read the HOW, the C/A-code must first be acquired to gain access to the navigation message. Unfortunately, the C/A-code is relatively susceptible to jamming, which would seriously impair the ability of a military receiver to acquire the P(Y) code. It would be far better if direct acquisition of a high-performance code were possible.

3.7.2 Elements of the Modernized GPS

3.7.2.1 Civil Spectrum Modernization The upper part of Fig. 3.21 outlines the current civil GPS signal spectrum and the additional codes and signal frequencies in the plans for modernization. The major elements are as follows:

1. *A New L_2 Civil Signal (L_{2C}) Modulated with a New Code Structure.* The L_{2C} signal, described in detail below, will offer civilian users the following improvements:

   (a) *Two-frequency ionospheric error correction becomes possible.* The $1/f^2$ dispersive delay characteristic of the ionosphere can be used to accurately estimate errors in propagation delay.

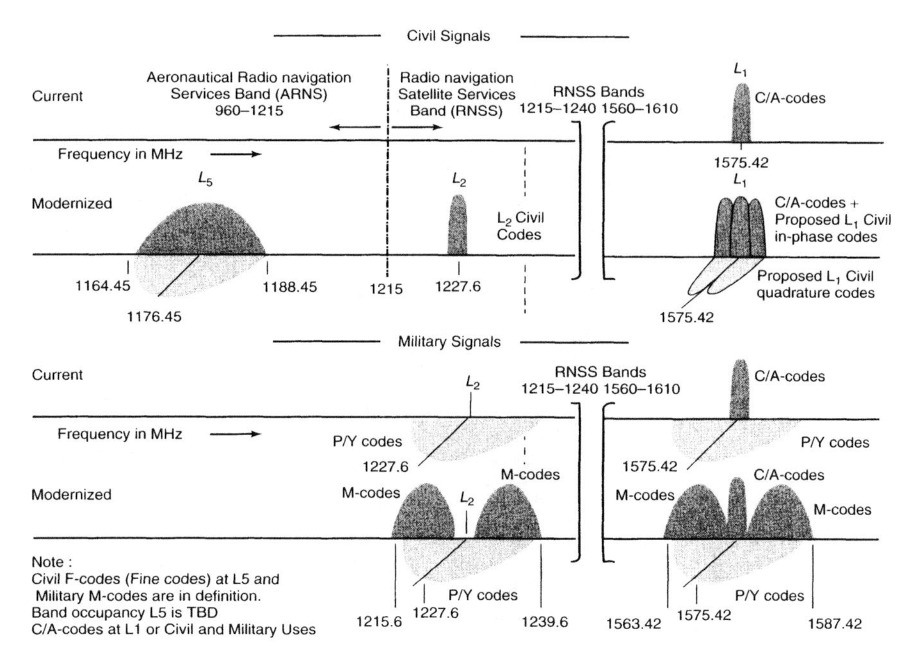

Fig. 3.21 Existing and modernized GPS signal spectrum.

(b) *Carrier phase ambiguity resolution will be significantly improved,* The accessibility of the L_1 and L_2 carriers provides "wide lane" phase measurements having ambiguities that are much easier to resolve.

(c) *The additional L_2 signal will improve robustness in acquisition and tracking and improve C/A code positioning accuracy.*

The existing C/A code at the L_1 frequency will be retained for legacy purposes.

2. *A New L_5 Signal Modulated by a New Code Structure.* Although the use of the L_1 and L_2 frequencies can satisfy most civil users, there are concerns that the L_2 frequency band may be subject to unacceptable levels of interference for applications involving public safety, such as aviation. The potential for interference arises because the International Telecommunications Union (ITU) has authorized this band on a coprimary basis with radiolocation services, such as high-power radars. As a result of FAA requests, the Department of Transportation and Department of Defense have determined a new civil GPS frequency, called L_5, in the Aeronautical Radio Navigation System band at 1176.45 MHz. To gain maximum performance, the L_5 spread-spectrum codes will have a higher chipping rate and longer period than do the C/A-codes. Proposed codes have a 10.23-megachip/s chipping rate and a period of 10,230 chips. Additionally, the plan is to transmit two signals in phase quadrature, one of which will

not carry data modulation. The L_5 signal will provide the following system improvements:

(a) *Ranging accuracy will improve.* Pseudorange errors due to random noise will be reduced below levels obtainable with the C/A-codes, due to the larger bandwidth of the proposed codes. As a consequence, both code-based positioning accuracy and phase ambiguity resolution performance will improve.

(b) *Errors due to multipath will be reduced.* The larger bandwidth of the new codes will sharpen the peak of the code autocorrelation function, thereby reducing the shift in the peak due to multipath signal components.

(c) *Carrier phase tracking will improve.* Weak-signal phase tracking performance of GPS receivers is severely limited by the necessity of using a Costas (or equivalent-type) PLL to remove carrier phase reversals of the data modulation. Such loops rapidly degrade below a certain threshold (about 25–30 dB-Hz) because truly coherent integration of the carrier phase is limited to the 20-ms data bit length. In contrast, the "data-free" quadrature component of the L_5 signal will permit coherent integration of the carrier for arbitrarily long periods, which will permit better phase tracking accuracy and lower tracking thresholds.

(d) *Weak-signal code acquisition and tracking will be enhanced.* The "data-free" component of the L_5 signal will also permit new levels of positioning capability with very weak signals. Acquisition will be improved because fully coherent integration times longer than 20 ms will be possible. Code tracking will also improve by virtue of better carrier phase tracking for the purpose of code rate aiding.

(e) *The L_5 signal will further support rapid and reliable carrier phase ambiguity resolution.* Because the difference between the L_5 and L_2 frequencies is only 51.15 MHz as opposed to the 347.82-MHz difference between the L_1 and L_2 frequencies, carrier phase ambiguity will be possible using an extrawide lane width of about 5.9 m instead of 0.86 m. The inevitable result will be virtually instantaneous ambiguity resolution, a critical issue in high-performance real-time kinematic modes of GPS positioning.

(f) *The codes will be better isolated from each other.* The longer length of the L_5 codes will reduce the size of cross-correlation between codes from different satellites, thus minimizing the probability of locking onto the wrong code during acquisition, even at the increased power levels of the modernized signals.

3. *Higher Transmitted Power Levels.* For safety, cost, and performance, many in the GPS community are advocating a general increase of 3–6 dB in the signal power at at all three civil frequencies.

4. *A Proposed New L_1 Civil Signal (L_{1C}) Using Binary Offset Carrier (BOC) Modulation.* Although a decision to use this signal has not yet been made, if

implemented it will be the first civilian signal to use BOC modulation. The characteristics of the L_{1C} signal are described in more detail below.

3.7.2.2 ***Military Spectrum Modernization*** The lower part of Fig. 3.21 shows the current and modernized spectrum used by the military community. The current signals consist of C/A-codes and P/Y-codes transmitted in quadrature in the L_1 band and only P/Y-codes in the L_2 band. The primary elements of the modernized spectrum are as follows:

1. *All existing signals will be retained for legacy purposes.*
2. *New M-codes will also be transmitted in both the L_1 and L_2 bands.* These codes are BOC(10,5) codes in which a 5.115 Mcps chipping sequence modulates a 10.23 MHz square wave subcarrier. The resulting spectrum has two lobes, one on each side of the band center, and for this reason the M-codes are sometimes called "split-spectrum codes". They will be transmitted in the same quadrature channel as the C/A-codes, that is, in phase quadrature with the P(Y)-codes. Civil use of these codes will be denied by as yet unannounced encryption techniques. The M-codes will provide the following advantages to military users:

   (a) *Direct acquisition of the M-codes will be possible.* The design of these codes will eliminate the need to first acquire the L_1 C/A-code with its relatively high vulnerability to jamming.
   (b) *Better ranging accuracy will result.* As can be seen in Fig. 3.21, the M-codes have significantly more energy near the edges of the bands, with a relatively small amount of energy near band center. Since most of the C/A-code power is near band center, potential interference between the codes is mitigated. The effective bandwidth of the M-codes is much larger than that of the P(Y)-codes, which concentrate most of their power near the L_1 or L_2 carrier. Because of the modulated subcarrier, the autocorrelation function of the M-codes has, not just one peak, but several peaks spaced one subcarrier period apart, with the largest at the center. The modulated subcarrier will cause the central peak to be significantly sharpened, significantly reducing pseudorange measurement error.
   (c) *Error due to multipath will be reduced.* The sharp central peak of the M-code autocorrelation function is less susceptible to shifting in the presence of multipath correlation function components.

3.7.3 Families of GPS Satellites

The families of satellites prior to modernization have been Block I (1978–1985), Block II (1989–1990), and Block IIA (1990–1997). These satellites all carry the standard L_1 C/A-, P-, and L_2 P-codes.

In 1997 a new family, the Block IIR satellites, began to replace the older Block II/IIA family. The Block IIR satellites have several improvements, including reprogrammable processors enabling problem fixes and upgrades in flight. Eight Block IIR satellites are being modernized to include the new military M-code signals on both the L_1 and L_2 frequencies, as well as the new L_{2C} signal on L_2. The first modernized Block IIR was launched in September 2005.

The Block IIF family is the next generation of GPS satellites, retaining all the capabilities of the previous blocks, but with many improvements, including an extended design life of 12 years, faster processors with more memory, and the inclusion of the new L_5 signal on a third L_5 frequency (1176.45 MHz). The first Block IIF satellite is scheduled to launch in 2007.

The GPS system of the future is the Block III family, still under development. The first satellite is scheduled to launch in 2010. Military improvements may include two high-power spot beams for the L_1 and L_2 military M-code signals, giving 20 dB higher received power over the earlier M-code signals. It is also likely that there will be two additional channels providing navigation signals for civilian use in local, regional and national safety-of-life applications for improved positioning, navigation, and timing, perhaps with higher power as well. Perhaps the proposed L_{1C} signal will be considered for this purpose. The entire Block III constellation is expected to remain operational through at least 2030.

3.7.4 Accuracy Improvements from Modernization

The progress in standalone positioning accuracy prior to and after the steps of modernization can be summarized as follows:

- 20–100 m with C/A code and with selective availability (SA) on (prior to May 2002)
- 10–20 m using C/A code with SA off (typical at the present time)
- 5–10 m by 2009 using L_1 C/A-code and L_{2C} code together for dual-frequency ionospheric correction
- 1–5 m by 2013 using L_1 C/A code, L_{2C} code, and L_5 code

3.7.5 Structure of the Modernized Signals

The bandwidths of all modernized GPS signals will be at least 24 MHz. Assuming equal received power and filtered bandwidth, the ranging performance (with or without multipath) on a GPS signal is determined by its spectral shape (or equivalently, the shape of the autocorrelation function), where fine structure is ignored. In this sense, the L_1 C/A-coded and L_2 civil signals are equivalent, as are the P/Y and L_5 civil signals. The military M-coded signal and the proposed civil L_1 signal use BOC modulation, but are not equivalent in performance because they use different subcarrier frequencies and chipping rates. However, they both place more of the signal power near the band edges, resulting in a multilobed autocorrelation function.

3.7.5.1 L_1 C/A-Coded and L_2 Civil Signals

The C/A-Coded Signal The GPS modernization program retains the C/A-code at the L_1 carrier frequency (1575.42 MHz) for legacy purposes, mostly for civilian users. These codes are maximal-length direct-sequence Gold codes, each consisting of a 1023-chip sequence transmitted at 1.023×106 chips/s, which repeats every 1 ms.

The L_2 Civil Signal Originally the modernization plan also called for the C/A-code at the L_2 carrier frequency (1227.60 MHz) to provide the civilian community with ionospheric correction capability as well as additional flexibility and robustness. However, late in the planning process it was realized that additional advantages could be obtained by replacing the planned L_2 C/A signal with a new L_2 civil signal (L_{2C}). The decision was made to use this new signal, and its structure was made public early in 2001. Both the L_{2C} and the new military M-code signal (to be described) will appear on the L_2 in-phase (I) channel, with the P/Y coded signal on the quadrature (Q) channel.

Like the C/A code, the L_{2C} code appears to be a 1.023×106 chip/s. sequence. However, it is generated by 2:1 time-division multiplexing of two independent subcodes, each having half the chipping rate, namely 511.5×103 chips/s. Each of these subcodes is made available to the receiver by demultiplexing. The first subcode (CM) has a moderate length of 10,230 chips, a 20-ms period, and is modulated with either a 25-s or a 50-bps (bits per second) navigation message. The moderate length of this code permits relatively easy acquisition of the signal although the 2:1 multiplexing results in a 3 dB acquisition and data demodulation loss. The second subcode (CL) has a length of 707,250 chips, a 1.5-s period, and is data-free. With no data there is no limit on coherent processing time, thereby permitting better code and carrier tracking performance, especially at low SNR. Full-cycle carrier tracking is possible, with a 6 dB improvement in the tracking threshold compared to that using only the CM code, where squaring loss is incurred in removing data phase changes. The relatively long CL code length also generates smaller correlation sidelobes as compared to the CM (or C/A) code.

Details on the L_2 civil signal are given by Fontana et al. [56].

3.7.5.2 P/Y-Coded and L_5 Civil Signals

The P/Y-Coded Signal For legacy purposes, GPS modernization will retain the P/Y-code on both the L_1 and L_2 frequencies. This code will be in phase quadrature with the C/A-code and the military M-code at the L_1 frequency, and at L_2 will be in quadrature with the new L_2 civil signal and the M-code. The P/Y-code is transmitted at 10.23×106 chips/s in either unencrypted (P-code) or encrypted (Y-code) form. The P-code sequence is publicly known and has a very long period of 1 week. The Y-code is formed by modulating the P-code with a slower sequence of encrypting chips, called a *W-code*, generated at 511.5×103 chips/sec.

The L_5 Civil Signal GPS modernization calls for a completely new civil signal at a carrier frequency of 1176.45 MHz with the total received signal power divided equally between in-phase (I) and quadrature (Q) components. Each component is modulated with a different but synchronized 10,230-chip direct sequence L_5 code transmitted at 10.23×10^6 chips/s, the same rate as the P/Y-code, but with a 1-ms period (the same as the C/A-code period). The I channel is modulated with a 100-symbol-per-second (sps) data stream, which is obtained by applying 1/2-rate, constraint length 7, forward error correction (FEC) convolutional coding to a 50-bps (bits per second) navigation data message that contains a 24-bit cyclic redundancy check (CRC). The Q-channel is unmodulated by data. However, both channels are further modulated by Neuman–Hoffman (NH) synchronization codes, which provide additional spectral spreading of narrowband interference, improve bit and symbol synchronization, and also improve cross-correlation properties between signals from different GPS satellites.

Compared to the C/A code, the 10-times larger chip count of the I- and Q-channel civil L_5-codes provides lower autocorrelation sidelobes, and the 10-times higher chipping rate substantially improves ranging accuracy, provides better interference protection, and substantially reduces multipath errors at longer path separations (far multipath). Additionally, these codes were selected to reduce as much as possible the cross-correlation between satellite signals. The absence of data modulation on the Q-channel permits longer coherent processing intervals in code and carrier tracking loops, with full-cycle carrier tracking in the latter. As a result, tracking capability and phase ambiguity resolution become more robust.

Further details on the civil L_5 signal structure can be found in Ref. 45.

3.7.5.3 The Proposed L_1 Civil (L_{1C}) Signal Although the current C/A code will remain on the L_1 frequency (1575.42 MHz), a new L_1 civil signal has recently been proposed. Like the L_5 civil signal, it will have a data-free quadrature component. However, it is unique among the civil signals in that it will use binary offset carrier (BOC) code modulation. The modulation candidates under consideration are BOC(1,1) and several versions which time-multiplex BOC(1,1) waveforms. These are collectively known as MBOC. The BOC(1,1) modulation consists of a 1.023-MHz square-wave subcarrier modulated by a 1.023-megachip/second (Mcps) spreading sequence. Each spreading chip subtends exactly one cycle of the subcarrier, with the rising edge of the first subcarrier half-cycle coincident with initiation of the spreading chip. The MBOC codes provide a larger RMS bandwidth compared to pure BOC(1,1).

Some GPS receiver manufacturers prefer the pure BOC(1,1) modulation for the following reasons:

1. The BOC(1,1) RMS bandwidth is smaller than that of MBOC modulation, permitting a lower digital sampling rate in the receiver, which is desirable to keep the receiver cost and power consumption as small as possible.

2. Although a low-cost, narrow bandwidth receiver can use any of the MBOC signal candidates, the higher frequency components of an MBOC signal will be lost, resulting in some loss of received signal power. This disadvantage is most serious in very high-sensitivity receivers designed for indoor use.

On the other hand, MBOC modulation provides better ranging accuracy and is inherently more robust against multipath, since its rms bandwidth is considerably greater than that of BOC(1,1).

Details of the proposed L_{1C} signal can be found in T. Stansell, "BOC or MBOC," *Inside GNSS*, Jul/Aug 2006, pub. by Gibbons Media & Research, Eugene, Oregon.

3.7.5.4 The M-Coded Signal New military M-coded signals will be transmitted on the L_1 and L_2 carriers, with the capability of using different codes on the two frequencies. The nominal received power level will be -158 dBW over the entire portion of the earth viewed by the satellite. The received L_1 M-code will appear in the I channel additively superimposed on the C/A-code, and the L_2 M-code will appear in the I channel superimposed on the civil L_2 code. However, in the fully modernized Block III satellites, the M-coded signal component can be radiated as a physically distinct signal from a separate antenna on the same satellite. This is done in order to enable optional transmission of a spot beam for greater antijam resistance within a selected local region on the earth. Spot beam nominal received power will be 20 dB greater than that of normal earth coverage.

The M-code, denoted a BOC(10,5) code, consists of a 10.23-MHz square-wave subcarrier modulated by a 5.115×106 chip/second spreading sequence. Each spreading chip subtends exactly two cycles of the subcarrier, with the rising edge of the first subcarrier cycle coincident with initiation of the spreading chip. The spectrum of the BOC(10,5) code has considerably more relative power near the edges of the signal bandwidth than any of the C/A-coded, L_2 civil, L_5 civil, or P/Y-coded signals. As a consequence, the M-coded signal not only offers the best pseudoranging accuracy and resistance to multipath but also has minimal spectral overlap with the other GPS transmitted signals, which permits transmission at higher power levels without mutual interference.

Details on the BOC(10,5) code can be found in a paper by Parker et al. [149].

PROBLEMS

3.1 An important signal parameter is the maximum Doppler shift due to satellite motion, which must be accommodated by a receiver. Find its approximate value by assuming that a GPS satellite has a circular orbit with a radius of 27,000 km, an inclination angle of 55°, and a 12-h period. Is the rotation rate of the earth significant? At what latitude(s) would one expect to see the largest possible Doppler shift?

3.2 Another important parameter is the maximum *rate* of Doppler shift in hertz per second that a phase-lock loop must be able to track. Using the orbital parameters of the previous problem, calculate the maximum rate of Doppler shift of a GPS signal one would expect, assuming that the receiver is stationary with respect to the earth.

3.3 Find the power spectrum of the 50-bps (bits per second) data stream containing the navigation message. Assume that the bit values are -1 and 1 with equal probability of occurrence, that the bits are uncorrelated random variables, and that the location of the bit boundary closest to $t = 0$ is a uniformly distributed random variable on the interval [-0.01 s, 0.01 s]. [*Hint:* First find the auto-correlation function $R(\tau)$ of the bit stream and then take its Fourier transform.]

3.4 In two-dimensional positioning, the user's altitude is known, so only three satellites are needed. Thus, there are three pseudorange equations containing two position coordinates (e.g., latitude and longitude) and the receiver clock bias term B. Since the equations are nonlinear, there will generally be more than one position solution, and all solutions will be at the same altitude. Determine a procedure that isolates the correct solution.

3.5 Some civil receivers attempt to extract the L_2 carrier by squaring the received waveform after it has been frequency-shifted to a lower IF. Show that the squaring process removes the P(Y)-code and the data modulation, leaving a sinusoidal signal component at twice the frequency of the original IF carrier. If the SNR in a 20-MHz IF bandwidth is -30 dB before squaring, find the SNR of the double-frequency component after squaring if it is passed through a 20-MHz bandpass filter. How narrow would the bandpass filter have to be to increase the SNR to 0 dB?

3.6 The relativistic effect in a GPS satellite clock which is compensated by a deliberate clock offset is about

(a) 4.5 parts per million

(b) 4.5 parts per 100 million

(c) 4.5 parts per 10 billion

(d) 4.5 parts per trillion

3.7 The following component of the ephemeris error contributes the most to the range error:

(a) Along-track error

(b) Cross-track error

(c) Both along-track and cross-track error

(d) Radial error

3.8 The differences between pseudorange and carrier phase observations are

(a) Integer ambiguity, multipath errors, and receiver noise

(b) Satellite clock, integer ambiguity, multipath errors, and receiver noise

(c) Integer ambiguity, ionospheric errors, multipath errors, and receiver noise

(d) Satellite clock, integer ambiguity, ionospheric errors, multipath errors, and receiver noise

3.9 GPS week number started incrementing from zero at

(a) Midnight of Jan. 5–6, 1980

(b) Midnight of. Jan. 5–6, 1995

(c) Midnight of Dec. 31–Jan. 1, 1994–1995

(d) Midnight of Dec. 31–Jan. 1, 1999–2000

3.10 The complete set of satellite ephemeris data comes once in every

(a) 6 s

(b) 30 s

(c) 12.5 s

(d) 12 s

3.11 Describe how the time of travel (from satellite to receiver) of the GPS signal is determined.

3.12 Describe how the receiver locks on via code correlation. (Use sketches if it helps.)

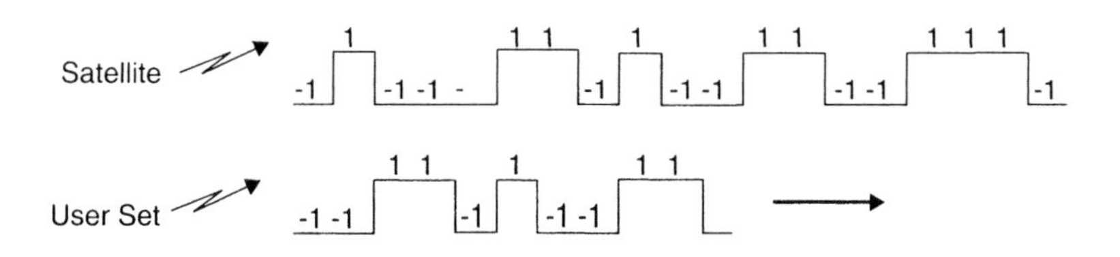

3.13 For high accuracy of the carrier phase measurements, the most suitable carrier tracking loop will be

(a) PLL with low loop bandwidth

(b) FLL with low loop bandwidth

(c) PLL with high loop bandwidth

(d) FLL with high loop bandwidth

3.14 Which of the following actions does *not* reduce the receiver noise (code)?

(a) Reducing the loop bandwidth

(b) Decreasing the predetection integration time

(c) Spacing the early–late correlators closer

(d) Increasing the signal strength

4

RECEIVER AND ANTENNA DESIGN

4.1 RECEIVER ARCHITECTURE

Although there are many variations in GPS receiver design, all receivers must perform certain basic functions. We will now discuss these functions in detail, each of which appears as a block in the diagram of the generic receiver shown in Fig. 4.1.

4.1.1 Radiofrequency Stages (Front End)

The purpose of the receiver front end is to filter and amplify the incoming GPS signal. As was pointed out earlier, the GPS signal power available at the receiver antenna output terminals is extremely small and can easily be masked by interference from more powerful signals adjacent to the GPS passband. To make the signal usable for digital processing at a later stage, RF amplification in the receiver front end provides as much as 35–55 dB of gain. Usually the front end will also contain passband filters to reduce out-of-band interference without degradation of the GPS signal waveform. The nominal bandwidth of both the L_1 and L_2 GPS signals is 20 MHz (± 10 MHz on each side of the carrier), and sharp-cutoff bandpass filters are required for out-of-band signal rejection. However, the small ratio of passband width to carrier frequency makes the design of such filters infeasible. Consequently, filters with wider skirts are commonly used as a first stage of filtering, which also helps prevent front-end overloading by strong interference, and the sharp-cutoff filters are used later after downconversion to intermediate frequencies (IFs).

Global Positioning Systems, Inertial Navigation, and Integration, Second Edition, by M. S. Grewal, L. R. Weill, and A. P. Andrews

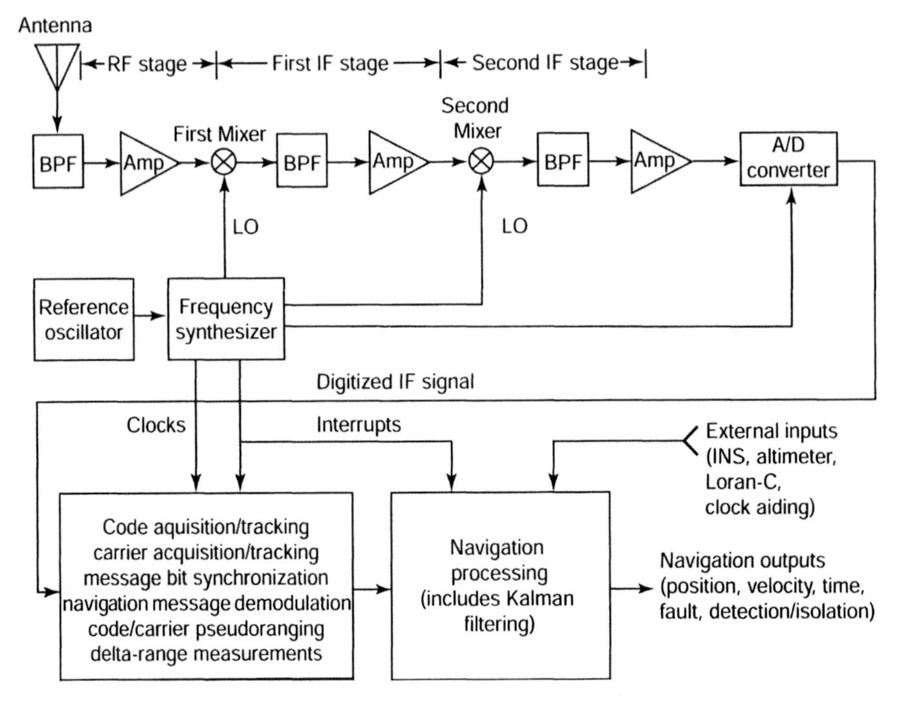

Fig. 4.1 Generic GPS receiver.

4.1.2 Frequency Downconversion and IF Amplification

After amplification in the receiver front end, the GPS signal is converted to a lower frequency called an *intermediate frequency* for further amplification and filtering. Downconversion accomplishes several objectives:

1. The total amount of signal amplification needed by the receiver exceeds the amount that can be performed in the receiver front end at the GPS carrier frequency. Excessive amplification can result in parasitic feedback oscillation, which is difficult to control. In addition, since sharp-cutoff filters with a GPS signal bandwidth are not feasible at the L-band, excessive front-end gain makes the end-stage amplifiers vulnerable to overloading by strong nearby out-of-band signals. By providing additional amplification at an IF different from the received signal frequency, a large amount of gain can be realized without the tendency toward oscillation.
2. By converting the signal to a lower frequency, the signal bandwidth is unaffected, and the increased ratio of bandwidth to center frequency permits the design of sharp-cutoff bandpass filters. These filters can be placed ahead of the IF amplifiers to prevent saturation by strong out-of-band signals. The filtering is often by means of surface acoustic wave (SAW) devices.

3. Conversion of the signal to a lower frequency makes the sampling of the signal required for digital processing much more feasible.

Downconversion is accomplished by multiplying the GPS signal by a sinusoid called the *local-oscillator signal* in a device called a *mixer*. The local-oscillator frequency is either larger or smaller than the GPS carrier frequency by an amount equal to the IF. In either case the IF signal is the difference between the signal and local-oscillator frequencies. Sum frequency components are also produced, but these are eliminated by a simple bandpass filter following the mixer. An incoming signal either above or below the local-oscillator frequency by an amount equal to the IF will produce an IF signal, but only one of the two signals is desired. The other signal, called the image, can be eliminated by bandpass filtering of the desired signal prior to downconversion. However, since the frequency separation of the desired and image signals is twice the IF, the filtering becomes difficult if a single downconversion to a low IF is attempted. For this reason downconversion is often accomplished in more than one stage, with a relatively high first IF (30–100 MHz) to permit image rejection.

Whether it is single-stage or multistage, downconversion typically provides a final IF that is low enough to be digitally sampled at feasible sampling rates without frequency aliasing. In low-cost receivers typical final IFs range from 4 to 20 MHz with bandwidths that have been filtered down to several MHz. This permits a relatively low digital sampling rate and at the same time keeps the lower edge of the signal spectrum well above 0 Hz to prevent spectral foldover. However, for adequate image rejection either multistage downconversion or a special single-stage image rejection mixer is required. In more advanced receivers there is a trend toward single conversion to a signal at a relatively high IF (30–100 MHz), because advances in technology permit sampling and digitizing even at these high frequencies.

4.1.2.1 Signal-to-Noise Ratio An important aspect of receiver design is the calculation of signal quality as measured by the signal-to-noise ratio (SNR) in the receiver IF bandwidth. Typical IF bandwidths range from about 2 MHz in low-cost receivers to the full GPS signal bandwidth of 20 MHz in high-end units, and the dominant type of noise is the thermal noise in the first RF amplifier stage of the receiver front end (or the antenna preamplifier if it is used). The noise power in this bandwidth is given by

$$N = kT_eB \tag{4.1}$$

where $k = 1.3806 \times 10^{-23}$ J/K, B is the bandwidth in Hz, and T_e is the effective noise temperature in degrees Kelvin. The effective noise temperature is a function of sky noise, antenna noise temperature, line losses, receiver noise temperature, and ambient temperature. A typical effective noise temperature for a GPS receiver is 513 K, resulting in a noise power of about −138.5 dBW in a 2-MHz bandwidth and −128.5 dBW in a 20-MHz bandwidth. The SNR is defined as the ratio of signal power to noise power in the IF bandwidth, or the difference of these powers when expressed in decibels. Using −154.6 dBW for the received signal power obtained in Section 3.3, the SNR in a 20-MHz bandwidth is seen to be −154.6 −(−128.5) = −26.1 dB. Although the GPS signal has a 20-MHz bandwidth, about 90% of the C/A-code power lies in a 2-MHz bandwidth, so there is only about 0.5 dB loss in signal power. Consequently the SNR in a 2-MHz bandwidth is (−154.6 − 0.5) − (−138.5) = −16.6 dB. In either case it is evident that the signal is completely masked by noise. Further processing to elevate the signal above the noise will be discussed subsequently.

4.1.3 Digitization

In modern GPS receivers digital signal processing is used to track the GPS signal, make pseudorange and Doppler measurements, and demodulate the 50-bps (bits per second) data stream. For this purpose the signal is sampled and digitized by an analog-to-digital converter (ADC). In most receivers the final IF signal is sampled, but in some the final IF signal is converted down to an analog baseband signal prior to sampling. The sampling rate must be chosen so that there is no spectral aliasing of the sampled signal; this generally will be several times the final IF bandwidth (2–20 MHz).

Most low-cost receivers use 1-bit quantization of the digitized samples, which not only is a very-low cost method of analog-to-digital conversion, but has the additional advantage that its performance is insensitive to changes in voltage levels. Thus, the receiver needs no automatic gain control (AGC). At first glance it would appear that 1-bit quantization would introduce severe signal distortion. However, the noise, which is Gaussian and typically much greater than the signal at this stage, introduces a dithering effect that, when statistically averaged, results in an essentially linear signal component. One-bit quantization does introduce some loss in SNR, typically about 2 dB, but in low-cost receivers this is an acceptable tradeoff. A major disadvantage of 1-bit quantization is that it exhibits a capture effect in the presence of strong interfering signals and is therefore quite susceptible to jamming.

Typical high-end receivers use anywhere from 1.5-bit (three-level) to 3-bit (eight-level) sample quantization. Three-bit quantization essentially eliminates the SNR degradation found in 1-bit quantization and materially improves performance in the presence of jamming signals. However, to gain the advantages of multibit quantization, the ADC input signal level must exactly match the ADC dynamic range. Thus the receiver must have AGC to keep the ADC input level constant. Some military receivers use even more than 3-bit quantization to extend the dynamic range so that jamming signals are less likely to saturate the ADC.

4.1.4 Baseband Signal Processing

Baseband signal processing refers to a collection of high-speed real-time algorithms implemented in dedicated hardware and controlled by software that acquire and track the GPS signal, extract the 50-bps (bits per second) navigation data, and provide measurements of code and carrier pseudoranges and Doppler.

4.1.4.1 Carrier Tracking Tracking of the carrier phase and frequency is accomplished by using feedback control of a numerically controlled oscillator (NCO) to frequency shift the signal to precisely zero frequency and phase. Because the shift to zero frequency results in spectral foldover of the signal sidebands, both in-phase (I) and a quadrature (Q) baseband signal components are formed in order to prevent signal information loss. The I component is generated by multiplying the digitized IF by the NCO output, and the Q component is formed by first introducing a 90° phase lag in the NCO output before multiplication. Feedback is accomplished by using the measured baseband phase to control the NCO so that this phase is driven toward zero. When this occurs, signal power is entirely in the I component, and the Q component contains only noise. However, both components are necessary in order to measure the phase error for feedback and to provide full signal information during acquisition when phase lock has not yet been achieved. The baseband phase θ_{baseband} is defined by

$$\theta_{\text{baseband}} = \text{atan2}(I, Q) \tag{4.2}$$

where atan2 is the four-quadrant arctangent function. The phase needed for feedback is recovered from I and Q after despreading of the signal. When phase lock has been achieved, the output of the NCO will match the incoming IF signal in both frequency and phase but will generally have much less noise due to lowpass filtering used in the feedback loop. Comparing the NCO phase to a reference derived from the receiver reference oscillator provides the phase measurements needed for carrier phase pseudoranging. Additionally, the cycles of the NCO output can be accumulated to provide the raw data for Doppler, delta-range, and integrated Doppler measurements.

4.1.4.2 Code Tracking and Signal Spectral Despreading The digitized IF signal, which has a wide bandwidth due to the C/A- (or P-) code modulation, is completely obscured by noise. The signal power is raised above the noise power by *despreading*, in which the digitized IF signal is multiplied by a receiver-generated replica of the code precisely time-aligned with the code on the received signal. Typically the individual baseband I and Q signals from the controlled NCO mixer are despread in parallel, as previously shown in Fig. 3.13. The despreading process removes the code from the signal, thus concentrating the full signal power into the approximately 50-Hz baseband bandwidth of the data modulation. Subsequent filtering (usually in the form of integration) can now be employed to dramatically raise the SNR to values permitting observation and measurement of the signal. As an example, recall that in a GPS receiver a typical SNR in a 2-MHz IF bandwidth is −16.6 dB. After despreading and 50-Hz lowpass filtering the total signal power is still about the same, but the bandwidth of the noise has been reduced from 2 MHz to about 50 Hz, which increases the SNR by the ratio $2 \times 10^6/50$, or 46 dB. The resulting SNR is therefore $-16.6 + 46.0 = 29.4$ dB.

4.2 RECEIVER DESIGN CHOICES

4.2.1 Number of Channels and Sequencing Rate

GPS receivers must observe the signal from at least four satellites to obtain three-dimensional position and velocity estimates. If the user altitude is known, three satellites will suffice. There are several choices as to how the signal observations from a multiplicity of satellites can be implemented. In early designs, reduction of hardware cost and complexity required that the number of processing channels be kept at a minimum, often smaller than the number of satellites observed. In this case, each channel must sequentially observe more than one satellite. As a result of improved lower-cost technology, most modern GPS receivers have a sufficient number of channels to permit one satellite to be continuously observed on each channel.

4.2.1.1 Receivers with Channel Time Sharing

Single-Channel Receivers In a single-channel receiver, all processing, such as acquisition, data demodulation, and code and carrier tracking, is performed by a single channel in which the signals from all observed satellites are time-shared. Although this reduces hardware complexity, the software required to manage the time-sharing process can be quite complex, and there are also severe performance penalties. The process of acquiring satellites can be very slow and requires a juggling act to track already-acquired satellites while trying to acquire others. The process is quite tricky when receiving ephemeris data from a satellite, since about 30 s of continuous reception is required. During this time the signals from other satellites are eclipsed, and resumption of reliable tracking can be difficult.

After all satellites have been acquired and their ephemeris data stored, two basic techniques can be used to track the satellite signals in a single-channel receiver. In *slow-sequencing* designs the signal from each satellite is observed for a duration (dwell time) on the order of 1 s. Since a minimum of four satellites must typically be observed, the signal from each satellite is eclipsed for an appreciable length of time. For this reason, extra time must be allowed for signal reacquisition at the beginning of each dwell interval. Continually having to reacquire the signal generally results in less reliable operation, since the probability of losing a signal is considerably greater as compared to the case of continuous tracking. This is especially critical in the presence of dynamics, in which unpredictable user platform motion can take place during signal eclipse. Generally the positioning and velocity accuracy is also degraded in the presence of dynamics.

If a single-channel receiver does not have to accurately measure velocity, tracking can be accomplished with only a frequency-lock loop (FLL) for carrier tracking. Since a FLL typically has a wider pull-in range and a shorter pull-in time than does a phase-lock loop (PLL), reacquisition of the signal is relatively fast and the sequencing dwell time can be as small as 0.25 s per satellite. Because loss of phase lock is not an issue, this type of receiver is also more robust in the presence of dynamics. On the other hand, if accurate velocity determination is required, a PLL must be used and the extra time required for phase lock during signal reacquisition pushes the dwell time up to about 1 – 1.5 s per satellite, with an increased probability of reacquisition failure due to dynamics.

A single-channel receiver requires relatively complex software for managing the satellite time-sharing process. A typical design employs only one pseudonoise (PN) code generator and one PPL in hardware. Typical tasks that the software must perform during the dwell period for a specific satellite are as follows:

1. Select the PN code corresponding to the satellite observed.
2. Compute the current state of the code at the start of the dwell based on the state at the end of the last dwell, the signal Doppler, and the eclipse time since the last dwell.
3. Load the code state into the code generator hardware.
4. Compute the initial Doppler frequency of the FLL/PLL reference.
5. Load the Doppler frequency into the FLL/PLL hardware.
6. Initiate the reacquisition process by turning on the code and carrier tracking loops.
7. Determine when reacquisition (code/frequency/phase lock) has occurred.
8. Measure pseudorange/carrier phase/carrier phase rate during the remainder of the dwell.

In addition to these tasks, the software must be capable of ignoring measurements from a satellite if the signal is momentarily lost and must permanently remove the satellite from the sequencing cycle when its signal becomes unusable, such as when the satellite elevation angle is below the mask angle. The software must also have the capability of acquiring new satellites and obtaining their ephemeris data as their signals become available while at the same time not losing the satellites already being tracked. A satellite whose ephemeris data are being recorded must have a dwell time (about 30 s) much longer than those of other satellites that are only being tracked, which causes a much longer eclipse time for the latter. The software must therefore modify the calculations listed above to take this into account.

Because current technology makes the hardware costs of a multichannel receiver almost as small as that for a single channel, the single-channel approach has been almost entirely abandoned in modern designs.

Another method of time sharing that can be used in single-channel receivers is *multiplexing*, in which the dwell time is much shorter, typically 5 – 10 ms per satellite. Because the eclipse time is so short, the satellites do not need to be reacquired at each dwell. However, a price is paid in that the effective SNR is significantly reduced in proportion to the number of satellites being tracked. Resistance to jamming is also degraded by values of 7 dB or more. Additionally, the process of acquiring new satellites without disruption is made more demanding because the acquisition search must be broken into numerous short time intervals. Because of the rapidity with which satellites are sequenced, a common practice with a two-channel receiver is to use a full complement of PN code generators that run all the time, so that high-speed multiplexing of a single code generator can be avoided.

Two-Channel Receivers The use of two channels permits the second channel to be a "roving" channel, in which new satellites can be acquired and ephemeris data collected while on the first channel satellites can be tracked without slowdown in position/velocity updates. However, the satellites must still be time-shared on the first channel. Thus the software must still perform the functions listed above and in addition must be capable of inserting/deleting satellites from the sequencing cycle. As with single-channel designs, either slow sequencing or multiplexing may be used.

Receivers with Three to Five Channels In either slow-sequencing or multiplexed receivers, additional channels will generally permit better accuracy and jamming immunity as well as more robust performance in the presence of dynamics. A major breakthrough in receiver performance occurs with five or more channels, because four satellites can be simultaneously tracked without the need for time sharing. The fifth channel can be used to acquire a new satellite and collect its ephemeris data before using it to replace one of the satellites being tracked on the other four channels.

Multichannel All-in-View Receivers The universal trend in receiver design is to use enough channels to receive all satellites that are visible. In most cases eight or fewer useful satellites are visible at any given time; for this reason modern receivers typically have no more than 10–12 channels, with perhaps several channels being used for acquisition of new satellites and the remainder for tracking. Position/velocity accuracy is materially improved because satellites do not have to be continually reacquired as is the case with slow sequencing, there is no reduction in effective SNR found in multiplexing designs, and the use of more than the minimum number of satellites results in an overdetermined solution. In addition, software design is much simpler because each channel has its own tracking hardware that tracks only one satellite and does not have to be time shared.

4.2.2 L_2 Capability

GPS receivers that can utilize the L_2 frequency (1227.60 MHz) gain several advantages over L_1-only receivers. Currently the L_2 carrier is modulated only with a military-encrypted P-code, called the *Y-code*, and the 50-bps (bits per second) data stream. Because of the encryption, civilians are denied the use of the P-code. However, it is still possible to recover the L_2 carrier, which can provide significant performance gains in certain applications.

4.2.2.1 Dual-Frequency Ionospheric Correction Because the pseudorange error caused by the ionosphere is inversely proportional to the square of frequency, it can be calculated in military receivers by comparing the P-code pseudorange measurements obtained on the L_1 and L_2 frequencies. After subtraction of the calculated error from the pseudorange measurements, the residual error due to the ionosphere is typically no more than a few meters as compared to an uncorrected error of 5–30 m. Although civilians do not have access to the P-code, in differential positioning applications the L_2 carrier phase can be extracted without decryption, and the ionospheric error can then be estimated by comparing the L_1 and L_2 phase measurements.

4.2.2.2 Improved Carrier Phase Ambiguity Resolution in High-Accuracy Differential Positioning High-precision receivers, such as those used in surveying, use carrier phase measurements to obtain very precise pseudoranges. However, the periodic nature of the carrier makes the measurements highly ambiguous. Therefore, solution of the positioning equations yields a grid of possible positions separated by distances on the order of one to four carrier wavelengths, depending on geometry. Removal of the ambiguity is accomplished by using additional information in the form of code pseudorange measurements, changes in satellite geometry, or the use of more satellites than is necessary. In general, ambiguity resolution becomes less difficult as the frequency of the carrier decreases. By using both the L_1 and L_2 carriers, a virtual carrier frequency of $L_1 - L_2 = 1575.42 - 1227.60 = 347.82$ MHz can be obtained, which has a wavelength of about 86 cm as compared to the 19 cm wavelength of the L_1 carrier. Ambiguity resolution can therefore be made faster and more reliable by using the difference frequency.

4.2.3 Code Selections: C/A, P, or Codeless

All GPS receivers are designed to use the C/A-code, since it is the only code accessible to civilians and is used by the military for initial signal acquisition. Most military receivers also have P-code capability to take advantage of the improved performance it offers. On the other hand, commercial receivers seldom have P-code capability because the government does not make the needed decryption equipment available to the civil sector. Some receivers, notably those used for precision differential positioning application, also incorporate a codeless mode that permits recovery of the L_2 carrier without knowledge of the code waveform.

4.2.3.1 The C/A-Code The C/A-code, with its 1.023-MHz chipping rate and 1-ms period, has a bandwidth that permits a reasonably small pseudorange error due to thermal noise. The code is easily generated by a few relatively small shift registers. Because the C/A-code has only 1023 chips per period, it is relatively easy to acquire. In military receivers direct acquisition of the P-code would be extremely difficult and time-consuming. For this reason these receivers first acquire the C/A-code on the L_1 frequency, allowing the 50-bps (bits per second) data stream to be recovered. The data contains a hand-over word that tells the military receiver a range in which to search for the P-code.

4.2.3.2 ***The P-Code*** The unencrypted P-code has a 10.23-MHz chipping rate and is known to both civilian and military users. It has a very long period of one week. The Y-code is produced by biphase modulation of the P-code by an encrypting code known as the *W-code*. The W-code has a slower chipping rate than does the P-code; there are precisely 20 P-code chips per W-code chip. Normally the W-code is known only to military users who can use decryption to recover the P-code, so that the civilian community is denied the full use of the L_2 signal. However, as will be indicated shortly, useful information can still be extracted from the L_2 signal in civilian receivers without the need for decryption. Advantages of the P-code include the following:

Improved Navigation Accuracy Because the P-code has 10 times the chipping rate of the C/A-code, its spectrum occupies a larger portion of the full 20-MHz GPS signal bandwidth. Consequently, military receivers can typically obtain 3 times better pseudoranging accuracy compared to that obtained with the C/A-code.

Improved Jamming Immunity The wider bandwidth of the P-code gives about 40 dB suppression of narrowband jamming signals as compared to about 30 dB for the C/A-code, which is of obvious importance in military applications.

Better Multipath Rejection In the absence of special multipath mitigation techniques, the P-code provides significantly smaller pseudorange errors in the presence of multipath as compared to the C/A-code. Because the P-code correlation function is approximately one-tenth as wide as that of the C/A-code, there is less opportunity for a delayed-path component of the receiver-generated signal correlation function to cause range error by overlap with the direct-path component.

4.2.3.3 ***Codeless Techniques*** Commercial receivers can recover the L_2 carrier without knowledge of the code modulation simply by squaring the received signal waveform or by taking its absolute value. Because the a priori SNR is so small, the SNR of the recovered carrier will be reduced by as much as 33 dB because the squaring of the signal greatly increases the noise power relative to that of the signal. However, the squared signal has extremely small bandwidth (limited only by Doppler variations), so that narrowband filtering can make up the difference.

4.2.4 Access to SA Signals

Selective Availability (SA) refers to errors that may be intentionally introduced into the satellite signals by the military to prevent full-accuracy capability by the civilian community. SA was suspended on May 1, 2000 but can be turned on again at the discretion of the DoD. The errors appear to be random, have a zero long-term average value, and typically have a standard deviation of 30 m. Instantaneous position errors of 50–100 m occur fairly often and are magnified by large position dilution of precision (PDOP) values. Part of the SA error is in the ephemeris data transmitted by the satellite, and the rest is accomplished by dithering of the satellite clock that controls the timing of the carrier and code waveforms. Civil users with a single receiver generally have no way to eliminate errors due to SA, but authorized users (mostly military) have the key to remove them completely. On the other hand, civilians can remove SA errors by employing differential operation, and a large network of differential reference stations has been spawned by this need.

4.2.5 Differential Capability

Differential GPS (DGPS) is a powerful technique for improving the performance of GPS positioning. This concept involves the use of not only the user's receiver (sometimes called *the remote or roving* unit) but also a *reference receiver* at an accurately known location within perhaps 200 km of the user. Because the location of the reference receiver is known, pseudorange errors common to the user and reference receivers can be measured and removed in the user's positioning calculations.

4.2.5.1 Errors Common to Both Receivers The major sources of error common to the reference and remote receivers, which can be removed (or mostly removed) by differential operation, are the following:

1. *Selective Availability Error.* As mentioned previously, these are typically about 30 m, 1σ.
2. *Ionospheric Delays.* Ionospheric signal propagation group delay, which is discussed further in Chapter 5, can be as much as 20–30 m during the day to 3–6 m at night. Receivers that can utilize both the L_1 and L_2 frequencies can largely remove these errors by applying the inverse square-law dependence of delay on frequency.
3. *Tropospheric Delays.* These delays, which occur in the lower atmosphere, are usually smaller and more predictable than ionospheric errors, and typically are in the 1–3 m range but can be significantly larger at low satellite elevation angles.
4. *Ephemeris Errors.* Ephemeris errors, which are the difference between the actual satellite location and the location predicted by satellite orbital data, are typically less than 3 m and will undoubtedly become smaller as satellite tracking technology improves.
5. *Satellite Clock Errors.* These are the difference between the actual satellite clock time and that predicted by the satellite data.

Differential operation can almost completely remove satellite clock errors, errors due to SA, and ephemeris errors. For these quantities the quality of correction has little dependence on the separation of the reference and roving receivers. However, because SA errors vary quite rapidly, care must be taken in time-synchronizing the corrections to the pseudorange measurements of the roving receiver. The degree of correction that can be achieved for ionospheric and tropospheric delays is excellent when the two receivers are in close proximity, say, up to 20 km. At larger separations the ionospheric/tropospheric propagation delays to the receivers become less correlated, and residual errors after correction are correspondingly larger. Nonetheless, substantial corrections can often be made with receiver separations as large as 100–200 km.

Differential operation is ineffective against errors due to multipath, because these errors are strictly local to each of the two receivers.

4.2.5.2 Corrections in the Measurement Domain versus the Solution Domain

In the broadest sense there are two ways that differential corrections can be made. In the *measurement domain*, corrections are determined for pseudorange measurements to each satellite in view of the reference receiver, and the user simply applies the corrections corresponding to the satellites the roving receiver is observing. On the other hand, in the *solution-domain* approach, the reference station computes the position error that results from pseudorange measurements to a set of satellites, and this is applied as a correction to the user's computed position. A significant drawback to the solution-domain approach is that the user and reference station must use exactly the same set of satellites if the position correction is to be valid. In most cases the reference station does not know which satellites can be received by the roving receiver (e.g., some might be blocked by obstacles) and therefore would have to transmit the position corrections for many possible sets of satellites. The impracticality of doing this strongly favors the use of the measurement-domain method.

4.2.5.3 Real-Time versus Postprocessed Corrections In some applications, such as surveying, it is not necessary to obtain differentially corrected position solutions in real time. In these applications it is common practice to obtain corrected positions at a later time by bringing together recorded data from both receivers. No reference-to-user data link is necessary if the recorded data from both receivers can be physically transported to a common computer for processing.

However, in the vast majority of cases it is imperative that corrections be applied as soon as the user has enough pseudorange measurements to obtain a position solution. When the user needs to know his or her corrected position in real time, current pseudorange corrections can be transmitted from the reference receiver to the user via a radio or telephone link, and the user can use them in the positioning calculations. This capability requires a user receiver input port for receiving and using differential correction messages. A standardized format of these messages has been recommended by Special Committee 104 (SC-104), established by the Radio Technical Commission for Maritime Service (RTCM) in 1983. Details on this format appear in Ref. 103.

4.2.6 Pseudosatellite Compatibility

Although differential GPS can improve the reliability, integrity, and accuracy of GPS navigation, it cannot overcome inherent limitations that are critical to successful operation in specific applications. A major limitation is poor satellite geometry, which can be caused by signal failure of one or more satellites, signal blockage by local objects and/or terrain, and occasional periods of high PDOP, which can occur even with a full constellation of satellites. Vertical positioning error is usually more sensitive to this effect, which is bad news for aviation applications. In some cases a navigation solution may not exist because not enough satellite signals can be received.

The use of *pseudolites* can solve these problems within a local area. A pseudolite is simply a ground-based transmitter that acts as an additional GPS satellite by transmitting a GPS-like signal. This signal can be utilized by a receiver for pseudoranging and can also convey messages to the receiver to improve reliability and signal integrity. The RTCM SC-104 was formed in 1983 to study pseudolite system and receiver design issues. The recommendations of SC-104 can be found in Ref. 180. The major improvements offered by pseudolites are the following:

1. *Improvement in Geometry.* Pseudolites, acting as additional satellites, can provide major improvements in geometry, hence in positioning accuracy, within their region of coverage. Vertical (VDOP) as well as horizontal (HDOP) dilution of precision can be dramatically reduced, which is of major importance to aviation. Experiments have shown that PDOP of about 3 over a region having a radius of 20–40 km can be obtained by using several pseudolites even when there are fewer than the minimum of four satellites that would otherwise be needed for a navigation solution.
2. *Improvement in Signal Availability.* Navigation solutions with fewer than the minimum required number of GPS satellites are made possible by using the additional signals provided by pseudolites.
3. *Inherent Transmission of Differential Corrections.* The GPS-like signals transmitted by a pseudolite include messaging capability that can be received directly by the GPS receiver, thus allowing the user to receive differential corrections without the need for a separate communications link.
4. *Self-Contained Failure Notification.* The additional signals provided by pseudolites permit users to perform their own failure assessments. For example, if pseudorange measurements from four satellites and one pseudolite are available, a problem can be detected by examining the consistency of the measurements. If two pseudolites are available, not only can the failure of a single signal be detected, but the offending signal can be identified

as well. These advantages are especially important in aviation, where pilot notification of signal failures must occur very rapidly (within 1–10 s).

5. *Solution of Signal Blockage Problems.* The additional signals from pseudolites can virtually eliminate problems due to blockage of the satellite signals by objects, terrain, or the receiving platform itself.

4.2.6.1 Pseudolite Signal Structure Ideally the pseudolite signal structure would permit reception by a standard GPS receiver with little or no modification of the receiver design. Thus it would seem that the pseudolite signal should have a unique C/A-code with the same characteristics as the C/A-codes used by the satellites. However, with this scheme it would be difficult to prevent a pseudolite signal from interfering with the reception of the satellite signals, even if its C/A-code were orthogonal to the satellite codes. The fundamental difficulty, which is called the *near–far problem*, occurs because of the inverse square-law dependence of received signal power with range. The near–far problem does not occur with the GPS satellite signals because variation in the user-to-satellite range is relatively small compared to its average value. However, with pseudolites this is not the case. The problem is illustrated by considering that the received signal strength of a pseudolite must be at least approximately that of a satellite. If the pseudolite signal equals that of a satellite when the user is, say, 50 km from the pseudolite, then that same signal will be 60 dB stronger when the user is 50 m from the pseudolite. At this close range the pseudolite signal would be so strong that it would jam the weaker GPS satellite signals.

Several solutions to the near–far problem involving both pseudolite signal design and receiver design have been proposed [180] for the 60-dB received signal dynamic range discussed above.

4.2.6.2 Pseudolite Signal Design Approaches

1. *Use of High-Performance Pseudorandom Codes.* The 60 dB of jamming protection would require the pseudolite to transmit a code much longer than a C/A-code and clocked at a much higher rate. This has been judged to be an impractical solution because it would reduce compatibility with the GPS signal structure and significantly increase receiver costs.
2. *Pseudolite Frequency Offset.* By moving the frequency of the pseudolite signal sufficiently far away from the 1575.42-MHz L_1 frequency, filters in the receiver could prevent the pseudolite signals from interfering with the satellite signals. Again, however, this approach would significantly increase receiver costs and reduce compatibility with the GPS signal structure.
3. *Low-Duty-Cycle Time-Division Multiplexing.* A preferred approach is for the pseudolite to transmit at the L_1 frequency using short, low-duty-cycle pulses that interfere with the satellite signals only a small fraction of the time. The impact on receiver design is minimal because modifications are primarily digital and low in cost. This approach retains compatibility with the GPS signal structure by using a new set of 51 pseudolite Gold codes

with the same chipping rate, period, and number of chips per period as the satellite C/A-codes and a 50-bps (bits per second) data stream. Although the codes run continuously in both the pseudolite and the user receiver, the pseudolite signal is gated on only during eleven 90.91-μs intervals in each 10-ms (half-data-bit) interval. Each of the 11 gate intervals transmits 93 new chips of the code, so that all 1023 chips get transmitted in 10 ms. However, the timing of the gate intervals is randomized in order to randomize the signal spectrum. Further details of the signal structure can be found in Ref. 180.

4.2.6.3 Pseudolite Characteristics

1. *Pseudolite Identification.* Identification of a pseudolite is accomplished by both its unique Gold code and its physical location, which appears in its 50-bps (bits per second) message. Since pseudolite signals are low power and thus can be received only within a relatively small coverage area, it is possible for pseudolites spaced sufficiently far apart to use the same Gold code. In this case correct identification is effected by noting the location transmitted by the pseudolite.
2. *Pseudolite Clock Offset.* Since the pseudolite can monitor GPS signals over extended time periods, it can determine GPS time. This permits the transmitted epochs of the pseudolite signal to be correct in GPS time and avoids the necessity of transmitting pseudolite clock corrections. The time reference for the differential pseudorange corrections transmitted by the pseudolite is also GPS time.
3. *Transmitted Signal Power.* The primary use of pseudolite signals is for aircraft in terminal areas, so that a typical maximum reception range is 50 km. At this range a half-hemisphere omnidirectional transmitting antenna fed with approximately 30 mW of signal power will provide a signal level comparable to that typical of a GPS satellite (−116 dBm). At a range of 50 m the signal level will be 60 dB larger (−56 dBm).
4. *Pseudolite Message Structure.* Although the pseudolite data stream is 50 bps (bits per second) to ensure compatibility with GPS receivers, its structure must be modified to transmit information that differs somewhat from that transmitted by the GPS satellites. A proposed structure can be found in Ref. 180.
5. *Minimum Physical Spacing of Pseudolites.* Placement of pseudolites involves considerations that depend on whether the pseudolites use the same or different Gold codes.

4.2.6.4 Separation of Pseudolites Using the Same Code One approach when two pseudolites use the same code is to synchronize the timing of the gated signals of the pseudolites and separate the pseudolites by a distance that guarantees that received transmissions from different pseudolites will not overlap. This requires that the pseudolites be separated by at least 130 km, which guarantees that a user 50 km from the desired pseudolite will be at least 80 km from the undesired pseudolite. The pulses from the latter will then travel at least 30 km further than those from the desired pseudolite, thus arriving at least 100 μs later. Since the width of pulses is 90.91 μs, pulses from two pseudolites will not overlap and interference is thereby avoided.

However, a more conservative approach is to separate two pseudolites by a distance that is sufficient to guarantee that when the user is at the maximum usable range from one pseudolite, the signal from the other is too weak to interfere. Suppose that each pseudolite is set to achieve a received signal level of −126 dBm at a maximum service radius of 50 km and that an undesired pseudolite signal must be at least 14 dB below the desired signal to avoid interference. A simple calculation involving the inverse square power law shows that this can be achieved with a minimum spacing of 300 km between the two pseudolites, so that the minimum distance to the undesired pseudolite will be 250 km when the user is 50 km from the desired pseudolite.

4.2.6.5 Separation of Pseudolites Using Different Codes When the user must receive several pseudolites simultaneously, separation of the signals from different pseudolites might be possible by using different timing offsets of the transmitted pulses. However, this would substantially complicate system design. A preferred approach is to use synchronous transmissions but space the pseudolites so that when the received pulses do overlap, they can still be recovered by using a suitable low-cost receiver design. The situation is clarified by considering the two pseudolites shown in Fig. 4.2, which are separated by at least 27.25 km, the distance traveled by a signal in the time required to transmit a single pulse. With synchronous pulse transmissions from the pseudolites there exists a central

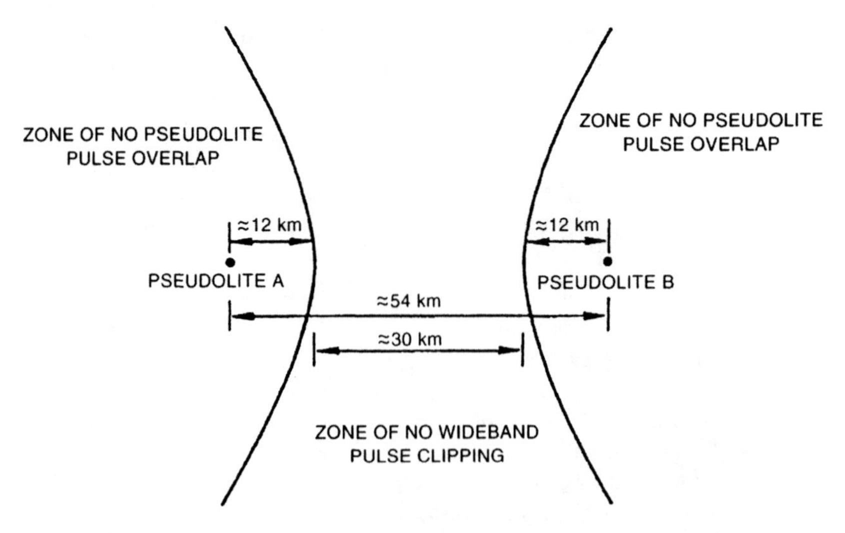

Fig. 4.2 Minimum spacing of pseudolites.

region bounded on the left and right by two hyperbolic curves 27.25 km apart along the baseline connecting the pseudolites. This distance is independent of the separation of the pseudolites, but the curvature of the hyperbolas decreases as the pseudolite separation increases. Outside the central region the received pulses will not overlap and can easily be recovered by the receiver. The difficulty of separating the overlapping pulses within the central region is a function of the pseudolite separation. Separation is most difficult when the receiver is located at the intersection of a hyperbola and the baseline where the stronger of the two signals has its largest value, thus having the potential to overpower the weaker signal. This problem can be avoided by adequate separation of the pseudolites, but the separation required is a function of receiver design.

It will be seen later that a typical receiver designed for pseudolite operation might clip the incoming signal at $\pm 2\sigma$ of the precorrelation noise power in order to limit the received power of strong pseudolite signals. Under this assumption and an assumed ± 1-MHz precorrelation bandwidth, the clipping threshold in a receiver with a 4-dB noise figure would be -104 dBm. Assuming that the pseudolites are designed to produce a -116-dBm power level at 50 km, a receiver receiving overlapping pulses would need to be at least 12.5 km from both pseudolites to avoid the capture effect in the clipping process. Thus, the two pseudolites in Fig. 4.2 should each be moved 12.5 km from the boundaries of the central region, resulting in a minimum distance of 52.25 km between them.

4.2.6.6 Receiver Design for Pseudosatellite Compatibility Major design issues for a GPS receiver that receives pseudosatellite signals (often called a *participating receiver*) are as follows:

1. *Continuous Reception.* Because the receiver must continuously recover the pseudolite data message, a channel must be dedicated to this task. For this reason a single-channel slow-sequencing receiver could not be used. This is really not a problem, since almost all modern receivers use parallel channels.
2. *Ability to Track Pseudolite Gold Codes.* The receiver must be capable of generating and tracking each of the 51 special C/A-codes specified for the pseudolite signals. These codes and their method of generation can be found in Ref. 61. Although the codes can be tracked with standard GPS tracking loops, optimum performance demands that the noise between pseudolite pulses be blanked to obtain a 10-dB improvement in SNR.
3. *Reduction of Pseudosatellite Interference to GPS Signal Channels.* In a GPS satellite channel a pseudolite signal appears as pulsed interference that can be 60 dB greater above the satellite signal level. The resulting degradation of the GPS satellite signal can be reduced to acceptable levels by properly designed wideband precorrelation signal clipping in the receiver. This approach, which generally improves with increasing precorrelation bandwidth and decreasing clipping level, typically results in a reduction in the GPS SNR of 1–2 dB. A somewhat more effective approach is to blank

the GPS signal ahead of the correlator during the reception of a pseudolite pulse, which results in a GPS SNR reduction of about 0.5 dB.

4. *Ability to Receive Overlapping Pseudolite Pulses.* A group of pseudolites designed to be utilized simultaneously must be located relatively close together, inevitably causing received pulse overlap in certain portions of the coverage area. Consequently, receiver design parameters must be chosen carefully to ensure that overlapping pulses from different pseudolites can be separated. The signal level from a nearby pseudolite often can be strong enough to overcome the approximately 24 dB of interference suppression provided by the cross-correlation properties of distinct Gold codes and also can obliterate a second overlapping signal by saturating the receiver amplifiers. Both of these problems can be solved by properly designed wideband precorrelation signal clipping, in which there are two conflicting requirements. Deep (severe) clipping significantly reduces the amount of interfering power from a strong signal but gives the stronger signal more ability to blank out the weaker one (capture effect). On the other hand, more modest clipping levels reduce the capture effect at the expense of passing more power from the stronger signal into the correlators. As a result, more stress is put on the Gold codes to separate the weaker pulses from the stronger ones in the correlation process. An acceptable compromise for most purposes is to clip the received signal at about ± 2 standard deviations of the precorrelation noise power.

4.2.7 Immunity to Pseudolite Signals

A receiver that is not designed to receive pseudolite signals (a so-called *nonparticipating receiver*) must be designed so that a pseudolite signal, which might be 60 dB stronger than a satellite signal, will not interfere with the latter. The importance of this requirement cannot be overstated, since it is expected that use of pseudolites will grow dramatically, especially near airports. Therefore, purchasers of nonparticipating receivers would be well advised to obtain assurances of immunity to jamming by pseudolites.

Pseudolite immunity in a nonparticipating receiver can be effected by designing the front-end amplifier circuits for quick recovery from overload in combination with precorrelation hard limiting of the signal. This approach is suitable for low-cost receivers such as handheld units. More sophisticated receivers using more than 1 bit of digital quantization to avoid quantization loss may still be designed to operate well if the clipping level is the same as that used in participating receivers. The design issues for obtaining immunity to pseudolite interference have been analyzed by RTCM SC 104 and can be found in Ref. 180.

4.2.8 Aiding Inputs

Although GPS can operate as a standalone system, navigation accuracy and coverage can be materially improved if additional information supplements the received

GPS signals. Basic sources include the following:

1. *INS Aiding.* Although GPS navigation is potentially very accurate, periods of poor signal availability, jamming, and high-dynamics platform environments often limit its capability. INSs are relatively immune to these situations and thus offer powerful leverage in performance under these conditions. On the other hand, the fundamental limitation of INS long-term drift is overcome by the inherent calibration capability provided by GPS. Incorporation of INS measurements is readily achieved through Kalman filtering.
2. *Aiding with Additional Navigation Inputs.* Kalman filtering can also use additional measurement data from navigation systems such as LORAN-C, vehicular wheel sensors, and magnetic compasses, to improve navigation accuracy and reliability.
3. *Altimeter Aiding.* A fundamental property of GPS satellite geometry causes the greatest error in GPS positioning to be in the vertical direction. Vertical error can be significantly reduced by inputs from barometric, radar, or laser altimeter data. Coupling within the system of positioning equations tends to reduce the horizontal error as well.
4. *Clock Aiding.* An external clock with high stability and accuracy can materially improve navigation performance. It can be continuously calibrated when enough satellite signals are available to obtain precise GPS time. During periods of poor satellite visibility it can be used to reduce the number of satellites needed for positioning and velocity determination.

4.3 HIGH-SENSITIVITY-ASSISTED GPS SYSTEMS (INDOOR POSITIONING)

The last decade (since 1996) has seen increasing interest in the development of very high-sensitivity GPS receivers for use in poor signal environments. More generally, such receivers can be designed for use with any global navigation satellite system (GNSS), such as the Russian GLONASS and European Galileo systems. A major application is incorporation of such receivers in cell phones, thus enabling a user to automatically transmit his location to rescue authorities in emergencies such as a 911 call. Such a receiver must be able to reliably operate deep within buildings or heavy vegetation, which severely attenuates the GPS signals.

In order to achieve the requisite reliable and rapid positioning for such applications, assisting data from a base station receiver (the server) at a location having good signal reception is sent to the user's receiver (the client). The assisting data can include base station location, satellite ephemeris data, the demodulated navigation data bit stream, frequency calibration data, and timing information. In addition the base station can provide pseudorange and/or carrier phase measurements that enable differential operation. The assisting data can be transmitted via a cell phone or other radiolink. In some case the assisting information can be transmitted over the Internet and relayed to a cell phone via a local-area wireless link. An example of an assisted GPS system can be found in Ref. 207.

Assisting data can not only increase the sensitivity of the client receiver, but also can significantly reduce the time required to obtain a position. A typical standalone GPS receiver can acquire signals down to about −145 dBm, and might require a minute or more to obtain a position from a cold start On the other hand, high-sensitivity assisted GPS receivers are currently being produced by a number of major manufacturers who are claiming sensitivities in the −155 to −165 dBm range and a cold start time to first fix (TTFF) under 10 seconds. To gain the required sensitivity and processing speed, assisted GPS receivers usually capture several seconds of received signal in a memory that can be accessed at high speed to facilitate the signal processing operations.

4.3.1 How Assisting Data Improves Receiver Performance

4.3.1.1 Reduction of Frequency Uncertainty To achieve rapid positioning, the range of frequency uncertainty in acquiring the satellites at the client receiver must be reduced as much as practicable in order to reduce the search time. Reducing the number of searched frequency bins also increases receiver sensitivity because the acquisition false-alarm rate is reduced. Two ways that frequency uncertainty can be reduced are as follows:

1. *Transmission of Doppler Information.* The server can accurately calculate signal Doppler shifts at its location and transmit them to the user. For best results, the user must either be reasonably close to the server's receiver or must know his or her approximate position to avoid excessive uncompensated differential Doppler shift between server and client.
2. *Transmission of a Frequency Reference.* If only Doppler information is transmitted to the user, the frequency uncertainty of the client receiver local oscillator still remains an obstacle to rapid acquisition. Today's technology can produce oscillators that have a frequency uncertainty on the order of 1 part per million at a cost low enough to permit incorporation into a consumer product such as a cell phone. Even so, 1 part per million translates into about ±1575 Hz of frequency uncertainty at the GPS L_1 frequency. Assuming that the coherent integration time during satellite search is 20 ms (the length of a navigation message data bit), the frequency bins in the search would have a 50-Hz spacing, and a total of $2 \times 1575/50 = 63$ frequency bins might have to be searched to find the first satellite. Once the first satellite is acquired, the local-oscillator offset can be determined, and the frequency uncertainty in searching for the remaining satellites can thereby be reduced to a small value. To remedy the problem of acquiring the first satellite in a sufficiently short time, some assisted GPS systems use an accurate frequency reference transmitted from the server to the client, in addition to satellite Doppler measurements. However, this requirement

significantly complicates the design of the server-to-client communication system, and is certainly undesirable when trying to use an existing communication system for assisting purposes. If the communication system is a cell phone network, every cell tower would need to transmit a precise frequency reference.

4.3.1.2 Determination of Accurate Time In order to obtain accurate pseudoranges, a conventional GPS receiver obtains time information from the navigation data message that permits the precise GPS time of transmission of any part of the received signal to be tracked at the receiver. When a group of pseudorange measurements is made, the time of transmission from each satellite is used for two purposes: (1) to obtain an accurate position of each satellite at the time of transmission and (2) to compute pseudorange by computing the difference between signal reception time (according to the receiver clock) and transmission time.

In order to obtain time information from the received GPS signal, a conventional receiver must go through the steps of acquiring the satellite signal, tracking it with a phase-lock loop to form a coherent reference for data demodulation, achieve bit synchronization, demodulate the data, achieve frame synchronization, locate the portion of the navigation message that contains the time information, and finally, continue to keep track of time (usually by counting C/A-code epochs as they are received).

However, it is desirable to avoid these numerous and time-consuming steps in a positioning system that must reliably obtain a position within several seconds of startup in a weak-signal environment. Because the navigation data message contains time information only once per 6-s subframe, the receiver may have to wait a minimum of 6 s to obtain it (additionally, more time is needed to phase-lock to the signal and achieve bit and frame synchronization). Furthermore, if the signals are below about -154 dBm, demodulation of the navigation data message has an error rate that precludes the reading of time from the signal.

If the approximate position of the client is known with sufficient accuracy (perhaps within 100 km), it is possible to resolve the difference in times of transmission. This is possible because the times of transmission of the C/A-code epochs are known to be integer multiples of 1 ms according to SV time (which can be corrected to GPS time using slowly changing time correction data sent from the server). This integer ambiguity in differences of time transmission is resolved by using approximate ranges to the satellites, which are calculated from the approximate position of the client and insertion of approximate time into satellite ephemeris data sent by the server to the client. For this purpose the accuracy of the approximated time needs to be sufficiently small to avoid excessive uncertainty in the satellite positions. Generally a time accuracy of better than 10 s will suffice.

Once the ambiguity of the differences in transmission times has been resolved, accurate positioning is possible if the positions of the satellites at transmission time are known with an accuracy comparable to the positioning accuracy desired.

However, since the satellites are moving at a tangential orbital velocity of approximately 3800 m/s, the accuracy in knowledge of signal transmission time for the purpose of locating the satellites must be significantly more accurate than that required for the ambiguity resolution previously described.

Most weak-signal assisted GPS rapid positioning systems obtain the necessary time accuracy for locating the satellites by using time information transmitted from the server. It is important to recognize that such time information must be in "real time"; that is, it must have a sufficiently small uncertainty in latency as it arrives at the client receiver. For example, a latency uncertainty of 0.1 s could result in a satellite position error of 380 m along its orbital path, causing a positioning error of the same order of magnitude.

Transmission of time from the server with small latency uncertainty has a major impact on the design of the server-to-client communication system, and is a major disadvantage in getting the providers of existing communication systems, such as cellular networks, to become involved in providing indoor assisted GPS positioning service.

4.3.1.3 Transmission of Satellite Ephemeris Data Due to the structure of the GPS navigation message, up to 30 s is required for a standalone GPS receiver to obtain the ephemeris data necessary to determine the position of a satellite. This delay is undesirable in emergency applications. Furthermore, in indoor operation the signal is likely to be too weak to demodulate the ephemeris data. The problem is solved if the server transmits the data to the client via a high-speed communication link. The Internet can even be used for this purpose if the client receiver has access to a high-speed internet connection.

4.3.1.4 Provision of Approximate Client Location Some servers (e.g., a cell phone network) can transmit the approximate position of the client receiver to the user. As mentioned previously, this information can be used to resolve the ambiguity in times of signal transmission from the satellites.

4.3.1.5 Transmission of the Demodulated Navigation Bit Stream The ultimate achievable receiver sensitivity is affected by the length of the signal capture interval and the presence of navigation data modulation on the GPS signal.

Fully Coherent Processing If the GPS signal were modulated only by the C/A-code and contained no navigation data modulation, maximum theoretically possible acquisition sensitivity would result from fully coherent delay and Doppler processing. In this form of processing the baseband signal in the receiver is frequency-shifted and precession-compensated in steps (Doppler bins), and for each step the signal is cross-correlated with a replica of the C/A-code spanning the entire signal observation interval. Alternatively, the 1-msec periods of the C/A-code could be synchronously summed prior to cross-correlation.

However, the presence of the 50-bps (bits per second) navigation data modulation precludes the use of fully coherent processing over signal capture intervals exceeding 20 ms unless some means is available to reliably strip the modulation from the signal. If the server can send the demodulated data bit stream to the client, the data modulation can be stripped from the user's received signal, thus enabling fully coherent processing. However, the timing of the bit stream must be known with reasonable accuracy (within approximately 1 ms); otherwise a search for time alignment must be made.

Partially Coherent Processing In the absence of a demodulated data bit stream from the server, a common method of dealing with the presence of data modulation is to coherently process the signal within each data bit interval, followed by noncoherent summation of the results. Assuming that the timing of the data bit boundaries is available from the server, the usual implementation of this technique is to first coherently sum the 20 periods of the complex baseband C/A coded signal within each data bit. For each data bit a waveform is produced that contains one 1-ms period of the C/A-code, with a processing gain of $10 \log(20) = 13$ dB. Each waveform is then cross-correlated with a replica of the C/A-code to produce a complex-valued cross-correlation function. The squared magnitudes of the cross-correlation functions are computed and summed to produce a single function spanning 1 ms, and the location of the peak value of the function is the signal delay estimate. We shall call this form of processing *partially coherent.*

When 1 second of signal is observed by the user, fully coherent processing gives a sensitivity approximately 3–4 dB over partially coherent processing. It is important to note that fully coherent processing has a major drawback-- many more delay/Doppler bins must be processed, which either dramatically slows down processing speed or requires a large amount of parallel processing to maintain that speed.

Data Detection and Removal by the Client Receiver An alternate method of achieving fully coherent processing is to have the client receiver detect the data bits and use them to homogenize the polarity of the signal, thus permitting coherent processing over the full signal capture interval. In order for this method to be effective, the signal must be strong enough to ensure reliable data bit detection. Furthermore, a phase reference is needed, and it should be estimated using the entire signal observation. A practical technique for estimating phase, which approaches theoretically optimum results, is the method of maximum likelihood (ML). We shall call this methodology *coherent processing with data stripping*, or simply *data stripping* for short. At low signal power levels (less than about -160 dBm) its performance approaches that of partially coherent processing, and at high signal levels its performance approaches that of fully coherent processing. At first glance, it seems that data stripping might give a worthwhile advantage over partially coherent processing. However, it shares a common disadvantage with fully coherent processing in that a larger number of delay/Doppler bins must be processed, and the cost is often prohibitive.

4.3.2 Factors Affecting High-Sensitivity Receivers

In a good signal environment, a certain amount of signal-to-noise ratio (SNR) implementation loss is tolerable. Typical standalone GPS receivers for outdoor use may have total losses as great as 3–6 dB. However in a high-sensitivity receiver the maintenance of every decibel of signal-to-noise ratio is important, thus requiring attention to minimizing losses that would otherwise not be of concern. The following are some of the more important issues that arise in high-sensitivity receiver design.

4.3.2.1 Antenna and Low-Noise RF Design A good antenna and a low-noise receiver front end are mandatory elements of a high-sensitivity receiver.

4.3.2.2 Degradation Due to Signal Phase Variations With fully coherent processing over long time intervals, performance is adversely affected by signal phase variations from sources including Doppler curvature due to satellite motion, receiver oscillator phase stability, and motion of the receiver. Doppler curvature can be partially predicted from assisting almanac or ephemeris data, but its accuracy depends on knowledge of the approximate position of the user. On the other hand, oscillator phase noise is random and unpredictable, and hence resistant to compensation (for this reason research efforts are currently underway to produce a new generation of low-cost atomic and optical frequency sources). Especially pernicious is motion of a GPS receiver in the user's hand. Because of the short wavelengths of GPS signals, such motion can cause phase variations of more that a full cycle during the time that the receiver is searching for a signal, thus seriously impairing acquisition performance.

4.3.2.3 Signal Processing Losses There are various forms of processing loss that must be minimized in a high-sensitivity GPS receiver.

Digitization Losses due to quantization of the analog-to-digital converter (ADC) digital output must be minimized. The 1-bit quantization often used in low-cost receivers causes almost 2 dB of SNR loss. Hence it is desirable to use an ADC with at least 2 bits in high-sensitivity applications.

Sampling Considerations The bandwidth of the receiver should be large enough to avoid SNR loss. However, this generally requires higher sampling rates with an attendant increase in power consumption and processing loads, a factor that is detrimental to low-cost, low-power consumer applications.

Correlation Losses Rapid signal acquisition drives the need for coarser quantization of correlator reference code phase during signal search. However, this causes correlation loss, and an acceptable tradeoff must be made. Correlation loss is further exacerbated if the receiver bandwidth is made small to reduce the required sampling rate.

Doppler Compensation Losses One source of these losses is "scalloping loss," caused by the discretization of the Doppler frequencies used in searching for the satellites. Scalloping loss can be as large as 2 dB in some receivers. Another source is phase quantization of the Doppler compensation, which can introduce a degradation of as much as 1 dB in the simplest designs.

4.3.2.4 Multipath Fading It is common in poor signal environments, especially indoors, for the signal to have large and/or numerous multipath components. In addition to causing pseudorange biases, multipath can significantly reduce receiver sensitivity when phase cancellation of the signal occurs.

4.3.2.5 Susceptibility to Interference and Strong Signals As receiver sensitivity is increased, so does the susceptibility to various forms of interference. Although this is seldone a problem with receivers of normal sensitivity, in a high-sensitivity receiver steps must be taken to prevent erroneous acquisition of lower-level PN code correlation sidelobes from both desired and undesired satellite signals.

4.3.2.6 The Problem of Time Synchronization In assisted GPS systems designed for rapid positioning (within a few seconds) using weak signals, the user's receiver does not have time to read unambiguous time from the received signal itself. The need for the base station to transmit to the user low-latency time accurate enough to establish the position of the satellites is a major disadvantage in getting the providers of existing communication networks, such as cellular networks, to provide this capability.

4.3.2.7 Difficulties in Reliable Sensitivity Assessment Realistic assessment of receiver sensitivity is a challenging task. At the extremely low signal levels for which a high-sensitivity receiver has been designed, laboratory signal generators often have signal leakage, which causes the signal levels to be higher than indicated by the generator. For this and other reasons, published sensitivity specifications should at least be regarded with healthy skepticism. A meaningful comparison of competing specifications can be a daunting task if there is not an adequate description of the conditions under which the sensitivity measurements are made.

4.4 ANTENNA DESIGN

Although there is a wide variety of GPS antennas, most are normally right-hand circularly polarized to match the incoming signal and the spatial reception pattern is nominally a hemisphere. Such a pattern permits reception of satellites in any azimuthal direction from zenith down to the horizon. The short wavelengths at the L_1 and L_2 frequencies permit very compact designs. In low-cost handheld receivers the antenna is often integrated with the receiver electronics in a rugged case. In more sophisticated applications it is often desirable that the antenna be separate from the receiver in order to site it more advantageously. In these situations the signal is fed from the antenna to the receiver via a low-loss coaxial cable. At L-band frequencies the cable losses are still quite large and can reach 1 dB for every 10 ft of cable. Thus, it is often necessary to use a low-noise preamplifier at the antenna (active antenna). Preamplifier gain is usually in the range of 20–40 dB, and DC power is commonly fed to the preamplifier via the coaxial cable itself, with appropriate decoupling filters to isolate the signal from the DC power voltage. The preamplifier sets the noise figure for the entire receiver system and typically has a noise figure of 1.5–3 dB.

4.4.1 Physical Form Factors

Figure 4.3 shows several common physical forms of GPS antennas.

Patch Antennas The patch antenna, the most common antenna type, is often used in low-cost handheld receivers. In typical designs the antenna elements are formed by etching the copper foil on a printed-circuit board, which forms a very rugged low-profile unit. This is advantageous in some aviation applications, because it is relatively easy to integrate the antenna into the skin of the aircraft.

Dome Antennas These antennas are housed in a bubblelike housing.

Blade Antennas The blade antenna, also commonly used in aviation applications, resembles a small airfoil protruding from its base.

Helical (Volute) Antennas Helical antennas contain elements that spiral along an axis that typically points toward the zenith. In some designs the helical elements

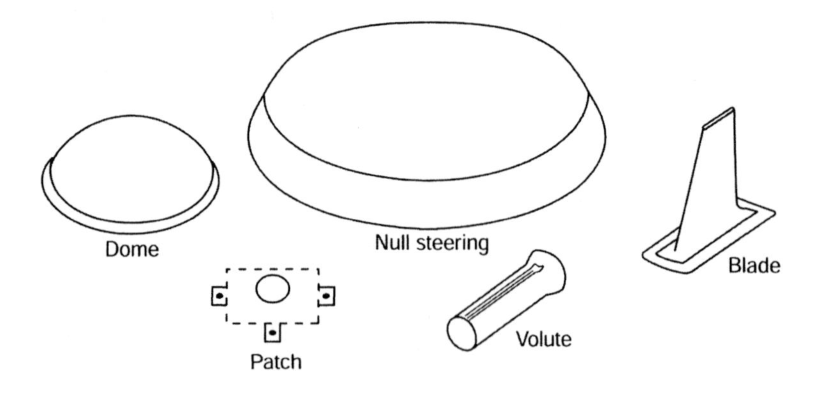

Fig. 4.3 Types of GPS antennas.

are etched from a cylindrical copper-clad laminate to reduce cost. Helical antennas are generally more complex and costly to manufacture than patch antennas but tend to be somewhat more efficient. Some handheld receivers use this type of antenna as an articulated unit that can be adjusted to point skyward while the case of the receiver can be oriented for comfortable viewing by the user. A popular design is the quadrifilar helix, which consists of four helixes symmetrically wound around a circular insulating core.

Choke-Ring Designs In precision applications, such as surveying, choke-ring antennas are used to reduce the effects of multipath signal components reflected from the ground. These antennas are usually of the patch or helical type with a groundplane containing a series of concentric circular troughs one-quarter wavelength deep that act as transmission lines shortcircuited at the bottom ends so that their top ends exhibit a very high impedance at the GPS carrier frequency. Low-elevation angle signals, including ground-reflected components, are nulled by the troughs, reducing the antenna gain in these directions. The size, weight, and cost of a choke-ring antenna are significantly greater than that of simpler designs.

Phased-Array Antennas Although most applications of GPS require a nominally hemispherical antenna pattern, certain applications (especially military) require that the antenna be capable of forming beams in specified directions to obtain better spatial gain or to form nulls in the direction of intentional jamming signals to reduce their effect on the desired GNSS signals. Principles of operation of phased-array antennas are outlined in Section 4.4.3.

Needless to say, phased-array antennas are much more costly than simpler designs and historically have only been used by the military. However, civilian applications have recently begun to emerge, primarily for the purpose of improving positioning performance in the presence of multipath. An introduction to multipath-mitigation antennas and a design example can be found in [41].

4.4.2 Circular Polarization of GPS Signals

An important design goal of GPS antennas is to obtain good performance with the right-hand circularly polarized (RHCP) electromagnetic field characteristic of these signals. It is also desirable for the antenna to have little or no response to multipath signal components in which the sense of polarization is typically changed to left-hand circular polarization (LHCP) by reflection from objects in the vicinity of the antenna.

Figure 4.4 is a somewhat oversimplified illustration of the response of an antenna designed for RHCP signals. The antenna consists of two orthogonal dipoles, which can be assumed to lie in a horizontal plane, with the incoming signal coming directly from above. The centers of the dipoles lie on the origin

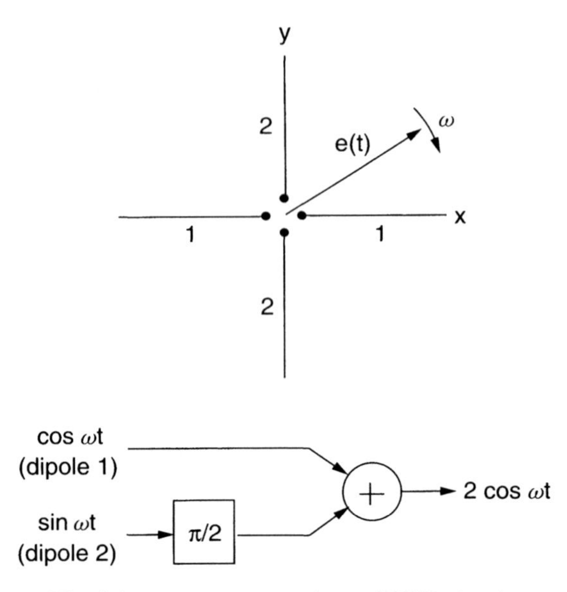

Fig. 4.4 Antenna responsive to RHCP signals.

of an $x-y$ coordinate system, with dipole 1 on the x axis and dipole 2 on the y axis. The positive directions of the x and y axes are respectively indicated by unit vectors **i** and **j**.

The normalized vector electrostatic field **e**(t) of an arriving RHCP signal in the x-y plane can be represented by

$$\mathbf{e}(t) = \mathbf{i} \cos(\omega t) - \mathbf{j} \sin(\omega t), \tag{4.3}$$

which is a unit vector rotating clockwise at angular rate ω, where ω is the frequency of the GPS carrier in radians per second (rad/s). Dipole 1 responds only to the x component of the arriving signal and dipole 2, only to the y component. The polarities of the dipole outputs are such that their respective normalized output voltages are

$$\begin{aligned} s_1(t) &= \cos \omega t, \\ s_2(t) &= \sin \omega t. \end{aligned} \tag{4.4}$$

The output of dipole 2 is phase-shifted by $\pi/2$ radians and summed with the output of dipole 1 to obtain the signal

$$\begin{aligned} s(t) &= \cos \omega t + \sin(\omega t + \pi/2) \\ &= 2 \cos \omega t. \end{aligned} \tag{4.5}$$

On the other hand, if the incoming signal is changed to LHCP, then

$$\left.\begin{aligned} \mathbf{e}(t) &= \mathbf{i}\cos(\omega t) + \mathbf{j}\sin(\omega t) \\ s_1(t) &= \cos\omega t \\ s_2(t) &= -\sin\omega t \\ s(t) &= \cos\omega t - \sin(\omega t + \pi/2) \\ &= 0, \end{aligned}\right\} \qquad (4.6)$$

and there is no response to the signal.

4.4.3 Principles of Phased-Array Antennas

Two general types of phased-array antennas are used with GNSS receivers: the single output adaptive nulling antenna and the multiple output beam-steering antenna, as illustrated in Fig. 4.5. Both types employ N antenna elements whose outputs are amplitude- and/or phase-weighted to produce the desired spatial reception pattern.

4.4.3.1 The Single-Output Adaptive Nulling Antenna This type of antenna is used to sense the presence of an interfering signal and adaptively place nulls in the direction of jamming signals. The objective of the adaptive algorithm is to adjust the weighting of the antenna elements so that the power in the sum of

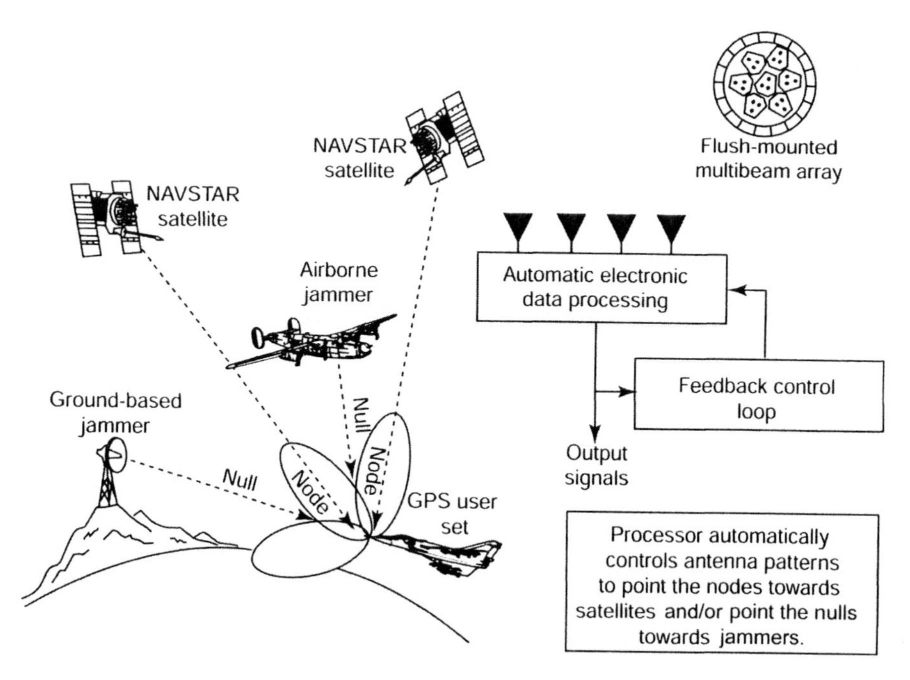

Fig. 4.5 Phased-array antenna operation.

the weighted signal outputs is minimized, subject to a constraint that prevents the minimized power from being zero. Typically the constraint is provided by fixing the weight of one of the antenna elements and allowing the other weights to be adjusted. In mathematical terms the weights are adjusted to minimize the average power $\widehat{|r|^2}$ of the sum

$$r(t) = \sum_{j=1}^{N} w_j s_j(t), \tag{4.7}$$

where the unweighted output signal from the jth antenna element is $s_j(t)$ and w_j is the complex weight (characterized by amplitude and phase) applied to the jth element. Without loss of generality it can be assumed that the constraint is applied by fixing the weight w_1 at unity. Since there are $N - 1$ degrees of freedom in adjusting the weights, as many as $N - 1$ nulls in the antenna spatial pattern can be generated. Because the direction of arriving GNSS signals is not taken into account, there is the possibility that the antenna gain might be low in some signal directions.

Since the single output adaptive nulling antenna provides only one output signal $r(t)$, it can be used with a standard receiver plus the hardware and software for performing the adaptive nulling.

4.4.3.2 The Multiple-Output Beamforming Antenna Instead of forming nulls to reject interfering signals, this type of antenna uses beamforming to produce M independent beams. Thus, there are M corresponding output signals that the receiver must be able to process independently. Each beam can be pointed toward a satellite to achieve a first-order spatial gain of $10 \log_{10} N$ dB relative to an isotropic antenna. A separate set of weights is required for each beam. Instead of being altered adaptively, the weights are typically computed on the basis of antenna attitude information and satellite ephemeris data, so that each beam points toward a satellite. The beamformer signal outputs are

$$r_k(t) = \sum_{j=1}^{N} w_{kj} s_j(t) \quad k = 1, 2, \ldots, M \tag{4.8}$$

where w_{kj} is the complex weight for the kth beam and the jth antenna element. For each k the weights w_{kj}, $j = 1, 2, \ldots, N$ are computed to generate a narrow beam in the direction of a specified unit vector $\mathbf{u}_k$.

If jamming signals arrive from directions sufficiently different from the directions of the satellites, the multiple-output beamsteering antenna can also provide reasonably good suppression of jammers.

4.4.3.3 Space–Frequency Processing The interference reduction capability of both types of phased-array antennas is best when the interfering signals are narrowband. Broadband jammers present a problem, because for a given set of weights the antenna spatial gain pattern varies with frequency. Thus, a set of weights optimal at one frequency will not be optimal at another frequency. Space–frequency processing solves this problem by dividing the frequency band into multiple narrow subbands, typically by using a fast Fourier transform. For each subband a set of optimal weights is used to obtain a corresponding desired antenna spatial pattern for that subband. By combining the nullformed or beamformed signals from all the subbands, an optimal wideband antenna spatial pattern is obtained. Similar results can be obtained by using a bank of narrowband filters in the time domain.

4.4.3.4 Relative Merits of Adaptive Nulling and Beamforming Antennas

- Adaptive nulling is much simpler and cheaper than beamforming, since only one output emerges from the process, enabling use with a standard GNSS receiver. A beamforming antenna produces one output for each beam, so a considerably more complex receiver is required to process each output independently.
- Beamforming can produce significant spatial gain in the direction of the GNSS satellites, while adaptive nulling makes no attempt to maximize gain in the desired directions.
- Jamming reduction is usually greater with an adaptive nulling antenna, because it can place deep nulls in the direction of jamming signals. Beamforming makes no attempt to do this, so its jamming performance is determined by the ratio of gain in the direction of a satellite to the gain in the direction of a jammer.
- Beamformers tend to have high spatial gain in the direction of desired signals and lower gain in other directions. Multipath arriving from a low-gain direction is therefore attenuated relative to the desired signal. Adaptive nulling is less effective against multipath because it does not emphasize signals arriving from a desired direction.
- Because the large physical extent of the antenna array, beamformers and nullers have the common problem of causing biases in signal delay caused by movement of the antenna phase center as a function of the weight values. In some cases biases of 100° in carrier phase and one meter in code phase can occur. For high-precision systems, these errors can be significant.

4.4.4 The Antenna Phase Center

GNSS positioning locates the antenna position, not the receiver position. However, since the antenna has physical extent, it is not simply a point in space. This is not significant in most applications in which positioning errors on the order of decimeters or more are usually encountered. However, in high-precision differential applications where accuracy at the centimeter level or below is needed, the definition of antenna location becomes important. The *phase center* of an antenna can be defined as the point in space where the electrostatic field of the signal exactly matches the signal emerging from the antenna terminals (or equivalently, matches the emerging signal except for a known delay).

The phase center can vary with the arrival direction of the signal, usually within the range of 1 cm or less. There are two basic methods of dealing with this problem. One method is to calibrate the phase center as a function of signal arrival direction. The calibration uses a physical point on the antenna as a reference point. Calibration is usually performed in an anechoic chamber with a very precise signal source that can be moved to different positions around the antenna. Another method is to use identical antennas oriented in the same way at the two receivers in a differential GNSS system.

PROBLEMS

4.1 An ultimate limit on the usability of weak GPS signals occurs when the bit error rate (BER) in demodulating the 50-bps (bits per second) navigation message becomes unacceptably large. Find the signal level in dBm at the output of the receiver antenna that will give a BER of 10^{-5}. Assume an effective receiver noise temperature of 513° K, and that all signal power has been translated to the baseband I channel with optimal demodulation (integration over the 20-ms bit duration followed by polarity detection).

4.2 Support the claim that a 1-bit analog-to-digital converter (ADC) provides an essentially linear response to a signal deeply buried in Gaussian noise by solving the following problem. Suppose that the input signal s_{in} to the ADC is a DC voltage embedded in zero-mean additive Gaussian noise $n(t)$ with standard deviation σ_{in} and that the power spectral density of $n(t)$ is flat in the frequency interval $[-W, W]$ and zero outside the interval. Assume that the 1-bit ADC is modeled as a hard limiter that outputs a value $v_{\text{out}} = 1$ if the polarity of the signal plus noise is positive and $v_{\text{out}} = -1$ if the polarity is negative. Define the output signal s_{out} by

$$s_{\text{out}} = E[v_{\text{out}}], \tag{4.9}$$

where E denotes expectation, and let σ_{out} be the standard deviation of the ADC output. The ADC input signal-to-noise ratio SNR_{in} can then be defined by

$$\text{SNR}_{\text{in}} = \frac{s_{\text{in}}}{\sigma_{\text{in}}} \tag{4.10}$$

and the ADC output signal-to-noise ratio SNR_{out} by

$$\text{SNR}_{\text{out}} = \frac{s_{\text{out}}}{\sigma_{\text{out}}}, \tag{4.11}$$

where s_{out} and σ_{out} are respectively the expected value and the standard deviation of the ADC output. Show that if $s_{\text{in}} << \sigma_{\text{in}}$, then $s_{\text{out}} = K s_{\text{in}}$, where K is a constant, and

$$\frac{\text{SNR}_{\text{out}}}{\text{SNR}_{\text{in}}} = \frac{2}{\pi}. \tag{4.12}$$

Thus, the signal component of the ADC output is linearly related to the input signal component, and the output SNR is about 2 dB less than that of the input.

4.3 Some GPS receivers directly sample the signal at an IF instead of using mixers for the final frequency shift to baseband. Suppose that you wish to sample a GPS signal with a bandwidth of 1 MHz centered at an IF of 3.5805 MHz. What sampling rates will not result in frequency aliasing? Assuming that a sampling rate of 2.046 MHz were used, show how a digitally sampled baseband signal could be obtained from the samples.

4.4 Instead of forming a baseband signal with I and Q components, a single-component baseband signal can be created simply by multiplying the incoming L_1 (or L_2) carrier by a sinusoid of the same nominal frequency, followed by lowpass filtering. Discuss the problems inherent in this approach. (*Hint*: Form the product of a sinusoidal carrier with a sinusoidal local-oscillator signal, use trigonometric identities to reveal the sum and difference frequency components, and consider what happens to the difference frequency as the phase of the incoming signal assumes various values.)

4.5 Write a computer program using C or another high-level language that produces the 1023-chip C/A-code used by satellite SV1. The code for this satellite is generated by two 10-stage shift registers called the *G1 and G2 registers*, each of which is initialized with all 1s. The input to the first stage of the G1 register is the exclusive OR of its 3rd and 10th stages. The input to the first stage of the G2 register is the exclusive OR of its 2nd, 3rd, 6th, 8th, 9th, and 10th stages. The C/A-code is the exclusive OR of stage 10 of G1, stage 2 of G2, and stage 6 of G2.

5

GLOBAL NAVIGATION SATELLITE SYSTEM DATA ERRORS

5.1 SELECTIVE AVAILABILITY ERRORS

Prior to May 1, 2000, Selective Availability (SA) was a mechanism adopted by the Department of Defense (DOD) to control the achievable navigation accuracy by nonmilitary GPS receivers. In the GPS SPS mode, the SA errors were specified to degrade navigation solution accuracy to 100 m (2D RMS) horizontally and 156 m (RMS) vertically.

In a press release on May 1, 2000, the President of the United States announced the decision to discontinue this intentional degradation of GPS signals available to the public. The decision to discontinue SA was coupled with continuing efforts to upgrade the military utility of systems using GPS and supported by threat assessments that concluded that setting SA to zero would have minimal impact on United States national security. The decision was part of an ongoing effort to make GPS more responsive to civil and commercial users worldwide.

The transition as seen from Colorado Springs, Colorado (USA) at the GPS Support Center is shown in Fig. 5.1. The figure shows the horizontal and vertical errors with SA, and after SA was suspended, midnight GMT (8 p.m. EDT), May 1, 2000. Figure 5.2 shows mean errors with and without SA, with satellite PRN numbers.

Aviation applications will probably be the most visible user group to benefit from the discontinuance of SA. However, precision approach will still require some form of augmentation to ensure that integrity requirements are met. Even though setting SA to zero reduces measurement errors, it does not reduce the need for and design of WAAS and LAAS ground systems and avionics.

Global Positioning Systems, Inertial Navigation, and Integration, Second Edition, by M. S. Grewal, L. R. Weill, and A. P. Andrews

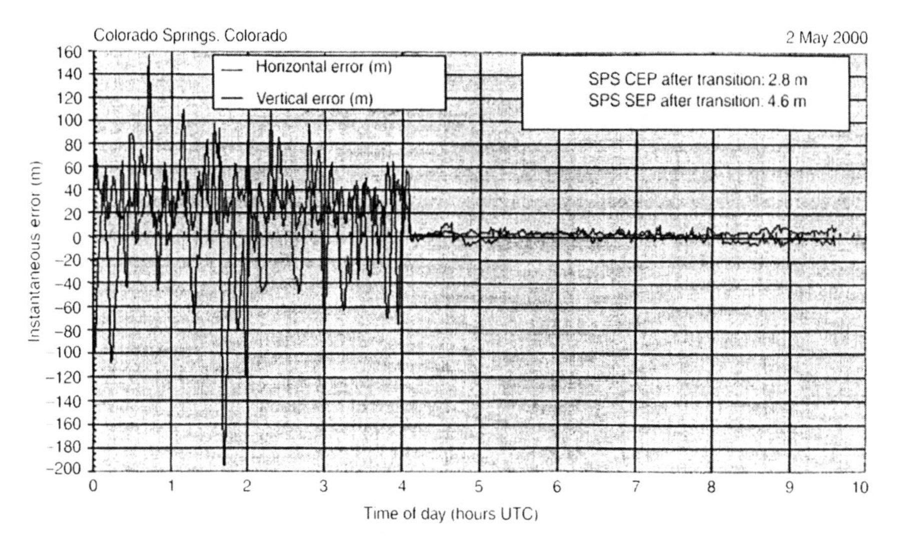

Fig. 5.1 Change in errors when SA is turned off.

Time and frequency users may see greater effects in the long term via communication systems that can realize significant future increases in effective bandwidth use due to tighter synchronization tolerances. The effect on vehicle tracking applications will vary. Tracking in the trucking industry requires accuracy only good enough to locate in which city the truck is, whereas public safety applications can require the precise location of the vehicle. Maritime applications have the potential for significant benefits. The personal navigation consumer will benefit from the availability of simpler and less expensive products, resulting in more extensive use of GPS worldwide.

Because SA could be resumed at any time, for example, in time of military alert, one needs to be aware of how to minimize these errors.

There are at least two mechanisms to implement SA. Mechanisms involve the manipulation of GPS ephemeris data and dithering the satellite clock (carrier frequency). The first is referred to as *epsilon-SA* (ε-SA), and the second as *clock-dither SA*. The clock-dither SA may be implemented by physically dithering the frequency of the GPS signal carrier or by manipulating the satellite clock correction data or both.

Although the mechanisms for implementation of SA and the true SA waveform are classified, a variety of SA models exist in the literature [e.g., [4, 24, 35, 212]]. These references show various models. One proposed by Braasch [24] appears

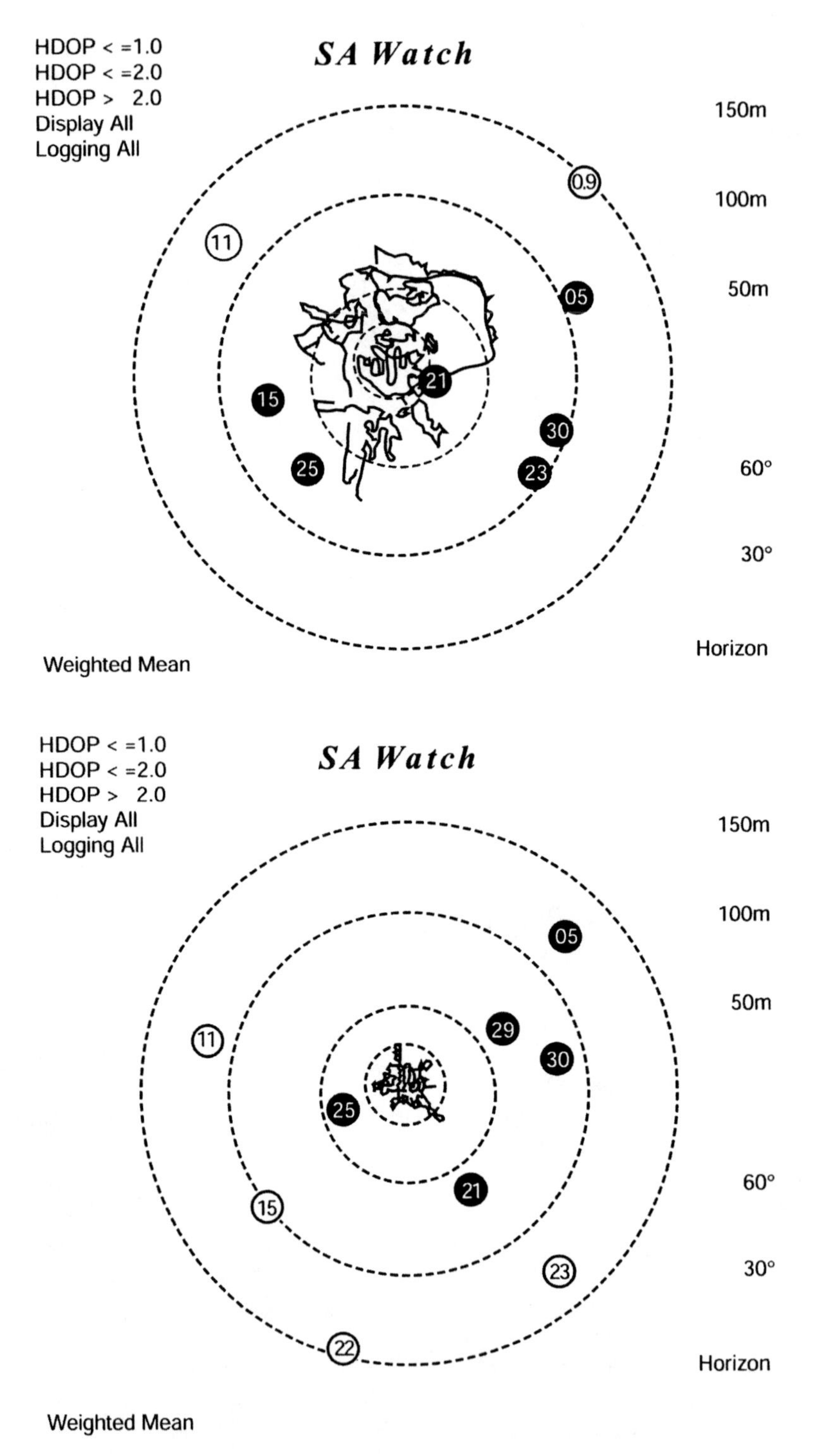

Fig. 5.2 Location errors with and without SA.

to be the most promising and suitable. Another used with some success for predicting SA is a Levinson predictor [10].

The Braasch model assumes that all SA waveforms are driven by normal white noise through linear system [autoregressive moving average (ARMA)] models (see Chapter 3 of Ref. 66). Using the standard techniques developed in system and parameter identification theory, it is then possible to determine the structure and parameters of the optimal linear system that best describes the statistical characteristics of SA. The problem of modeling SA is estimating the model of a random process (SA waveform) based on the input/output data.

The technique used to find an SA model involves three basic elements:

The observed SA

A model structure

A criterion for determination of the best model from the set of candidate models

There are three choices of model structures:

1. An ARMA model of order (p,q), which is represented as ARMA(p,q)
2. An ARMA model of order $(p,0)$ known as the moving-average MA(p) model
3. An ARMA model of order $(q,0)$, the auto regression AR(q) model

Selection from these three models is performed with physical laws and past experience.

5.1.1 Time-Domain Description

Given observed SA data, the identification process repeatedly selects a model structure and then calculates its parameters. The process is terminated when a satisfactory model, according to a certain criterion, is found.

We start with the general ARMA model. Both the AR and MA models can be viewed as special cases of an ARMA model. An ARMA(p, q) model is mathematically described by

$$a_1 y_k + a_2 y_{k-1} + \cdots + a_q y_{k-q+1} = b_1 x_k + b_2 x_{k-1} + \cdots + b_p x_{k-p+1} + e_k, \tag{5.1}$$

or in a concise form by

$$\sum_{i=1}^{q} a_i y_{k-i+1} = \sum_{j=1}^{p} b_j x_{k-j+1} + e_k, \tag{5.2}$$

where $a_i, i = 1, 2, \ldots, q$ and $b_j, j = 1, 2, \ldots, p$ are the sets of parameters that describe the model structure, x_k and y_k are the input and output to the model at any time k for $k = 1, 2, \ldots$, and e_k is the noise value at time k. Without loss of generality, it is always assumed that $a_l = 1$.

Once the model parameters a_i and b_j are known, calculation of y_k for an arbitrary k can be accomplished by

$$y_k = -\sum_{i=2}^{q} a_i y_{k-i+1} + \sum_{j=1}^{p} b_j x_{k-j+1} + e_k. \tag{5.3}$$

It is noted that when all the a_i in Eq. 5.3 take the value of 0, the model is reduced to the MA $(p,0)$ model or simply MA(p). When all of the b_j take the value of 0, the model is reduced to the AR$(0,q)$ model or AR(q). In the latter case, y_k is calculated by

$$y_k = -\sum_{i=2}^{q} a_i y_{k-i+1} + e_k. \tag{5.4}$$

5.1.1.1 Model Structure Selection Criteria Two techniques, known as *Akaike's final prediction error* (FPE) criterion and the closely related *Akaike information-theoretic criterion* (AIC), may be used to aid in the selection of model structure. According to Akaike's theory, in the set of candidate models, the one with the smallest values of FPE or AIC should be chosen. The FPE is calculated as

$$\text{FPE} = \frac{1 + n/N}{1 - n/N} V, \tag{5.5}$$

where n is the total number of parameters of the model to be estimated, N is the length of the data record, and V is the loss function for the model under consideration. Here, V is defined as

$$V = \sum_{i=1}^{n} e_i^2, \tag{5.6}$$

where e is as defined in Eq. 5.2. The AIC is calculated as

$$\text{AIC} = \log\left[(1 + 2n/N)\, V\right]. \tag{5.7}$$

In the following, an AR(12) model was chosen to characterize SA. This selection was based primarily on Braasch's recommendation [24]. As such, the resulting model should be used with caution before the validity of this model structure assumption is further studied using the above criteria.

5.1.1.2 Frequency-Domain Description The ARMA models can be equivalently described in the frequency domain, which provides further insight into model behavior. Introducing a one-step delay operator Z^{-1}, Eq. 5.2 can be rewritten as

$$A\left(Z^{-1}\right) y_k = B\left(Z^{-1}\right) x_k + e_k, \tag{5.8}$$

where

$$A(Z^{-1}) = \sum_{i=1}^{q} a_i Z^{-i+l}, \tag{5.9}$$

$$B(Z^{-1}) = \sum_{i=1}^{p} b_i Z^{i+1}, \tag{5.10}$$

and

$$Z^{-1} y_k = y_{k-1}. \tag{5.11}$$

It is noted that $A(Z^{-1})$ and $B(Z^{-1})$ are polynomials of the timeshift operator Z^{-1} and normal arithmetic operations may be carried out under certain conditions. Defining a new function $H(Z^{-1})$ as $B(Z^{-1})$ divided by $A(Z^{-1})$ and expanding the resulting $H(Z^{-1})$ in terms of operator Z^{-1}, we have

$$H\left(Z^{-1}\right) = \frac{B\left(Z^{-1}\right)}{A\left(Z^{-1}\right)} = \sum_{i=1}^{\infty} h_i Z^{i+1}. \tag{5.12}$$

The numbers of $\{h_i\}$ are the impulse responses of the model. It can be shown that h_i is the output of the ARMA model at time $i = 1, 2, \ldots$ when the model input x_i takes the value of zero at all times except for $i = 1$. The function $H(Z^{-1})$ is called the *frequency function* of the system. By evaluating its value for $Z^{-1} = e^{j\omega}$, the frequency response of the model can be calculated directly. Note that this process is a direct application of the definition of the discrete Fourier transform (DFT) of h_i.

5.1.1.3 AR Model Parameter Estimation The parameters of an AR model with structure

$$A\left(Z^{-1}\right) y_k = e_k, \tag{5.13}$$

may be estimated using the least-squares (LS) method. If we rewrite Eq. 5.13 in matrix format for $k = q, q+1, \ldots, n$, we get

$$\begin{bmatrix} y_n & y_{n-l} & \cdots & y_{n-q+1} \\ y_{n-l} & y_{n-2} & \cdots & y_{n-q} \\ \vdots & \vdots & \vdots & \vdots \\ y_q & y_{q-1} & \cdots & y_1 \end{bmatrix} \begin{bmatrix} a_1 \\ a_2 \\ \vdots \\ a_q \end{bmatrix} = \begin{bmatrix} e_n \\ e_{n-1} \\ \vdots \\ e_q \end{bmatrix}, \tag{5.14}$$

or

$$H \cdot A = E, \tag{5.15}$$

where

$$H = \begin{bmatrix} y_n & y_{n-l} & \cdots & y_{n-q+1} \\ y_{n-l} & y_{n-2} & \cdots & y_{n-q} \\ \vdots & \vdots & \vdots & \vdots \\ y_q & y_{q-1} & \cdots & y_1 \end{bmatrix}, \tag{5.16}$$

$$A = \left[a_1 a_2 a_3 \cdots a_q\right]^T, \tag{5.17}$$

and

$$E = [e_n \ e_{n-1} \ e_{n-2} \ \cdots \ e_q]^T. \tag{5.18}$$

The LS estimation of the parameter matrix A can then be obtained by

$$A = \left(H^T H\right)^{-1} H^T E. \tag{5.19}$$

5.1.2 Collection of SA Data

To build effective SA models, samples of true SA data must be available. This requirement cannot be met directly as the mechanism of SA generation and the actual SA waveform are classified. The approach we take is to extract SA from flight test data. National Satellite Test Bed (NSTB) flight tests recorded the pseudorange measurements at all 10 RMS (reference monitoring station) locations. These pseudorange measurements contain various clock, propagation, and receiver measurement errors, and they can, in general, be described as

$$\text{PR}_M = \rho + \Delta T_{\text{sat}} + \Delta T_{\text{rcvr}} + \Delta T_{\text{iono}} + \Delta T_{\text{trop}} + \Delta T_{\text{multipath}} + \text{SA} + \Delta t_{\text{noise}}, \tag{5.20}$$

where ρ is the true distance between the GPS satellite and the RMS receiver; ΔT_{sat} and ΔT_{rcvr} are the satellite and receiver clock errors; ΔT_{iono} and ΔT_{trop} are the ionosphere and troposphere propagation delays, $\Delta T_{\text{multipath}}$ is the multipath error; SA is the SA error; and Δt_{noise} is the receiver measurement noise.

To best extract SA from PR_M, values of the other terms were estimated. The true distance ρ is calculated by knowing the RMS receiver location and the precise orbit data available from the National Geodetic Survey (NGS) bulletin board. GIPSY[1]/OASIS analysis (GOA) was used for this calculation, which recreated the precise orbit and converted all relevant data into the same coordinate system. Models for propagation and satellite clock errors have been built into GOA, and these were used to estimate ΔT_{sat}, ΔT_{iono}, and ΔT_{trop}. The receiver clock errors were estimated by the NSTB algorithm using data generated from GOA

[1]GPS Positioning System.

for the given flight test conditions. From these, a simulated pseudorange PR_{sim} was formed

$$PR_{sim} = \rho_{sim} + \Delta T_{sat_{sim}} + \Delta T_{rcvr_{sim}} + \Delta T_{iono_{sim}} + \Delta T_{trop_{sim}}, \tag{5.21}$$

where $\Delta T_{sat_{sim}}$, $\Delta T_{rcvr_{sim}}$, $\Delta T_{iono_{sim}}$, and $\Delta T_{trop_{sim}}$ are, respectively, the estimated values of ΔT_{sat}, ΔT_{rcvr}, ΔT_{iono}, and ΔT_{trop} in the simulation.

From Eqs. 5.20 and 5.21, pseudorange residuals are calculated

$$\Delta PR = PR_M - PR_{sim} = SA + \Delta T_{multipath} + \Delta t_{noise} + \Delta T_{models}, \tag{5.22}$$

where ΔT_{models} stands for the total modeling error, given by

$$\begin{aligned} \Delta T_{models} = (\rho - \rho_{sim}) + \left(\Delta T_{sat} - \Delta T_{sat_{sim}}\right) + \left(\Delta T_{rcvr} - \Delta T_{rcvr_{sim}}\right) \\ + \left(\Delta T_{iono} - \Delta T_{iono_{sim}}\right) + \left(\Delta T_{trop} - \Delta T_{trop_{sim}}\right). \end{aligned} \tag{5.23}$$

It is noted that the terms $\Delta T_{multipath}$ and Δt_{noise} should be significantly smaller than SA, although it is not possible to estimate their values precisely. The term ΔT_{models} should also be negligible compared to SA. It is, therefore, reasonable to use ΔPR as an approximation to the actual SA term to estimate SA models. Examination of all available data show that their values vary between ±80 m. These are consistent with previous reports on observed SA and with the DoD's specification of SPS accuracy.

5.2 IONOSPHERIC PROPAGATION ERRORS

The ionosphere, which extends from approximately 50 to 1000 km above the surface of the earth, consists of gases that have been ionized by solar radiation. The ionization produces clouds of free electrons that act as a dispersive medium for GPS signals in which propagation velocity is a function of frequency. A particular location within the ionosphere is alternately illuminated by the sun and shadowed from the sun by the earth in a daily cycle; consequently the characteristics of the ionosphere exhibit a diurnal variation in which the ionization is usually maximum late in midafternoon and minimum a few hours after midnight. Additional variations result from changes in solar activity.

The primary effect of the ionosphere on GPS signals is to change the signal propagation speed as compared to that of free space. A curious fact is that the signal modulation (the code and data stream) is delayed, while the carrier phase is advanced by the same amount. Thus the measured pseudorange using the code is larger than the correct value, while that using the carrier phase is equally smaller. The magnitude of either error is directly proportional to the total electron content (TEC) in a tube of 1 m^2 cross section along the propagation path. The TEC varies spatially, due to spatial nonhomogeneity of the ionosphere. Temporal variations are caused not only by ionospheric dynamics but also by rapid changes in the propagation path due to satellite motion. The path delay for a satellite at zenith typically varies from about 1 m at night to 5–15 m during late afternoon. At low elevation angles the propagation path through the ionosphere is much longer, so the corresponding delays can increase to several meters at night and as much as 50 m during the day.

Since ionospheric error is usually greater at low elevation angles, the impact of these errors could be reduced by not using measurements from satellites below a certain elevation mask angle. However, in difficult signal environments, including blockage of some satellites by obstacles, the user may be forced to use low-elevation satellites. Mask angles of 5°–7.5° offer a good compromise between the loss of measurements and the likelihood of large ionospheric errors.

The L_1-only receivers in nondifferential operation can reduce ionospheric pseudorange error by using a model of the ionosphere broadcast by the satellites, which reduces the uncompensated ionospheric delay by about 50% on the average. During the day errors as large as 10 m at midlatitudes can still exist after compensation with this model and can be much worse with increased solar activity. Other recently developed models offer somewhat better performance. However, they still do not handle adequately the daily variability of the TEC, which can depart from the modeled value by 25% or more.

The L_1/L_2 receivers in nondifferential operation can take advantage of the dependence of delay on frequency to remove most of the ionospheric error. A relatively simple analysis shows that the group delay varies inversely as the square of the carrier frequency. This can be seen from the following model of the code pseudorange measurements at the L_1 and L_2 frequencies:

$$\rho_i = \rho \pm \frac{k}{f_i^2}, \tag{5.24}$$

where ρ is the error-free pseudorange, ρ_i is the measured pseudorange, and k is a constant that depends on the TEC along the propagation path. The subscript $i = 1, 2$ identifies the measurement at the L_1 or L_2 frequencies, respectively, and the plus or minus sign is identified with respective code and carrier phase pseudorange measurements. The two equations can be solved for both ρ and k. The solution for ρ for code pseudorange measurements is

$$\rho = \frac{f_1^2}{f_1^2 - f_2^2}\rho_1 - \frac{f_2^2}{f_1^2 - f_2^2}\rho_2, \tag{5.25}$$

where f_1 and f_2 are the L_1 and L_2 carrier frequencies, respectively, and ρ_1 and ρ_2 are the corresponding pseudorange measurements.

An equation similar to Eq. 5.25 can be obtained for carrier phase pseudorange measurements. However, in nondifferential operation the residual carrier phase pseudorange error can be greater than either an L_1 or L_2 carrier wavelength, making ambiguity resolution difficult.

With differential operation ionospheric errors can be nearly eliminated in many applications, because ionospheric errors tend to be highly correlated when the base and roving stations are in sufficiently close proximity. With two L_1-only receivers separated by 25 km, the unmodeled differential ionospheric error is typically at the 10–20-cm level. At 100 km separation this can increase to as much as a meter. Additional error reduction using an ionospheric model can further reduce these errors by 25–50%.

5.2.1 Ionospheric Delay Model

J. A. Klobuchar's model [54, 111] for ionospheric delay in seconds is given by

$$T_g = \mathrm{DC} + A\left[1 - \frac{x^2}{2} + \frac{x^4}{24}\right] \quad \text{for } |x| \leq \frac{\pi}{2}, \tag{5.26}$$

where

$$\begin{aligned} x &= \frac{2\pi(t - T_p)}{P}\,\mathrm{rad} \\ \mathrm{DC} &= 5\ \text{ns (constant offset)} \\ T_p &= \text{phase} \\ &= 50{,}400\ \text{s} \\ A &= \text{amplitude} \\ P &= \text{period} \\ t &= \text{local time of the earth subpoint of the signal} \\ &\quad\ \text{intersection with mean ionospheric height (s)} \end{aligned}$$

The algorithm assumes this latter height to be 350 km. The DC and phasing T_p are held constant at 5 ns and 14 h (50,400 s) local time.

Amplitude (A) and period (P) are modeled as third-order polynomials:

$$A = \sum_{n=0}^{3} \alpha_n \phi_m^n \quad \text{(s)},$$

$$P = \sum_{n=0}^{3} \beta_n \phi_m^n \quad \text{(s)},$$

where ϕ_m is the geomagnetic latitude of the ionospheric subpoint and α_n,β_n are coefficients selected (from 370 such sets of constants) by the GPS master control station and placed in the satellite navigation upload message for downlink to the user.

For Southbury, Connecticut, we obtain

$$\alpha_n = \left[0.8382 \times 10^{-8}, -0.745 \times 10^{-8}, \ -0.596 \times 10^{-7}, 0.596 \times 10^{-7}\right],$$

$$\beta_n = \left[0.8806 \times 10^{5}, -0.3277 \times 10^{5}, \ -0.1966 \times 10^{6}, 0.1966 \times 10^{6}\right].$$

The parameter ϕ_m is calculated as follows:

1. Subtended earth angle (EA) between user and satellite is given by the approximation

$$\text{EA} \approx \left(\frac{445}{\text{el} + 20}\right)^{-4} \quad \text{(deg)}$$

where el is the elevation of the satellite and with respect to the user equals 15.5°.

2. Geodetic latitude (lat) and longitude (long) of the ionospheric subpoint are found using the approximations

$$\text{Iono lat } \phi_I = \phi_{\text{user}} + \text{EA} \cos \text{AZ} \quad \text{(deg)},$$

$$\text{Iono long } \lambda_I = \lambda_{\text{user}} + \frac{\text{EA} \cos \text{AZ}}{\cos \phi_I} \quad \text{(deg)},$$

where ϕ_{user} is geodetic latitude $= 41^\circ$, λ_{user} is geodetic longitude $= -73^\circ$, and AZ is azimuth of the satellite with respect to the user $= 112.5^\circ$.

3. The geodetic latitude is converted to a geomagnetic coordinate system using the approximation

$$\phi_m \approx \phi_I + 11.6^\circ \cos\left(\lambda_I - 291^\circ\right) \quad \text{(deg)}$$

4. The final step in the algorithm is to account for elevation angle effect by scaling with an obliquity scale factor (SF):

$$\text{SF} = 1 + 2\left[\frac{96^\circ - \text{el}}{90^\circ}\right]^3 \quad \text{(unitless)}.$$

With scaling, time delay due to ionospheric becomes

$$T_g = \begin{cases} \text{SF(DC)} + A\left(1 - \frac{x^2}{2} + \frac{x^4}{24}\right) & |x| < \frac{\pi}{2}, \\ \text{SF(DC)}, & |x| \geq \frac{\pi}{2}, \end{cases}$$

$$T_G = C T_g$$

$$C = \text{ speed of light}$$

$$t = \frac{\lambda_I}{15} + \text{UTC},$$

where T_g is in seconds and T_G is in meters.

The MATLAB programs `Klobuchar fix.m` and `Klobuchar(PRN)` for computing ionospheric delay (for PRN = satellite number) are described in Appendix A.

5.2.2 GPS Ionospheric Algorithms

The ionospheric correction computation algorithms (ICC) enable the computation of the ionospheric delays applicable to a signal on L_1 and to the GPS and WRS (Wide-Area Reference Station) L_1 and L_2 interfrequency biases. These algorithms also calculate GIVEs (grid Ionospheric vertical errors), empirically derived error bounds for the broadcast ionospheric corrections. The ionospheric delays are employed by the SBAS user to correct the L_1 measurements, as well as internally to correct the WRSs' L_1 GEO measurement for orbit determination if dual-frequency corrections are not available from GEOs. The interfrequency biases are needed internally to convert the dual-frequency-derived SBAS corrections to single-frequency corrections for the SBAS users. The vertical ionospheric delay and GIVE information is broadcast to the SBAS user via Message Types 18 and 26. See MOPs for details on the content and usage of the SBAS messages [167].

The algorithms used to compute ionospheric delays and interfrequency biases are based on those originated at the Jet Propulsion Laboratory [130]. The ICC models assume that ionospheric electron density is concentrated on a thin shell of height 350 km above the mean earth surface. The estimates of interfrequency biases and ionospheric delays are derived using a pair of Kalman filters, herein referred to as the L_1L_2 and *iono filters*. The purpose of the L_1 L_2 filter is to estimate the interfrequency biases, while the purpose of the IONO filter is to estimate the ionospheric delays. The inputs to both filters are leveled WRS receiver slant delay measurements (L_2 minus L_1 differential delay), which are output from the data. Both filters perform their calculations in total electron count units (TECU) (1m of L_1 ranging delay = 6.16 TECU, and 1 m of $L_1 - L_2$ differential delay = 9.52 TECU). Conceptually, the measurement equation is (neglecting the noise term):

$$\tau_{\text{TECU}} = 9.52 \times \tau_m \tag{5.27}$$

$$= 9.52 \times \left(t_{L_2,\,m} - t_{L_1,\,m}\right) \tag{5.28}$$

$$= 9.52 \times \left(b^r_m + b^s_m\right) + \text{TEC}_{\text{TECU}} \tag{5.29}$$

$$= b^r_{\text{TECU}} + b^s_{\text{TECU}} + \text{TEC}_{\text{TECU}}, \tag{5.30}$$

where τ is differential delay, b^r and b^s are the interfrequency biases of the respective receiver and satellite, and TEC is the ionospheric delay. The subscripts m (meters) and TECU denote the corresponding units of each term. The ionospheric delay in meters for a signal on the L_1 frequency is

$$\tau^{L_1}_m = 1.5457 \times \frac{1}{9.52}\text{TEC}_{\text{TECU}} \tag{5.31}$$

$$= \frac{1}{6.16}\text{TEC}_{\text{TECU}}. \tag{5.32}$$

Both Kalman filters contain the vertical delays at the vertices of a triangular spherical grid of height 350 km fixed in the solar-magnetic coordinate frame as states. The L_1 L_2 filter also contains interfrequency biases as states. In contrast, the IONO filter does not estimate the interfrequency biases, but instead they are periodically forwarded to the IONO filter, along with the variances of the estimates, from the L_1 L_2 filter. Each slant measurement is modeled as a linear combination of the vertical delays at the three vertices surrounding the corresponding measurement pierce point (the intersection of the line of sight and the spherical grid), plus the sum of the receiver and satellite biases, plus noise. The ionospheric delays computed in the IONO filter are eventually transformed to a latitude–longitude grid that is sent to the SBAS users via Message Type 26. Because SBAS does not have any calibrated ground receivers, the interfrequency bias estimates are all relative to a single receiver designated as a reference, whose L_1 L_2 interfrequency bias filter covariance is initialized to a small value, and to which no process noise is applied.

The major algorithms making up the ICC discussed here are

- *Initialization*—the L_1 L_2 and IONO filters are initialized using either the Klobuchar model or using previously recorded data.
- *Estimation*—the actual computation of the interfrequency biases and ionospheric delays involves both the L_1 L_2 and IONO filters.
- *Thread switch*—the measurements from a WRS may come from an alternate WRS receiver. In this case, the ICC must compensate for the switch by altering the value of the respective receiver's interfrequency bias state in the L_1 L_2 filter. In the nominal case, an estimate of the L_1 L_2 bias difference is available.
- *Anomaly processing*—the L_1 L_2 filter contains a capability to internally detect when a bias estimate is erroneous. Both thread switch and anomaly processing algorithms may also result in the change of the reference receiver.

5.2.2.1 L_1L_2 Receiver and Satellite Bias and Ionospheric Delay Estimations

System Model The ionospheric delay estimation Kalman filter uses a random-walk system model. A state of the Kalman filter at time t_k is modeled to be equal to that state at the previous time t_{k-1}, plus a random process noise representing the uncertainty in the transition from time t_{k-1} to time t_k; that is

$$\mathbf{x}_k = \mathbf{x}_{k-1} + \mathbf{w}_k ,$$

where $\mathbf{x}_k$ is the state vector of the Kalman filter at time t_k and $\mathbf{w}_k$ is a white process noise vector with known covariance Q. The state vector $\mathbf{x}_k$ consists of three subgroups of states: the ionospheric vertical delays at triangular tile vertices, the satellite L_1L_2 biases, and the receiver L_1/L_2 biases; that is

$$\mathbf{x}_k = \begin{bmatrix} x_{1,k} \\ \vdots \\ x_{NV,k} \\ x_{NV+1,k} \\ \vdots \\ x_{NV+NS,k} \\ x_{NV+NS+1,k} \\ \vdots \\ x_{NV+NS+NR,k} \end{bmatrix},$$

where NV is the number of triangular tile vertices, NS is the number of GPS satellites, and NR is the number of WRSs. The values of NV, NS, and NR must be adjusted to fit the desired configuration. In simulations, one can use 24 GPS satellites in the real orbits generated by GIPSY using ephemeris data downloaded from the GPS bulletin board. The number of WRSs is 25 and these WRSs are placed at locations planned for SBAS operations.

Observation Model The observation model or measurement equation establishes the relationship between a measurement and the Kalman filter state vector. For any GPS satellite in view, there is an ionospheric slant delay measurement corresponding to each WRS–satellite pair. Ionospheric slant delay measurement is converted to the vertical delay at its corresponding pierce point through an obliquity factor. At any time t_k, there are approximately 80–200 pierce points and hence the same number of ionospheric vertical delay measurements that can be used to update the Kalman filter state vector.

Denote the ionospheric vertical delay measurement at t_k for the ith satellite and jth WRS as z_{ijk}. Thus

$$z_{ijk} = i_{ijk} + \frac{b_{si}}{q_{ijk}} + \frac{b_{sj}}{q_{ijk}} + v_{ijk}$$

where i_{ijk} is the vertical ionospheric delay at the piece point corresponding to satellite i and WRS j, b_{si} and b_{sj} are the L_1/L_2 interfrequency biases for satellite i and WRS j, respectively, q_{ijk} is the obliquity factor, and v_{ijk} is the receiver measurement noise, white with covariance R.

To establish an observation model, we need to relate i_{ijk} , b_{si}, and b_{sj} to the state vector of the ionospheric delay estimation Kalman filter. Note that b_{si} and b_{sj} are the elements of the state vector labeled $NV + i$ and $NV + NS + j$, respectively. The relationship between i_{ijk} and the state vector is established below. The value i_{ijk} is modeled as a linear combination of the vertical delay values at the three vertices of the triangular tile in which the piece point is located, as shown in Fig. 5.3.

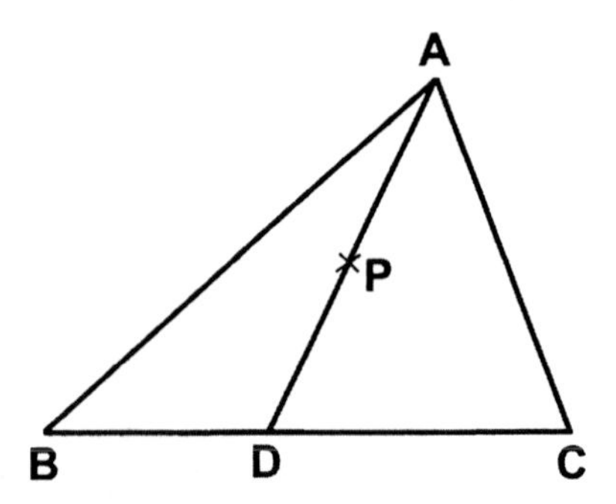

Fig. 5.3 Bilinear interpolation.

In Fig. 5.3, assume a pierce point P is located arbitrarily in the triangular tile ABC. The ionospheric delay at pierce point P is obtained from the vertical delay values at vertices A, B, and C using a bilinear interpolation as follows. Draw a line from point A to point P and find the intersection point D between this line and the line BC. The bilinear interpolation involves two simple linear interpolations—the first yields the vertical delay value at point D from points B and C; the second yields the vertical delay value at point P from points D and A. The result can be summarized as

$$I_P = w_A I_A + w_B I_B + w_C I_C,$$

where I_P, I_A, I_B, and I_C are the ionospheric vertical delay values at points P, A, B, and C, respectively, and w_A, w_B, and w_C are the bilinear weighting coefficients from points A, B, and C, respectively, to point P. The values of w_A, w_B, and w_C can be readily calculated from the geometry involved. It is recognized that I_A, I_B, and I_C are three elements of the Kalman filter state vector. In summary, the measurement equation can be written as

$$z_{ijk} = \mathbf{h}_{ijk}\, \mathbf{x}_k + v_{ijk},$$

where h_{ijk} is the measurement matrix and v_{ijk} is the measurement noise, respectively, for the pierce point measurement for the satellite with index i and WRS with index j at time t_k . Here, $\mathbf{h}_{ijk}$ is an $(NV + NS + NR)$ dimension row vector with all elements equal to zeros except five elements. The first three of these five nonzero elements correspond to the vertices of the tile that contains the pierce point under consideration, and the other two correspond to the ith satellite and jth WRS, which yields the ionospheric slant delay measurement z_{ijk}.

UDU^T Kalman Filter (See Chapter 8) As noted previously, there are approximately 180–200 pierce points at any time t_k . Each pierce point corresponds to one of the possible combinations of a satellite and a WRS, which further corresponds to an ionospheric vertical delay measurement at that pierce point. The

ionospheric estimation Kalman filter is designed so that its state vector is updated upon the reception of each ionospheric vertical delay measurement.

SM (Solar Magnetic)-to-ECEF Transformation At the end of each 5-min interval (Kalman filter cycle), the ionospheric vertical delays at the vertices of all tiles are converted from the SM coordinates to the ECEF coordinates. This conversion is completed by first transforming the SBAS IGPs from the ECEF coordinates to the SM coordinates. For each IGP converted to SM coordinates, the triangular tile which contains this IGP is found. A bilinear interpolation identical to the one described in Fig. 5.3 is then used to calculate the ionospheric vertical delay values at this IGP. (Transformations are given in Appendix C.)

In new GEOs (3rd, PRN 135, at 133° longitude; 4th, PRN 138, at 107° longitude) will have L_1/L_5 frequencies (see Chapter 6). Ionospheric delays can be calculated at the WRSs directly instead of using ionospheric delay provided by ionospheric grids from SBAS broadcast messages.

5.2.2.2 Kalman Filter In estimating the ionospheric vertical delays in the SM coordinate system by the Kalman filter, there are three types of estimation errors:

1. Estimation error due to ionospheric slant delay measurement noise error
2. Estimation error due to the temporal variation of the ionosphere
3. Estimation error due to nonlinear spatial variation of the ionosphere

Each of the three sources of error can be individually minimized by adjusting the values of the covariances **Q** and **R**. However, the requirements to minimize the errors due to noise and temporal variations are often in conflict.

Intuitively, to minimize the measurement noise implies that we want the **Q** and **R** values to result in a Kalman gain that averages out the measurement noise. That is, we want the Kalman gain to take values so that for each new measurement, the value of innovation is small, such that a relatively large noise component of the measurement results in a relatively small estimation error. On the other hand, if we want to minimize the estimation error due to temporal variations, then we want to have a Kalman gain that can produce a large innovation, so that the component in the measurement that represents the actual ionospheric delay variation with time can be quickly reflected in the new state estimate. This suggests that we usually need to compromise in selecting the values of **Q** and **R** when a conventional nonadaptive Kalman filter is used.

Although the Kalman filter estimation error is the dominant source of error, it is not the only source. The nonlinear spatial variation introduces additional error when converting the ionospheric vertical delay estimated by the Kalman filter in the SM coordinate system to the SBAS IGP in the ECEF coordinate system. This is because bilinear interpolation is used and there is an implicit assumption that interpolation is a strictly valid procedure. However, if the actual value of the vertical delay was measured at some location, it would not be equal to the value found by interpolation. Violation of this assumption results in interpolation error during the transformation. It can be shown by simulations that, under certain conditions, this conversion error can be significant and non-negligible.

In order to isolate the sources of errors and understand how the algorithm responds to various conditions, consider seven scenarios, with each testing one aspect of the possible estimation error, and all their possible combinations.

Scenario 1: Measurement Noise. In this scenario, the ionospheric vertical delay is assumed to be a time-invariant constant anywhere over the earth's surface. The Kalman filter estimation errors due to temporal and spatial variations are zero. For each of the ionospheric slant delay measurements, a zero mean white Gaussian noise is added. The magnitude of the noise is characterized by its variance. The measurement noise is added to the slant delay rather than the vertical delay because this is where the actual measurement noise is introduced by a GPS receiver.

Scenario 2: Temporal Variation. In this scenario, the ionosphere is assumed to be uniformly distributed spatially, but its TEC values change with time; that is, the ionospheric vertical delays vary with time, but these variations are identical everywhere. Various time variation functions, such as a sinusoidal function, a linear ramp, a step function, or an impulse function, can be used to study this scenario. In a simulation using a sinusoidal time variation function, the sinusoidal function is characterized by two parameters—its amplitude and frequency. The values of these two parameters are chosen to produce a time variation that is similar in magnitude to the ionospheric delay variation data published in the literature. The measurement noise is zero. Kalman filter estimation errors due to both the measurement noise and spatial variation are fixed at zero (for this scenario).

Scenario 3: Spatial variation. In this scenario, the ionosphere is assumed to be a constant at any fixed location when observed in the SM coordinate system. The ionospheric delays at different locations in the SM coordinate system, however, are different. Various spatial variation functions can be used to study this scenario. Here, we use a three-dimensional surface constructed from two orthogonal sinusoidal functions of varying amplitude and frequency to model the values of ionospheric vertical delays over the earth. The values of the parameters of the two sinusoidal functions are chosen to produce gradients in TEC similar in magnitude to the ionospheric delay variation data published in the literature. The measurement noise is zero. Kalman filter estimation errors due to both the measurement noise and temporal variations are fixed at zero for this scenario.

Scenario 4: Noise + Temporal. Scenarios 1 and 2 are combined, and the Kalman filter estimation error due to spatial variation is zero.

Scenario 5: Noise + Spatial. Scenarios 1 and 3 are combined. In this scenario, the Kalman filter estimation error due to temporal variation is zero.

Scenario 6: Temporal + Spatial. Here, the Kalman filter estimation error due to measurement noise is zero. The combined values of temporal and spatial variations define the "truth ionosphere" in the simulation.

Scenario 7: Noise + Temporal + Spatial. In this scenario, the parameters that define the "true ionosphere" and "measurement noise" can be configured to mimic any ionospheric conditions.

In the simulations, the GPS satellite orbits used are the precise orbits generated by GIPSY using GPS satellite ephemeris data downloaded from the GPS bulletin board. The WRS locations used are those currently recommended by the FAA. These locations may be adjusted to evaluate the impact of other WRS locations or additions WRSs.

***5.2.2.3 Selection of* Q *and* R** Theoretically, a Kalman filter yields optimal estimation of the states of a system, given a knowledge of the system dynamics and measurement equations, when both the system process noise and measurement noise are zero-mean Gaussian at each epoch and white in time and their variances are known. However, in practice, the system dynamics are often unknown and system modeling errors are introduced when the actual system dynamics differ from the assumptions. In addition, the system process noise and the measurement noise are often non-Gaussian and their variances are not known precisely. To ensure a stable solution, a relatively large value of Q is often used, sacrificing estimation accuracy. Careful selection of Q and R values impacts the performance of the Kalman filter in practical applications, including the SBAS ionospheric estimation filter.

In each phase of the validation, many parameters are tuned. The procedures and rationale involved in selecting the final values of these parameters include an effort to distinguish those parameters for which the performance is particularly sensitive. For many parameters, performance is not particularly sensitive. Table 5.1 shows typical values of the parameters used in two Kalman filters. The L_1 L_2 filter can be eliminated and thereby use the IONO filter, including the satellite and receiver biases, may be sufficient to estimate the biases and IONO delays. This reduces the computational load and simplifies the process.

The algorithms must be validated to ensure that the estimation accuracy is good enough to ultimately support downstream precision-approach requirements. Convergence properties of the estimation algorithms must be examined, and the logic associated with restarting the estimation using recorded data must be analyzed. The capabilities to perform thread switches and detect anomalies must be examined, and the special cases necessitating a change of reference receiver. In each phase of validation, the critical test is whether there is any significant degradation in accuracy as compared with nominal performance, and whether the nominal performance itself is adequate.

5.2.2.4 Calculation of Ionospheric Delay Using Pseudoranges The problem of calculating ionospheric propagation delay from P-code and C/A-code can be

TABLE 5.1. Representative Kalman Filter Parameter Values

Parameter Term	Value	Units
L_1L_2 filter bias process noise update interval	300	s
L_1L_2 filter TEC process noise	0.05	TECU/s$^{1/2}$
L_1L_2 filter TEC process noise update interval	300	s
Iono filter process noise	0.05	TECU/s$^{1/2}$
Iono filter process noise update interval	300	s
Iono meas floor	9	TECU^2
Iono meas scale	0	
L_1L_2 filter bias process noise	4.25×10^{-4}	TECU/s$^{1/2}$
L_1L_2 next bias distribution time interval	300	s
L_1L_2 cold start bias distribution time interval	300	s
L_1L_2 cold start time interval	86,400	s
Iono a priori covariance matrix	$400 = 20^2$	TECU^2
L_1L_2bias a priori covariance matrix	$10,000 = 100^2$	TECU^2
(ref receiver)	10^{-10}	TECU^2
Maximum initial TEC	1000	TECU
Nominal initial TEC	25	TECU

formulated in terms of the following measurement equalities:

$$P_{\mathrm{RL1}} = \rho + \mathrm{L}_{1\mathrm{iono}} + c\tau_{\mathrm{RX1}} + c\tau_{\mathrm{GD}}, \tag{5.33}$$

$$P_{\mathrm{RL2}} = \rho + \frac{\mathrm{L}_{1\mathrm{iono}}}{\left(f_{\mathrm{L2}}/f_{\mathrm{L1}}\right)^2} + c\tau_{\mathrm{RX2}} + c\frac{\tau_{\mathrm{GD}}}{\left(f_{\mathrm{L2}}/f_{\mathrm{L1}}\right)^2}, \tag{5.34}$$

where

$$\left.\begin{array}{lcl}
P_{\mathrm{RL1}} & = & \mathrm{L}_1 \text{ pseudorange} \\
P_{\mathrm{RL2}} & = & \mathrm{L}_2 \text{ pseudorange} \\
\rho & = & \text{geometric distance between GPS satellite} \\
 & & \text{transmitter and GPS receiver, including} \\
 & & \text{nondispersive contributions such as} \\
 & & \text{tropospheric refraction and clock drift} \\
f_{\mathrm{L1}} & = & \mathrm{L}_1 \text{ frequency} \\
 & = & 1575.42 \text{ MHz} \\
f_{\mathrm{L2}} & = & \mathrm{L}_2 \text{ frequency} \\
 & = & 1227.6 \text{ MHz} \\
\tau_{\mathrm{RX1}} & = & \text{receiver noise as manifested in code} \\
 & & \text{(receiver and calibration biases) at } \mathrm{L}_1 \text{ (ns)} \\
\tau_{\mathrm{RX2}} & = & \text{receiver noise as manifested in code} \\
 & & \text{(receiver and calibration biases) at } \mathrm{L}_2 \text{ (ns)} \\
\tau_{\mathrm{GD}} & = & \text{satellite group delay (interfrequency bias)} \\
c & = & \text{speed of light} \\
 & = & 0.299792458 \text{ m/ns}
\end{array}\right\}. \tag{5.35}$$

Subtracting Eq. 5.34 from Eq. 5.33, we get

$$L_{1\,\mathrm{iono}} = \frac{P_{\mathrm{RL1}} - P_{\mathrm{RL2}}}{1 - \left(f_{\mathrm{L1}}/f_{\mathrm{L2}}\right)^2} - \frac{c\left(\tau_{\mathrm{RX1}} - \tau_{\mathrm{RX2}}\right)}{1 - \left(f_{\mathrm{L1}}/f_{\mathrm{L2}}\right)^2} - c\tau_{\mathrm{GD}}. \tag{5.36}$$

What is actually measured in the ionospheric delay is the sum of receiver bias and interfrequency bias. The biases are determined and taken out from the ionospheric delay calculation. These biases may be up to 10 ns (3 m) [50, 142].

However, the presence of ambiguities N_1 and N_2 in carrier phase measurements of L_1 and L_2 preclude the possibility of using these in the daytime by themselves. At night, these ambiguities can be calculated from the pseudoranges and carrier phase measurements may be used for ionospheric calculations.

The MATLAB program `Iono_delay(PRN#)` (described in Appendix A) uses pseudorange and carrier phase data from L_1 and L_2 signals.

5.3 TROPOSPHERIC PROPAGATION ERRORS

The lower part of the earth's atmosphere is composed of dry gases and water vapor, which lengthen the propagation path due to refraction. The magnitude of the resulting signal delay depends on the refractive index of the air along the propagation path and typically varies from about 2.5 m in the zenith direction to 10–15 m at low satellite elevation angles. The troposphere is nondispersive at the GPS frequencies, so that delay is not frequency dependent. In contrast to the ionosphere, tropospheric path delay is consequently the same for code and carrier signal components. Therefore, this delay cannot be measured by utilizing both L_1 and L_2 pseudorange measurements, and either models and/or differential positioning must be used to reduce the error.

The refractive index of the troposphere consists of that due to the dry-gas component and the water vapor component, which respectively contribute about 90% and 10% of the total. Knowledge of the temperature, pressure, and humidity along the propagation path can determine the refractivity profile, but such measurements are seldom available to the user. However, using standard atmospheric models for dry delay permits determination of the zenith delay to within about 0.5 m and with an error at other elevation angles that approximately equals the zenith error times the cosecant of the elevation angle. These standard atmospheric models are based on the laws of ideal gases and assume spherical layers of constant refractivity with no temporal variation and an effective atmospheric height of about 40 km. Estimation of dry delay can be improved considerably if surface pressure and temperature measurements are available, bringing the residual error down to within 2–5% of the total.

The component of tropospheric delay due to water vapor (at altitudes up to about 12 km) is much more difficult to model, because there is considerable spatial and temporal variation of water vapor in the atmosphere. Fortunately, the wet delay is only about 10% of the total, with values of 5–30 cm in continental midlatitudes. Despite its variability, an exponential vertical profile model can reduce it to within about 2–5 cm.

In practice, a model of the standard atmosphere at the antenna location would be used to estimate the combined zenith delay due to both wet and dry components. Such models use inputs such as the day of the year and the latitude and altitude of the user. The delay is modeled as the zenith delay multiplied by a factor that is a function of the satellite elevation angle. At zenith, this factor is unity, and it increases with decreasing elevation angle as the length of the propagation path through the troposphere increases. Typical values of the multiplication factor are 2 at 30° elevation angle, 4 at 15°, 6 at 10°, and 10 at 5°. The accuracy of the model decreases at low elevation angles, with decimeter level errors at zenith and about 1 m at 10° elevation.

Much research has gone into the development and testing of various tropospheric models. Excellent summaries of these appear in the literature [84, 96, 177].

Although a GPS receiver cannot measure pseudorange error due to the troposphere, differential operation can usually reduce the error to small values by taking advantage of the high spatial correlation of tropospheric errors at two points within 100–200 km on the earth's surface. However, exceptions often occur when storm fronts pass between the receivers, causing large gradients in temperature, pressure, and humidity.

5.4 THE MULTIPATH PROBLEM

Multipath propagation of the GPS signal is a dominant source of error in differential positioning. Objects in the vicinity of a receiver antenna (notably the ground) can easily reflect GPS signals, resulting in one or more secondary propagation paths. These secondary-path signals, which are superimposed on the desired direct-path signal, always have a longer propagation time and can significantly distort the amplitude and phase of the direct-path signal.

Errors due to multipath cannot be reduced by the use of differential GPS, since they depend on local reflection geometry near each receiver antenna. In a receiver without multipath protection, C/A-code ranging errors of 10 m or more can be experienced. Multipath can not only cause large code ranging errors but also severely degrade the ambiguity resolution process required for carrier phase ranging such as that used in precision surveying applications.

Multipath propagation can be divided into two classes: static and dynamic. For a stationary receiver, the propagation geometry changes slowly as the satellites move across the sky, making the multipath parameters essentially constant for perhaps several minutes. However, in mobile applications there can be rapid fluctuations in fractions of a second. Therefore, different multipath mitigation techniques are generally employed for these two types of multipath environments. Most current research has been focused on static applications, such as surveying, where greater demand for high accuracy exists. For this reason, we will confine our attention to the static case.

5.5 HOW MULTIPATH CAUSES RANGING ERRORS

To facilitate an understanding of how multipath causes ranging errors, several simplifications can be made that in no way obscure the fundamentals involved. We will assume that the receiver processes only the C/A-code and that the received signal has been converted to complex (i.e., analytic) form at baseband (nominally zero frequency), where all Doppler shift has been removed by a carrier tracking phase-lock loop. It is also assumed that the 50-bps (bits per second) GPS data modulation has been removed from the signal, which can be achieved by standard techniques. When no multipath is present, the received waveform is represented by

$$r(t) = ae^{j\phi}c(t - \tau) + n(t), \tag{5.37}$$

where $c(t)$ is the normalized, undelayed C/A-code waveform as transmitted, τ is the signal propagation delay, a is the signal amplitude, ϕ is the carrier phase, and $n(t)$ is Gaussian receiver thermal noise having flat power spectral density. Pseudoranging consists of estimating the delay parameter τ. As we have previously seen, an optimal estimate (i.e., a minimum-variance unbiased estimate) of τ can be obtained by forming the cross-correlation function

$$R(\tau) = \int_{T_1}^{T_2} r(t)c_r(t - \tau)\,dt, \tag{5.38}$$

of $r(t)$ with a replica $c_r(t)$ of the transmitted C/A-code and choosing as the delay estimate that value of τ that maximizes this function. Except for an error due to receiver thermal noise, this occurs when the received and replica waveforms are in time alignment. A typical cross-correlation function without multipath for C/A-code receivers having a 2-MHz precorrelation bandwidth is shown by the solid lines Fig. 5.4 (these plots ignore the effect of noise, which would add small random variations to the curves).

If multipath is present with a single secondary path, the waveform of Eq. 5.37 changes to

$$r(t) = ae^{j\phi_1}c(t - \tau_1) + be^{j\phi_2}c(t - \tau_2) + n(t), \tag{5.39}$$

where the direct and secondary paths have respective propagation delays τ_1 and τ_2, amplitudes a and b, and carrier phases ϕ_1 and ϕ_2. In a receiver not designed expressly to handle multipath, the resulting cross-correlation function will now have two superimposed components, one from the direct path and one from the secondary path. The result is a function with a distortion depending on the relative amplitude, delay, and phase of the secondary-path signal, as illustrated at the top of Fig. 5.4 for an in-phase secondary path and at the bottom of the figure for an out-of-phase secondary path. Most importantly, the location of the peak of the function has been displaced from its correct position, resulting in a pseudorange error.

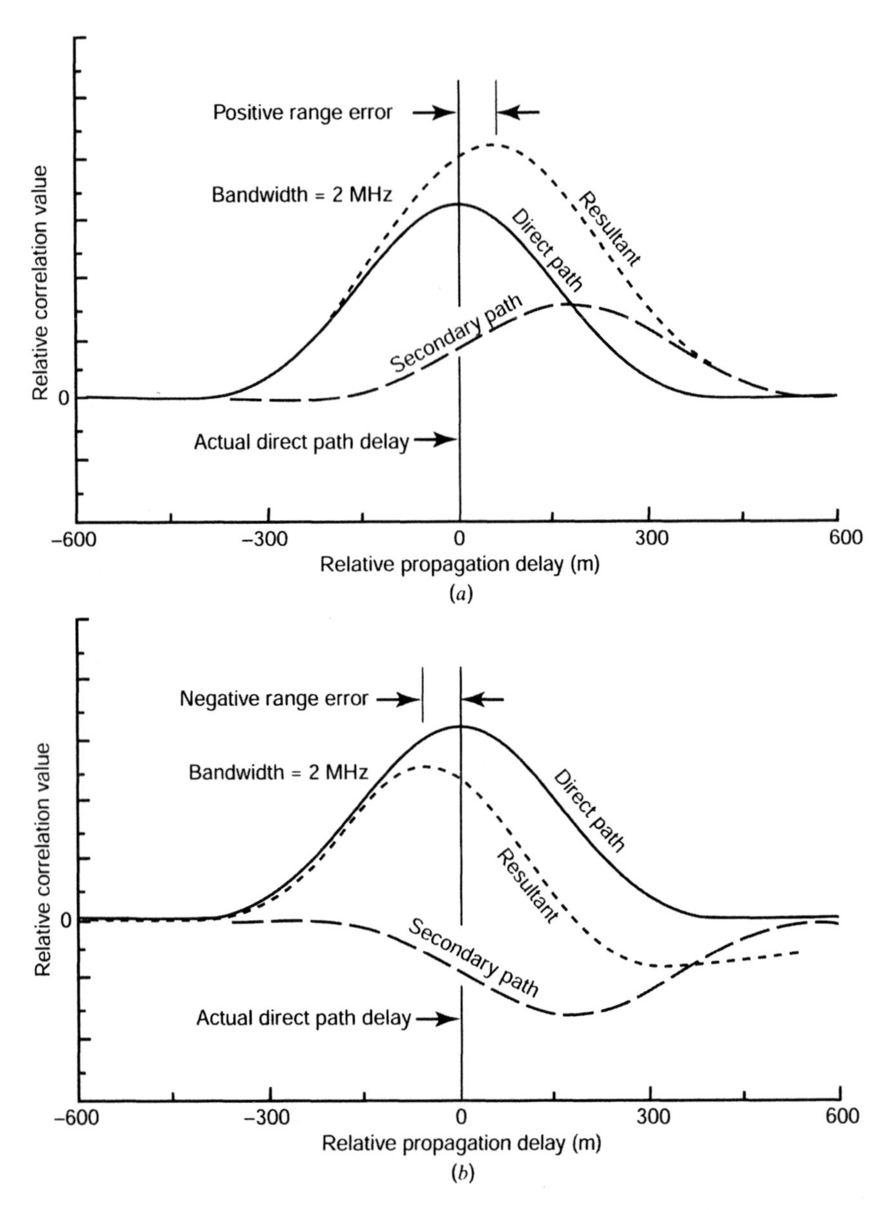

Fig. 5.4 Effect of multipath on C/A-code cross-correlation function.

In vintage receivers employing standard code tracking techniques (early and late codes separated by one C/A-code chip), the magnitude of pseudorange error caused by multipath can be quite large, reaching 70–80 m for a secondary-path signal one-half as large as the direct-path signal and having a relative delay of approximately 250 m. Further details can be found in Ref. 78.

5.6 METHODS OF MULTIPATH MITIGATION

Processing against slowly changing multipath can be broadly separated into two classes: spatial processing and time-domain processing. Spatial processing uses antenna design in combination with known or partially known characteristics of signal propagation geometry to isolate the direct-path received signal. In contrast, time domain processing achieves the same result by operating only on the multipath-corrupted signal within the receiver.

5.6.1 Spatial Processing Techniques

5.6.1.1 Antenna Location Strategy Perhaps the simplest form of spatial processing is to locate the antenna where it is less likely to receive reflected signals. For example, to obtain the position of a point near reflective objects, one can first use GPS to determine the position of a nearby point "in the clear" and then calculate the relative position of the desired point by simple distance and/or angle measurement techniques. Another technique that minimizes ever-present ground signal reflections is to place the receiver antenna directly at ground level. This causes the point of ground reflection to be essentially coincident with the antenna location so that the secondary path has very nearly the same delay as the direct path. Clearly such antenna location strategies may not always be possible but can be very effective when feasible.

5.6.1.2 Groundplane Antennas The most common form of spatial processing is an antenna designed to attenuate signals reflected from the ground. A simple design uses a metallic groundplane disk centered at the base of the antenna to shield the antenna from below. A deficiency of this design is that when the signal wavefronts arrive at the disk edge from below, they induce surface waves on the top of the disk that then travel to the antenna. The surface waves can be eliminated by replacing the groundplane with a *choke ring*, which is essentially a groundplane containing a series of concentric circular troughs one-quarter wavelength deep. These troughs act as transmission lines shorted at the bottom ends so that their top ends exhibit a very high impedance at the GPS carrier frequency. Therefore, induced surface waves cannot form, and signals that arrive from below the horizontal plane are significantly attenuated. However, the size, weight, and cost of a choke-ring antenna is significantly greater than that of simpler designs. Most importantly, the choke ring cannot effectively attenuate secondary-path signals arriving from above the horizontal, such as those reflecting from buildings or other structures. Nevertheless, such antennas have proven to be effective when signal ground bounce is the dominant source of multipath, particularly in GPS surveying applications.

5.6.1.3 Directive Antenna Arrays A more advanced form of spatial processing uses antenna arrays to form a highly directive spatial response pattern with high gain in the direction of the direct-path signal and attenuation in directions from which secondary-path signals arrive. However, inasmuch as signals from different satellites have different directions of arrival and different multipath geometries, many directivity patterns must be simultaneously operative, and each must be capable of adapting to changing geometry as the satellites move across the sky. For these reasons, highly directive arrays seldom are practical or affordable for most applications.

5.6.1.4 Long-Term Signal Observation If a GNSS signal is observed for sizable fractions of an hour to several hours, one can take advantage of changes in multipath geometry caused by satellite motion. This motion causes the relative delays between the direct and secondary paths to change, resulting in measurable variations in the received signal. For example, a periodic change in signal level caused by alternate phase reinforcement and cancellation by the reflected signals is often observable. Although a variety of algorithms have been proposed for extracting the direct-path signal component from measurements of the received signal, the need for long observation times rules out this technique for most applications. However, it can be an effective method of multipath mitigation at a fixed site, such as at a differential GNSS base station. In this case, it is even possible to observe the same satellites from one day to the next, looking for patterns of pseudorange or phase measurements that repeat daily.

Multipath Calculation from Long-Term Observations Delays can be computed as follows by using pseudoranges and carrier phases over long signal observations (one day to next). This technique may be ruled out for most applications. Ambiguities and cycle slips have been eliminated or mitigated.

Let

$$\left.\begin{array}{rcl}
\lambda_1 & = & 19.03 \text{ cm, wavelength of } L_1 \\
\lambda_2 & = & 24.42 \text{ cm, wavelength of } L_2 \\
\phi_{L1} & = & \text{carrier phase for } L_1 \\
\phi_{L2} & = & \text{carrier phase for } L_2 \\
f_{L1} & = & L_1 \text{frequency} = 1575.42 \text{ MHz} \\
f_{L2} & = & L_2 \text{ frequency} = 1227.6 \text{ MHz} \\
\rho & = & \text{geometrical pseudorange} \\
P_{RL1} & = & \text{pseudorange } L_1 \\
P_{RL2} & = & \text{pseudorange } L_2 \\
I & = & \text{ionospheric delay} \\
I_{L1} & = & \text{Ionospheric delay in } L_1 \\
MP_{L1} & = & \text{multipath in } L_1 \\
MP_{L2} & = & \text{multipath in } L_2
\end{array}\right\} . \tag{5.40}$$

For dual-frequency GNSS receivers, one obtains

$$\lambda_1 \phi_{L1} = \rho - \frac{I}{(f_{L1})^2}, \tag{5.41}$$

$$\lambda_2 \phi_{L2} = \rho - \frac{I}{(f_{L2})^2}. \tag{5.42}$$

Subtracting Eq. 5.42 from Eq. 5.41, one can obtain

$$\lambda_1 \phi_{L1} - \lambda_2 \phi_{L2} = \frac{I\,(f_{L1})^2 - I\,(f_{L2})^2}{(f_{L1})^2\,(f_{L2})^2},$$

$$I_{L1} = \frac{(\lambda_1 \phi_{L1} - \lambda_2 \phi_{L2})\,(f_{L2})^2}{(f_{L1})^2 - (f_{L2})^2},$$

$$K = \frac{(f_{L2})^2}{(f_{L1})^2 - (f_{L2})^2},$$

$$I_{L1} = K\,(\lambda_1 \phi_{L1} - \lambda_2 \phi_{L2}), \tag{5.43}$$

$$P_{RL1} = \rho + \frac{I}{(f_{L1})^2}. \tag{5.44}$$

Subtracting Eq. 5.41 from Eq. 5.44, one obtains the multipath as

$$MP_{L1} = P_{RL1} - \lambda_1 \phi_{L1} - 2I_{L1}, \tag{5.45}$$

where

$$I_{L1} = \frac{I}{(f_{L1})^2}.$$

Substitute Eq. 5.43 into Eq. 5.45 to obtain

$$\begin{aligned} MP_{L1} &= P_{RL1} - \lambda_1 \phi_{L1} - 2K\,(\lambda_1 \phi_{L1} - \lambda_2 \phi_{L2}) \\ &= P_{RL1} - [(1 + 2K)\,\lambda_1 \phi_{L1} + 2K\,\lambda_2 \phi_{L2}\,], \end{aligned} \tag{5.46}$$

the multipath solution for L_1.

5.6.2 Time-Domain Processing

Although time-domain processing against GPS multipath errors has been the subject of active research for at least two decades, there is still much to be learned, both at theoretical and practical levels. Most of the practical approaches have been developed by receiver manufacturers, who are often reluctant to explicitly reveal their methods. Nevertheless, enough information about multipath processing exists to gain insight into its recent evolution.

5.6.2.1 Narrow-Correlator Technology (1990–1993) The first significant means to reduce GPS multipath effects by receiver processing made its debut in the early 1990s. Until that time, most receivers had been designed with a 2-MHz precorrelation bandwidth that encompassed most, but not all, of the GPS spread-spectrum signal power. These receivers also used one-chip spacing between the early and late reference C/A-codes in the code tracking loops. However, the 1992 paper [195] makes it clear that using a significantly larger bandwidth combined with much closer spacing of the early and late reference codes would dramatically improve the ranging accuracy both with and without multipath. It is somewhat surprising that these facts were not recognized earlier by the GPS community, given that they had been well known in radar circles for many decades.

A 2-MHz precorrelation bandwidth causes the peak of the direct-path cross-correlation function to be severely rounded, as illustrated in Fig. 5.4. Consequently, the sloping sides of a secondary-path component of the correlation function can significantly shift the location of the peak, as indicated in the figure. The result of using an 8-MHz bandwidth is shown in Fig. 5.5, where it can be noted that the sharper peak of the direct-path cross-correlation function is less easily shifted by the secondary-path component. It can also be shown that at larger bandwidths the sharper peak is more resistant to disturbance by receiver thermal noise, even though the precorrelation signal-to-noise ratio is increased.

Another advantage of a larger precorrelation bandwidth is that the spacing between the early and late reference codes in a code tracking loop can be made smaller without significantly reducing the gain of the loop; hence the term *narrow correlator*. It can be shown that this causes the noises on the early and late

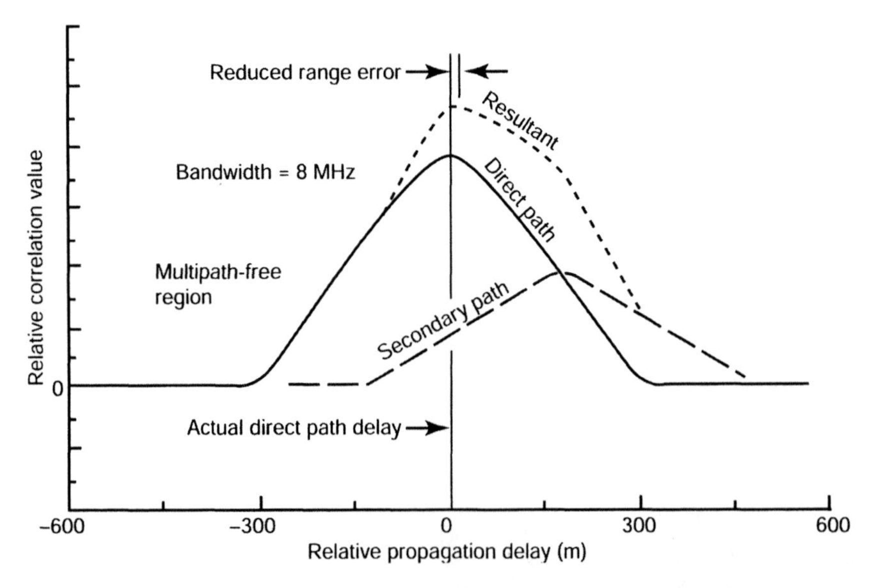

Fig. 5.5 Reduced multipath error with larger precorrelation bandwidth.

correlator outputs to become more highly correlated, resulting in less noise on the loop error signal. An additional benefit is that the code tracking loop will be affected only by the multipath-induced distortions near the peak of the correlation function.

5.6.2.2 ***Leading-Edge Techniques*** Because the direct-path signal always precedes secondary-path signals, the leading (left-hand) portion of the correlation function is uncontaminated by multipath, as is illustrated in Fig. 5.5. Therefore, if one could measure the location of just the leading part, it appears that the direct-path delay could be determined with no error due to multipath. Unfortunately, this seemingly happy state of affairs is illusory. With a small direct-/secondary-path separation, the uncontaminated portion of the correlation function is a minuscule piece at the extreme left, where the curve just begins to rise. In this region, not only is the signal-to-noise ratio relatively poor, but the slope of the curve is also relatively small, which severely degrades the accuracy of delay estimation.

For these reasons, the leading-edge approach best suits situations with a moderate to large direct-/secondary-path separation. However, even in these cases there is the problem of making the delay measurement insensitive to the slope of the correlation function leading edge, which can vary with signal strength. Such a problem does not occur when measuring the location of the correlation function peak.

5.6.2.3 ***Correlation Function Shape-Based Methods*** Some GPS receiver designers have attempted to determine the parameters of the multipath model from the shape of the correlation function. The idea has merit, but for best results many correlations with different values of reference code delay are required to obtain a sampled version of the function shape. Another practical difficulty arises in attempting to map each measured shape into a corresponding direct-path delay estimate. Even in the simple two-path model (Eq. 5.39) there are six signal parameters, so that a very large number of correlation function shapes must be handled. An example of a heuristically developed shape-based approach called the *early—late slope* (ELS) *method* can be found in Ref. 190, while a method based on maximum-likelihood estimation called the multipath-estimating delay-lock loop (MEDLL) is described in Ref. 191.

5.6.2.4 ***Modified Correlator Reference Waveforms*** A relatively new approach to multipath mitigation alters the waveform of the correlator reference PRN code to provide a cross-correlation function with inherent resistance to errors caused by multipath. Examples include the strobe correlator [58], the use of special code reference waveforms to narrow the correlation function developed in Refs. 202 and 203, and the gated correlator developed in Ref. 138. These techniques take advantage of the fact that the range information in the received signal resides primarily in the chip transitions of the C/A-code. By using a correlator reference waveform that is not responsive to the flat portions of the C/A-code, the resulting correlation function can be narrowed down to the width of a chip transition,

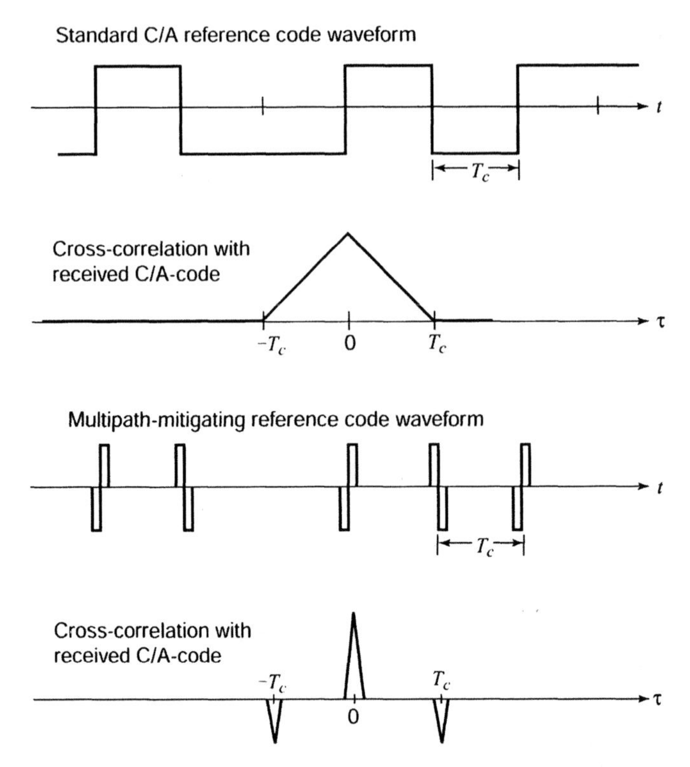

Fig. 5.6 Multipath-mitigating reference code waveform.

thereby being almost immune to multipath having a primary/secondary-path separation greater than 30–40 m. An example of such a reference waveform and the corresponding correlation function are shown in Fig. 5.6.

5.6.3 MMT Technology

The latest approach to time-domain multipath mitigation is called multipath mitigation technology (MMT), and is incorporated a number of GPS receivers manufactured by NovAtel Corporation of Canada. The MMT technique not only reaches theoretical performance limits described in Section 5.7 for both code and carrier phase ranging but also, compared to existing approaches, has the advantage that its performance improves as the signal observation time is lengthened. A description of MMT follows in Section 5.6.3 and also appears in a patent [208].

5.6.3.1 Description MMT is based on maximum-likelihood (ML) estimation. Although the theory of ML estimation is well-developed, its application to GPS multipath mitigation has not been feasible until now, due to the large amount of computation required. However, recent mathematical breakthroughs have solved this problem. Before introducing the MMT algorithm we first briefly describe the process of ML estimation in the context of the multipath problem.

5.6.3.2 Maximum-Likelihood Multipath Estimation Maximum likelihood estimation (MLE) is described in detail in Chapter 8. Its application to multipath mitigation is described below.

5.6.3.3 The Two-Path ML Estimator (MLE) The simplest ML estimator designed for multipath is based on a two-path model (one direct path and one secondary delayed path). For simplicity in describing MMT we consider only this model, although generalization to additional paths is straightforward, and the MMT algorithm can be implemented for such cases. It is assumed that the received signal has been frequency-shifted to baseband, and the navigation data have been stripped off. The two-path signal model is

$$r(t) = A_1 e^{j\phi_1} m(t - \tau_1) + A_2 e^{j\phi_2} m(t - \tau_2) + n(t). \tag{5.47}$$

In this model the parameters A_1, ϕ_1, τ_1 are respectively the direct path signal amplitude, phase, and delay, and the parameters A_2, ϕ_2, and τ_2 are the corresponding parameters for the secondary path. The code modulation is denoted by $m(t)$, and the noise function $n(t)$ is an additive zero-mean complex Gaussian noise process with a flat power spectral density. It will be convenient to group the multipath parameters into the vector

$$\overline{\theta} = [A_1, \phi_1, \tau_1, A_2, \phi_2, \tau_2]. \tag{5.48}$$

Observation of the received signal $r(t)$ is accomplished by sampling it on the time interval $[0,T]$ to produce a complex observed vector $\overline{r}$.

The ML estimate of the multipath parameters is the vector $\hat{\overline{\theta}}$ of parameter values that maximizes the likelihood function $p(\overline{r}\,|\overline{\theta})$, which is the probability density of the received signal vector conditioned on the values of the multipath parameters. In this maximization the vector $\overline{r}$ is held fixed at its observed value. Within the vector $\hat{\overline{\theta}}$ the estimates $\hat{\tau}_1$ and $\hat{\phi}_1$ of direct-path delay and carrier phase are normally the only ones of interest. However, the ML estimate of these parameters requires that the likelihood function $p(\overline{r}\,|\overline{\theta})$ be maximized over the six-dimensional (6D) space of *all* multipath parameters (components of $\overline{\theta}$). For this reason the unwanted parameters are called *nuisance parameters.*

Since the natural logarithm is a strictly increasing function, the maximization of $p(\overline{r}\,|\overline{\theta})$ is equivalent to maximization of $L\left(\overline{r};\overline{\theta}\right) = \ln p\left(\overline{r}\,|\overline{\theta}\right)$, which is called the *log-likelihood function*. The log-likelihood function is often simpler than the likelihood function itself, especially when the noise in the observations is additive and Gaussian. In our application this is the case.

Maximization of $L\left(\overline{r};\overline{\theta}\right)$ by standard techniques is a daunting task. A brute-force approach is to find the maximum by a search over the 6D multipath parameter space, but it takes too long to be of practical value. Reliable gradient-based or hill climbing methods are too slow to be useful. Finding the maximum using differential calculus is difficult, because of the nonlinearity of the resulting equations and the possibility of local maxima that are not global maxima. Iterative solution techniques are often difficult to analyze and may not converge to the correct solution in a timely manner, if they converge at all. As we shall see, the MMT algorithm solves these problems by reducing the dimensionality of the search space.

5.6.3.4 Asymptotic Properties of ML Estimators ML estimation is used by MMT not only because it can be made computationally simple enough to be practical but also because ML estimators have desirable asymptotic[2] properties:

1. The ML estimate of a parameter asymptotically converges in probability to the true parameter value.
2. The ML estimate is asymptotically efficient, that is, the ratio of the variance of the estimation error to the Cramer–Rao bound approaches unity.
3. The ML estimate is asymptotically Gaussian.

5.6.3.5 The MMT Multipath Mitigation Algorithm The MMT algorithm uses several mathematical techniques to solve what would otherwise be intractable computational problems. The first of these is a nonlinear transformation on the multipath parameter space to permit rapid computation of a log-likelihood function that has been partially maximized with respect to all of the multipath parameters except for the path delays. Thus, final maximization requires a search in only two dimensions for the two-path case, aided by acceleration techniques.

A new method of signal compression, described in Section 5.6.3.10, is used to transform the received signal into a very small vector on which MMT can operate very rapidly.

A major advantage of the MMT algorithm is that its performance improves with increasing E/N_0, the ratio of signal energy E to noise power spectral density. This is not true for most GNSS multipath mitigation methods, because their estimation error is in the form of an irreducible bias. Additionally, the MMT algorithm provides ML estimates of *all* parameters in the multipath model, and can utilize known bounds on the magnitudes of the secondary paths, if available, to improve performance.

5.6.3.6 The MMT Baseband Signal Model In the complex baseband signal $r(t)$ given by Eq. (5.47) it is assumed that the signal has been Doppler-compensated and stripped of the 50-bps (bits per second) navigation data modulation. In developing the MMT algorithm it is useful to separate $r(t)$ into its real

[2]*Asymptotic* refers to the behavior of an estimator when the error becomes small. In GNSS this occurs when E/N_0 is sufficiently large.

component $x(t)$ and imaginary component $y(t)$:

$$\begin{aligned} x(t) &= A_1 \cos\phi_1 \; m(t-\tau_1) \\ &\quad + A_2 \cos\phi_2 \; m(t-\tau_2) + n_x(t) , \\ y(t) &= A_1 \sin\phi_1 \; m(t-\tau_1) \\ &\quad + A_2 \sin\phi_2 \; m(t-\tau_2) + n_y(t) , \end{aligned} \tag{5.49}$$

where $n_x(t)$ and $n_y(t)$ are independent, real-valued, zero-mean Gaussian noise processes with flat power spectral density.

5.6.3.7 *Baseband Signal Vectors* The real and imaginary signal components are synchronously sampled on $[0,T]$ at the Nyquist rate $2W$, corresponding to the lowpass baseband bandwidth W, to produce the vectors

$$\begin{aligned} \overline{x} &= (x_1, x_2, \ldots, x_M) , \\ \overline{y} &= (y_1, y_2, \ldots, y_M) , \end{aligned} \tag{5.50}$$

in which the noise components of distinct samples are essentially uncorrelated (hence independent, since the noise is Gaussian).

5.6.3.8 *The Log-Likelihood Function* The ML estimates of the six parameters in the vector $\overline{\theta}$ given by (5.48) are obtained by maximizing the log-likelihood function with respect to these parameters. For MMT the log-likelihood function is

$$\begin{aligned} L\left(\overline{x}, \overline{y} \,|\overline{\theta}\right) &= \ln\left[p\left(\overline{x}, \overline{y} \,|\overline{\theta}\right)\right] \\ &= \ln C_1 \\ &\quad - C_2 \sum_{k=1}^{M} \left[\begin{array}{r} x_k - A_1 \cos\theta_1 m_k(\tau_1) \\ - A_2 \cos\theta_2 m_k(\tau_2) \end{array} \right]^2 \\ &\quad - C_2 \sum_{k=1}^{M} \left[\begin{array}{r} y_k - A_1 \sin\theta_1 m_k(\tau_1) \\ - A_2 \sin\theta_2 m_k(\tau_2) \end{array} \right]^2 , \end{aligned} \tag{5.51}$$

where

$$\begin{aligned} C_1 &= \left(\frac{1}{\sqrt{2\pi}\sigma}\right)^M \\ C_2 &= \frac{1}{2\sigma^2} \\ \sigma^2 &= \text{noise variance of } x(t) \text{ and } y(t) \\ m_k(\tau_1) &= k\text{th sample of } m(t-\tau_1) \\ m_k(\tau_2) &= k\text{th sample of } m(t-\tau_2) . \end{aligned} \tag{5.52}$$

Replacing the summations in (5.51) by integrals and utilizing the fact that C_1 and $-C_2$ are negative constants that do not depend on the multipath parameters, maximization of Eq.5.51 is equivalent to *minimization* of

$$\Gamma = \int_0^T \left[\begin{array}{r} x(t) - A_1 \cos\phi_1 \, m(t-\tau_1) \\ - A_2 \cos\phi_2 \, m(t-\tau_2) \end{array} \right]^2 dt + \int_0^T \left[\begin{array}{r} y(t) - A_1 \sin\phi_1 \, m(t-\tau_1) \\ - A_2 \sin\phi_2 \, m(t-\tau_2) \end{array} \right]^2 dt. \tag{5.53}$$

with respect to the six multipath parameters. This is a highly coupled, nonlinear minimization problem on the 6D space spanned by the parameters A_1, ϕ_1, τ_1, A_2, ϕ_2, and τ_2. Standard minimization techniques such as a gradient search on this space or ad hoc iterative approaches are either unreliable or too slow to be useful.

However, a major breakthrough results by using the invertible transformation

$$\left. \begin{array}{ll} a = A_1 \cos\phi_1 & c = A_1 \sin\phi_1 \\ b = A_2 \cos\phi_2 & d = A_2 \sin\phi_2 \end{array} \right\}. \tag{5.54}$$

When this transformation is applied and the integrands in (5.53) are expanded, the problem becomes one of minimizing

$$\begin{aligned} \Gamma = & \int_0^T \left[x^2(t) + y^2(t)\right] dt \\ & + \left(a^2 + b^2 + c^2 + d^2\right) R_{mm}(0) \\ & - 2a R_{xm}(\tau_1) - 2b R_{xm}(\tau_2) + 2ab R_{mm}(\tau_1 - \tau_2) \\ & - 2c R_{ym}(\tau_1) - 2d R_{ym}(\tau_2) + 2cd R_{mm}(\tau_1 - \tau_2). \end{aligned} \tag{5.55}$$

Note that Γ in Eq. 5.55 is quadratic in a, b, c, and d, and uses the correlation functions

$$\begin{aligned} R_{xm}(\tau) &= \int_0^T x(t)\, m(t-\tau)\, dt \\ R_{ym}(\tau) &= \int_0^T y(t)\, m(t-\tau)\, dt \\ R_{mm}(\tau) &= \int_0^T m(t)\, m(t-\tau)\, dt. \end{aligned} \tag{5.56}$$

Thus, minimization of Eq. 5.55 with respect to a, b, c, and d can be accomplished by taking partial derivatives with respect to these parameters, resulting in the linear system

$$\begin{aligned}
0 &= \frac{\partial \Gamma}{\partial a} = 2aR_{mm}(0) - 2R_{xm}(\tau_1) + 2bR_{mm}(\tau_1 - \tau_2) \\
0 &= \frac{\partial \Gamma}{\partial b} = 2bR_{mm}(0) - 2R_{xm}(\tau_2) + 2aR_{mm}(\tau_1 - \tau_2) \\
0 &= \frac{\partial \Gamma}{\partial c} = 2cR_{mm}(0) - 2R_{ym}(\tau_1) + 2dR_{mm}(\tau_1 - \tau_2) \\
0 &= \frac{\partial \Gamma}{\partial d} = 2dR_{mm}(0) - 2R_{ym}(\tau_2) + 2cR_{mm}(\tau_1 - \tau_2).
\end{aligned} \tag{5.57}$$

For each pair of values of τ_1 and τ_2 this linear system can be explicitly solved for the minimizing values of a, b, c, and d. Thus the space to be searched for a minimum of (5.55) (i.e., Eq. 5.55) is now 2D instead of 6D. The minimization procedure is as follows. Search the (τ_1, τ_2) domain. At each point (τ_1, τ_2) compute the values of the correlation functions in system (5.57) and then solve the system to find the values of a, b, c, and d that minimize Γ at that point. Identify the point $(\hat{\tau}_1, \hat{\tau}_2)_{ML}$ where the smallest of all such minima is obtained, as well as the associated minimizing values of a, b, c, and d. Transform these values of a, b, c, and d back to the estimates $\hat{A}_{1ML}$, $\hat{A}_{2ML}$, $\hat{\phi}_{1ML}$, $\hat{\phi}_{2ML}$ by using the inverse of transformation (5.54), which is

$$\left.\begin{array}{ll}
A_1 = \sqrt{a^2 + c^2} & A_2 = \sqrt{b^2 + d^2} \\
\phi_1 = \arctan 2(a, c) & \phi_2 = \arctan 2(b, d)
\end{array}\right\}. \tag{5.58}$$

5.6.3.9 Secondary-Path Amplitude Constraint In the majority of multipath scenarios, the amplitudes of secondary-path signals are smaller than that of the direct path. The multipath mitigation performance of MMT can be significantly improved by minimizing Γ in (5.55) subject to the constraint

$$\frac{A_2}{A_1} \le \alpha, \tag{5.59}$$

where α is a positive constant (a typical value is 0.7). The constraint in terms of the transformed parameters a, b, c, and d is

$$b^2 + d^2 \le \alpha^2 \left(a^2 + c^2\right). \tag{5.60}$$

The constrained minimization of (5.55) uses the method of Lagrange multipliers.

5.6.3.10 Signal Compression In the MMT algorithm the correlation functions $R_{xm}(\tau)$, $R_{ym}(\tau)$, and $R_{mm}(\tau)$ defined by (5.56) and appearing in (5.57) are computed very rapidly by first using a process called *signal compression*, in which the large number of signal samples (on the order of 10^8–10^9) that would normally be involved is reduced to only a few tens of samples (the exact number depends on which type of GPS signal is being processed). This processing is easily done in real time.

The correlation functions appearing in (5.56) have the form

$$R(\tau) = \int_0^T r(t)\, m(t-\tau)\, dt, \tag{5.61}$$

where $r(t)$ is a given function and $m(t)$ is a replica of the code modulation, which includes the effects of filtering in the satellite and receiver. The calculation of $R(\tau)$ in a conventional receiver is ordinarily not computationally difficult because in such receivers $m(t)$ can be an ideal chipping sequence with only the values ± 1, and the multiplications of samples of the integrand of (5.61) then become trivial. Furthermore, conventional receivers track only the peak of the correlation function so that $R(\tau)$ needs to be computed for only a few values of τ (usually for early, punctual, and late correlations). However, the MMT algorithm cannot employ these simplifications. The function $m(t)$ used by MMT must include the aforementioned effects of filtering, thus requiring multibit multiplications (typically numbering in the millions) in the calculation of $R(\tau)$. Furthermore, $R(\tau)$ must be calculated for many values of τ to obtain high resolution for accurate estimation of direct-path delay in the presence of multipath.

These difficulties are circumvented by using signal compression. To simplify its description, we assume that the correlation function $R(\tau)$ in (5.61) is a cyclic correlation over one period T of the replica code $m(t)$ in which $m(t-\tau)$ is a rotation by τ (right for positive τ and left for negative τ). However, compression can be accomplished over an arbitrary interval of observation of the function $r(t)$ in which many periods of a received PN code occur, and furthermore the correlation function need not be cyclic.

A single period of replica code can be written as

$$m(t) = \sum_{k=0}^{N-1} \varepsilon_k c(t - kT_c), \tag{5.62}$$

where T_c is the duration of each chip, ε_k is the chip polarity (either $+1$ or -1), and N is the number of chips in one period of the code. The function $c(t)$ is the response of the combined satellite and receiver filtering to a single ideal chip of the code. This ideal chip has a constant value of 1 on the interval $0 \le t \le T_c$. Because the filtering is linear and time-invariant, it follows that $m(t)$ is the filter response to the entire code sequence. The index k identifies the individual chips of the code, where $k = 0$ identifies the epoch chip, defined as the first chip of the chipping sequence.

The *compressed signal* $\tilde{r}(t)$ is defined by

$$\tilde{r}(t) = \sum_{k=0}^{N-1} \varepsilon_k r(t + kT_c). \quad (5.63)$$

In this expression $\varepsilon_k r(t + kT_c)$ is $r(t)$ weighted by ε_k and left-rotated by kT_c. In GPS applications the compressed signal has the very nice property that essentially all of its energy (excluding noise) is concentrated into a pulse of one filtered chip in duration. This is made evident by noting that the received signal $r(t)$ without multipath can be expressed as

$$r(t) = am(t - \tau_0) + n(t) = a\left[\sum_{j=0}^{N-1} \varepsilon_j c(t - \tau_0 - jT_c)\right] + n(t), \quad (5.64)$$

where a is the signal amplitude, τ_0 is the signal delay, $n(t)$ is noise, and all timeshifts are rotations (i.e., cyclic over one code period). Substitution of this expression into (5.63) gives

$$\begin{aligned}
\tilde{r}(t) &= \sum_{k=0}^{N-1} \varepsilon_k r(t + kT_c) \\
&= \sum_{k=0}^{N-1} \varepsilon_k \left\{\left[\sum_{j=0}^{N-1} \varepsilon_j c(t - \tau_0 + kT_c - jT_c)\right] + n(t + kT_c)\right\} \\
&= \sum_{k=0}^{N-1}\sum_{j=0}^{N-1} \varepsilon_k \varepsilon_j c\left[t - \tau_0 + (k - j)T_c\right] + \sum_{k=0}^{N-1} \varepsilon_k n(t + kT_c) \\
&= \sum_{k=0}^{N-1}\sum_{j=0}^{N-1} \varepsilon_k \varepsilon_j c\left[t - \tau_0 + (k - j)T_c\right] + \tilde{n}(t),
\end{aligned} \quad (5.65)$$

where the double summation is the compressed signal component, and the single summation is the *compressed noise* function $\tilde{n}(t)$. The terms in the double summation can be grouped into N groups such that each group contains N terms having the same value of $k - j$ modulo N. Thus, $\tilde{r}(t)$ will be the summation of N group sums plus $\tilde{n}(t)$. The group sum corresponding to particular value p of $k - j$ modulo N is $c\left[t - \tau_0 + pT_c\right]$ weighted by the sum of terms $\varepsilon_j \varepsilon_k$, which satisfy $k - j = p$ modulo N. Since T_c is the duration of $c(t)$ before filtering, it can be seen that $\tilde{r}(t)$ consists of a concatenation of N weighted and translated copies of $c(t)$ which do not overlap, except for a trailing transient from each copy due to filtering.

5.6.3.11 ***Properties of the Compressed Signal*** If the number of chips N is sufficiently large (on the order of 10^3 or more), the autocorrelation function of the GPS chipping sequence has the property that the group sums in which $k - j \neq 0$

modulo N are negligible compared to the group sum in which $k - j = 0$ modulo N. Furthermore, the sum of all of these small group sums is also negligible because the translations of the weighted copies of $c(t)$ prevent the small group sums from accumulating to large values. Thus, to a very good approximation, the double summation in (5.65) is just the sum of the terms where $k - j = 0$ modulo$N$:

$$\tilde{r}(t) \cong \left[\sum_{k=0}^{N-1} \varepsilon_k^2 c(t - \tau_0) \right] + \tilde{n}(t) = Nc(t - \tau_0) + \tilde{n}(t). \qquad (5.66)$$

This is a very significant result, because it tells us that the compressed received signal is essentially just the single weighted filtered chip $Nc(t - \tau_0)$ plus noise, with small "sidelobe" chips to either side. Furthermore, the compression process provides a processing gain of $10 \log N$ dB. Since a receiver can measure the delay τ_0, a window can be constructed that need be long enough only to contain $Nc(t - \tau_0)$, and the sidelobe chips as well as all noise outside this window can be rejected. The required length of the window is $T_c + \delta$, where δ is large enough to accommodate the measurement uncertainty of τ_0, the trailing transient due to filtering, and any multipath components with delays larger than τ_0 (almost certainly the only multipath components having significant amplitude are found within 1 chip of the direct path delay) . Thus the window length is somewhat larger than one chip duration of the code, a quantity much smaller than the length T of the observed signal $r(t)$, which must include all N chips of the code. It is because of this result that $\tilde{r}(t)$ can justifiably be called a *compressed signal*. An illustration of the compressed signal is shown in Fig. 5.7.

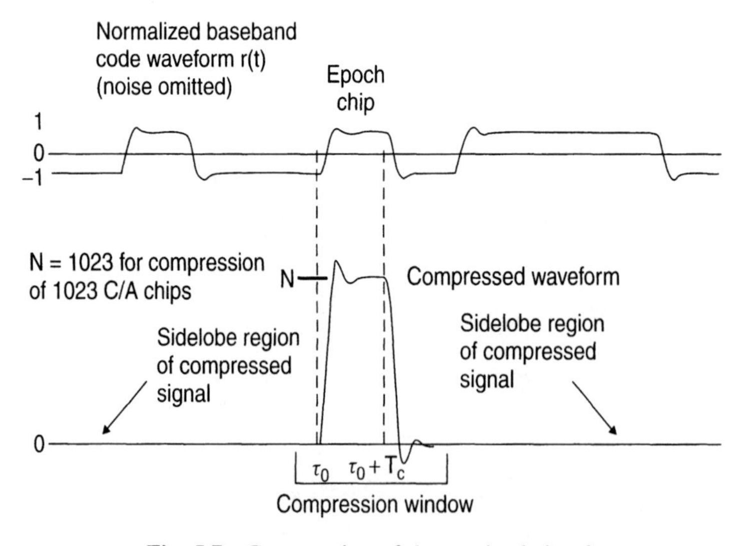

Fig. 5.7 Compression of the received signal.

If N is sufficiently large, the processing gain is great enough to make the compressed signal within the window visible with very little noise, so that small subtleties in the chip waveshape due to multipath or other causes can easily be seen. This property is very beneficial for signal integrity monitoring. It has been put to practical use in GPS receivers sold by the NovAtel Corporation, which calls its implementation the *Vision Correlator*.

The compressed signal also enjoys a *linearity property*: If $r(t) = a_1 r_1(t) + a_2 r_2(t)$, then $\tilde{r}(t) = a_1 \tilde{r}_1(t) + a_2 \tilde{r}_2(t)$. The linearity property is essential for the MMT to properly process a multipath corrupted signal.

5.6.3.12 The Compression Theorem Most importantly, the compressed signal can be used to drastically reduce the amount of computation of the correlation function $R(\tau)$ in (5.61). The basis for this assertion is the following theorem:

The correlation function

$$R(\tau) = \int_0^T r(t)\, m(t-\tau)\, du, \tag{5.67}$$

can be computed by the alternate method

$$R(\tau) = \int_0^T \tilde{r}(t)\, c(t-\tau)\, du. \tag{5.68}$$

Proof:

$$\left.\begin{aligned}
R(\tau) &= \int_0^T r(t)\, m(t-\tau)\, dt \\
&= \int_0^T r(t) \left[\sum_{k=0}^{N-1} \varepsilon_k c(t - kT_c - \tau) \right] dt \\
&= \sum_{k=0}^{N-1} \int_0^T \varepsilon_k r(t)\, c(t - kT_c - \tau)\, dt \\
&= \sum_{k=0}^{N-1} \int_0^T \varepsilon_k r(u + kT_c)\, c(u-\tau)\, du \ (\text{using } u = t - kT_c) \\
&= \int_0^T \left[\sum_{k=0}^{N-1} \varepsilon_k r(u + kT_c) \right] c(u-\tau)\, du \\
&= \int_0^T \tilde{r}(u)\, c(u-\tau)\, du
\end{aligned}\right\} . \tag{5.69}$$

This theorem shows that $R(\tau)$ can be computed by cross-correlating the compressed signal $\tilde{r}(t)$ with the very short function $c(t)$. Furthermore, since we have already noted that the significant portion of $\tilde{r}(t)$ also spans a short time interval, the region surrounding the correlation peak of $R(\tau)$ can be obtained with far less computation than the original correlation (5.67). The bottom line is that the cross-correlations in (5.56) used by MMT can be calculated very efficiently by using the compressed versions of the signals $x(t)$, $y(t)$, and $m(t)$.

5.6.4 Performance of Time-Domain Methods

5.6.4.1 Ranging with the C/A-Code Typical C/A-code ranging performance curves for several multipath mitigation approaches are shown in Fig. 5.8 for the case of an in-phase secondary path with amplitude one-half that of the direct path. Even with the best available methods (other than MMT), peak range errors of 3–6 m are not uncommon. It can be observed that the error tends to be largest for "close-in" multipath, where the separation of the two paths is less that 20–30 m. Indeed, this region poses the greatest challenge in multipath mitigation research because the extraction of direct-path delay from a signal with small direct/secondary-path separation is an ill-conditioned parameter estimation problem.

A serious limitation of most existing multipath mitigation algorithms is that the residual error is mostly in the form of a bias that cannot be removed by further filtering or averaging. On the other hand, the above mentioned MMT algorithm overcomes this limitation and also appears to have significantly better performance than other published algorithms, as is indicated by curve F of Fig. 5.8.

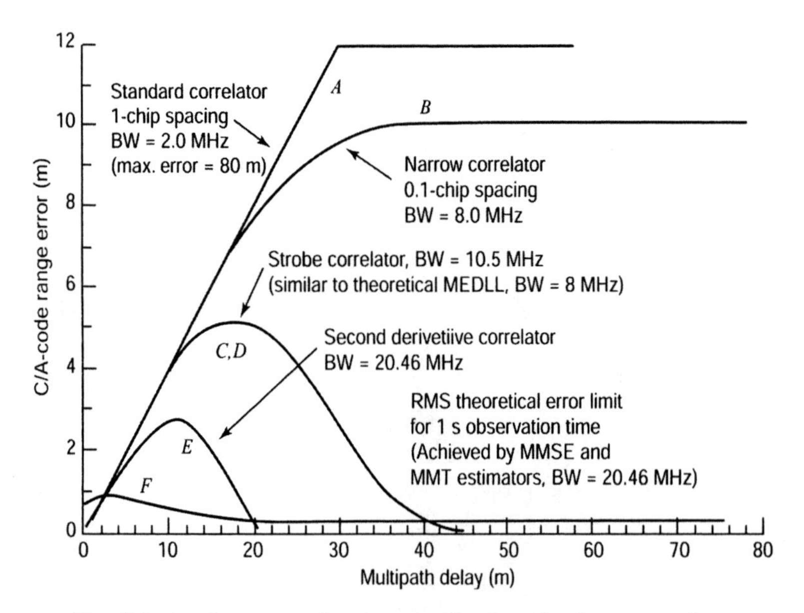

Fig. 5.8 Performance of various multipath mitigation approaches.

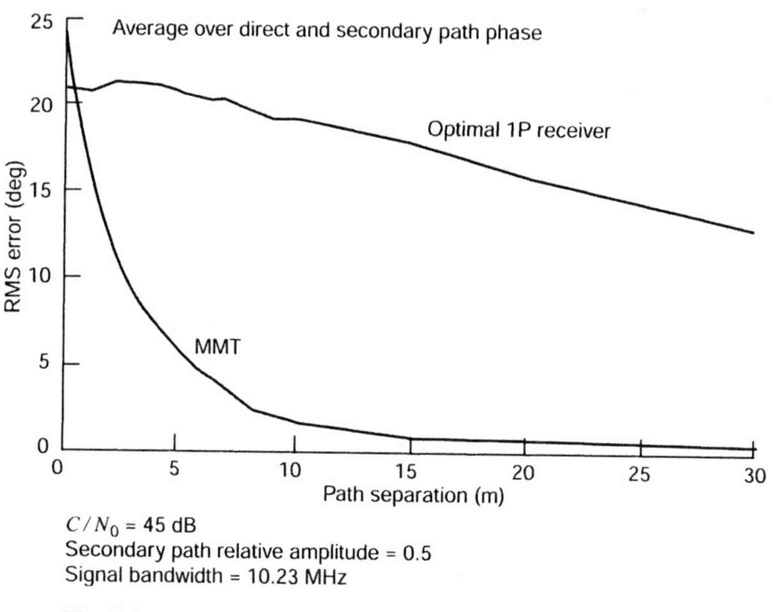

Fig. 5.9 Residual multipath phase error using MMT algorithm.

5.6.4.2 Carrier Phase Ranging The presence of multipath also causes errors in estimating carrier phase, which limits the performance in surveying and other precision applications, particularly with regard to carrier phase ambiguity resolution. Not all current multipath mitigation algorithms are capable of reducing multipath-induced phase error. The most difficult situation occurs at small separations between the direct and secondary paths (less than a few meters). It can be shown that under such conditions essentially no mitigation is theoretically possible. Typical phase error curves for the MMT algorithm, which appears to have the best performance of published methods, is shown in Fig. 5.9.

5.6.4.3 Testing Receiver Multipath Performance Conducting meaningful tests of receiver multipath mitigation performance on either an absolute or a comparative basis is no easy matter. There are often two conflicting goals. On the one hand, the testing should be under strictly controlled conditions, so that the signal levels and true multipath parameters are precisely known; otherwise the measured performance cannot be linked to the multipath conditions that actually exist. Generally this will require precision signal simulators and other ancillary equipment to generate accurately characterized multipath signals.

On the other hand, receiver end users place more credence on how well a receiver performs in the field. However, meaningful field measurements pose a daunting challenge. It is extremely difficult to know the amount and character of the multipath, and great difficulty can be experienced in isolating errors caused by multipath from those of other sources. To add to these difficulties, it is not clear that either the receiver manufacturers or the users have a good feel for the range of multipath parameter values that represent typical operation in the field.

5.7 THEORETICAL LIMITS FOR MULTIPATH MITIGATION

5.7.1 Estimation-Theoretic Methods

Relatively little has been published on multipath mitigation from the fundamental viewpoint of statistical estimation theory, despite the power of its methods and its ability to reach theoretical performance limits in many cases. Knowledge of such limits provides a valuable benchmark in receiver design by permitting an accurate assessment of the potential payoff in developing techniques that are better than those in current use. Of equal importance is the revelation of the signal processing operations that can reach performance bounds. Although it may not be feasible to implement the processing directly, its revelation often leads to a practical method that achieves nearly the same performance.

5.7.1.1 Optimality Criteria In discussing theoretical performance limits, it is important to define the criterion of optimality. In GPS the optimal range estimator is traditionally considered to be the minimum-variance unbiased estimator (MVUE), which can be realized by properly designed receivers. However, in Ref. 204 it is shown that the standard deviation of a MVUE designed for multipath becomes infinite as the primary-to-secondary-path separation approaches zero. For this reason it seems that a better criterion of optimality would be the minimum RMS error, which can include both random and bias components. Unfortunately, it can be shown that *no* estimator exists having minimum RMS error for *every* combination of true multipath parameters.

5.7.2 MMSE Estimator

There is an estimator that can be claimed optimal in a weaker sense. The *minimum-mean-square-error* (MMSE) estimator has the property that no other estimator has a uniformly smaller RMS error. In other words, if some other estimator has smaller RMS error than the MMSE estimator for some set of true multipath parameter values, then that estimator must have a *larger* RMS error than the MMSE estimator for some *other* set of values.

The MMSE estimator also has an important advantage not possessed by most current multipath mitigation methods in that the RMS error decreases as the length of the signal observation interval is increased.

5.7.3 Multipath Modeling Errors

Although a properly designed estimation-theoretic approach such as the MMSE estimator will generally outperform other methods, the design of such estimators requires a mathematical model of the multipath-contaminated signal containing parameters to be estimated. If the actual signal departs from the assumed model, performance degradation can occur. For example, if the model contains only two signal propagation paths but in reality the signal is arriving via three or more paths, large bias errors in range estimation can result. On the other hand, poorer performance (usually in the form of random error cause by noise) can also occur if the model has too many degrees of freedom. Striking the right balance in the number of parameters in the model can be difficult if little information exists about the multipath reflection geometry.

5.8 EPHEMERIS DATA ERRORS

Small errors in the ephemeris data transmitted by each satellite cause corresponding errors in the computed position of the satellite (here we exclude the ephemeris error component of SA, which is regarded as a separate error source). Satellite ephemerides are determined by the master control station of the GPS ground segment based on monitoring of individual signals by four monitoring stations. Because the locations of these stations are known precisely, an "inverted" positioning process can calculate the orbital parameters of the satellites as if they were users. This process is aided by precision clocks at the monitoring stations and by tracking over long periods of time with optimal filter processing. Based on the orbital parameter estimates thus obtained, the master control station uploads the ephemeris data to each satellite, which then transmits the data to users via the navigation data message. Errors in satellite position when calculated from the ephemeris data typically result in range errors less than 1 m. Improvements in satellite tracking will undoubtedly reduce this error further.

5.9 ONBOARD CLOCK ERRORS

Timing of the signal transmission from each satellite is directly controlled by its own atomic clock without any corrections applied. This time frame is called *space vehicle* (SV) *time*. A schematic of a rubidium atomic clock is shown in Fig. 5.10. Although the atomic clocks in the satellites are highly accurate, errors can be large enough to require correction. Correction is needed partly because it would be difficult to directly synchronize the clocks closely in all the satellites. Instead, the clocks are allowed some degree of relative drift that is estimated by ground station observations and is used to generate clock correction data in the GPS navigation message. When SV time is corrected using this data, the result is called *GPS time*. The time of transmission used in calculating pseudoranges must be in GPS time, which is common to all satellites.

The onboard clock error is typically less than 1 ms and varies slowly. This permits the correction to be specified by a quadratic polynomial in time whose

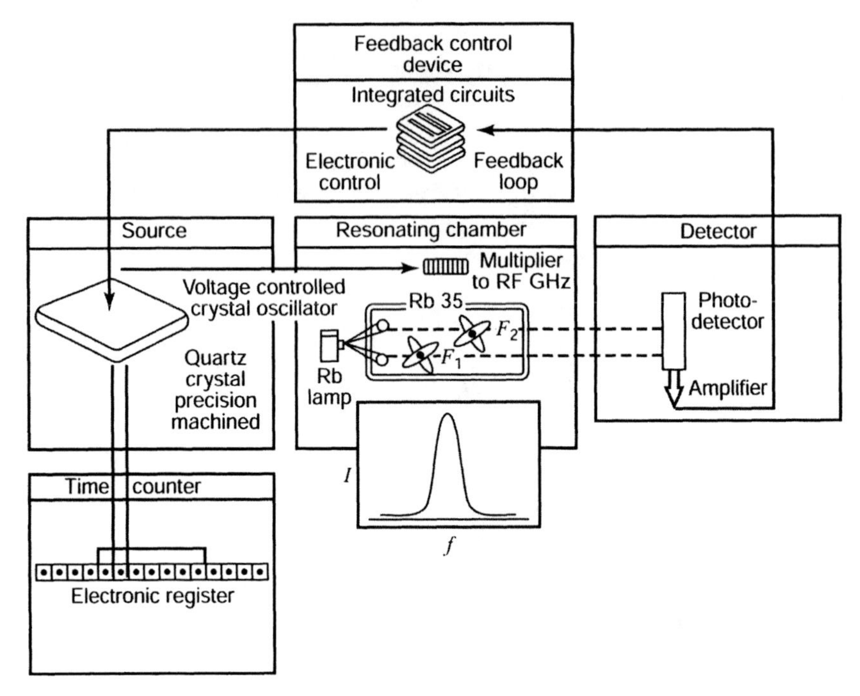

Fig. 5.10 Schematic of a rubidium atomic clock.

coefficients are transmitted in the navigation message. The correction has the form

$$\Delta t_{sv} = a_{f0} + a_{f1}(t_{sv} - t_{0c}) + a_{f2}(t_{sv} - t_{0c})^2 + \Delta t_R, \tag{5.70}$$

with

$$t_{\text{GPS}} = t_{sv} - \Delta t_{sv}, \tag{5.71}$$

where a_{f0}, a_{f1}, a_{f2} are the correction coefficients, t_{sv} is SV time, and Δt_R is a small relativistic clock correction caused by the orbital eccentricity. The clock data reference time t_{0c} in seconds is broadcast in the navigation data message. The stability of the atomic clocks permits the polynomial correction given by Eq. 5.70 to be valid over a time interval of 4–6 h. After the correction has been applied, the residual error in GPS time is typically less than a few nanoseconds, or about 1 m in range. Complete calculations of GPS time are given as exercises in Problems 3.6–3.10 (in Chapter 3).

5.10 RECEIVER CLOCK ERRORS

Because the navigation solution includes a solution for receiver clock error, the requirements for accuracy of receiver clocks is far less stringent than for the

GPS satellite clocks. In fact, for receiver clocks short-term stability over the pseudorange measurement period is usually more important than absolute frequency accuracy. In almost all cases such clocks are quartz crystal oscillators with absolute accuracies in the 1–10 ppm range over typical operating temperature ranges. When properly designed, such oscillators typically have stabilities of 0.01–0.05 ppm over a period of a few seconds.

Receivers that incorporate receiver clock error in the Kalman filter state vector need a suitable mathematical model of the crystal clock error. A typical model in the continuous-time domain is shown in Fig. 5.11, which is easily changed to a discrete version for the Kalman filter. In this model the clock error consists of a bias (frequency) component and a drift (time) component. The frequency error component is modeled as a random walk produced by integrated white noise. The time error component is modeled as the integral of the frequency error after additional white noise (statistically independent from that causing the frequency error) has been added to the latter. In the model the key parameters that need to be specified are the power spectral densities of the two noise sources, which depend on characteristics of the specific crystal oscillator used.

The continuous time model has the form

$$\dot{x}_1 = w_1, \tag{5.72}$$

$$\dot{x}_2 = x_1 + w_2, \tag{5.73}$$

where $w_1(t)$ and $w_2(t)$ are independent zero-mean white-noise processes with known variances.

The equivalent discrete-time model has the state vector

$$\mathbf{x} = \begin{bmatrix} x_1 \\ x_2 \end{bmatrix}, \tag{5.74}$$

and the stochastic process model

$$\mathbf{x}_k = \begin{bmatrix} 1 & 0 \\ \Delta t & 1 \end{bmatrix} \mathbf{x}_{k-1} + \begin{bmatrix} w_{1,k-1} \\ w_{2,k-1} \end{bmatrix}, \tag{5.75}$$

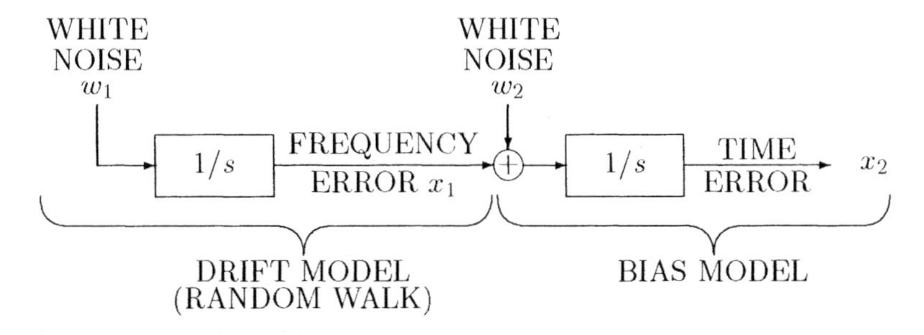

Fig. 5.11 Crystal clock error model.

where Δt is the discrete-time step and $\{w_{1,k-1}\}$, $\{w_{2,k-1}\}$ are independent zero-mean white-noise sequences with known variances.

5.11 ERROR BUDGETS

For purposes of analyzing the effects of the errors discussed above, it is convenient to convert each error into an equivalent range error experienced by a user, which is called the *user-equivalent range error* (UERE). In general, the errors from different sources will have different statistical properties. For example, satellite clock and ephemeris errors tend to vary slowly with time and appear as biases over moderately long time intervals, perhaps hours. On the other hand, errors due to receiver noise and quantization effects may vary much more rapidly, perhaps within seconds. Nonetheless, if sufficiently long time durations over many navigation scenarios are considered, all errors can be considered as zero-mean random processes that can be combined to form a single UERE. This is accomplished by forming the root sum square of the UERE errors from all sources:

$$\text{UERE} = \sqrt{\sum_{i=1}^{n} (\text{UERE})_i^2}. \tag{5.76}$$

Figure 5.12 depicts the various GPS UERE errors and their combined effect for both C/A-code and P(Y)-code navigation at the 1-σ level.

When SA is on, the UERE for the C/A-code user is about 36 m and reduces to about 19 m when it is off. Aside from SA, it can be seen that for such a user the dominant error sources in nondifferential operations are multipath, receiver noise/resolution, and ionospheric delay (however, recent advances in receiver technology have in some cases significantly reduced receiver noise/resolution errors). On the other hand, the P(Y)-code user has a significantly smaller UERE of about 6 m, for the following reasons:

1. Errors due to SA can be removed, if present. The authorized user can employ a key to eliminate them.
2. The full use of the L_1 and L_2 signals permits significant reduction of ionospheric error.
3. The wider bandwidth of the P(Y)-codes greatly reduces errors due to multipath and receiver noise.

5.12 DIFFERENTIAL GNSS

Differential GNSS (DGNSS) is a technique for improving the performance of GNSS positioning. The basic idea of DGNSS is to compute the spatial

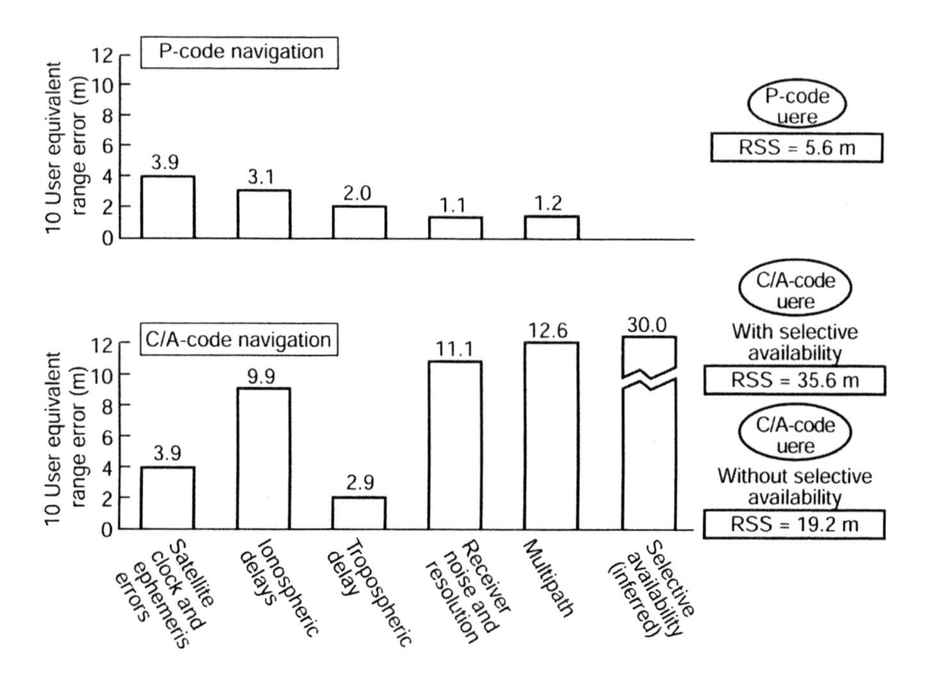

Fig. 5.12 GPS UERE budget.

displacement vector of the user's receiver (sometimes called the *roving* or *remote* receiver) relative to another receiver (usually called the *reference receiver* or *base station*). In most DGNSS applications the coordinates of the reference receiver are precisely known from highly accurate survey information; thus, the accurate location of the roving receiver can be determined by vector addition of the reference receiver coordinates and the reference-to-rover displacement vector.

The positioning accuracy of DGNSS depends on the error in estimating the reference-to-rover displacement vector. This error can be made considerably smaller than the positioning error of a standalone receiver, because major components of pseudorange measurement errors are common to the roving and reference receivers and can be canceled out by using the difference between the reference and rover measurements to compute the displacement vector.

There are basically two ways that errors common to the roving and reference receiver can be canceled. The first method is called the *measurement* or *solution-domain* technique, in which both receivers individually compute their positions and the reference-to-rover displacement vector is simply the difference of these positions. However, the two receivers must use exactly the same set of satellites for this method to be effective. Since this requirement is often impossible to fulfill (e.g., due to blockage of signals at the roving receiver), this method is seldom used. A far better method, which offers more flexibility, is to use only difference the measurements from the set of satellites which are viewed in common by both receivers. Therefore, only this method will be described.

The two primary types of differential measurements are code measurements and carrier phase measurements.

5.12.1 Code Differential Measurements

To obtain code differential measurements, the roving and reference receivers each make a pseudorange measurement of the following form for each satellite

$$\rho_M = \rho + c\,dt - c\,dT + d_{\text{ION}} + d_{\text{TROP}} + d_{\text{EPHEM}} + d_\rho, \tag{5.77}$$

where ρ_M is the measured pseudorange in meters, ρ is the true receiver-to-satellite geometric range in meters, c is the speed of light in meters per second, dt is satellite clock error in seconds, dT is receiver clock error in seconds, d_{ION} is ionospheric delay error in meters, d_{TROP} is tropospheric delay error in meters, d_{EPHEM} is delay error in meters due to satellite ephemeris error, and d_ρ represents other pseudorange errors in meters, such as multipath, interchannel receiver biases, thermal noise, and selective availability (when turned on). The pseudorange measurements made by both receivers must occur at a common GPS time, or if not, corrections must be applied to extrapolate the measurements to a common time.

5.12.1.1 First Difference Observations A code first difference observation is determined by subtracting an equation of the form (5.77) for the reference receiver from a similar equation for the roving receiver, where both equations relate to the same satellite. The result is

$$\Delta\rho_M = \Delta\rho - c\Delta dT + \Delta d_{\text{ION}} + \Delta d_{\text{TROP}} + \Delta d_{\text{EPHEM}} + \Delta d_\rho, \tag{5.78}$$

where the symbol Δ denotes the difference between the corresponding terms in the two equations of the form (5.77). Note that the term $c\Delta dt$ representing the satellite clock error has disappeared, since the satellite clock error is the same for the pseudorange measurements made by each receiver. Furthermore, if the distance between the roving and reference receivers is sufficiently small (say, $<$ 20 km), the terms Δd_{ION}, Δd_{TROP}, and Δd_{EPHEM} will be nearly canceled out, since errors due to the ionosphere, troposphere, and ephemeredes vary slowly with position.

5.12.1.2 Second Difference Observations A code second difference measurement is formed by subtraction of the first difference observation of the form (5.78) for one satellite from a similar first difference observation for another satellite. Thus, if there are N first difference observations corresponding to N satellites, there will be $N - 1$ independent second difference observations that can be formed. The second difference observations have the form

$$\nabla\Delta\rho_M = \nabla\Delta\rho + \nabla\Delta d_{\text{ION}} + \nabla\Delta d_{\text{TROP}} + \nabla\Delta d_{\text{EPHEM}} + \nabla\Delta d_\rho, \tag{5.79}$$

where the symbol ∇ denotes difference between the corresponding difference terms in the two equations of the form (5.78). Note that the double difference error term $c\nabla\Delta dT$ involving receiver clock error has been cancelled out, since receiver clock error is constant across all satellite measurements in both the reference and roving receivers. Furthermore, for a sufficiently small distance between the rover and reference receivers, the first difference errors Δd_{ION}, Δd_{TROP}, and Δd_{EPHEM} are so small that the corresponding double difference errors $\nabla\Delta d_{\text{ION}}$, $\nabla\Delta d_{\text{TROP}}$, and $\nabla\Delta d_{\text{EPHEM}}$ can be neglected. In this case the second difference observations become

$$\nabla\Delta\rho_M \cong \nabla\Delta\rho + \nabla\Delta d_\rho, \tag{5.80}$$

which can result positioning accuracies that are often in the submeter range.

Although DGPS is effective in removing satellite and receiver clock errors, ionospheric and tropospheric errors, ephemeris errors, and selective availability errors, it cannot remove errors due to multipath, receiver interchannel biases, and thermal noise, since these errors are not common to the roving and reference receivers.

5.12.2 Carrier Phase Differential Measurements

Because carrier phase pseudorange measurements have a significantly smaller error than do those using the code, positioning accuracy is potentially much more accurate. However, since only the fractional and not the integer part of a carrier cycle can be observed, some method of finding the integer part must be employed. This is the classic *ambiguity resolution* problem.

First and second difference observations can be obtained from carrier phase pseudorange measurements having the form

$$\lambda\phi_M = \rho + cdt - cdT + \lambda N - d_{\text{ION}} + d_{\text{TROP}} + d_{\text{EPHEM}} + d_\phi, \tag{5.81}$$

where the new variables are the carrier wavelength λ (0.1903 m for L_1 and 0.2442 m for L_2), the measured carrier phase ϕ_M in cycles, the carrier phase ambiguity N in cycles, and other errors d_ϕ in meters. Because the ionospheric group delay for the carrier is opposite that of the code, the ionospheric error is reversed in sign in (5.81).

In some cases triple differences of carrier phase measurements are used, as will be described subsequently.

5.12.2.1 First Difference Observations Each carrier phase first difference observation is determined in the same manner as for PN code by subtracting an equation of the form (5.81) for the reference receiver from a similar equation for the roving receiver, where both equations relate to the same satellite. The result is

$$\Delta\lambda\phi_M = \Delta\rho - c\Delta dT + \lambda\Delta N - \Delta d_{\mathrm{ION}} + \Delta d_{\mathrm{TROP}} + \Delta d_{\mathrm{EPHEM}} + \Delta d_{\phi}, \tag{5.82}$$

where, as before, the satellite clock error term has disappeared and the terms Δd_{ION}, Δd_{TROP}, and $\Delta d_{\mathrm{EPHEM}}$ are small for small distances between the rover and reference receivers.

5.12.2.2 Second Difference Observations A second difference carrier phase measurement is formed by subtraction of the first difference observation of the form (5.82) for one satellite from a similar first difference observation for another satellite. The second difference observations have the form

$$\nabla\Delta\lambda\phi_M = \nabla\Delta\rho + \lambda\nabla\Delta N - \nabla\Delta d_{\mathrm{ION}} + \nabla\Delta d_{\mathrm{TROP}} + \nabla\Delta d_{\mathrm{EPHEM}} + \nabla\Delta d_{\phi}, \tag{5.83}$$

Again, the receiver clock error term has disappeared, and the terms $\nabla\Delta d_{\mathrm{ION}}$, $\nabla\Delta d_{\mathrm{TROP}}$, and $\nabla\Delta d_{\mathrm{EPHEM}}$ are usually small. However, the value of N in the phase ambiguity term $\lambda\nabla\Delta N$ must be determined by some method of ambiguity resolution.

5.12.2.3 Third Difference Observations Triple difference carrier observations are sometimes used in DGPS, primarily to detect and correct cycle slips during carrier tracking. These observations have the form

$$\delta\nabla\Delta\lambda\phi_M = \delta\nabla\Delta\rho - \delta\nabla\Delta d_{\mathrm{ION}} + \delta\nabla\Delta d_{\mathrm{TROP}} + \delta\nabla\Delta d_{\mathrm{EPHEM}} + \delta\nabla\Delta d_{\phi}, \tag{5.84}$$

where δ is the time difference between two successive double difference observations. Cycle slips can be detected by observing the deviation of successive triple difference observations from their predicted values as the carrier is tracked.

5.12.2.4 Combinations of L_1 and L_2 Carrier Phase Observations Summing the L_1 and L_2 second difference carrier observations results in higher-resolution phase measurements than can be obtained at either frequency alone. Such *narrow-lane* measurements result in more precision, but place greater demands on phase ambiguity resolution. On the other hand, it is easier to resolve the phase ambiguity of *wide-lane* measurements obtained by differencing the L_1 and L_2 observations at the expense of reduced resolution.

5.12.3 Positioning Using Double-Difference Measurements

5.12.3.1 Code-Based Positioning The linearized matrix equation for positioning using code double-difference measurements from four satellites has the form

$$\overbrace{\delta Z_{\nabla\Delta\rho}}^{3\times 1} = \overbrace{H^{[1]}_{\nabla\Delta\rho}}^{3\times 3} \overbrace{\delta x}^{3\times 1} + \overbrace{\nu_\rho}^{3\times 1}, \tag{5.85}$$

which is the same form as shown in Section 2.3.3. However, because the double-difference measurements have eliminated receiver clock error as an unknown, the unknowns are simply the x, y, and z coordinates of the roving receiver, constituting the components of the 3×1 vector δx. Thus, the measurement matrix $H^{[1]}$ is 3×3 and the partial derivatives in it are partial derivatives of the double differences $\nabla\Delta\rho_M$ with respect to user position coordinates x,y,z instead of partial derivatives of the pseudorange measurements.. Accordingly, the measurement vector $\delta Z_\rho$and the measurement noise vector ν_ρ are 3×1. As indicated in Section 2.3.3, a solution for position can be found by computing the measurement vector associated with an assumed initial position x, finding the difference δZ_ρ between the computed and actual measurement vectors, solving (5.85) (omitting the measurement noise vector) for the position correction δx, and obtaining the new value $x + \delta x$ for x. Iteration of this process is used to produce a sequence of positions which converges to the position solution.

5.12.3.2 Carrier Phase–Based Positioning For positioning using carrier phase double difference measurements the linearized matrix equation from four satellites used for iterative position solution has the form

$$\overbrace{\delta Z_{\nabla\Delta\rho}}^{3\times 1} = \overbrace{H^{[1]}_{\nabla\Delta\rho}}^{3\times 3} \overbrace{\delta x}^{3\times 1} + \overbrace{\nu_\rho}^{3\times 1}, \tag{5.86}$$

where the measurement matrix $H^{[1]}$ contains the partial derivatives of the double differences $\nabla\Delta\lambda\phi_M$ with respect to user position coordinates x,y,z. As compared to code-based positioning, the measurement noise term ν_ρ is much smaller, often in the centimeter range. However, the major difference is that the ambiguity in the phase measurements can cause convergence to any one of many possible positions in a spatial grid of points. Only one of these points is the correct position. Various techniques for resolving the ambiguity have been developed. A simple method is to use the position solution from the code second difference measurements as the initial position x in the carrier phase iterative position solution. If this initial position is sufficiently accurate, convergence to the correct solution will be obtained.

5.12.3.3 Real-Time Processing versus Postprocessing Since double differencing combines measurements made in the roving and reference receivers, these measurements must be brought together for processing. Often the processing site is at the roving receiver, although in other applications it can be at the reference station or at another off-site location. In real-time processing measurements are transmitted to the processing site using wireless communication or a telephone link. In post processing the data can be physically carried to the processing site in a storage medium such as a floppy disk or CD-ROM. Another post processing option is to transmit the data via the Internet.

5.13 GPS PRECISE POINT POSITIONING SERVICES AND PRODUCTS

The cost and inconvenience of setting up one's own DGPS system can be eliminated because there are numerous services and software packages available to the user, some of which are free. There are too many to describe completely, so only a few of them are described in this section.

The International GNSS Service (IGS) Many of the DGPS services are subsumed under the IGS, which is a voluntary federation of more than 200 worldwide agencies that pool resources and permanent GPS and GLONASS station data to generate precise DGPS positioning services. The IGS is committed to providing the highest quality data and products as the standard for global navigation satellite systems (GNSSs) in support of earth science research, multidisciplinary applications, and education. The IGS also intends to incorporate future GNSS systems, such as Galileo, as they become operational.

Continuously Operating Reference Stations (CORSs) The National Geodetic Survey (NGS), an office of NOAA's National Ocean Service, manages two networks of CORS: the National CORS network and the Cooperative CORS network. These networks consist of numerous base stations containing DGPS reference receivers that operate continuously to generate pseudorange and other DGPS data for postprocessing. The data is disseminated to a wide variety of users. Surveyors, GIS/LIS professionals, engineers, scientists, and others can apply CORS data to their own GPS measurements to obtain positioning accuracies approaching a few centimeters relative to the National Spatial Reference System, both horizontally and vertically. The CORS program is a multipurpose cooperative endeavor involving more than 130 government, academic, and private organizations, each of which operates at least one CORS site. In particular, it includes all existing National Differential GPS (NDGPS) sites and all existing FAA Wide-Area Augmentation System (WAAS) sites. New sites are continually being evaluated according to established criteria.

Typical uses of CORS include land management, coastal monitoring, civil engineering, boundary determination, mapping and geographic information systems (GISs), geophysical and infrastructure modeling, as well as future improvements to weather prediction and climate modeling.

All national CORS data are available from NGS at their original sampling rate for 30 days, after which the data are decimated to a 30-s sampling rate. Cooperative CORS data are available from a large number of participating organizations that operate individual sites. Most of the CORS data are available on the internet.

GPS-Inferred Positioning System (GIPSY) and Orbit Analysis Simulation Software (OASIS) The GIPSY-OASIS II (GOA II) package consists of extremely versatile software that can be used for GPS positioning and satellite orbit analysis. Developed by the Caltech Jet Propulsion Laboratory (JPL), it can provide centimeter-level DGPS positioning accuracy over short to intercontinental baselines. It is capable of unattended, automated, low-cost operation in near real time for precise positioning and time transfer in ground, sea, air, and space applications.

GOA II also includes many force models useful for orbit determination, such as earth/sun/moon/planet (and tidal) gravity perturbations, solar pressure, thermal radiation, and drag, which make it useful in non-GPS satellite positioning applications. To augment its potential accuracy, models are included for earth characteristics, such as tides, ocean/atmospheric loading, and crustal plate motion.

Parameter estimation for positioning and time transfer is state-of-the-art. A general estimator can be used for GPS and non-GPS data. Matrix factorization is used to maintain robustness of solutions, and the estimator can intelligently identify, correct, or exclude questionable data. A general and flexible noise model is included.

Australia's Online GPS Processing System (AUPOS) AUPOS provides users with the facility to submit via the Internet dual-frequency geodetic quality GPS RINEX data observed in a "static" mode and receive rapid-turnaround precise position coordinates. The service is free and provides both International Terrestrial Reference Frame (ITRF) and Geocentric Datum of Australia (GDA94) coordinates. This Internet service takes advantage of both IGS products and the IGS GPS network and can handle GPS data collected anywhere on earth.

Scripps Coordinate Update Tool (SCOUT) SCOUT, managed by the Scripps Institute of Oceanography, is also a system which provides precise positioning for users who submit GPS RINEX data from their receiver via the Internet. The reference stations are by default the three nearest sites for which data have been collected and are available for the specific day the user's data are taken. However, the user can specify the reference stations if desired. Station maps are provided to assist the user in specifying nearby reference sites. When SCOUT has finished determining a DGPS position solution, it sends a report of the results to the user via the Internet. The report contains both Cartesian and geodetic coordinates, standard deviations, and the locations of the reference sites that were used. The reported Cartesian coordinates are referenced to the International Terrestrial Reference Frame 2000 (ITRF2000), and the geodetic coordinates are referenced to both ITRF2000 and the World Geodetic System 1984 (WGS84) ellipsoid.

The Online Positioning User Service (OPUS) The National Geodetic Survey (NGS) operates OPUS as a means to provide GPS users easier access to the National Spatial Reference System (NSRS). OPUS users submit their GPS data files to the NGS Internet site. The NGS computers and software determine a position by using reference receivers from three CORS sites. The position is reported back to the user by email in both ITRF and NAD83 (North American Datum 1983) coordinates, as well as Universal Transverse Mercator (UTM), and State Plain Coordinates (SPC) northing and easting. Results are typically obtained within a few minutes. OPUS is intended for use in the coterminous United States and most U.S. territories. It is NGS policy not to publish geodetic coordinates outside the United States without the agreement of the affected countries.

PROBLEMS

5.1 Using the values provided for Klobuchar's model in Section 5.2, for Southbury, Connecticut, calculate the ionospheric delay and plot the results.

5.2 Assume that a direct-path GPS L_1 C/A-code signal arrives with a phase such that all of the signal power lies in the baseband I channel, so that the baseband signal is purely real. Further assume an infinite signal bandwidth so that the cross-correlation of the baseband signal with an ideal C/A reference code waveform will be an isosceles triangle 600 m wide at the base.

(a) Suppose that in addition to the direct-path signal there is a secondary-path signal arriving with a relative time delay of precisely 250 L_1 carrier cycles (so that it is in phase with the direct-path signal) and with an amplitude one-half that of the direct path. Calculate the pseudorange error that would result, including its sign, under noiseless conditions. Assume that pseudorange is measured with a delay-lock loop using 0.1-chip spacing between the early and late reference codes. (*Hint*: The resulting cross-correlation function is the superposition of the cross-correlation functions of the direct- and secondary-path signals.)

(b) Repeat the calculations of part (a) but with a secondary-path relative time delay of precisely $250\frac{1}{2}$ carrier cycles. Note that in this case the secondary-path phase is 180° out of phase with the direct-path signal, but still lies entirely in the baseband I channel.

5.3 (a) Using the discrete matrix version of the receiver clock model given by Eq. 5.75, find the standard deviation σ_{w_1} of the white-noise sequence $w_{1,k}$ needed in the model to produce a frequency standard deviation σ_{x_1} of 1Hz after 10 min of continuous oscillator operation. Assume that the initial frequency error at $t = 0$ is zero and that the discrete-time step Δt is 1 s.

(b) Using the assumptions and the value of σ_{w_1} found in part (a), find the standard deviation σ_{x_2} of the bias error after 10 min. Assume that $\sigma_{w_2} = 0$.

(c) Show that σ_{x_1} and σ_{x_2} approach infinity as the time t approaches infinity. Will this cause any problems in the development of a Kalman filter that includes estimates of the clock frequency and bias error?

5.4 The peak electron density in the ionosphere occurs in a height range of

(a) 50–100 km

(b) 250–400 km

(c) 500–700 km

(d) 800–1000 km

5.5 The refractive index of the gaseous mass in the troposphere is

(a) Slightly higher than unity

(b) Slightly lower than unity

(c) Unity

(d) Zero

5.6 If the range measurements for two simultaneously tracking satellites in a receiver are differenced, then the differenced measurement will be free of

(a) Receiver clock error only

(b) Satellite clock error and orbital error only

(c) Ionospheric delay error and tropospheric delay error only

(d) Ionospheric delay error, tropospheric delay error, satellite clock error, and orbital error only.

5.7 Zero baseline test (code) can be performed to estimate

(a) Receiver noise and multipath

(b) Receiver noise

(c) Receiver noise, multipath, and atmospheric delay errors

(d) None of the above

5.8 What are the purposes of selective availability (SA) and antispoofing (AS)?

5.9 What are GNSS

(a) Single difference?

(b) Double difference?

(c) Triple difference?

(d) Wide lane?

(e) Narrow lane?

5.10 What is a CORS site?

5.11 Derive the multipath formula equivalent to Eq. 5.46 for L_2, using the same notation as in Eq. 5.40.

5.12 Calculate the ionospheric delay using dual-frequency carrier phases.

6

DIFFERENTIAL GNSS

6.1 INTRODUCTION

Differential global navigation satellite system (differential GNSS; abbreviated DGNSS) is a technique for reducing the error in GPS-derived positions by using additional data from a reference GNSS receiver at a known position. The most common form of DGNSS involves determining the combined effects of navigation message ephemeris and satellite clock errors [including the effects of propagation] at a reference station and transmitting corrections, in real time, to a user's receiver. The receiver applies the corrections in the process of determining its position [94]. They include:

- Corrections for (1) selective availability (if present) (2) satellite ephemeris and clock errors, and (3) ionospheric delay errors.
- Still other error sources cannot be corrected with DGNSS: (1) multipath errors, (2) user receiver errors, and (3) tropospheric delay error.

6.2 DESCRIPTIONS OF LADGPS, WADGPS, AND SBAS

6.2.1 Local-Area Differential GPS (LADGPS)

LADGPS is a form of DGPS in which the user's GPS receiver receives real-time pseudorange and, possibly, carrier phase corrections from a reference receiver

Global Positioning Systems, Inertial Navigation, and Integration, Second Edition, by M. S. Grewal, L. R. Weill, and A. P. Andrews

generally located within the line of sight. The corrections account for the combined effects of navigation message ephemeris and satellite clock errors (including the effects of SA) and, usually, atmospheric propagation delay errors at the reference station. With the assumption that these errors are also common to the measurements made by the user's receiver, the application of the corrections will result in more accurate coordinates [117].

6.2.2 Wide-Area Differential GPS (WADGPS)

WADGPS is a form of DGPS in which the user's GPS receiver receives corrections determined from a network of reference stations distributed over a wide geographic area. Separate corrections are usually determined for specific error sources, such as satellite clock, ionospheric propagation delay, and ephemeris. The corrections are applied in the user's receiver or attached computer in computing the receiver's coordinates. The corrections are typically supplied in real time by way of a geostationary communications satellite or through a network of ground-based transmitters. Corrections may also be provided at a later date for postprocessing collected data [117].

6.2.3 Space-Based Augmentation Systems (SBAS)

6.2.3.1 Wide-Area Augmentation System (WAAS) WAAS enhances the GPS SPS and is available over a wide geographic area. The WAAS being developed by the Federal Aviation Administration, together with other agencies, will provide WADGPS corrections, additional ranging signals from geostationary (GEO) satellites, and integrity data on the GPS and GEO satellites [117].

Each GEO uplink subsystem includes a closed-loop control algorithm and special signal generator hardware. These ensure that the downlink signal to the users is controlled adequately to be used as a ranging source to supplement the GPS satellites in view.

The primary mission of WAAS is to provide a means for air navigation for all phases of flight in the National Airspace System (NAS) from departure, en route, arrival, and through approach. GPS augmented by WAAS offers the capability for both nonprecision approach (NPA) and precision approach (PA) within a specific service volume. A secondary mission of the WAAS is to provide a WAAS network time (WNT) offset between the WNT and Coordinated Universal Time (UTC) for nonnavigation users.

WAAS provides improved en route navigation and PA capability to WAAS-certified avionics. The safety critical WAAS system consists of the equipment and software necessary to augment the Department of Defense (DOD)-provided GPS SPS. WAAS provides a signal in space (SIS) to WAAS-certified aircraft avionics using the WAAS for any FAA-approved phase of flight. The SIS provides two services: (1) data on GPS and GEO satellites and (2) a ranging capability.

The GPS satellite data is received and processed at widely dispersed wide-area reference stations (WRSs), which are strategically located to provide coverage

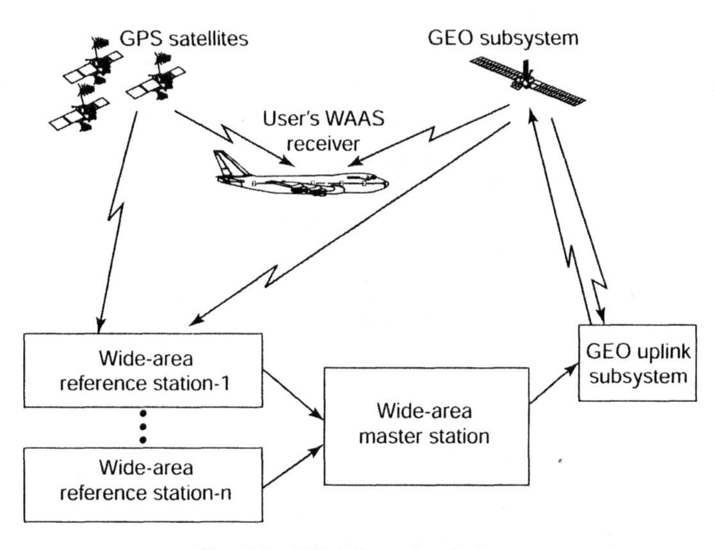

Fig. 6.1 WAAS top-level view.

over the required WAAS service volume. Data are forwarded to wide-area master stations (WMSs), which process the data from multiple WRSs to determine the integrity, differential corrections, and residual errors for each monitored satellite and for each predetermined ionospheric grid point (IGP). Multiple WMSs are provided to eliminate single-point failures within the WAAS network. Information from all WMSs is sent to each GEO uplink subsystem (GUS) and uplinked along with the GEO navigation message to GEO satellites. The GEO satellites downlink these data to the users via the GPS SPS L-band ranging signal (L_1) frequency with GPS-type modulation. Each ground-based station/subsystem communicates via a terrestrial communications subsystem (TCS). (See Fig. 6.1).

In addition to providing augmented GPS data to the users, WAAS verifies its own integrity and takes any necessary action to ensure that the system meets the WAAS performance requirements. WAAS also has a system operation and maintenance function that provides status and related maintenance information to FAA airway facilities (AFs) NAS personnel.

WAAS has a functional verification system (FVS) that is used for early development test and evaluation (DT&E), refinement of contractor site installation procedures, system-level testing, WAAS operational testing, and long-term support for WAAS.

Correction and Verification (C&V) processes data from all WRSs to determine integrity, differential corrections, satellite orbits, and residual error bounds for each monitored satellite. It also determines ionospheric vertical delays and their residual error bounds at each of the IGPs. C&V schedules and formats WAAS messages and forwards them to the GUSs for broadcast to the GEO satellites.

C&V's capabilities are as follows:

1. Control C&V Operations and Maintenance (COM) supports the transfer of files, performs remotely initiated software configuration checks, and accepts requests to start and stop execution of the C&V application software.
2. Control C&V Modes manage mode transitions in the C&V subsystem while the application software is running.
3. Monitor C&V (MCV) reports line replaceable unit (LRU) faults and configuration status. In addition, it monitors software processes and provides performance data for the local C&V subsystems.
4. Process Input Data (PID) selects and monitors data from the wide-area reference equipment (WREs). Data that passes PID screening is repackaged for other C&V capabilities. PID performs clock and L_1 GPS Precision Positioning Service L-band ranging signal (L_2) receiver bias calculations, cycle slip detection, outlier detection, data smoothing, and data monitoring. In addition, PID calculates and applies the windup correction to the carrier phase, accumulates data to estimate the pseudorange to carrier phase bias, and computes the ionosphere corrected carrier phase and measured slant delay.
5. Satellite Orbit Determination (SOD) determines the GPS and GEO satellite orbits and clock offsets, WRE receiver clock offsets, and troposphere delay.
6. Ionosphere Correction Computation (ICC) determines the L_1 IGP vertical delays, grid ionosphere vertical error (GIVE) for all defined IGPs, and L_1–L_2 interfrequency bias for each satellite transmitter and each WRS receiver.
7. Satellite Correction Processing (SCP) determines the fast and long-term satellite corrections, including the user differential range error (UDRE). It determines the WNT and the GEO and WNT clock steering commands [154].
8. Independent Data Verification (IDV) compares satellite corrections, GEO navigation data, and ionospheric corrections from two independent computational sources, and if the comparisons are within limits, one source is selected from which to build the WAAS messages. If the comparisons are not within limits, various responses may occur, depending on the data being compared, all the way from alarms being generated to the C&V being faulted.
9. Message Output Processing (MOP) transmits messages containing independently verified results of C&V calculations to the GUS processing (GP) for broadcast.
10. C&V Playback (PLB) processes the playback data that has been recorded by the other C&V capabilities.
11. Integrity Data Monitoring (IDM) checks both the broadcast and the to-be-broadcast UDREs and GIVEs to ensure that they are properly bounding their errors. In addition, it monitors and validates that the broadcast

messages are sent correctly. It also performs the WAAS time-to-alarm validation [1, 154].

WRS Algorithms Each WRS collects raw pseudorange (PR) and accumulated delta range (ADR) measurements from GPS and GEO satellites selected for tracking. Each WRS performs smoothing on the measurements and corrects for atmospheric effects, that is, ionospheric and tropospheric delays. These smoothed and atmospherically corrected measurements are provided to the WMS.

WMS Foreground (Fast) Algorithms The WMS foreground algorithms are applicable to real-time processing functions, specifically the computation of fast correction, determination of satellite integrity status and WAAS message formatting. This processing is done at a 1-Hz rate.

WMS Background (Slow) Algorithms The WMS background processing consists of algorithms that estimate slowly varying parameters. These algorithms consist of WRS clock error estimation, grid ionospecific delay computation, broadcast ephemeris computation, satellite orbit determination, satellite ephemeris error computation, and satellite visibility computation.

Independent Data Verification and Validation Algorithms This includes a set of WRS and at least one WMS, which enable monitoring the integrity status of GPS and the determination of wide-area DGPS correction data. Each WRS has three dual-frequency GPS receivers to provide parallel sets of measurement data. The presence of parallel data streams enables independent data verification and validation (IDV&V) to be employed to ensure the integrity of GPS data and their corrections in the WAAS messages broadcast via one or more GEOs. With IDV&V active, the WMS applies the corrections computed from one stream to the data from the other stream to provide verification of the corrections prior to transmission. The primary data stream is also used for the validation phase to check the active (already broadcast) correction and to monitor their SIS performance. These algorithms are continually being improved. The latest versions can be found in references the literature [68, 151, 152, 154, 217; 153, pp. 397–425].

6.2.3.2 European Global Navigation Overlay System (EGNOS) EGNOS is a joint project of the European Space Agency, the European Commission, and the European Organization for the Safety of Air Navigation (Eurocontrol). Its primary service area is the ECAC (European Civil Aviation Conference) region. However, several extensions of its service area to adjacent and more remote areas are under study. An overview of the EGNOS system architecture is presented in Fig. 6.2, where

[RIMS] are the Ranging and Integrity Monitoring Stations

[MCC] is the Mission and Control Center

[NLES] are the Navigation Land Earth Stations

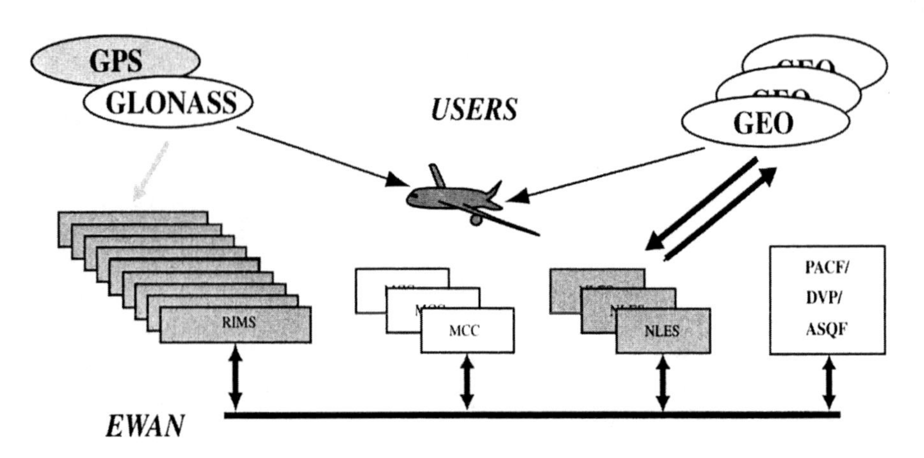

Fig. 6.2 European Global Navigation Overlay System architecture.

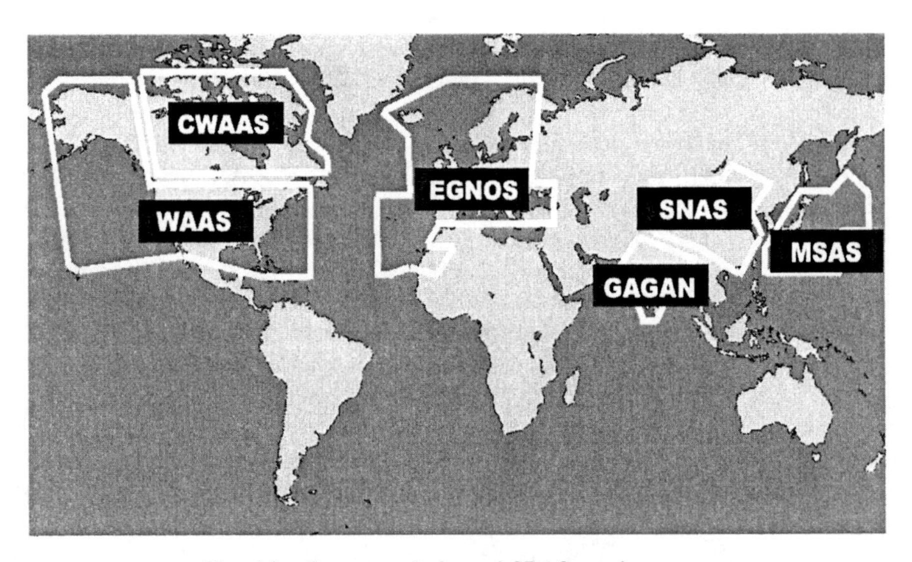

Fig. 6.3 Current and planned SBAS service areas.

[PACF] is the Performance Assessment and Checkout Facility
[DVP] is the Development and Verification Platform
[ASQF] is the Application Specific Qualification Facility
[EWAN] is the EGNOS Wide Area (communication) Network

6.2.3.3 Other SBAS Service areas of current and future SBAS systems are mapped in Fig, 6.3, and the acronyms are listed in Table 6.1.

TABLE 6.1. Worldwide SBAS System Coverages

Country	Acronym	Title
USA	WAAS	Wide-Area Augmentation System
Europe	EGNOS	European Geostationary Navigation Overlay System
Japan	MSAS	MTSAT Satellite-based Augmentation System
Canada	CWAAS	Canadian Wide Area Augmentation System
China	SNAS	Satellite Navigation Augmentation System
India	GAGAN	GPS & GEO Augmented Navigation

6.3 GROUND-BASED AUGMENTATION SYSTEM (GBAS)

6.3.1 Local-Area Augmentation System (LAAS)

The Local-Area Augmentation System (near airports) will be designed to provide differential GPS corrections in support of navigation and landing systems. The system provides monitoring functions via LAAS Ground Facility (LGF) and includes individual measurements, ranging sources, reference receivers, navigation data, data broadcast, environment sensors, and equipment failures. Each identified monitor has a corresponding system response including alarms, alerts and service alerts. (See Fig. 6.4.)

6.3.2 Joint Precision Approach Landing System (JPALS)

The Joint Precision Approach Landing System (JPALS) is in the "concept and technology development" phase by the Department of Defense. Concept exploration led to the determination that the Local-Area Differential Global Positioning System (LADGPS) was the best precision approach and landing solution.

The objective of upcoming "component advanced development" (CAD) work effort is to provide sufficient evidence that key technical risks for LADGPS have been reduced and to help define the JPALS technical architecture. The CADevelopment effort will operate under the belief that a fully developed LADGPS will give both conventional manned aircraft and possible unmanned aerial vehicles (UAVs) fully coupled (automatic) landing capability in any weather, and in any mission environment.

LADGPS-complementing technologies are also being explored. These include: GPS-based local- and wide-area augmentation systems (LAAS and WAAS) that will enhance civil interoperability, and autonomous landing capability (ALC), which will greatly improve a pilot's visibility by filtering out meteorological conditions like snow and fog.

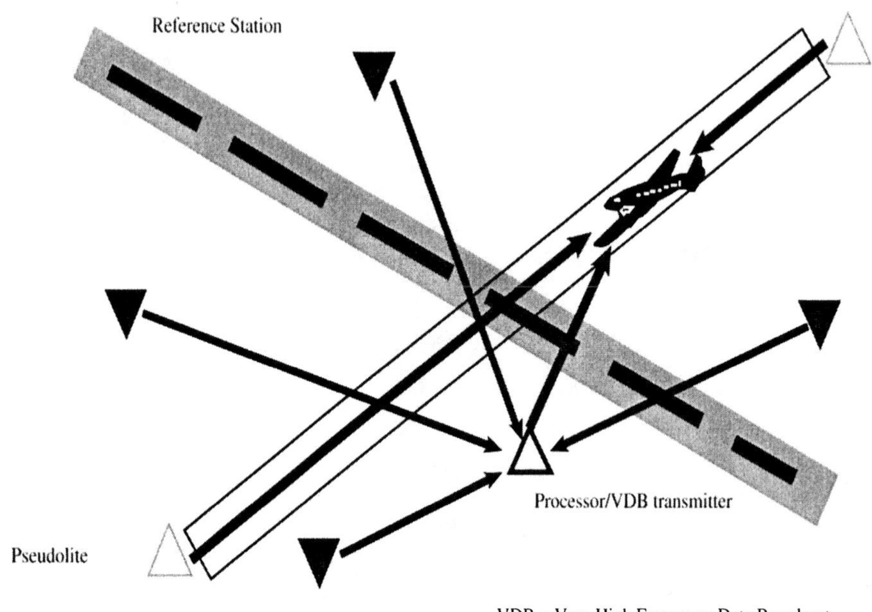

Fig. 6.4 Local-Area Augmentation System (LAAS).

6.3.3 LORAN-C

Long-range navigation (LORAN) uses signal phase information from three or more long-range navigation signal sources positioned at fixed, known locations. The LORAN-C system relies upon a plurality of ground-based signal towers, spaced 100-300 km apart. Antenna towers transmit distinguishable electromagnetic signals that are received and processed by a LORAN signal antenna and LORAN signal receiver/processor that are analogous to GPS antenna and receiver/processor.

6.4 GEO UPLINK SUBSYSTEM (GUS)

Corrections from the WMS are sent to the ground uplink subsystem (GUS) for uplink to the GEO. The GUS receives integrity and correction data and WAAS specific messages from the WMS, adds forward error correction (FEC) encoding, and transmits the messages via a C-band uplink to the GEO satellites for broadcast to the WAAS user. The GUS signal uses the GPS standard positioning service waveform (C/A-code, BPSK modulation); however, the data rate is higher (250 (bits per second)). The 250 (bits per second) of data are encoded with a one-half rate convolutional code, resulting in a transmission rate of 500 symbols per second (sps).

Each symbol is modulated by the C/A-code, a 1.023×10^6-chips/s pseudo random sequence to provide a spread-spectrum signal. This signal is then BPSK-modulated by the GUS onto an IF carrier, upconverted to a C-band frequency, and uplinked to the GEO. It is the C/A-code modulation that provides the ranging capability if its phase is properly controlled.

Control of the carrier frequency and phase is also required to eliminate uplink Doppler and to maintain coherence between code and carrier. The GUS monitors the C-band and L_1 downlinks from the GEO to provide closed-loop control of the PRN code and L_1 carrier coherency. WAAS short-and long-term code–carrier coherence requirements are met.

6.4.1 Description of the GUS Algorithm

The GUS control loop algorithm "precorrects" the code phase, carrier phase, and carrier frequency of the GEO uplink signal to maintain GEO broadcast code–carrier coherence. The uplink effects such as ionospheric code–carrier divergence, uplink Doppler, equipment delays, and frequency offsets must be corrected in the GUS control loop algorithm.

Figure 6.5 provides an overview of the functional elements of the GUS control loop. The control loop contains algorithm elements (shaded boxes) and hardware elements that either provide inputs to the algorithm or are controlled or affected by outputs from the algorithm. The hardware elements include a WAAS GPS receiver, GEO satellite, and GUS signal generator.

Downlink ionospheric delay is estimated in the ionospheric delay and rate estimator using pseudorange measurements form the WAAS GPS receiver on L_1 and L_2 (downconverted from the GEO C-band downlink at the GUS). This is a two-state Kalman filter that estimates the ionospheric delay and delay rate.

At each measurement interval, a range measurement is taken and fed into the range, rate, and acceleration estimator. This measurement is the average between the reference pseudorange from the GUS signal generator PR_{sign} and the received pseudorange from the L_1 downlink as measured by the WAAS GPS receiver (PR_{geo}) and adjusted for estimated ionospheric delay (PR_{iono}). The equation for the range measurement is then

$$z = \frac{1}{2}[(PR_{\text{geo}} - PR_{\text{iono}}) + PR_{\text{sign}}] - T_{C\text{up}} - T_{L1\text{dwn}S},$$

where $T_{C\text{up}}$ is C-band uplink delay (m) and $T_{L1\text{dwn}S}$ is L_1 receiver delay of the GUS (m).

The GUS signal generator is initialized with a pseudorange value from satellite ephemeris data. This is the initial reference from which corrections are made.

The range, rate and acceleration estimator is a three-state Kalman filter that drives the frequency and code control loops.

The code control loop is a second-order control system. The error signal for this control system is the difference between the WAAS pseudorange (P_{rsign}) and

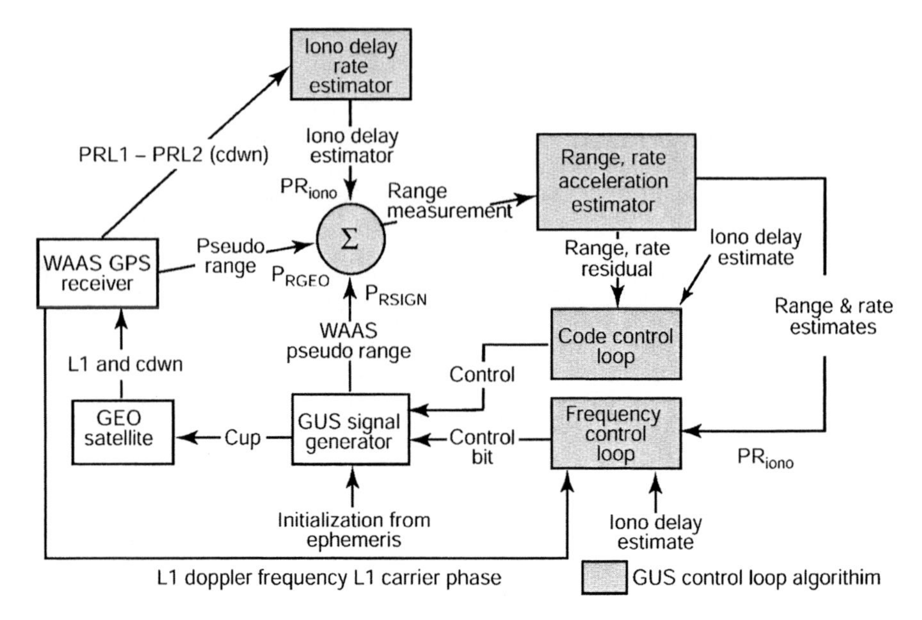

Fig. 6.5 GUS control loop block diagram.

the estimated pseudorange from the Kalman filter. The loop output is the code rate adjustments to the GUS signal generator.

The frequency control loop has two modes. First, it adjusts the signal generator frequency to compensate for uplink Doppler effects. This is accomplished using a first-order control system. The error signal input is the difference between the L_1 Doppler frequency from the WAAS GPS receiver and the estimated range rate (converted to a Doppler frequency) from the Kalman filter.

Once the frequency error is below a threshold value, the carrier phase is controlled. This is accomplished using a second-order control system. The error signal input to this system is the difference between the L_1 carrier phase and a carrier phase estimate based on the Kalman filter output. This estimated range is converted to carrier cycles using the range estimate at the time carrier phase control starts as a reference. Fine-tuning adjustments are made to the signal generator carrier frequency to maintain phase coherence [52, 67, 68, 69, 145].

6.4.2 In-Orbit Tests (IOT)

Two separate series of in-orbit tests (IOTs) were conducted, one at the COMSAT GPS Earth Station (GES) in Santa Paula, California with Pacific Ocean Region (POR) and Atlantic Ocean Region West (AOR-W) Inmarsat-3 (I-3) satellites and

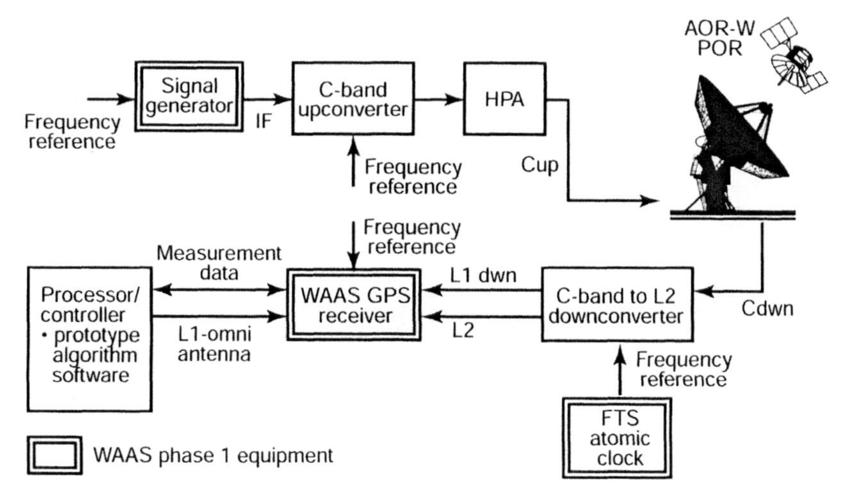

Fig. 6.6 IOT test GUS setup.

the other at the COMSAT GES in Clarksburg, Maryland, using AOR-W. The IOTs were conducted to validate a prototype version of the GUS control loop algorithm. Data were collected to verify the ionospheric estimation and code–carrier coherence performance capability of the control loop and the short–term carrier frequency stability of the Inmarsat-3 satellites with a prototype ground station. The test results were also used to validate the GUS control loop simulation.

Figure 6.6 illustrates the IOT setup at a high level. Prototype ground station hardware and software were used to assess algorithm performance at two different ground stations with two different Inmarsat-3 satellites.

6.4.3 Ionospheric Delay Estimation

The GUS control loop estimates the ionospheric delay contribution of the GEO C-band uplink to maintain code–carrier coherence of the broadcast SIS. Figure 6.7 shows the delay estimates for POR using the Santa Paula AOR-W. The plot shows the estimated ionospheric delay (output of the two-state Kalman filter) versus the calculated delay using the L_1 and C pseudorange data from a WAAS GPS receiver. Calculated delay is noisier and varying about 1 m/s, whereas the estimated delay by the Kalman filter is right in middle of the measured delay, as shown in Fig. 6.7. Delay measurements were calculated using the equation

$$\text{Ionospheric delay} = \frac{P_{\text{RL1}} - P_{\text{RC}} - \tau \text{L}_1 + \tau C1 - [\text{L}_1 \text{ freq}]^2}{1 - [\text{L}_1 \text{freq}]^2 / [C \text{ freq}]^2}$$

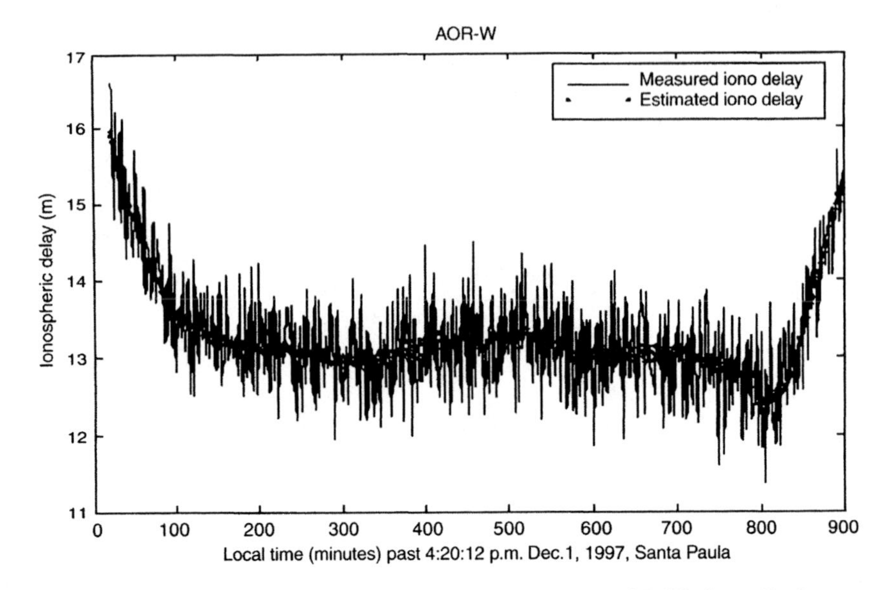

Fig. 6.7 Measured and estimated ionospheric delay, AOR-W, Santa Paula.

where

$$P_{\mathrm{RL1}} = \mathrm{L_1 \ pseudorange \ (m)}$$

$$P_{\mathrm{RC}} = C \ \mathrm{pseudorange \ (m)}$$

$$\tau \mathrm{L_1} = \mathrm{L_1 \ downlink \ delay \ (m)}$$

$$\tau \mathrm{C} = C \ \mathrm{downlink \ delay \ (m)}$$

$$\mathrm{L_1 \ freq} = \mathrm{L_1 \ frequency} = 1575.42 \ \mathrm{MHz}$$

$$\mathrm{Cfreq} = C \ \mathrm{frequency} = 3630.42 \ \mathrm{MHz}$$

The ionosphere during the IOTs was fairly benign with no high levels of solar activity observed. Table 6.2 provides the ionospheric delay statistics (in meters) between the output of the ionospheric Kalman filter in the control loop, and the calculated delay from the WAAS GPS receiver's L_1 and L_2 pseudoranges. The statistics show that the loop's ionospheric delay estimation is very close (low

TABLE 6.2. Observed RMS WAAS Ionospheric Correction Errors

In-Orbit Test	RMS Error (m)
Santa Paula GES, Oct. 10, 1997, POR	0.20
Santa Paula GES, Dec. 1, 1997, AOR-W	0.45
Clarksburg GES, March 20, 1998, AOR-W	0.34

RMS) to the ionospheric delay calculated using the measured pseudorange from the WAAS GPS receiver.

6.4.4 Code–Carrier Frequency Coherence

The GEO's broadcast code–carrier frequency coherence requirement is specified in the WAAS System Specification and Appendix A of Ref. 165. It states:

> The lack of coherence between the broadcast carrier phase and the code phase shall be limited. The short term fractional frequency difference between the code phase rate and the carrier frequency will be less than 5×10^{-11}. That is,
>
> $$\left| \frac{f_{\text{code}}}{1.023\ \text{MHz}} - \frac{f_{\text{carrier}}}{1575.42\ \text{MHz}} \right| < 5 \times 10^{-11}$$
>
> Over the long term, the difference between the broadcast code phase (1/1540) and the broadcast carrier phase will be within one carrier cycle, one sigma. This does not include code–carrier divergence due to ionospheric refraction in the downlink propagation path.

For the WAAS program, *short term* is defined as less than 10 s and *long term*, less than 100 s. Pseudorange minus the ionospheric estimates averaged over τ seconds is expressed as

$$F_{\text{PR}} = \frac{PR_{\text{L1}}(t) - \text{ionoestimate}\ (t)}{\tau} \qquad \text{m/s}.$$

Carrier phase minus the ionospheric estimate average over τ seconds is expressed as

$$F_{\text{PH}} = \frac{-\phi_{\text{L1}}(t) + (\text{ionoestimate}\ (t)/\lambda_{\text{L1}})}{\tau} \qquad \text{cycles/s}.$$

For long-term code–carrier coherence calculations, a τ of 60 s was chosen to mitigate receiver bias errors in the pseudorange and carrier phase measurements of the WAAS GPS receiver. For short-term code–carrier coherence a shorter 30-s averaging time was selected. The code–carrier coherence requirement is specified at the output of the GEO and not the receiver, so data averaging has to be employed to back out receiver effects such as multipath and noise. Each averaging time was based on analyzing GPS satellite code–carrier coherence data and selecting the minimum averaging time required for GPS to meet the WAAS code–carrier coherence requirements.

For long-term code–carrier coherence calculations, the difference between the pseudorange and the phase measurements is given by

$$\Delta_{\text{PR-PH}} = \frac{F_{\text{PR}}}{\lambda_{\text{L1}}} - F_{\text{PH}} \quad \text{cycles/s},$$

TABLE 6.3. Code–Carrier Coherence Test Results

		Carrier Coherence Requirements	
Test Location	Test Date	Short-Term[a] (10 s) $<5 \times 10^{-11}$	Long-Term[b] (100 s) <1 cycle
Santa Paula prototyping	POR, Oct. 10, 1997	1.89×10^{-11}	0.326
	AOR-W, Dec. 1, 1997	1.78×10^{-11}	0.392
Clarksburg prototyping	AOR-W, March 20, 1998	1.92×10^{-11}	0.434

[a] Data averaging 30 s for short term.
[b] Data averaging 60 s for long term.

where λ_{L1} is the wavelength of the L_1 carrier frequency and *long-term coherence* equals $|\Delta_{PR-PH}(t + 100) - \Delta_{PR-PH}(t)|$ cycles.

For short-term code–carrier coherence calculations, the difference between the pseudorange and the phase measurements is given by

$$\delta_{PR-PH} = \frac{F_{PR} - F_{PH}}{10 \times c \text{ (speed of light)}},$$

and *short-term coherence* is $|\delta_{PR-PH}(t + 10) - \delta_{PR-PH}(t)|$.

The IOT long- and short-term code–carrier results from Santa Paula and Clarksburg are shown in Table 6.3. The results indicate that the control loop algorithm performance meets the long- and short-term code–carrier requirements of WAAS with the Inmarsat-3 satellites.

6.4.5 Carrier Frequency Stability

Carrier frequency stability is a function of both the uplink frequency standard, GUS signal generator, and Inmarsat-3 transponder. The GEO's short-term carrier frequency stability requirement is specified in the WAAS System Specification and Appendix A of Ref. 165. It states: "The short term stability of the carrier frequency (square root of the Allan variance) at the input of the user's receiver antenna shall be better than 5×10^{-11} over 1 to 10 seconds, excluding the effects of the ionosphere and Doppler."

The Allan variance [2] is calculated on the second difference of L_1 phase data divided by the center frequency over 1–10 s. Effects of smoothed ionosphere and Doppler are compensated for in the data prior to this calculation. Test results in Table 6.4 show that the POR and AOR-W Inmarsat-3 GEOs, in conjunction with WAAS ground station equipment, meet the short-term carrier frequency stability requirement of WAAS.

TABLE 6.4. Carrier Frequency Stability Requirements Satisfied

	Requirements for L_1:	$<5 \times 10^{-11}$ (1 s)	$<5 \times 10^{-11}$ (10 s)
Santa Paula prototyping	Oct. 10, 1997, POR	4.52×10^{-11}	5.32×10^{-12}
	Dec. 1, 1997, AOR-W	3.93×10^{-11}	4.5×10^{-12}
Clarksburg prototyping	March 20, 1998	4.92×10^{-11}	4.73×10^{-12}

6.5 GUS CLOCK STEERING ALGORITHMS

The local oscillator (cesium frequency standard) at the GUS is not perfectly stable with respect to WAAS network time (WNT). Even though the cesium frequency standard is very stable, it has inherent drift. Over a long period of operation, as in the WAAS scenario, this slow drift will accumulate and result in an offset so large that the value will not fit in the associated data fields in the WAAS Type 9 message. This is why a clock steering algorithm is necessary at the GUS. This drifting effect will cause GUS time and WNT to slowly diverge. The GUS can compensate for this drift by periodically re-synchronizing the receiver time with the WNT using the estimated receiver clock offset $[a_0(t_k)]$. This clock offset is provided by the WMS in WAAS Type 9 messages. (See Fig. 6.8.)

GUS steering algorithms for the primary and backup GEO uplink subsystems [66, 158] are discussed in the next section.

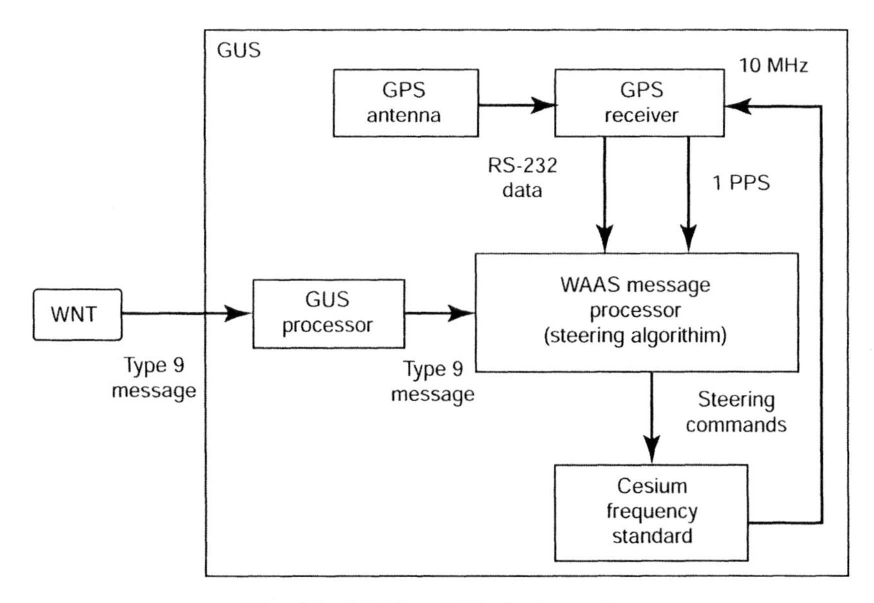

Fig. 6.8 WMS-to-GUS clock steering.

The primary GUS clock steering is closed loop via the signal generator, GEO, WRS, WMS, to the GUS processor. The backup GUS clock steering is an open-loop system, because the backup does not uplink to the GEO. The clock offset is calculated using the estimated range and the range calculated from the C&V provided GEO positions.

The GUS also contains the WAAS clock steering algorithm. This algorithm uses the WAAS Type 9 messages from the WMS to align the GEO's epoch with the GPS epoch. The WAAS Type 9 message contains a term referred to as a_0, or clock offset. This offset represents a correction, or time difference, between the GEOs epoch and WNT. WNT is the internal time reference scale of WAAS and is required to track the GPS timescale, while at the same time providing the users with the translation to UTC. Since GPS master time is not directly obtainable, the WAAS architecture requires that WNT be computed at multiple WMSs using potentially differing sets of measurements from potentially differing sets of receivers and clocks (WAAS reference stations). WNT is required to agree with GPS to within 50 ns. At the same time, the WNT-to-UTC offset must be provided to the user, with the offset being accurate to 20 ns. The GUS calculates local clock adjustments. In accordance with these clock adjustments, the frequency standard can be made to speed up or slow the GUS clock. This will keep the total GEO clock offset within the range allowed by the WAAS Type 9 message so that users can make the proper clock corrections in their algorithms.

6.5.1 Primary GUS Clock Steering Algorithm

The GUS clock steering algorithm calculates the fractional frequency control adjustment required to slowly steer the GUS's cesium frequency standard to align the GEO's epoch. These frequency control signals are very small so normal operation of the code and frequency control loops of any user receiver is not disturbed. Figure 6.8 shows the primary GUS's closed-loop control system block diagram. The primary GUS is the active uplink dedicated to either the AOR-W or POR GEO satellite. If this primary GUS fails, then the hot "backup GUS" is switched to primary.

The clock steering algorithm is designed using a proportional and integral (PI) controller. This algorithm allows one to optimize by adjusting the parameters a, b, and T. Values of a and b are optimized to 0.707 damping ratio.

The value $\overline{a}_0(t_k)$ is the range residual for the primary GUS:

$$\overline{a}_0(t_k) = \frac{1}{N} \sum_{n=1}^{N} a_0(t_{k-n}).$$

The value $f_c(t_k)$ is the frequency control signal to be applied at time t_k to the GUS cesium frequency standard:

$$f_c(t_k) = -\left[\frac{\alpha}{T} \overline{a}_0(t_k) + \frac{\beta}{T^2} \int_0^{t_k} \overline{a}_0(t)\, dt \right],$$

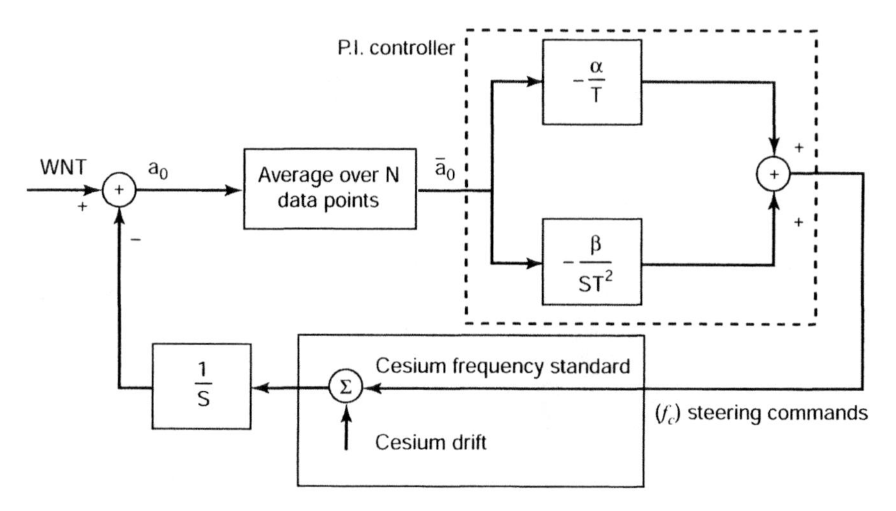

Fig. 6.9 Clock steering block diagram.

where

$$
\begin{aligned}
T &= \text{large time constant} \\
\alpha, \beta &= \text{control parameters} \\
N &= \text{number of data points within period } t \\
t &= \text{time of averaging period} \\
t_k &= \text{time when the frequency control signal is applied to} \\
&\qquad \text{the cesium frequency standard} \\
a_0(t_k) &= \text{time offset for GEO at time} t_k \text{ provided by WMS} \\
&\qquad \text{for primary GUS} \\
S &= \text{Laplace transform variable(see Fig. 6.9)}
\end{aligned}
$$

6.5.2 Backup GUS Clock Steering Algorithm

The backup GUS must employ a different algorithm for calculating the range residual. Since the backup GUS is not transmitting to the satellite, the WMS cannot model the clock drift at the backup GUS, and therefore an a_0 term is not present in the WAAS Type 9 message. In lieu of the a_0 term provided by the WMS, the backup GUS calculates an equivalent a_0 parameter.

The range residual $a_0(t_k)$ for the backup GUS is calculated as follows [64]:

$$a_0(t_k) = \frac{B_{\text{RE}} - R_{\text{WMS}}}{c - S(t_k)},$$

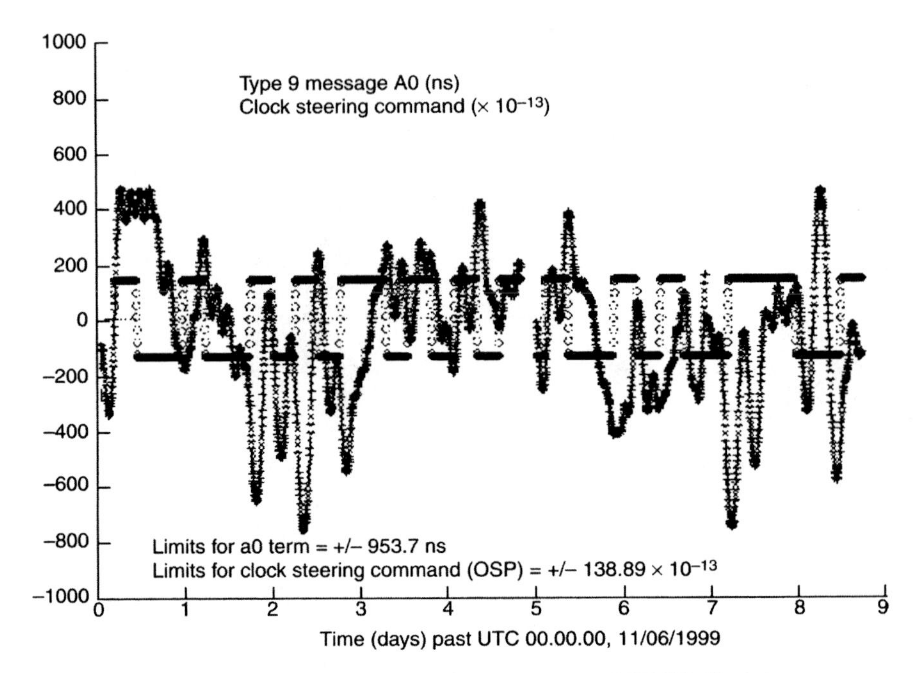

Fig. 6.10 Primary GP clock steering parameters, AOR-W, Clarksburg.

where $B_{\rm RE}$ is range estimate in the backup range estimator, $R_{\rm WMS}$ is range estimate calculated from the GEO position supplied by WMs Type 9 message, c is the speed of light, and $S(t_k)$ is Sagnac effect correction in an inertial frame.

The backup GUS uses the same algorithm $f_c(t_k)$ as the primary GUS.

6.5.3 Clock Steering Test Results Description

6.5.3.1 AOR-W Primary (Clarksburg, MD) Figure 6.10 shows the test results for the first 9 days. The first two to three days had cold-start transients and WMS switch overs (LA to DC and DC to LA). From the third to the sixth day, the clock stayed within ±250 ns. At the end of the sixth day, a maneuver took place and caused a small transient and the clock offset went to −750 ns. On the eighth day, the primary GUS was switched to Santa Paula, and another transient was observed. Clock steering command limits are $\pm 138.89 \times 10^{-13}$. Limits on the clock offset from the WAAS Type 9 messages are ±953.7 ns.

6.5.3.2 POR Backup (Santa Paula, CA) Figure 6.11 shows cold-start transients. After initial transients, the backup GUS stayed within ±550 ns for 9 days.

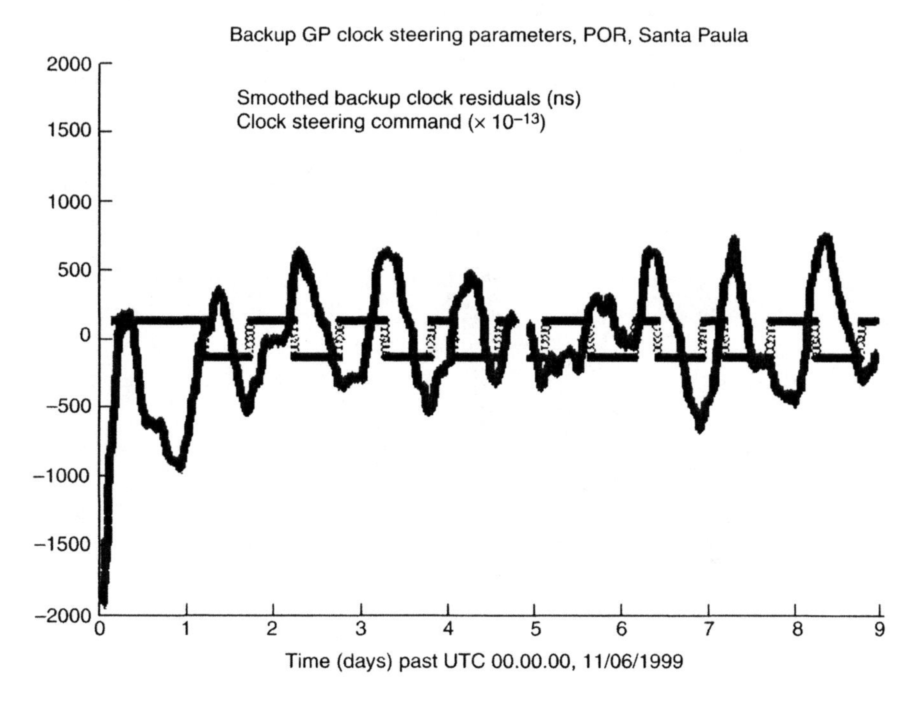

Fig. 6.11 Backup GPS clock steering parameters for POR (Santa Paula, CA).

The clock offsets in all four cases are less than ±953.7 ns (limit on WAAS Type 9 Message) for 9 days.

6.6 GEO WITH L_1/L_5 SIGNALS

The space-based augmentation system (SBAS) uses geostationary earth orbit (GEO) satellites to relay correction and integrity information to users. A secondary use of the GEO signal is to provide users with a GPS-like ranging source. The ranging signal is generated on the ground and provided via C-band uplink to the GEO, where the navigation payload translates the uplinked signal to an L_1 downlink frequency. The GEO incorporates an additional C-band downlink to provide ionospheric delay observations to the GEO uplink ground station. The GEO Communication and control segment (GCCS) will add new L_1 L_5 GEOs and ground stations to SBAS.

A key feature of GCCS is the addition of a second independently generated and controlled uplink signal. In contrast to SBAS, which uplinks and controls a single C-band signal, GCCS uplinks two independent C-band signals, which are translated to L_1 and L_5 downlink signals. Closed-loop control of the GEO's L_1 and L_5 broadcast signals in space is necessary to ensure that the algorithms compensate for various sources of uplink divergence between the code and carrier, including uplink ionospheric delay, uplink Doppler, and divergence due to carrier frequency translation errors induced by the GEO's transponder. Use of two independent broadcast signals creates unique challenges in estimating biases and maintaining coherency between the two signals.

Raytheon Company is developing a subsystem for GCCS uplink signal generation under a subcontract to Lockheed Martin. GCCS will add new GEOs to the space-based augmentation system (SBAS), which Raytheon is developing under contract with the Federal Aviation Administration (FAA). SBAS is a GPS-based navigation system that is intended to become the primary navigational aid for aviation during all phases of flight.

The SBAS makes use of a network of wide-area reference stations (WRSs) distributed throughout the United States. These reference stations collect pseudorange measurements and send them to the SBAS wide-area master stations (WMSs). The master stations process the data to provide correction and integrity information for each geostationary earth orbit (GEO) and GPS satellite in view. The corrections information includes satellite ephemeris errors, clock bias, and ionospheric estimation data. The corrections from the WMS are sent to the GEO uplink subsystem (GUS) for uplink to the GEO.

The GUS receives SBAS messages from the WMS, adds forward error correction (FEC) encoding, and transmits the messages via a single C-band uplink to the GEO satellite for broadcast to SBAS users. The GUS uplink signal uses the GPS standard positioning service waveform (C/A code, BPSK modulation); however, the data rate is higher (250 (bits per second)). The 250 bits of data are encoded with a one-half rate convolutional code, resulting in a 500 symbols per second transmission rate.

New GEO satellites will be added to SBAS under the SBAS *GEO Communication and Control Segment* (GCCS) contract. Raytheon is currently developing control algorithms for GCCS under a subcontract to Lockheed Martin for the FAA. A key feature of GCCS is that satellite broadcasts will be available at both the GPS L_1 and L_5 frequencies. Unlike current SBAS broadcasts, which utilize a single uplink signal frequency translated into two downlinks, the future GEOs will uplink two independent C-band signals, which the transponder will frequency translate and broadcast as independent L_1 and L_5 downlink signals. Figure 6.12 provides a top-level view of the GCCS architecture.

For L_1 loop, each symbol is modulated by the C/A code, a 1.023×106 chips-per-second (cps) pseudorandom sequence to provide a spread spectrum signal. This signal is then BPSK modulated by the GUS onto an intermediate frequency (IF) carrier, upconverted to a C-band frequency, and uplinked to the GEO. The satellite's navigation transponder translates the signal in frequency to both L-band (GPS L_1) and C-band downlink frequencies. The GUS monitors the L_1 and C-band downlink signals from the GEO to provide closed-loop control of the code and L_1 carrier. When properly controlled, the SBAS GEO provides ranging signals, as well as GPS corrections and integrity data, to end users.

The L_5 spread-spectrum signal will be generated by modulating each message symbol with a 10.23×106-cps pseudorandom code, which is an order of magnitude longer than that of the L_1 C/A-code. As with L_1, the L_5 signal will then be BPSK modulated onto an IF carrier, upconverted to a C-band frequency, and uplinked to the GEO. The GEO transponder will independently translate the two uplink signals to L band for broadcast to SBAS end users. Use of two independent broadcast signals creates unique challenges in estimating biases and maintaining coherency between the two signals.

An important aspect of the downlink signals is coherence between the code and carrier frequency. To ensure code carrier coherency, closed-loop control algorithms, implemented in the safety computer's SBAS message processors (WMPs), are used to maintain the code chipping rate and carrier frequency of the received L_1 signal at a constant ratio of 1:1540. The C-band downlink is used by the control algorithms to estimate and correct for ionospheric delay on the uplink signal. Control algorithms also correct for other uplink effects such as Doppler, equipment delays, and transponder offsets in order to maintain the correct Doppler and ionospheric divergence as observed by the user.

Closed loop control of each signal is required to maintain coherence between its code and carrier frequency, as described above. With two independent signal paths, it is also required that coherence between the two carriers be maintained for correct ionospheric delay estimation. As before, the control-loop algorithms "pre-correct" the code phase, carrier phase, and carrier frequency of the L_1 and L_5 signals to remove uplink effects such as ionospheric delays, uplink Doppler, equipment delays, and frequency offsets. In addition, differential biases between the L_1 and L_5 signals must be estimated and corrected.

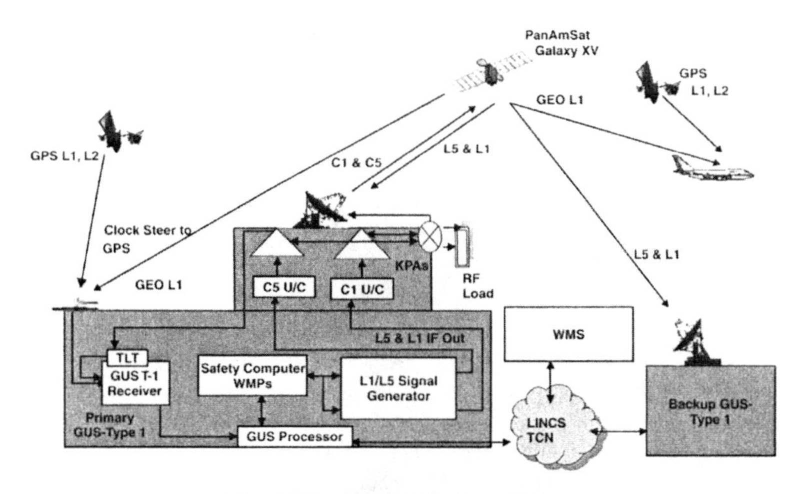

Fig. 6.12 GCCS Top Level View.

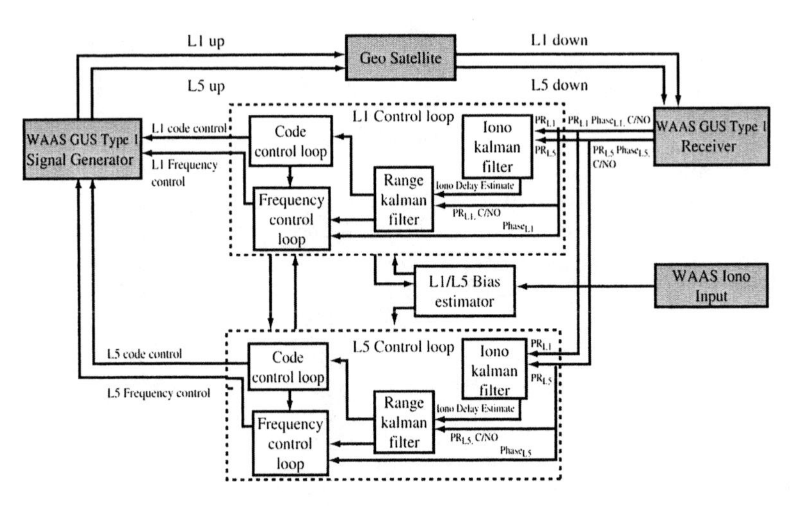

Fig. 6.13 Primary GUST control loop functional block diagram.

Each control algorithm contains two Kalman filters and two control loops. One Kalman filter estimates the ionospheric delay and its rate of change from L_1 and L_5 pseudorange measurements. The second Kalman filter estimates range, range rate, range acceleration, and acceleration rate from raw pseudorange measurements. Range estimates are adjusted for ionospheric delay, as estimated by the first Kalman filter. Each code control loop generates a code chip rate command and chip acceleration command to compensate for uplink ionospheric delay and for the uplink Doppler effect. Each frequency control loop generates a carrier frequency command and a frequency rate command. A final estimator is used to calculate bias between the L_1 and L_5 signals.

Results of laboratory tests utilizing live L_1/L_5 hardware elements and simulated satellite effects follow.

6.6.1 GEO Uplink Subsystem Type 1 (GUST) Control Loop Overview

The primary GUST control loop functional block diagram is shown in Figure 6.13. The backup GUST control loop is similar to the primary GUST control loop except that the uplink signal is radiated into a dummy load. The operation of the backup GUST control loop is different from the primary GUST because of the latter.

Each of the L_1 and L_5 control loops in the primary GUST consists of an iono Kalman filter, a range Kalman filter, a code control function, and a frequency control function. In addition, there is an L_1/L_5 bias estimation function. These control loop functions reside inside the safety computer. The external inputs to the control loop algorithm are the pseudorange, carrier phase, Doppler, and carrier-to-noise ratio from the receiver.

6.6.1.1 Ionospheric Kalman Filters The L_1 and L_5 ionospheric (iono) Kalman filters are two-state filters.

$$\mathbf{x} = \begin{bmatrix} \text{iono delay} \\ \text{iono delay rate} \end{bmatrix}.$$

During every 1-s timeframe in the safety computer, the ionospheric Kalman filter states and the covariance are propagated. The equations for Kalman filter propagation are given in Table 8.1 (in Chapter 8).

The L_1 filter measurement is formulated as follows:

$$z = \frac{(PR_{\mathrm{L1}} - d_{\mathrm{L1}}) - (PR_{\mathrm{L5}} - d_{\mathrm{L5}})}{(1 - \mathrm{L_1 freq})^2 / (\mathrm{L_5 freq})^2},$$

where PR_{L1} is the L_1 pseudorange, PR_{L5} is the L_5 pseudorange, d_{L1} and d_{L5} are the predetermined L_1 and L_5 downlink path hardware delays, L_5freq is the L_1 nominal frequency of 1575.42 MHz, and L_5freq is the L_5 nominal frequency of 1176.45 MHz.

The L_5 ionospheric Kalman filter design is similar to that for L_1, with the filter measurement as follows:

$$z = \frac{(PR_{\mathrm{L1}} - d_{\mathrm{L1}}) - (PR_{\mathrm{L5}} - d_{\mathrm{L5}})}{(\mathrm{L_5 freq})^2 / (\mathrm{L_1 freq})^2 - 1}. \tag{6.1}$$

6.6.1.2 Range Kalman Filter The L_1 and L_5 range Kalman filters use four state variables:

$$\mathbf{x} \stackrel{\text{def}}{=} \begin{bmatrix} \text{range} \\ \text{range rate} \\ \text{acceleration} \\ \text{acceleration rate} \end{bmatrix}. \tag{6.2}$$

During every 1-s timeframe in the safety computer, the range Kalman filter states and their covariance of uncertainty are propagated (predicted) in the Kalman filter.

After the filter propagation, if L_1 pseudorange is valid, the L_1 range estimate and covariance are updated in the Kalman filter using the L_1 pseudorange measurement correction. Likewise, if L_5 pseudorange is valid, the L_5 range estimate and covariance are updated in the Kalman filter using the L_5 pseudorange measurement correction.

The L_1 range Kalman filter measurement is

$$\begin{aligned} Z_1 = {} & \mathrm{L_1}\ \text{pseudorange} \\ & -\text{predetermined } \mathrm{L_1} \text{ downlink path hardware delay} \\ & -\mathrm{L_1}\ \text{iono delay estimate}. \end{aligned}$$

Likewise, the L_5 range Kalman filter measurement is

$$Z_5 = L_5 \text{ pseudorange}$$
$$-\text{predetermined } L_5 \text{ downlink path hardware delay}$$
$$-L_5 \text{ iono delay estimate.}$$

The required Kalman filter equations are given in Table 8.1.

6.6.1.3 Code Control Function The L_1 and L_5 code control functions compute the corresponding code chip rate commands, and the chip acceleration commands to be sent to the signal generator. The signal generator adjusts its L_1 and L_5 chip rates according to these commands. The purpose of code control is to compensate for any initial GEO range estimation error, the iono delay on the uplink C-band signal, and the Doppler effects due to the GEO movement on the uplink signal code chip rate. This compensation will ensure that the GEO signal code phase deviation is within the required limit.

The receiver and signal generator timing [1-pulse-per-second (pps)] errors also affect the GEO signal code phase deviation. These errors are compensated separately by the clock steering algorithm [72].

Measurement errors in the predetermined hardware delays of the two signal paths (both uplink and downlink) will result in additional code phase deviation for the GEO signal due to the closed-loop control. This additional code phase deviation will be interpreted as GEO satellite clock error by the master station's GEO orbit determination. Since the clock steering algorithm will use the SBAS broadcast Type 9 message GEO clock offset as part of the input to the clock steering controller [72], the additional code phase deviation due to common measurement errors will be compensated for by the clock steering function.

There are several inputs to the code control function: the uplink range, the projected range of the GEO for the next one-second timeframe, the estimated iono delay, and so on. The uplink range is the integration of the commanded chip rate, and this integration is performed in the safety computer. The commanded chip acceleration is computed on the basis of the estimated acceleration from the Kalman filter (see Table 8.1).

6.6.1.4 Frequency Control Function The L_1 and L_5 frequency control functions compute the corresponding carrier frequency commands and the frequency change rate (acceleration) commands to be sent to the signal generator. The signal generator adjusts the L_1 and L_5 IF outputs according to these commands. The purpose of frequency control is to compensate for the Doppler effects due to the GEO movement on the carrier of the uplink signal, the effect of iono rate on the uplink carrier, and the frequency offset of the GEO transponders. This function also continuously estimates the GEO transponder offset, which could drift during the lifetime of the GEO satellite.

6.6.1.5 L_1/L_5 Bias Estimation Function This function estimates the bias between the L_1 and L_5 that is due to differential measurement errors in the predetermined hardware delays of the two signal paths. If not estimated and compensated, the bias between L_1 and L_5 will be indistinguishable from iono delay, as shown in the equations below. L_1 and L_5 pseudorange can be expressed as

$$PR_{L1} = R + I_{L1} + \text{true } d_{L1} + \text{clock error} + \text{tropo delay}, \quad (6.3)$$

$$PR_{L5} = R + I_{L5} + \text{true } d_{L5} + \text{clock error} + \text{tropo delay}, \quad (6.4)$$

where R is the true range, I_{L1} is the true L_1 iono delay, I_{L5} is the true L_5 iono delay, true d_{L1} is the true L_1 downlink path hardware delay, and true d_{L5} is the true L_5 downlink path hardware delay.

This becomes

$$PR_{L1} - d_{L1} = R + I_{L1} + \text{true } d_{L1} + \text{clock error} + \text{tropo delay} - d_{L1}, \quad (6.5)$$

$$PR_{L5} - d_{L5} = R + I_{L5} + \text{true } d_{L5} + \text{clock error} + \text{tropo delay} - d_{L5}, \quad (6.6)$$

where d_{L1} is the predetermined (measured) L_1 downlink path hardware delay and d_{L5} is the predetermined (measured) L_5 downlink path hardware delay.

Let $\Delta d_{L1} = \text{true } d_{L1} - d_{L1}$ and $\Delta d_{L5} = \text{true } d_{L5} - d_{L5}$. The measurement for the L_1 iono Kalman filter becomes

$$z = \frac{(PR_{L1} - d_{L1}) - (PR_{L5} - d_{L5})}{(1 - L_1\text{freq})^2/(L_5\text{freq})^2}, \quad (6.7)$$

$$= I_{L1} + \frac{(\Delta d_{L1} - \Delta d_{L5})}{(1 - L_1\text{freq})^2/(L_5\text{freq})^2}. \quad (6.8)$$

The term $(\Delta d_{L1} - \Delta d_{L5})$ /(1- $_{L1}$ freq)^2/ L_5 freq)^2 is the differential L_1/L_5 bias term, and it becomes an error in the L_1 iono delay estimation. The L_5 iono Kalman filter is similarly affected by the L_1/L_5 bias term.

6.7 NEW GUS CLOCK STEERING ALGORITHM

Presently, the SBAS wide-area master station (WMS) calculates SBAS network time (WNT) and estimates clock parameters (offset and drift) for each satellite. The GEO uplink system (GUS) clock is an independent free running clock. However, the GUS clock must track WNT (GPS time) to enable accurate ranging from the GEO signal in space (SIS). Therefore, a clock steering algorithm is necessary. The GUS clock steering algorithms reside in the SBAS Message Processor (WMP). The SBAS Type 9 message (GEO navigation message) is used as input to the GUS WMP, provided by the WMS.

In the new algorithm, the GUS clock is steered to the GPS time epoch (see also Fig. 6.14). The GUS receiver clock error is the deviation of its one-second pulse

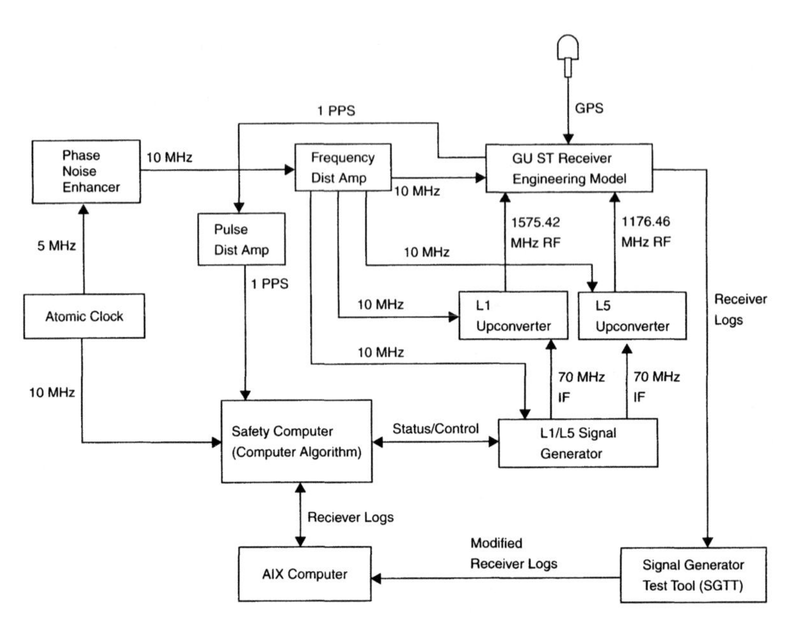

Fig. 6.14 Control loop test setup.

from the GPS epoch. The clock error is computed in the GUS processor by calculating the user position error by combining (in the least-squares sense, weighted with expected error statistics) multiple satellite data (pseudorange residuals called *MOPS*[1] *residuals*) into a position error estimate with respect to surveyed GUS position. The clock steering algorithm is initialized with the SBAS Type 9 message (GEO navigation message). This design keeps the GUS receiver clock 1 pulse per second synchronized with the GPS time epoch. Since the 10-MHz frequency standard is the frequency reference for the receiver, its frequency output needs to be controlled so that the 1 pps is adjusted. A proportional, integral and differential (PID) controller has been designed to synchronize to the GPS time at GUS locations. Two sets of prototype clock steering results are shown. Clock adjustment commands as applied to the frequency standard to null the MOPS clock offset are given.

This new algorithm also decouples the GUS clock from orbit errors and increases the observability of orbit errors in the orbit determination filter in the correction processor of the WMS. It also synchronizes GUS clocks at all GUS locations to GPS time. This section shows the algorithm prototype results of GUS clock steering to GPS time. It also shows the improvements in the GEO (AOR-W) orbits when the GEO clock state is known, thereby making the GEO a valid ranging source. The GEO range errors with the known clock solution are found to be up to a factor of ~5 better than those for the fielded SBAS with real

[1]Reference 165 specifies Minimum Operational Performance Standards (MOPS) for GPS/WAAS airborne equipment, including pseudorange residuals.

data using correction verification simulation. This will increase the availability and continuity of SBAS services since the signal is already processed by the user.

Raytheon Company is currently proposing to upgrade the SBAS to achieve *full operational capability* (FOC) for the Federal Aviation Administration (FAA). One of the new features will be to steer the GUS clock to GPS time. SBAS is a GPS-based navigation system that is intended to become the primary navigational aid for aviation during all phases of flight, from en route through lateral and vertical navigation (LNAV/VNAV) approach. SBAS makes use of a network of wide-area reference stations (WRSs) distributed throughout the U.S. National Airspace System. Figure 6.1 provides a top-level view of the SBAS architecture.

In the L_1 path, the GUS receives integrity and correction data and SBAS specific messages from the WMS, adds forward error correction (FEC) encoding, and transmits the messages via a C-band uplink to the GEO satellites for broadcast to the SBAS user. The GUS uplink signal uses the GPS standard positioning service waveform (C/A-code, BPSK modulation); however, the data rate is higher (250 bits per second). The 250 bits of data are encoded with a one-half rate convolutional code, resulting in a 500-(symbols per second) transmission rate. Each symbol is modulated by the C/A code, a 1.023×10^6-cps pseudorandom sequence, to provide a spread-spectrum signal. This signal is then binary phase-shift keying (BPSK) modulated by the GUS onto an intermediate-frequency (IF) carrier, upconverted to a C-band frequency, and uplinked to the GEO.

The existing GUS in SBAS contains a clock steering algorithm. This algorithm uses SBAS Type 9 messages from the WMS to align the GEO's epoch with the GPS epoch. The SBAS Type 9 message contains a term referred to as a_{Gf0} or clock offset. This offset represents a correction, or time difference, between the GEO's epoch and SBAS network time (WNT). WNT is the internal time reference scale of SBAS and is required to track the GPS timescale, while at the same time providing users with the translation to Universal Time, Coordinated (UTC). Since GPS master time is not directly obtainable, the SBAS architecture requires that WNT be computed at multiple WMSs using potentially differing sets of measurements from potentially differing sets of receivers and clocks (SBAS reference stations). WNT is required to agree with GPS to within 50 ns. At the same time, the WNT to UTC offset must be provided to the user, with the offset being accurate to 20 ns. The GUS calculates local clock adjustments. On the basis of these clock adjustments, the frequency standard can be made to speed up, or slow the GUS clock. This will keep the total GEO clock offset within the range allowed by the SBAS Type 9 message so that users can make the proper clock corrections in their algorithms [33, 192].

The new algorithm in the GUS will use the clock steering method described above during the initial 24 h after it becomes primary. Once the GUS clock is synchronized with WNT, a second steering method of clock steering is used. The algorithm now uses the composite of the MOPS [165] solution for the receiver clock error, and the average of the a_{Gf0}, and the average of the MOPS solution as the input to the clock steering controller.

6.7.1 Receiver Clock Error Determination

Determination of receiver clock error is based on the user position solution algorithm described in the SBAS MOPS. The clock bias (C_b) is a resultant of the MOPS weighted least-squares solution.

Components of the weighted least-squares solution are the observation matrix (**H**), the measurement weighting matrix (**W**) and the MOPS residual column vector ($\Delta\rho$). The weighted gain matrix (**K**) is calculated using **H** and **W** (see Eq. 2.37):

$$\mathbf{K} = (\mathbf{H}^T\mathbf{W}\mathbf{H})^{-1}\mathbf{H}^T W. \tag{6.9}$$

From this, the column vector for the user position error and the clock bias solution is:

$$\begin{aligned}\mathbf{\Delta X} &= \mathbf{K}\mathbf{\Delta}\boldsymbol{\rho} && (6.10)\\ &= (\mathbf{H}^T\mathbf{W}\mathbf{H})^{-1}\mathbf{H}^T\mathbf{W}\mathbf{\Delta}\boldsymbol{\rho}, && (6.11)\end{aligned}$$

where

$$\mathbf{\Delta X} = \begin{bmatrix} \Delta X(U) \\ \Delta X(E) \\ \Delta X(N) \\ C_b \end{bmatrix} \tag{6.12}$$

and $\Delta X(U)$ is the *up* error, $\Delta X(E)$ is the *east* error, $\Delta X(N)$ is the *north* error, and C_b is the clock bias or receiver clock error.

The $n \times 4$ observation matrix (**H**) is computed in up–east–north (UEN) reference frame using the line-of-sight (LOS) azimuth (Az_i) and LOS elevation (El_i) from the GUS omni antenna to the space vehicle (SV). The value n is the number of satellites in view. The formula for calculating the observation matrix is

$$\mathbf{H} = \begin{bmatrix} \cos(\mathrm{El}_1)\cos(\mathrm{Az}_1) & \cos(\mathrm{E1}_1)\sin(\mathrm{Az}_1) & \sin(\mathrm{E1}_1) & 1 \\ \cos(\mathrm{El}_2)\cos(\mathrm{Az}_2) & \cos(\mathrm{E1}_2)\sin(\mathrm{Az}_2) & \sin(\mathrm{E1}_2) & 1 \\ \vdots & \vdots & \vdots & \vdots \\ \cos(\mathrm{El}_n)\cos(\mathrm{Az}_n) & \cos(\mathrm{E1}_n)\sin(\mathrm{Az}_n) & \sin(\mathrm{E1}_n) & 1 \end{bmatrix}. \tag{6.13}$$

The $n \times n$ weighting matrix (**W**) is a function of the total variance (σ_i^2) of the individual satellites in view. The inverse of the weighting matrix is

$$\mathbf{W}^{-1} = \begin{bmatrix} \sigma_1^2 & 0 & 0 & \vdots & 0 \\ 0 & \sigma_2^2 & 0 & \vdots & 0 \\ 0 & 0 & \sigma_3^2 & \vdots & 0 \\ \vdots & \vdots & \vdots & \ddots & \vdots \\ 0 & 0 & 0 & 0 & \sigma_n^2 \end{bmatrix}. \tag{6.14}$$

The equation to calculate the total variance (σ_i^2) is

$$\sigma_i^2 = \left(\frac{\mathrm{UDRE}_i}{3.29}\right)^2 + \left(\frac{F_{\mathrm{pp}_i} \times \mathrm{GIVE}_i}{3.29}\right)^2 + \sigma^2_{\mathrm{L1},\,nmp,\,i} + \frac{\sigma^2_{\mathrm{tropo},\,i}}{\sin^2(\mathrm{El}_i)}. \tag{6.15}$$

The algorithms for calculating user differential range error (UDRE_i), user grid ionospheric vertical error (GIVE_i), LOS obliquity factor (F_{ppi}), standard deviation of uncertainty for the vertical troposphere delay model ($\sigma_{\mathrm{tropo},i}$) and the standard deviation of noise and multipath on the L_1 omni pseudorange ($\sigma_{L1,nmp,i}$) are found in the SBAS MOPS [165].

The MOPS residuals ($\boldsymbol{\Delta\rho}$) are the difference between the smoothed MOPS measured pseudorange ($PR_{M,i}$) and the expected pseudorange ($PR_{\mathrm{corr},\,i}$):

$$\boldsymbol{\Delta\rho} = \begin{bmatrix} PR_{M,1} - PR_{\mathrm{corr},1} \\ PR_{M,2} - PR_{\mathrm{corr},2} \\ PR_{M,3} - PR_{\mathrm{corr},3} \\ \vdots \end{bmatrix}. \tag{6.16}$$

The MOPS measured pseudorange ($PR_{M,i}$) in earth-centered earth-fixed (ECEF) reference is corrected for earth rotation, for SBAS clock corrections, for ionospheric effects and for tropospheric effects. The equation to calculate $\mathrm{PR}_{M,i}$ is

$$PR_{M,i} = PR_{L,i} + \Delta PR_{\mathrm{CC},i} + \Delta PR_{\mathrm{FC},i} + \Delta PR_{\mathrm{ER},i} - \Delta PR_{T,i} - \Delta PR_{I,i}. \tag{6.17}$$

The algorithms used to calculate smoothed L_1 omni pseudorange ($PR_{L,i}$), pseudorange clock correction ($\Delta PR_{\mathrm{CC},i}$), pseudorange fast correction ($\Delta PR_{\mathrm{FC},i}$), pseudorange earth rotation correction ($\Delta PR_{\mathrm{ER},i}$), pseudorange troposphere correction ($\Delta PR_{T,i}$), and pseudorange ionosphere correction ($\Delta PR_{I,i}$) are found in the SBAS MOPS [165].

Expected pseudorange ($PR_{\mathrm{corr},i}$), ECEF, at the time of GPS transmission is computed from broadcast ephemeris corrected for fast and long term corrections. The calculation is

$$PR_{\mathrm{corr},i} = \sqrt{(X_{\mathrm{corr},i} - X_{\mathrm{GUS}})^2 + (Y_{\mathrm{corr},i} - Y_{\mathrm{GUS}})^2 + (Z_{\mathrm{corr},i} - Z_{\mathrm{GUS}})^2}. \tag{6.18}$$

The fixed-position parameters of the WRE (X_{GUS}, Y_{GUS}, Z_{GUS}) are site-specific.

6.7.2 Clock Steering Control Law

In the primary GUS, the clock steering algorithm is initialized with SBAS Type 9 message (GEO navigation message). After the initialization, composite of MOPS solution and Type 9 message for the receiver clock error is used as the input to the control law (see Fig. 6.15). For the backup GUS, the MOPS solution for the receiver clock error is used as the input to the control law (see Fig. 6.16).

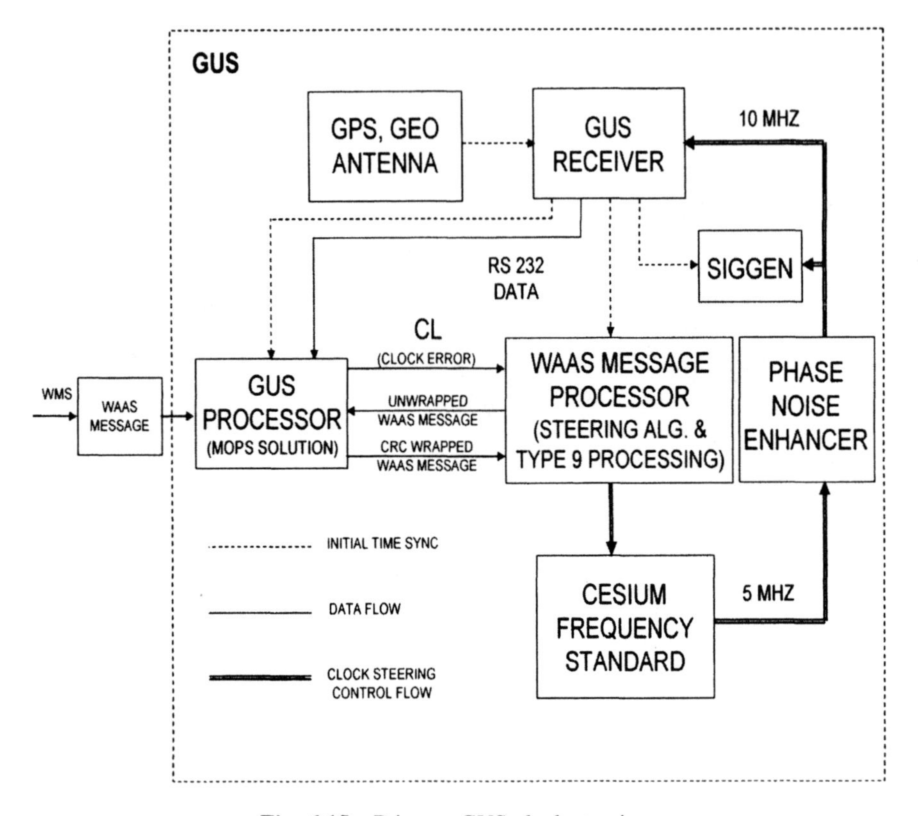

Fig. 6.15 Primary GUS clock steering.

For both the primary and backup clock steering algorithm, the control law is a proportional, integral, and differential (PID) controller. The output of the control law will be the frequency adjustment command. This command is sent to the frequency standard to adjust the atomic clock frequency. The output frequency to the receiver causes the 1 pps to approach the GPS epoch. Thus, a closed loop control of the frequency standard is established.

6.8 GEO ORBIT DETERMINATION

The purpose of WAAS is to provide pseudorange and ionospheric corrections for GPS satellites to improve the accuracy for the GPS navigation user and to protect the user with “integrity.” Integrity is the ability to provide timely warnings to the user whenever any navigation parameters estimated using the system are outside tolerance limits. WAAS may also augment the GPS constellation by providing additional ranging sources using GEO satellites that are being used to broadcast the WAAS signal.

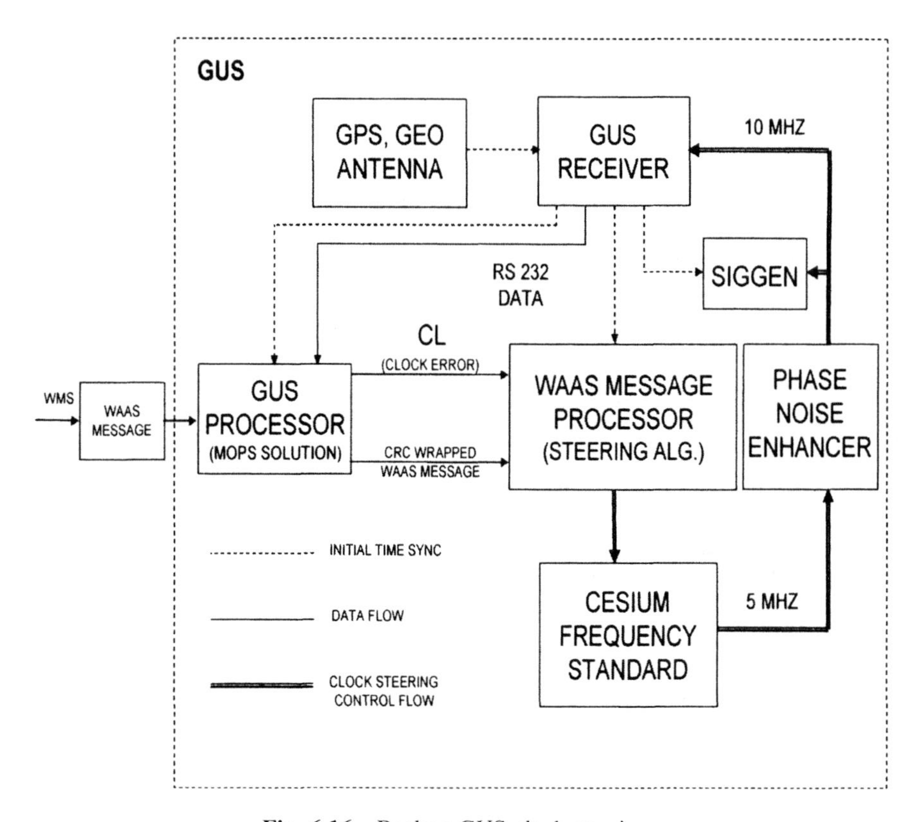

Fig. 6.16 Backup GUS clock steering.

The two parameters having the most influence on the integrity bounds for the broadcast data are user differential ranging error (UDRE) for the pseudorange corrections and grid ionospheric vertical error (GIVE) for the ionospheric corrections. With these, the onboard navigation system estimates the horizontal protection limit (HPL) and the vertical protection limit (VPL), which are then compared to the horizontal alert limit (HAL) and the vertical alert limit (VAL) requirements for the particular phase of flight involved, that is, oceanic/remote, en route, terminal, nonprecision approach, and precision approach. If the estimated protection limits are greater than the alert limits, the navigation system is declared unavailable. Therefore, the UDRE and GIVE values obtained by the WAAS (in concert with the GPS and GEO constellation geometry and reliability) essentially determine the degree of availability of the WADGPS navigation service to the user.

The WAAS algorithms calculate the broadcast corrections and the corresponding UDREs and GIVEs by processing the satellite signals received by the network of ground stations. Therefore, the expected values for UDREs and GIVEs are dependent on satellite and station geometries, satellite signal and clock performance, receiver performance, environmental conditions (such as multipath, ionospheric storms, etc.) and algorithm design [71, 147].

6.8.1 Orbit Determination Covariance Analysis

A full WAAS algorithm contains three Kalman filters—an orbit determination filter, an ionospheric corrections filter, and a fast corrections filter. The fast corrections filter is a Kalman filter that estimates the GEO, GPS, and ground station clock states every second. In this section, we derive an estimated lower bound of the GEO UDRE for a WAAS algorithm that contains only the orbit determination Kalman filter, called the $UDRE(OD)$, where OD refers to orbit determination.

A method is proposed to approximate the UDRE obtained for a WAAS including both the orbit determination filter and the fast corrections filter from UDRE(OD). From case studies of the geometries studied in the previous section, we obtain the essential dependence of UDRE on ground station geometry.

A covariance analysis on the orbit determination is performed using a simplified version of the orbit determination algorithms. The performance of the ionospheric corrections filter is treated as perfect, and therefore, the ionospheric filter model is ignored. The station clocks are treated as if perfectly synchronized using the GPS satellite measurements. Therefore, the station clock states are ignored. This allows the decoupling of the orbit determinations for all the satellites from each other, simplifying the orbit determination problem to that for one satellite with its corresponding ground station geometry and synchronized station clocks. Both of these assumptions are liberal; therefore, the UDRE(OD) obtained here is a lower bound for the actual UDRE(OD). Finally, we consider only users within the service volume covered by the stations and, therefore, ignore any degradation factors depending on user location.

To simulate the Kalman filter for the covariance matrix P, the following four matrices are necessary (Table 8.1):

$$\begin{aligned}\Phi &= \text{state transition matrix,}\\ \mathbf{H} &= \text{measurement sensitivity matrix,}\\ \mathbf{Q} &= \text{process noise covariance matrix,}\\ \mathbf{R} &= \text{measurement noise covariance matrix.}\end{aligned}$$

The methods used to determine these matrices are described below.

The state vector for the satellite is

$$\mathbf{x} = \begin{bmatrix} \mathbf{r} \\ \dot{\mathbf{r}} \\ C_b \end{bmatrix},$$

where

$$\mathbf{r} \equiv [x\ y\ z]^T$$

is the satellite position in the ECI frame,

$$\dot{\mathbf{r}} \equiv [\dot{x},\ \dot{y},\ \dot{z}]^T$$

is the satellite velocity in the ECI frame, and C_b is the satellite clock offset relative to the synchronized station clocks. Newton's second and third (gravitational) laws provide the equations of motion for the satellite:

$$\ddot{\mathbf{r}} \equiv \frac{d^2r}{dt^2} = -\frac{\mu_E \mathbf{r}}{|\mathbf{r}|^3} + M,$$

where $\ddot{\mathbf{r}}$ is the acceleration in the ECI frame, μ_E is the gravitational constant for the earth, and M is the total perturbation vector in the ECI frame containing all the perturbing accelerations. For this analysis, only the perturbation due to the oblateness of the earth is included. The effect of this perturbation on the behavior of the covariance is negligible, and therefore higher-order perturbations are ignored. (Note that although the theoretical model is simplified, the process noise covariance matrix Q is chosen to be consistent with a far more sophisticated orbital model.)

Therefore

$$M = -\frac{3}{2} J_2 \frac{\mu_E}{|\mathbf{r}|^3} \frac{a_E^2}{|\mathbf{r}|^2} [\mathbf{I}_{3\times 3} + 2\hat{\mathbf{z}}\hat{\mathbf{z}}^T]\mathbf{r},$$

where a_E is the semimajor axis of the earth-shape model, J_2 is the second zonal harmonic coefficient of the earth-shape model, and $\hat{\mathbf{z}} \equiv [0,\ 0,\ 1]^{\mathrm{T}}$ [11].

The second-order differential equation of motion can be rewritten as a pair of first-order differential equations

$$\dot{\mathbf{r}}_1 = \mathbf{r}_2, \qquad \dot{\mathbf{r}}_2 = \frac{\mu_E r_1}{|\mathbf{r}|^3} + M, \tag{6.19}$$

where $\boldsymbol{r}_1$ and $\boldsymbol{r}_2$ are vectors, which therefore gives a system of six first-order equations.

The variational equations are differential equations describing the rates of change of the satellite position and velocity vectors as functions of variations in the components of the estimation state vector. These lead to the state transition matrix Φ used in the Kalman filter. The variational equations are

$$\ddot{Y}(t) = A(t)Y(t) + B(t)\dot{Y}(t), \tag{6.20}$$

where

$$Y(t_k)_{3\times 6} \equiv \left[\left(\frac{\partial \mathbf{r}(t_k)}{\partial \mathbf{r}(t_{k-1})} \right)_{3\times 3} \left(\frac{\partial \mathbf{r}(t_{k)}}{\partial \dot{\mathbf{r}}(t_{k-1})} \right)_{3\times 3} \right], \tag{6.21}$$

$$\dot{Y}(t_k)_{3\times 6} \equiv \left[\left(\frac{\partial \dot{\mathbf{r}}(t_k)}{\partial \mathbf{r}(t_{k-1})} \right)_{3\times 3} \left(\frac{\partial \mathbf{r}(t_{k)}}{\partial \dot{\mathbf{r}}(t_{K-1})} \right)_{3\times 3} \right], \tag{6.22}$$

$$\begin{aligned} A(t)_{3\times 3} &\equiv \frac{\partial \ddot{\mathbf{r}}}{\partial \mathbf{r}} \\ &= \frac{-\mu_E}{|\mathbf{r}|^3}[\mathbf{I}_{3\times 3} - 3\hat{\mathbf{r}}\hat{\mathbf{r}}^{\mathrm{T}}] - \frac{3}{2}J_2\frac{\mu E}{|\mathbf{r}|^3}\frac{a_E^2}{|\mathbf{r}|^2}, \\ &\quad \times \left[\mathbf{I}_{3\times 3} + 2\hat{\mathbf{z}}\hat{\mathbf{z}}^{\mathrm{T}} - 10(\hat{\mathbf{r}}^{\mathrm{T}}\hat{\mathbf{z}}^{\mathrm{T}})(\hat{\mathbf{z}}\hat{\mathbf{r}}^{\mathrm{T}} + \hat{\mathbf{r}}\hat{\mathbf{z}}^{\mathrm{T}}) \right. \\ &\quad \left. +(10(\hat{\mathbf{r}}_{\mathrm{T}}\hat{\mathbf{z}})^2 - 5)(\hat{\mathbf{r}}\hat{\mathbf{r}}^{\mathrm{T}})\right], \end{aligned} \tag{6.23}$$

$$B(t)_{3\times 3} \equiv \frac{\partial \ddot{\mathbf{r}}}{\partial \dot{\mathbf{r}}} = \mathbf{0}_{3\times 3}, \tag{6.24}$$

where $\hat{\mathbf{r}} = \mathbf{r}/|\mathbf{r}|$.

Equations 6.21–6.24 are substituted into Eq. 6.20 and Eq. 6.19, and the differential equations are solved using the fourth-order Runge–Kutta method. The time step used is a 5-min interval. The initial conditions for the GEO are specified for the particular case given and propagated forward for each time step, whereas the initial conditions for the Y terms are

$$Y(t_{k-1})_{3\times 6} = [\mathbf{I}_{3\times 3} \quad \mathbf{0}_{3\times 3}], \qquad \dot{Y}(t_k)_{3\times 6} = [\mathbf{0}_{3\times 3} \quad \mathbf{I}_{3\times 3}]$$

and reset for each timestep. This is due to the divergence of the solution of the differential equation used in this method to calculate the state transition matrix for the Kepler problem.

This gives the state $\mathbf{x}^T = \left[\mathbf{r}^T_1 \quad \mathbf{r}^T_2\right]$ and the state transition matrix

$$\Phi_{k,k-1_{7\times 7}} = \begin{bmatrix} Y(t_k)_{3\times 6} & \mathbf{0}_{3\times 1} \\ \dot{Y}(t_k)_{3\times 6} & \mathbf{0}_{3\times 1} \\ \mathbf{0}_{1\times 6} & \mathbf{I}_{1\times 1} \end{bmatrix} \tag{6.25}$$

for the Kalman filter.

The measurement sensitivity matrix is given by

$$H_{N\times 7} \equiv \frac{\partial \rho}{\partial x} = \left[\left(\frac{\partial \rho}{\partial \mathbf{r}} \right)_{N\times 3} \left(\frac{\partial \rho}{\partial \dot{\mathbf{r}}} \right)_{N\times 3} \left(\frac{\partial \rho}{\partial (ct)} \right)_{N\times 1} \right],$$

where ρ is the pseudorange for a station and N is the number of stations in view of the satellite. Note that this is essentially the same H as in the previous section. Ignoring relativistic corrections and denoting the station position by the vector $r_S \equiv [x_S \; y_S \; z_S]^{\mathrm{T}}$, the matrices above are given by

$$\frac{\partial \rho}{\partial \mathbf{r}} = \frac{[\mathbf{r} - r_S]^{\mathrm{T}}}{|\mathbf{r} - r_S|} \frac{\partial \mathbf{r}(t_k)}{\partial \mathbf{r}(t_{k-1})},$$
$$\frac{\partial \rho}{\partial \dot{\mathbf{r}}} = \frac{[\mathbf{r} - r_S]^{\mathrm{T}}}{|\mathbf{r} - r_S|} \frac{\partial \mathbf{r}(t_k)}{\partial \dot{\mathbf{r}}(t_{k-1})},$$

and

$$\frac{\partial \rho}{\partial (ct)} = 1 \,. \tag{6.26}$$

The station position is calculated with the WGS84 model for the earth and converted to the ECI frame using the J2000 epoch. (See Appendix C.)

These are then combined with the measurement noise covariance matrix, R and the process noise covariance matrix Q to obtain the Kalman filter equations for the covariance matrix P, as shown in Table 8.1.

The initial condition, $P_0(+)$, and Q are chosen to be consistent with the WAAS algorithms. The value of R is chosen by matching the output of the GEO covariance for AOR-W with $R = \sigma^2 \mathbf{I}$ and is used as the input R for all other satellites and station geometries (note that this therefore gives approximate results). This corresponds to carrier phase ranging for the stations. The results corresponding to the value of R for code ranging are also presented.

From this covariance, the lower bound on the UDRE is obtained by

$$\text{UDRE} \geq \text{EMRBE} + K_{SS}\sqrt{\text{tr}(P)},$$

where EMRBE is the estimated maximum range and bias error. To obtain the .999 level of bounding for the UDRE with EMRBE $= 0$, $K_{SS} = 3.29$. Finally, since the message is broadcast every second, $\Delta t = 1$, so the trace can be used for the velocity components as well.

Figure 6.17 shows the relationship between UDRE and GDOP for various GEO satellites and WRS locations. Table 6.5 describes the various cases considered in this analysis.

The numerical values used for the filter are as follows [all units are Système International (SI)]:

- Earth parameters:

$$\mu_E = 3.98600441 \times 10^{14}, \qquad J_2 = 1082.63 \times 10^{-6},$$
$$a_E = 6,378,137.0, \qquad b_E = 6,356,752.3142.$$

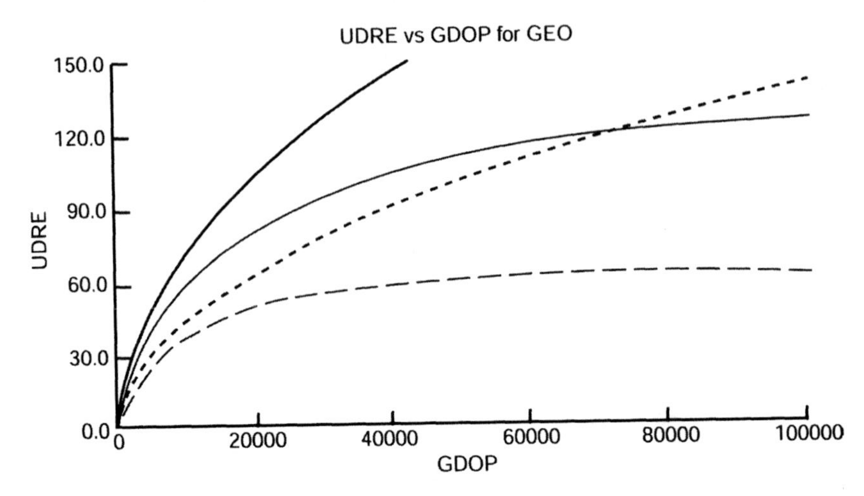

Fig. 6.17 Relationship between UDRE and GDOP.

TABLE 6.5. Cases Used in Geometry-per-Station Analysis

Case	UDRE	GDOP	Satellite	Geometry
1	17.9	905	AOR-W	WAAS stations (25),21 in view
2	45.8	2,516	AOR-W	4 WAAS stations (CONUS)[a]
3	135	56,536	AOR-W	4 WAAS stations (NE)[b]
4	4.5	254	AOR-W	WAAS stations + Santiago
5	5.8	212	AOR-W	WAAS stations + London
6	4.0	154	AOR-W	WAAS stations + Santiago + London
7	7.5	439	AOR-W	4 WAAS stations (CONUS) + Santiago
8	8.6	337	AOR-W	4 WAAS stations (CONUS) + London
9	6.6	271	AOR-W	4 WAAS stations (CONUS) + Santiago + London
10	47.7	2,799	AOR-W	4 WAAS stations (NE) + Santiago
11	21.5	1,405	AOR-W	4 WAAS stations (NE) + London
12	16.4	1,334	AOR-W	4 WAAS stations (NE) + Santiago + London
13	28.5	1,686	POR	WAAS stations (25), 8 in view
14	45.4	3,196	POR	WAAS stations, Hawaii
15	31.1	1,898	POR	WAAS stations, Cold Bay
16	55.0	4,204	POR	WAAS stations, Hawaii, Cold Bay
17	6.7	257	POR	WAAS stations + Sydney
18	8.3	338	POR	WAAS stations + Tokyo
19	6.7	257	POR	WAAS stations + Sydney + Tokyo
20	21.0	1,124	MTSAT	MSAS stations, 8 in view

TABLE 6.5. ***(continued)***

Case	UDRE	GDOP	Satellite	Geometry
21	22.0	1,191	MTSAT	MSAS stations–Hawaii
22	24.9	1,407	MTSAT	MSAS stations–Australia
23	54.6	4,149	MTSAT	MSAS stations –Hawaii, Australia
24	22.0	1,198	MTSAT	MSAS stations –Ibaraki
25	29.0	1,731	MTSAT	MSAS stations –Ibaraki, Australia
26	54.8	4,164	MTSAT	MSAS stations –Ibaraki, Australia, Hawaii
27	13.2	609	MTSAT	MSAS stations + Cold Bay
A		139	TEST	$\theta = 75°$
B		422	TEST	$\theta = 30°$
C		3,343	TEST	$\theta = 10°$
D		13,211	TEST	$\theta = 5°$
E		67	TEST	41 stations
F		64	TEST	41 + 4 stations

[a]4 WAAS stations (CONUS) are Boston, Miami, Seattle, and Los Angeles.
[b]4 WAAS stations (NE) are Boston, New York, Washington DC, and Cleveland.

- Filter parameters:

$$\begin{aligned} Q_{\text{pos}} &= 0, & Q_{\text{vel}} &= 0.75 \times 10^6, & Q_{ct} &= 60, \\ P_{0,\text{pos}} &= 144.9, & P_{0,\text{vel}} &= 1 \times 10^{-4}, & P_{0,\text{ct}} &= 100.9, \\ \sigma_R &= 0.013. \end{aligned}$$

- Curve fit parameters:

$$\sigma_{Q,\text{fit}} = 6.12, \qquad \varepsilon_{\phi,\text{fit}} = 0.0107.$$

PROBLEMS

6.1 Determine the code–carrier coherence at the GUS location using L_1 code and carrier.

6.2 Determine the frequency stability of the AOR and POR transponder using Allan variance for the L_1 using 1–10-s intervals.

7

GNSS AND GEO SIGNAL INTEGRITY

7.1 RECEIVER AUTONOMOUS INTEGRITY MONITORING (RAIM)

Navigation system integrity refers to the ability of the system to provide timely warnings to users when the system should not be used for navigation. The basic GPS (as described in Chapter 3) provides integrity information to the user via navigation message, but this may not be timely enough for some applications, such as civil aviation. Therefore, additional methods of providing integrity are necessary.

Two different methods will be discussed—GPS-only receiver (TSO-C129-compliant) autonomous integrity monitoring (RAIM) and use of ground monitoring stations to monitor the health of the satellites, as is done via SBAS and GBAS (TSO-C145-compliant receivers). Three RAIM methods have been proposed in recent papers on GPS integrity [150, 151, 152, 153]:

1. Range comparison method
2. Least-squares residual method
3. Parity method

We will briefly discuss the RAIM methods, then discuss SBAS and GBAS integrity design.

Global Positioning Systems, Inertial Navigation, and Integration, Second Edition, by M. S. Grewal, L. R. Weill, and A. P. Andrews

7.1.1 Range Comparison Method of Lee [121]

For the GNSS navigation problem described in Chapter 2, Section 2.5, there are four unknowns (three position coordinates $[x, y, z]$ and clock bias Cb) and more than four satellites in view (e.g., six satellites). One can solve the position and time equations for the first four satellites, ignoring noise, and find the user position. This solution can then be used to predict the remaining two pseudorange measurements, and the predicted values could be compared with actual measured values. If the two differences (residuals) are small, we have near consistency in the measurements and the detection algorithm can declare "no failure." It only remains to quantify what we mean by "small" or "large," and then assess the decision rule performance on actual data.

7.1.2 Least-Squares Method [151]

The basic measurement equation with noise (Eq. 2.32 from Chapter 2) is

$$\delta Z_\rho = \mathbf{H}\,\delta\mathbf{x} + v_\rho, \tag{7.1}$$

where the additive white noise $v_\rho \in N\left(0, \sigma^2\right)$.

Let us suppose six satellites are in view and four unknowns, as in Section 7.1.1, and solve for the four unknowns by the least-squares method.

The least-squares solution is given by Eq. 2.37:

$$\widehat{\delta\mathbf{x}} = \left(\mathbf{H}^{\mathrm{T}}\mathbf{H}\right)^{-1}\mathbf{H}^{\mathrm{T}}\delta Z_\rho. \tag{7.2}$$

The least-squares solution can be used to predict the six measurements, in accordance with

$$\widehat{\delta Z}_\rho(\text{predicted}) = \mathbf{H}\,\widehat{\delta\mathbf{x}}. \tag{7.3}$$

We can get a formula for the sum-squared residual error S by substituting δx from Eq. 7.3 into Eq. 7.2:

$$\Delta Z_\rho = \delta Z_\rho - \widehat{\delta Z}_\rho \text{ (residual error)} \tag{7.4}$$

$$= \left[\mathbf{I} - \mathbf{H}\left(\mathbf{H}^{\mathrm{T}}\mathbf{H}\right)^{-1}\mathbf{H}^{\mathrm{T}}\right]\delta Z_\rho \tag{7.5}$$

$$S = \Delta Z_\rho^T \Delta Z_\rho, \text{ the sum-squared error.} \tag{7.6}$$

This sum of squared error has three properties that are important in the decision rule:

1. S is a nonnegative scalar quantity. Choose a threshold value τ of S such that $S < \tau$ will be considered safe and $S \geq \tau$ will be declared a failure.

2. If the v_ρ have the same independent zero-mean Gaussian distribution, then the statistical distribution of S is completely independent of the satellite geometry for any number of satellites (n). Thresholds are precalculated, that results in the desired alarm rate for the various anticipated values of n. Then the real-time algorithm sets the threshold appropriately for the number of satellites in view at the moment.
3. With the v_ρ, from above, S has an unnormalized chi-square (χ^2) distribution with ($n - 4$) degrees of freedom. Parkinson and Axelrad [150] use $\sqrt{S/n-4}$ as the test statistic. Calculating the test statistic involves the same matrix manipulation, but these are no worse than calculating DOP.

7.1.3 Parity Method [182, 183]

The parity RAIM method is somewhat similar to the range comparison method, except that the way in which the test statistic is formed is different. In the parity method, perform a linear transformation on the measurement vector δZ_ρ as follows:

$$\begin{bmatrix} \delta x \\ p \end{bmatrix} = \begin{bmatrix} (\mathbf{H}^{\mathrm{T}}\mathbf{H})^{-1}\mathbf{H}^{\mathrm{T}} \\ \mathbf{P} \end{bmatrix} \delta Z_\rho. \tag{7.7}$$

The lower portion of Eq. 7.7, which yields p, is the result of operating on δZ_ρ with the special $(n-4)\times n$ matrix $\mathbf{P}$, whose rows are mutually orthogonal, unity magnitude and orthogonal to the columns of $\mathbf{H}$.

Under the same assumptions about the noise v_ρ as above, the following statements can be made:

$$\left.\begin{array}{lcl} \mathrm{E}\langle p\rangle & = & 0 \\ \mathrm{E}\langle pp^T\rangle & = & \sigma^2\mathbf{I}\ \text{(covariance of } p) \end{array}\right\}, \tag{7.8}$$

where σ^2 is the variance associated with v_ρ. Use p as the test statistic in this method. For detection, obtain all the information needed about p from its magnitude or magnitude squared. Thus, in the parity method, the test statistic for detection reduces to a scalar, as in the least squares method.

7.2 SBAS AND GBAS INTEGRITY DESIGN

The objectives of the space-based augmentation system (SBAS) and the ground-based augmentation system (GBAS) are to provide integrity, accuracy, availability and continuity for GPS, GLONASS, and Galileo Standard Positioning Service (SPS). Integrity is defined as the ability of the system to provide timely warnings to the user when individual corrections or certain satellites should not be used for navigation, specifically, the prevention of hazardously misleading information (HMI) data transmission to the user. The system should not be used for navigation when hardware, software or environmental errors directly pose a threat to the user or indirectly pose a threat by obscuring HMI from the integrity monitors. SBAS integrity is based on the premise that errors not detected or corrected in the operational environment can become threats to integrity and, if not mitigated, can become hazards to the user.

An SBAS design should mitigate the majority of these data errors with corrections that are proved to bound the integrity hazard to an acceptable level. The leftover data errors (referred to as *residual errors*) are mitigated by the transmission of residual error bounding information. The threat of potential underbounding of integrity information is mitigated by integrity monitors. This chapter examines both the faulted and unfaulted cases and mitigation strategies for these cases. These SBAS corrections improve the accuracy of satellite signals. The integrity data ensure that the residual errors are bounded. The SBAS integrity monitors help ensure that the integrity data have not been corrupted by SBAS failures.

The chapter addresses the data errors, error detection and correction pitfalls, and how such threats can become HMI to the user, as well as fault conditions, failure conditions, threats, and mitigation, and how safety integrity requirements are satisfied. Safety integrity assurance rules will be evaluated. Results from real SIS (signal in space) data, a high-level overview of the required SBAS safety architecture, and a data processing path protection approach are included.

This chapter provides information that defines how a safety-of-life-critical SBAS system should be designed and implemented in order to ensure mitigation of the entire International Civil Aviation Organization (ICAO) threat space to the required level less than 10^{-7}. It provides as an example, the rationale, background, and references to show that the SBAS can be used as a trusted navigational aid to augment the Global Positioning System (GPS) for lateral positioning with vertical guidance (LPV).

Rail integrity is one of the most stringent operational requirements, as evidenced by ERTMS (European Rail Traffic Management Systems) required integrity levels, which are in the order of 10^{-11}. Train detection will require an equally high level of positive integrity.

The chapter addresses the hazardous/severe–major integrity failure condition using LPV as an illustrative example. The ICAO integrity requirement is based on the premise that errors not detected or corrected in the operational environment system can become threats to the integrity and, if not mitigated, can become hazards to the user. These errors in the operational environment (referred to as *data errors*) can affect both the user and the SBAS system. *Integrity* in this context is defined as the ability of the system to provide timely warning to users when individual corrections or satellites should not be used for navigation, that is, the prevention of hazardously misleading information (HMI) data transmission to the user. The system should not be used for navigation when data errors in the environment, such as the ionosphere, and data processing, such as multipath, render the integrity data erroneous. The user must be protected from residual errors that can become threats to the integrity data that could result in HMI being transmitted to the user [171, 200].

An SBAS design mitigates the majority of these data errors with "corrections." The leftover data errors (referred to as *residual errors*) are mitigated by the transmission of residual error bounding information. The threat of potential under-bounding of the integrity information is mitigated by integrity monitors and point design features that protect the integrity of the information within the SBAS system. Additionally, analytic safety analyses are required to provide evidence and proof that the residual errors are acceptable (i.e., that the probability of HMI transmission to the user is sufficiently low).

Table 7.1 list the SBAS error sources. Mitigation of these errors when they become integrity threats are presented in Section 7.2.8. Section 7.3 gives an application of these techniques to SBAS for threat mitigation. Section 7.11 shows the conclusions. GPS Integrity Channel (GIC) is discussed in Section 7.5.

TABLE 7.1. List of SBAS Error Sources

GPS satellite	Integrity bound associated
GEO satellites	Message uplink
Reference receiver	Environment (ionosphere and troposphere)
Estimation	

7.2.1 SBAS Error Sources and Integrity Threats

The SBAS operational environment contains data errors. The SBAS ensures that these data errors do not become threats to the integrity data, so that HMI is not broadcasted to the user with a P_{HMI} greater than 10^{-7}.

The data used by an SBAS to calculate the correction and/or integrity data are assumed to contain errors, such as, GPS satellite clock offset, which must be sufficiently mitigated. The errors discussed are inherent in any SBAS design that utilizes GPS, Galileo, GLONASS, or GEO satellites; reference receivers; corrections; and integrity bounds. Depending on the system architecture, other error sources may also exist. Table 7.1 summarizes the error sources that every SBAS system must address [65].

The integrity threats associated with each of these error sources generally have two cases, shown in Fig. 7.1. The *fault-free case* addresses the nominal errors associated with each error source and the *faulted case* represents the errors when one or more of the system's components cause errors. The defining quality of an SBAS system that meets the ICAO standards is the mitigation of the faulted case and the fault-free case.

7.2.2 GPS-Associated Errors

GPS error sources are mitigated in an SBAS system by using corrections and integrity bounds. Generally, the SBAS system corrects the errors as well as possible, and then bounds the residual errors with integrity bounds that are broadcast

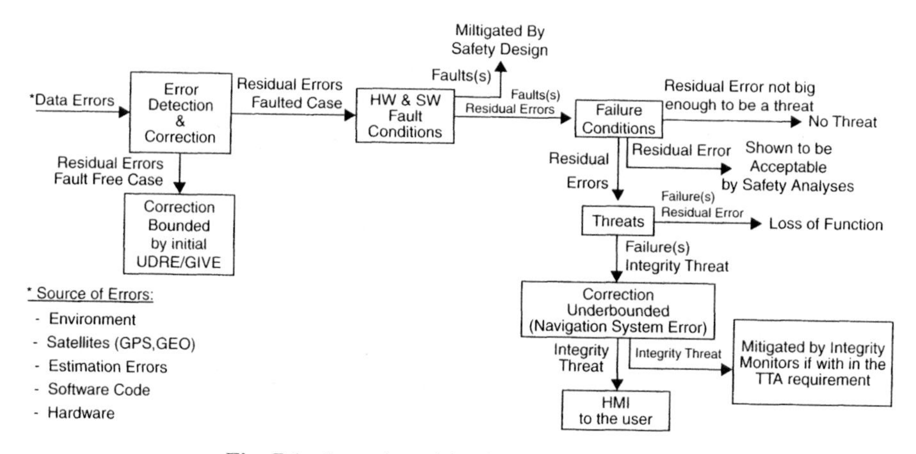

Fig. 7.1 Integrity mitigation within an SBAS.

to the user. The nominal GPS satellite errors are well understood. The literature includes many techniques for mitigating these errors. GPS failure modes are not as well understood and often require careful study to define the threat, which must be accounted for in threat models.

7.2.2.1 ***GPS Clock Error*** Each GPS satellite broadcasts a navigation data message containing an estimate of its clock offset (relative to GPS time) and drift rate. The GPS satellite clock value is utilized to correct the satellite's pseudorange, the measurement used to calculate the distance (range) from the satellite to the receiver (either the user's receiver or the reference receiver).

Under fault-free conditions the SBAS can accurately compute these corrections and mitigate this error source. Simple statistical techniques can be used to characterize these errors. The SBAS must also address satellite failures that cause the clock to rapidly accelerate, rendering the corrections suddenly invalid. As a result, the error bounds may not be bounding the residual error in the corrections. These types of failures have been observed many times in the history of GPS.

7.2.2.2 ***GPS Ephemeris Error*** Each GPS also broadcasts a navigation data message containing a prediction of its orbital parameters—Keplerian orbital parameters. The satellite's ephemeris data enable determination of the satellite position and velocity. Any difference between the satellite's calculated position and velocity and the true position is a potential source of error.

Under fault-free conditions the SBAS provides corrections relative to the GPS broadcast ephemeris data. The SBAS can accurately compute these corrections and mitigate this error source using standard statistical techniques. The satellite may experience an unexpected maneuver, rendering the corrections suddenly invalid. This threat includes geometric constraints that may be insufficient for the SBAS to adequately detect the orbit error.

7.2.2.3 ***GPS Code and Carrier Incoherence*** The GPS signal consists of a radiofrequency carrier encoded with a pseudorandom spread-spectrum code. The user's receiver performs smoothing of its pseudorange measurements using the carrier phase measurements. If the code and carrier are not coherent, there will be an error in this pseudorange smoothing process. This error is caused by a satellite failure. Incoherence between the code and carrier phase can increase the range error, ultimately resulting in the user incorrectly determining the code/carrier ambiguity.

7.2.2.4 ***GPS Signal Distortion*** A satellite may fail in a manner that distorts the pseudorange portion (PRN encoding) of the GPS transmission. This causes an error in the user's pseudorange measurements that may be different from the error that the SBAS receiver experiences. In 1993, the SV-19 GPS satellite experienced a failure that fits into this category. This error is caused by a satellite failure. If a satellite experiences this type of failure, the SBAS may not be able to estimate the satellite clock corrections that are aligned with the user's measurements, which could result in HMI.

7.2.2.5 ***GPS L_1L_2 Bias*** The GPS L_1 and L_2 signals are utilized together to compute the ionospheric delay so the delay can be removed from the range calculations. The satellite has separate signal paths for these two frequencies; therefore, the signals can have different delays. The difference in the delays must be modeled accurately to be able to properly calibrate and use L_1 and L_2 signals together.

Under nominal conditions, the SBAS estimation process is very accurate, and this error is easily modeled with standard statistical techniques. If a satellite experiences a fault, the L_1L_2 bias can suddenly change, resulting in a large estimation error. A large estimation error can lead to excessive errors in correction processing.

7.2.2.6 ***Environment Errors: Ionosphere*** As the GPS L_1 and L_2 signals propagate through the ionosphere, the signals are delayed by charged particles. The density of the charged particles, and therefore the delay, varies with location, time of day, angle of transmission through the ionosphere, and solar activity. This delay will cause an error in range measurements and must be corrected and properly accounted for in the SBAS measurement error models. As discussed earlier in Chapter 5, during calm ionospheric conditions, modeling errors are well understood and can be handled using standard statistical techniques. Ionospheric storms pose a multitude of threats for SBAS users. The model used in the error estimation may become invalid. The user may experience errors that are not observable to the SBAS, due to the geometry of the reference station pierce points. The error in the corrections may increase over time due to rapid fluctuations in the ionosphere.

7.2.2.7 Environment Errors: Troposphere As the GPS L_1 and L_2 signals propagate through the troposphere, the signals are delayed. This delay is dependent on temperature, humidity, angle of transmission through the atmosphere, and atmospheric pressure. This delay will cause an error in range measurements and must be corrected and properly accounted for in the measurement error models. Tropospheric modeling errors manifest themselves in the algorithms that generate the corrections. The user utilizes a separate tropospheric model that may have errors due to tropospheric modeling.

7.2.3 GEO-Associated Errors

7.2.3.1 GEO Code and Carrier Incoherence The GEO signal consists of a radiofrequency carrier encoded with a pseudorandom spread-spectrum code. The user's receiver performs smoothing of its GEO pseudorange measurements using the carrier phase measurements. If the code and carrier are not coherent, there will be an error in this pseudorange smoothing process. Under fault free conditions, some incoherence is possible (due to environmental effects). This will be a very small error that is easily modeled by the ground system. Under faulted conditions, severe divergence and potentially large errors are theoretically possible if the GEO uplink subsystem fails.

7.2.3.2 Environment Errors: Ionosphere As the GEO L_1 signal propagates through the ionosphere, the signal is delayed by charged particles. The density of the charged particles, and therefore the delay, varies with location, time of day, angle of transmission through the ionosphere, and solar activity. GEO satellites that are available today broadcast single-frequency (L_1) signals that do not allow a precise determination of the ionospheric delay at a reference station. Without dual-frequency measurements, uncertainty in the calculated ionospheric delay estimates bleed into the corrections. New GEO satellites [PRN 135,138] have two frequencies L_1, L_5. Ionospheric delay is calculated using those frequencies. (see Chapter 5.)

7.2.3.3 Environment Errors: Troposphere Like GPS satellite signals, the GEO L_1 signal is delayed as it propagates through the troposphere. This delay will cause an error in range measurements and must be corrected and properly accounted for in the measurement error models.

7.2.4 Receiver and Measurement Processing Errors

Measurement errors affect an SBAS system in two ways. They can corrupt or degrade the accuracy of the corrections. They can also mask other system errors and result in HMI slipping through to the user. The errors given below must all be mitigated and residual errors bounded.

7.2.4.1 Receiver Measurement Error The receiver outputs pseudorange and carrier phase measurements for all satellites that are in view. The receiver and antenna characteristics limit the measurement accuracy. Under fault-free conditions these errors can be addressed using well-documented processes. A receiver could fault and output measurement data that are in error for any or all of the satellites in view. A latent common-mode failure in the receiver firmware could cause all measurements in the system to simultaneously fail. Erroneous measurements pose two threats. They cannot only result in correction errors; they can also fool the integrity monitors and let HMI slip through to the user.

7.2.4.2 Intercard Bias For receiver designs that include multiple correlators, the internal delays in the subreceivers are different. This creates a different apparent clock for each subreceiver, called an *intercard bias*. Under nominal conditions, the intercard bias estimate is extremely accurate and the intercard bias error is easily accounted for. Any failure condition in the receiver or the algorithm computing the bias will result in an increase in the measurement data error.

7.2.4.3 Multipath Under nominal conditions, the dominant source of noise is multipath. Multipath is caused by reflected signals arriving at the receiver delayed relative to the direct signal. The amount of error is dependent on the delay time and the receiver correlator type. (See discussion of *multipath mitigation methods* in Chapter 5.)

7.2.4.4 L_1L_2 Bias The GPS L_1 and L_2 signals are utilized together to compute the ionospheric delay so that the delay can be removed from the range calculations. The receivers and antenna will experience different delays in the electronics when monitoring these two frequencies. The difference in the delays must be accurately modeled to be able to remove the bias and use the L_1 and L_2 signals together. If a receiver fails, the L_1/L_2 bias can suddenly change, resulting in large estimation errors. Under nominal conditions, the estimation process is very accurate and the error is not significant.

7.2.4.5 Receiver Clock Error A high-quality receiver generally utilizes a (cesium) frequency standard that provides a long-term stable time reference (clock). This clock does drift. If a receiver fails, the clock bias can suddenly change, resulting in a large estimation error. Under nominal conditions, the SBAS is able to accurately account for receiver clock bias and drift. If a receiver fails, the clock may accelerate, introducing errors into the corrections and the integrity monitoring algorithms.

7.2.4.6 Measurement Processing Unpack/Pack Corruption The measurement processing software that interfaces with the receiver needs to unpack and repack the GPS ephemeris. A software failure or network transmission failure could corrupt the GPS ephemeris data and result in the SBAS using an incorrect ephemeris.

7.2.5 Estimation Errors

The SBAS system provides corrections to improve the accuracy of the GPS measurements and mitigate the GPS/GEO error sources. Estimation of parameters and corrections described in Sections 7.2.5.1–7.2.5.4 cause these errors, which must be accounted for.

7.2.5.1 Reference Time Offset Estimation Error The difference between the SBAS and GPS reference time must be less than 50 nano-seconds. If the user is en route and mixing SBAS corrected satellite data with non-SBAS-corrected satellite data, then the offset (error) between the SBAS reference time and the GPS reference time could affect the user's receiver position solution. Under fault-free conditions, this error varies slowly. If one or more GPS satellites fail, the offset between GPS time and the SBAS reference time could vary rapidly.

7.2.5.2 Clock Estimation Error The SBAS system must compute estimates of the reference receiver clocks and GPS/GEO satellite clock errors. An error in this estimation results in errors in the user's position solution. The error in the estimation process must be accounted for in the integrity bounds.

7.2.5.3 Ephemeris Correction Error The SBAS computes estimates of each satellite's orbit (ephemeris) and then uses these estimates to compute corrections. Error in the orbit (ephemeris) estimation process will result in erroneous corrections. Sources of error include measurement noise, troposphere modeling error, and orbital parameter modeling error. The error in the estimation process must be accounted for in the integrity bounds.

7.2.5.4 L_1/L_2 WRE and GPS Satellite Bias Estimation Error The L_1/L_2 bias of the satellites and the receivers is used to generate the SBAS corrections. SBAS users utilize single-frequency corrections while corrections are generated using dual-frequency measurements that are unaffected by ionospheric delay errors. An error in the estimation process will result in erroneous corrections. Sources of estimation error include measurement error, time in view, ionospheric storms, and receiver/satellite malfunctions. The error in the estimation process must be accounted for in the integrity bounds.

7.2.6 Integrity-Bound Associated Errors

The integrity monitoring functionality in an SBAS system ensures that the system meets the allocated integrity requirement. This processing includes functionality that must be performed on a "trusted" platform with software developed to the proper RTCA/ DO178-B safety level.

The ICAO HMI hazard has been evaluated to be a "Hazardous/severe–major" failure condition. This requires all software that be responsible for preventing HMI to be developed using a process that meets all the RTCA/DO-178B Level B objectives.

A critical aspect of mitigating an integrity threat is the determination of the threat model. Threats originating in the RTCA/DO-178B Level B software can be characterized using observed performance, provided all the inputs originate from Level B software and the algorithms have been designed in an analytic methodology.

7.2.6.1 Ionospheric Modeling Errors The SBAS system uses an underlying characterization to transmit ionospheric corrections to the user. During periods of high solar activity the ionospheric decorrelation can be quite rapid and large and the true delay variation around the grid point may not match the underlying characterization. In this case, the SBAS-estimated delay measurement and the associated error bound may not be accurate or the SBAS may not sample a particular ionospheric event that is affecting a user.

7.2.6.2 Fringe Area Ephemeris Error Errors may be present in the SBAS GPS position estimates that are not observable from the reference receivers. These errors could cause position errors in a user's position solution that are not observable to the reference receivers.

7.2.6.3 Small-Sigma Errors It is possible that any quantity of satellites could contain small or medium-sized errors that combine in such a manner that creates an overall position error that is unbounded to a user.

7.2.6.4 Missed Message—Old But Active Data (OBAD) The user could have missed one or more messages and is allowed to use old corrections and integrity data. The use of these old data could result in an increased error compared to users that have not missed messages.

7.2.6.5 TTA Exceeded If there is an underbound condition, the SBAS is required to correct that condition within a specified period of time. This is called the *time to alarm* (TTA). This alarm is a series of messages that contain the new information, such as an increased error bound or new corrections that are needed to correct the situation and prevent HMI. Different types of failure, such as hardware, software, or network transmission delay, could occur and cause the alarm messages to be delayed in excess of the required time.

7.2.7 GEO Uplink Errors

Errors caused by the uplink system can also be a source of HMI to the user.

7.2.7.1 GEO Uplink System Fails to Receive SBAS Message Any hardware or software along the path to the satellite could fault, causing the message to be delayed or not broadcast at all.

7.2.8 Mitigation of Integrity Threats

This section describes some approaches that may be used to eliminate and minimize data errors, mitigate integrity threats, and satisfy the safety integrity requirements.

Safety design and safety analyses are utilized to protect the data transmission path into the integrity monitors and out to the user through the Geostationary satellite.

Such integrity monitors written to DO-178B Level B standards to provide adjustments to the integrity bounds, must test the associated integrity data, user differential range error (UDRE) or grid ionospheric vertical error (GIVE) in an analytically tractable manner. The test prevents HMI by either passing the integrity data with no changes, increasing the integrity data to bound the residual error in the corrections, or setting the integrity data to "not monitored" or "don't use". Each integrity monitor must carefully account for the uncertainty in each component of a calculation. Noisy measurements or poor quality corrections will result in large integrity bounds.

The examples given in this section are for a system that utilizes either a "calculate then monitor" or "monitor then calculate" design. Both techniques are used in the examples to fully illustrate the types of mitigation needed to meet the general SBAS integrity requirements. Under the "calculate then monitor" design, corrections and error bounds are computed assuming that the inputs to the system follow some observed or otherwise predetermined model. A monitoring system then verifies the validity of these corrections and error bounds against the integrity threats. With the "monitor then calculate design" the measurements inputs to the monitor are carefully screened and forced to meet strict integrity requirements. The corrections and the error bounds are then computed in an analytically tractable manner and no further testing is required. Both designs must address all of the errors associated with an SBAS system in an analytically tractable manner.

7.2.8.1 Mitigation of GPS Associated Errors

GPS Clock Error

Fault-free case—the clock corrections are computed in a Kalman filter. The broadcast UDRE should be constructed using standard statistical techniques to ensure that the nominal errors in the fast corrections and long-term clock corrections are bounded.

Faulted case—a monitor is designed to ensure that the probability of a large fast correction error and/or long-term clock correction error is less than the allocation on the fault tree. The monitor must use measurements that are independent of the measurements used to compute the corrections. Error models for each input into the monitor must be determined and validated. The monitor either passes the UDRE or increases the UDRE or sets it to "not monitored" or "don't use" depending on the size of the GPS clock error.

GPS Ephemeris Error

Fault-free case—the orbit corrections are computed in a Kalman filter. The broadcast UDRE would be constructed using standard statistical techniques to ensure that the nominal errors in the long-term position corrections are bounded.

Faulted case—clock errors are easily observed by a differential GPS system. The ability of an SBAS to observe orbit errors is dependent on the location of the system's reference stations. The SBAS can generate a covariance matrix and package it in SBAS Message Type 28. This message provides a location-specific multiplier for the broadcast UDRE. The covariance matrix must take into account the quality of the measurements from the reference stations and the quality of the ephemeris corrections broadcast from the SBAS. When the GPS ephemeris is grossly in error, the SBAS must either detect and correct the problem or increase the uncertainty in the UDRE. Under faulted conditions, the SBAS must account for the situation where clock error cancels with the ephemeris error at one or more of the reference stations.

GPS Code and Carrier

Fault-free case—GPS code–carrier divergence results from a failure on the GPS satellite and errors do not need to be mitigated in the fault free case.

Faulted case—a monitor must be developed to detect and alarm if the GPS code and carrier phase become incoherent. The monitor must account for differences in the SBAS measurement smoothing algorithm and the user's measurement smoothing algorithm. The most difficult threat to detect and mitigate is one where the code–carrier divergence occurs shortly (within seconds) after the user acquires the satellite. In this case, the error has an immediate effect on the user and a gradual effect on the SBAS.

GPS Signal Distortion

Fault-free case—GPS signal distortion results from a failure on the GPS satellite and errors do not need to be mitigated in the fault free case.

Faulted case—a monitor can be developed to mitigate the errors from GPS signal distortion. The measurement error incurred from signal distortion is receiver-dependent. The monitor must mitigate the errors regardless of the type of equipment the user is employing.

GPS L_1L_2 Bias

Fault-free case—L_1L_2 bias errors can be computed with a Kalman filter. These corrections are not sent to the user, but used in the other monitors. Nominal error bounds are computed with standard statistical techniques.

Faulted case—if the SBAS design utilizes the L_1L_2 bias corrections in the integrity monitors, then they must account for the faulted case. The L_1L_2 bias can suddenly change due to an equipment failure on board the GPS satellite. The SBAS must be designed so that this type of failure does not "blind" the monitors. One approach to this design is to form a single-frequency integrity monitor that tests the corrections without using the L_1L_2 bias corrections.

Environment (Ionosphere) Errors

Fault-free case—under calm ionospheric conditions, the GIVE is computed in a fashion that accounts for measurement uncertainty, L_1L_2 bias errors, and nominal fluctuations in the ionosphere.

Faulted case—the integrity monitors must ensure that an ionospheric storm cannot cause HMI. One approach to this problem is to create an ionospheric storm detector that is sensitive to spatial and/or temporal changes in the ionospheric delay. Proving such a detector mitigates HMI is a difficult endeavor since the ionosphere is unpredictable during ionospheric storms. It is possible for ionospheric storms to exist in regions where the SBAS does not sample the event. An additional factor can be added to the GIVE to account for unobservable ionospheric storms. In some cases (when a reference receiver is out or the grid point on the edge of the service volume) this term can be quite large. The GIVE must also account for rapid fluctuations in the ionosphere between ionospheric correction updates. One way to mitigate such errors is to run the monitor frequently and send alarm messages if such an event occurs.

Environment (Troposphere) Errors

Both cases—tropospheric delay errors are built into many of the SBAS corrections. The SBAS must determine error bounds on the tropospheric delay error and build them into the UDRE.

7.2.8.2 *Mitigation of GEO-Associated Errors*

GEO Code and Carrier and Environment Errors For GEO code-associated errors, fault-free and faulted, see Section 7.2.8.1, subsection "GPS Code and Carrier."

Fault-free case—since GEO measurements are single-frequency, the dual-frequency techniques utilized for GPS integrity monitoring have to be modified. One approach to working with single-frequency measurements is to compensate for the iono delay using the broadcast ionospheric grid delays. The uncertainty of the iono corrections (GIVE) needs to be accounted for in the integrity monitors.

Faulted case—during ionospheric storms, the GIVE is likely to be substantially inflated. The inflated values will "blind" the other integrity monitors from detecting small GEO clock and ephemeris errors, resulting in a large GEO UDRE.

For both faulted and fault-free cases, of environment (troposphere) errors, see Section 7.2.8.1, subsection "Environment (Troposphere) Errors," *Both cases*.

7.2.8.3 Mitigation of Receiver and Measurement Processing Errors

Receiver Measurement Error

Fault-free case—The integrity monitors must account for the noise in the reference station measurements. A bound on the noise can be computed and utilized in the integrity monitors. In the "calculate then monitor" approach, integrity monitors must use measurements that are uncorrelated with the measurements used to compute the corrections. Otherwise, error cancellation may occur.

Faulted case—in the faulted case, one or more receivers may be sending out erroneous measurements. An integrity monitor must be built to detect such events and ensure that erroneous measurements do not blind the integrity monitors.

Intercard Bias *Both cases*—Intercard bias errors appear to be measurement errors and are mitigated by the methods discussed in Section 7.2.4.1.

Code Noise and Multipath (CNMP)

Fault-free case—Small multipath errors are accounted for in the receiver measurement error discussed in Section 7.2.8.3.

Faulted case—Large multipath errors must be detected and screened from the integrity monitors or accounted for in the measurement noise error bounds.

WRE L_1/L_2 Bias

Fault-free case—The WRE L_1L_2 bias can be computed in a manner similar to that for the GPS L_1L_2 bias. The nominal errors in this computation must be bounded and accounted for in the integrity monitors.

Faulted case—A receiver can malfunction causing the L_1/L_2 bias to suddenly change. The L_1L_2 bias is used in the correction and integrity monitoring functions and such a change must be detected and corrected to prevent HMI. As discussed in Section 4.1.5, a single-frequency monitor can be created that tests the corrections without using L_1/L_2 bias as an input.

WRE Clock Error

Fault-free case—The receiver clock error can be computed using a Kalman filter. Standard statistical techniques can be used to determine the error in the WRE clock estimates. This error bound can be utilized by the integrity monitors.

Faulted case—if bad data is received in the Kalman filter, erroneous WRE clock corrections could result. An integrity monitor can be built that does not utilize the WRE clock estimates from the Kalman filter to test the corrections when the WRE clock estimates are bad.

7.2.8.4 Mitigation of Estimation Errors

Reference Time Offset Estimation Error

Fault-free case—in the fault-free case, the difference between the GPS reference time and the SBAS reference time are accounted for by the user, provided the difference is less than 50 ns.

Faulted case—in the faulted case, due to some system fault or GPS anomaly, the difference in the SBAS reference time and the GPS reference time exceeds 50 ns. A simple monitor can be constructed to measure the difference between the two references. The monitor would respond to a large offset by setting all satellites not monitored, stopping the user from mixing corrected and uncorrected satellites.

Clock Estimation Error, Ephemeris Correction Error, L_1/L_2 WRE, and GPS Satellite Bias Estimation Error See Section 7.2.8.1, "GPS Clock Error," "GPS Ephemeris Error," and "GPS L_1/L_2 Bias," and Section 7.2.8.3 "WRE L_1L_2 Bias."

7.2.8.5 Mitigation of Integrity-Bound-Associated Errors

Ionospheric Modeling Error

Fault-free case—extensive testing of the models used in the SBAS will provide assurance that the iono model error is properly bounding under quiet ionospheric conditions.

Faulted case—during an ionospheric storm, the validity of the model is in question. A monitor can be constructed to test the validity of the model and increase the GIVE when the model is in question.

Fringe Area Ephemeris Error

Fault-free case—this error is mitigated by Message Type 28 as discussed in Section 7.2.8.1, "GPS Ephemeris Error."

Faulted case—special considerations must be taken to ensure that the integrity monitors are sensitive to satellite ephemeris errors on the fringe of coverage. Errors in the satellite ephemeris are not well viewed by the SBAS on the edge of the service region. A specific proof of the monitors' sensitivity to errors of this nature is required. Additional inflation factors may be needed to adjust the UDRE for this error.

Small-Sigma Errors

Fault-free case—tests can easily be performed on individual corrections; the user, however, must be protected from the combination of all error sources. An analysis can be performed to demonstrate that any combination of errors observed in the fault-free case is bounded by the broadcast integrity bounds. An example of this analysis is discussed in Ref. 10.

Faulted case—under faulted conditions, small biases may occur which can "add" in the user position solution to cause HMI. This threat can be mitigated by monitoring the accuracy of the user position solution at the reference stations.

Missed Message—OBAD

Fault-free case—the old but active data deprivation factors broadcast by the SBAS account for aging data.

Faulted case—the integrity monitors must ensure that every combination of active SBAS messages meets the integrity requirements. Two methods are suggested for this threat. First, the integrity monitors can run on every active set of broadcast messages to check their validity after broadcast. If a large error is detected, an alarm will be sent. A second, preferable, approach is to test the messages against every active data set before broadcast and adjust the corrections/integrity bounds accordingly.

TTA Exceeded

Fault-free case—the system is designed to meet the time-to-alarm requirement by continually monitoring the satellite signals and responding to integrity faults with alarms.

Faulted case—A monitor can be designed to test the "loop back" time in the system and continually ensure that the time to alarm requirement is met. The monitor sends a test message every minute and measures the time it takes for message to loop back through the system.

7.3 SBAS EXAMPLE

The process for identifying, characterizing, and mitigating a failure condition is illustrated by the following SBAS example.

SBAS broadcasts corrections to compensate for range errors incurred as the signal passes through the ionosphere. The uncertainty in these corrections is computed and sent to the user along with the corrections. HMI would result if the SBAS broadcasts erroneous integrity data (error bounds) and does not alert the user to the erroneous integrity data within a specified time limit. This time limit is referred to as the *time to alarm* (TTA).

1. *Identify error conditions that can cause HMI.* Error conditions can be caused by internal or external hardware or software failures or fluctuations in environmental conditions. The onset of an ionospheric storm represents a failure condition that could result in large errors in the ionospheric corrections, ultimately resulting in an increased probability of HMI.
2. *Precisely characterize the threat.* On days with nominal ionospheric behavior, the ionospheric threats are well understood and reasonably easy to quantify. Scientists are not yet able to characterize the ionosphere during storm conditions. For these reasons, SBAS has generated specific threat models for the ionosphere based on real data collected during the worst ionospheric activity from the solar maximum period (an 11-year solar cycle). An important aspect of this model is the ionospheric irregularity detector, which assures the validity of the model and inflates the error bounds if the validity of the model is in question.
3. *Identify error detection mechanisms.* In the SBAS, errors in ionospheric corrections are mitigated by a monitor located in a "safety processor" and a special detector called the "ionospheric irregularity detector."
4. *Analytically determine that the threat is mitigated.* It's tempting to take an RMA (reliability, maintainability, availability) approach to dealing with ionospheric storms.
   (a) Ionospheric storms are "infrequent events."
   (b) "We haven't seen them cause HMI yet . . . ."
   (c) "They don't last very long."
   (d) "The system has other margins . . . "

The a priori probability of a storm is not the mitigation of the threat. SBAS must meet its 10^{-7} integrity allocation during ionospheric storms. The analysis must account for worst-case events, like storms that are not well sampled by the ground system. Furthermore, it is not necessarily the storms with the highest magnitude that are the hardest to detect or most likely to cause HMI. Extensive analysis is needed to characterize the threat.

In general, every requirement in a system's specification is tested by some type of formal demonstration. Most of the SBAS system-level requirements fall into this category; however 'the SBAS integrity requirement does not. Testing fault-tree allocations of 10^{-7} and smaller requires on the order of 100,000,000 independent points (1 sample every 5 min for 950 years). Integrity can only be demonstrated where reference stations exist. Integrity must be proved for every satellite/user geometry. Every user at every point in space must be protected at all times. Demonstrations cannot be conducted where data are not available. In addition, every satellite geometry (subset) must be tested. Since GPS orbits repeat, then, if at a specific airport a satellite/user geometry exists with an increased probability of HMI, the situation will repeat every day at the same time until the constellation changes. It is because of these considerations that analytic proofs are required to satisfy integrity requirements.

The identification, characterization, and mitigation of a threat to the SBAS user should be carefully scrutinized by a panel of experts in the SBAS field. The analysis supporting claims is formally documented, scrutinized, and approved by this panel. This four-step process should be completed for every error identified in the system [200].

7.4 CONCLUSIONS

The data used by an SBAS to calculate the corrections and integrity data are assumed to contain errors which have been sufficiently mitigated. The errors discussed are inherent in any SBAS design that utilizes GPS satellites. An SBAS design mitigates the majority of these errors with "corrections," thereby making it a trusted navigation aid. The leftover errors, referred to as residual errors, are mitigated by the transmission of residual error bounding information. The threat of potential underbounding of integrity information is mitigated by integrity monitors. Both faulted and unfaulted cases are examined and mitigation strategies are discussed in this chapter. These SBAS corrections improve the accuracy of satellite signals. The integrity data ensure that the residual errors are bounded. The SBAS integrity monitors ensure that the integrity data have not been corrupted by SBAS failures. Following the integrity design guidelines given in this chapter is an important factor in obtaining certification and approval for use of the SBAS system.

SBAS integrity concepts may be applied to GBAS. In GBAS, the integrity will be broadcast from the ground.

7.5 GPS INTEGRITY CHANNEL (GIC)

This GPS data integrity channel will be provided in the next generation of GPS satellites such as GPS IIF and GPS III.

8

KALMAN FILTERING

8.1 INTRODUCTION

Kalman's paper introducing his now-famous filter was first published in 1960 [104], and its first practical implementation was for integrating an inertial navigator with airborne radar aboard the C5A military aircraft [137].

The application of interest here is quite similar. We want to integrate an onboard inertial navigator with a different electromagnetic ranging system (GPS). There are many ways to do this [18], but nearly all involve Kalman filtering.

The purpose of this chapter is to familiarize you with theoretical and practical aspects of Kalman filtering that are important for GPS/INS integration, and the presentation is primarily slanted toward this application. We have also included a brief derivation of the Kalman gain matrix, based on the maximum-likelihood estimation (MLE) model. Broader treatments of the Kalman filter are presented in Refs. 6, 30, 59, and 101; more basic introductions can be found in Refs. 48 and 218, more mathematically rigorous derivations can be found in Ref. 99; and more extensive coverage of the practical aspects of Kalman filtering can be found in Refs. 29 and 66.

8.1.1 What Is a Kalman Filter?

The *Kalman filter* is an extremely effective and versatile procedure for combining *noisy sensor outputs* to estimate the *state* of a *system* with *uncertain dynamics,* where

Global Positioning Systems, Inertial Navigation, and Integration, Second Edition, by M. S. Grewal, L. R. Weill, and A. P. Andrews

The *noisy sensors* could be just *GPS receivers* and *inertial navigation systems,* but may also include subsystem-level sensors (e.g., GPS clocks or INS accelerometers and gyroscopes) or auxiliary sensors such as speed sensors (e.g., wheel speed sensors for land vehicles, water speed sensors for ships, air speed sensors for aircraft, or Doppler radar), magnetic compasses, altimeters (barometric or radar), or radionavigation aids (e.g., DME, VOR, LORAN).

The *system state* in question may include the *position, velocity, acceleration, attitude*, and *attitude rate* of a *vehicle* on land, at sea, in the air, or in space. The system state may also include ancillary "nuisance variables" for modeling *time-correlated noise sources* such as ionospheric propagation delays of GPS signals, and *time-varying parameters* of the sensors, GPS receiver clock frequency and phase, or scale factors and output biases of accelerometers or gyroscopes.

Uncertain dynamics includes unpredictable disturbances of the host vehicle, whether caused by a human operator or by the medium (e.g., winds, surface currents, turns in the road, or terrain changes), but it may also include unpredictable changes in the sensor parameters.

8.1.2 How it Works

8.1.2.1 Estimates and Uncertainties The Kalman filter maintains two types of variables:

1. *An estimate* $\hat{\mathbf{x}}$ *of the state vector* $\mathbf{x}$. The components of the estimated state vector include the following:
   (a) The variables of interest (i.e., what we want or need to know, such as position and velocity).
   (b) "Nuisance variables" that are of no intrinsic interest, but may be necessary to the estimation process. These nuisance variables may include, for example, the effective propagation delay errors in GPS signals. We generally do not wish to know their values but may be obliged to calculate them to improve the receiver estimate of position.

   The Kalman filter state variables for a specific application ordinarily include all those system dynamic variables that are measurable by the sensors used in that application. For example, A Kalman filter for an INS containing accelerometers and rate gyroscopes might include accelerations and rotation rates to which these instruments respond. However, simplified INS models might ignore the accelerometers and angular rate sensors and model the INS itself as a position-only sensor, or as a position and velocity sensor. In similar fashion, the Kalman filter state variables for GPS-only navigation might include the pseudoranges between the satellites and the receiver antenna, or they might contain the position coordinates of the receiver antenna. Position could be represented by geodetic latitude, longitude, and altitude with respect to a reference ellipsoid, or geocentric latitude, longitude, and altitude with respect to a reference sphere, or ECEF Cartesian coordinates, or ECI coordinates, or any equivalent coordinates.

2. *An estimate of estimation uncertainty.* Uncertainty is modeled by a *covariance matrix*

$$\mathbf{P} \stackrel{\text{def}}{=} \mathrm{E}\left\langle \left(\hat{\mathbf{x}} - \mathbf{x}\right)\left(\hat{\mathbf{x}} - \mathbf{x}\right)^{\mathrm{T}}\right\rangle \tag{8.1}$$

of estimation error $\left(\hat{\mathbf{x}} - \mathbf{x}\right)$, where $\hat{\mathbf{x}}$ is the estimated state vector, $\mathbf{x}$ is the actual state vector and E is the expectancy operator. The equations used to propagate the covariance matrix (collectively called the *Riccati equation*) model and manage *uncertainty*, taking into account how sensor noise and dynamic uncertainty contribute to uncertainty about the estimated system state.

By maintaining an estimate of its own estimation uncertainty and the relative uncertainty in the various sensor outputs, the Kalman filter is able to combine all sensor information "optimally," in the sense that the resulting estimate minimizes any linear quadratic loss function of estimation error, including the mean-squared value of any linear combination of state estimation errors. The *Kalman gain* is the optimal weighting matrix for combining new sensor data with a prior estimate to obtain a new estimate. The Kalman gain is usually obtained as a partial result in the solution of the Riccati equation.

8.1.2.2 Prediction Updates and Correction Updates The Kalman filter is a two-step process, the steps of which we call "prediction" and "correction." The filter can start with either step.

The *correction step* updates the estimate and estimation uncertainty, based on new information obtained from sensor measurements. It is also called the *observational update* or *measurement update,* and the Latin prepositional phrase *a posteriori* is often used for the corrected estimate and its associated uncertainty.

The *prediction step* updates the estimate and estimation uncertainty, taking into account the effects of uncertain system dynamics over the times between measurements. It is also called the *temporal update,* and the Latin phrase *a priori* is often used for the predicted estimate and its associated uncertainty.

8.2 KALMAN GAIN

The Kalman gain matrix $\overline{\mathbf{K}}$ is the crown jewel of Kalman filtering. All the effort of solving the matrix Riccati equation is for the purpose of computing the "optimal"

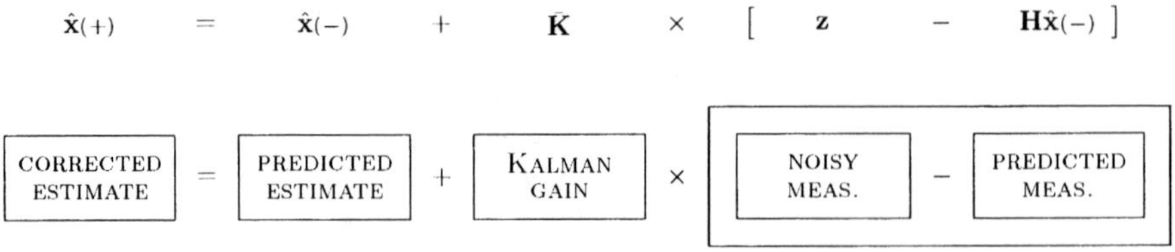

$$\hat{\mathbf{x}}(+) = \hat{\mathbf{x}}(-) + \overline{\mathbf{K}} \times [\mathbf{z} - \mathbf{H}\hat{\mathbf{x}}(-)]$$

CORRECTED ESTIMATE = PREDICTED ESTIMATE + KALMAN GAIN × [NOISY MEAS. − PREDICTED MEAS.]

Fig. 8.1 Estimate correction using kalman gain.

value of the gain matrix $\overline{\mathbf{K}}$ used as shown in Fig. 8.1 for correcting an estimate $\hat{\mathbf{x}}$, based on a measurement

$$\mathbf{z} = \mathbf{H}\mathbf{x} + \text{noise} \tag{8.2}$$

that is a linear function of the vector variable $\mathbf{x}$ to be estimated, plus additive noise with known statistical properties.

8.2.1 Approaches to Deriving the Kalman Gain

Kalman's original derivation of his gain matrix made no assumptions about the underlying probability distributions, but this requires a level of mathematical rigor that is a bit beyond standard engineering mathematics. As an alternative, it has become common practice to derive the formula for the Kalman gain matrix $\overline{\mathbf{K}}$ based on an analogous filter called the *Gaussian maximum-likelihood estimator*. It uses the analogies shown in Fig. 8.2 between concepts in Kalman filtering, Gaussian probability distributions, and maximum-likelihood estimation.

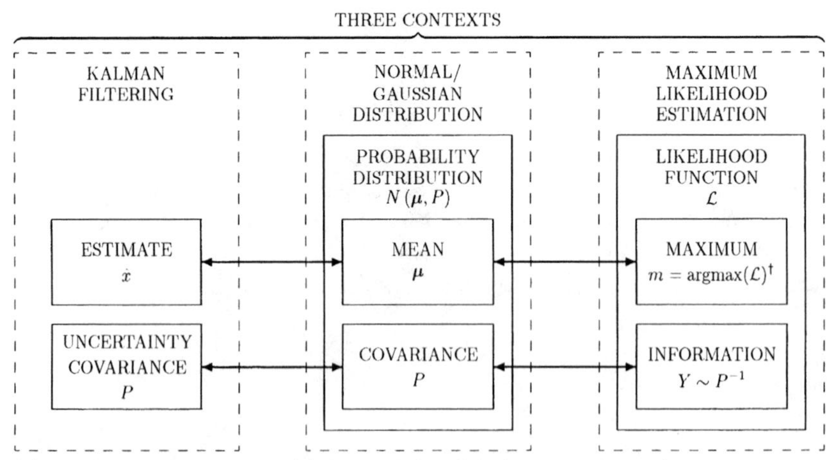

Fig. 8.2 Analogous concepts in three different contexts.

This derivation is given in the following subsections. It begins with some background information on the properties of Gaussian probability distributions and Gaussian likelihood functions, then development of models for noisy sensor outputs and a derivation of the associated *maximum-likelihood estimate* (MLE) for combining prior estimates with noisy sensor measurements.

8.2.2 Gaussian Probability Density Functions

Probability density functions (PDFs) are nonnegative integrable functions whose integral equals *unity* (i.e., 1). The density functions of *Gaussian* probability distributions all have the form

$$p(\mathbf{x}) = \frac{1}{\sqrt{(2\pi)^n \det \mathbf{P}}} \exp\left(-\tfrac{1}{2}[\mathbf{x}-\boldsymbol{\mu}]^T \mathbf{P}^{-1}[\mathbf{x}-\boldsymbol{\mu}]\right), \tag{8.3}$$

where n is the dimension of P (i.e., P is an $n \times n$ matrix) and the parameters

$$\boldsymbol{\mu} \stackrel{\text{def}}{=} E_{x \in N(\boldsymbol{\mu}, P)} \langle x \rangle \tag{8.4}$$

$$\stackrel{\text{def}}{=} \int_{x_1} dx_1 \cdots \int_{x_n} dx_n \; p(x)\, x \tag{8.5}$$

$$\mathbf{P} \stackrel{\text{def}}{=} E_{x \in N(\boldsymbol{\mu}, P)} \langle (x-\boldsymbol{\mu})(x-\boldsymbol{\mu})^{\mathrm{T}} \rangle \tag{8.6}$$

$$\stackrel{\text{def}}{=} \int_{x_1} dx_1 \cdots \int_{x_n} dx_n \; p(x)(x-\boldsymbol{\mu})(x-\boldsymbol{\mu})^T. \tag{8.7}$$

The parameter $\boldsymbol{\mu}$ is the *mean* of the distribution. It will be a column vector with the same dimensions as the variate $\mathbf{x}$.

The parameter $\mathbf{P}$ is the *covariance matrix* of the distribution. By its definition, it will always be an $n \times n$, *symmetric*, *nonnegative definite* matrix. However, because its determinant appears in the denominator of the square root and its inverse appears in the exponential function argument, it must be *positive definite* as well; that is, its eigenvalues must be real and positive for the definition to work.

The constant factor $1/\sqrt{(2\pi)^n \det \mathbf{P}}$ in Eq. 8.3 is there to make the integral of the probability density function equal to unity, a necessary condition for all probability density functions.

The operator $E\langle \cdot \rangle$ is the *expectancy operator*, also called the *expected-value operator*.

The notation $\mathbf{x} \in N(\boldsymbol{\mu}, \mathbf{P})$ denotes that the *variate* (i.e., random variable) $\mathbf{x}$ is drawn from the Gaussian distribution with mean $\boldsymbol{\mu}$ and covariance $\mathbf{P}$. Gaussian distributions are also called *normal* (the source of the "N" notation) or *Laplace* distributions.

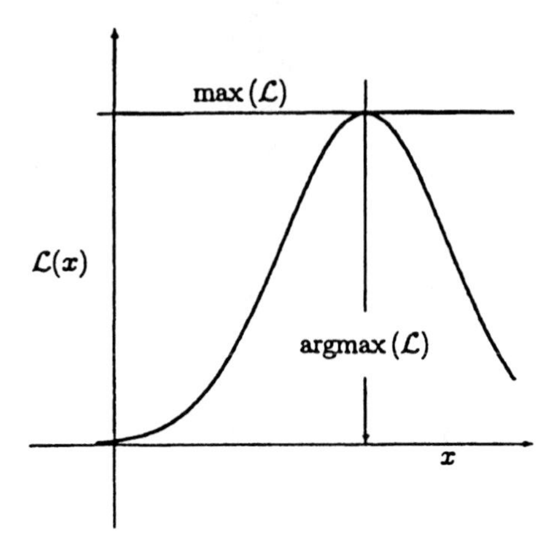

Fig. 8.3 Maximum-likelihood estimate.

8.2.3 Properties of Likelihood Functions

Likelihood functions are similar to probability density functions, except that their integrals are not constrained to equal unity, or even required to be finite. They are useful for comparing *relative* likelihoods and for finding a value

$$\mathbf{m} \in \text{argmax}\ [\mathcal{L}(\mathbf{x})] \tag{8.8}$$

of the unknown independent variable $\mathbf{x}$ at which the likelihood function $\mathcal{L}$ achieves its maximum,[1] as illustrated in Fig. 8.3.

8.2.3.1 Gaussian Likelihood Functions Gaussian likelihood functions have the form

$$\mathcal{L}(\mathbf{x}, \mathbf{m}, \mathbf{Y}) = c \exp\left\{-\frac{1}{2}[\mathbf{x} - \mathbf{m}]^{\mathrm{T}}\mathbf{Y}[\mathbf{x} - \mathbf{m}]\right\}, \tag{8.9}$$

where $c > 0$ is an arbitrary scaling variable and $\mathbf{m}$ (defined in Eq. 8.8) is a value of $\mathbf{x}$ at which $\mathcal{L}$ achieves its maximum value.

Information Matrices The parameter $\mathbf{Y}$ in Eq. 8.9 is called the *information matrix* of the likelihood function. It replaces $\mathbf{P}^{-1}$ in the Gaussian probability density function. If the information matrix $\mathbf{Y}$ is nonsingular, then its inverse $\mathbf{Y}^{-1} = \mathbf{P}$, a covariance matrix. However, *an information matrix is not required to be nonsingular*. This property of information matrices is important for representing the information from a set of measurements (sensor outputs) with incomplete

[1] It is possible that a likelihood function will achieve its maximum value at more than one value of $\mathbf{x}$, but that will not matter in the derivation.

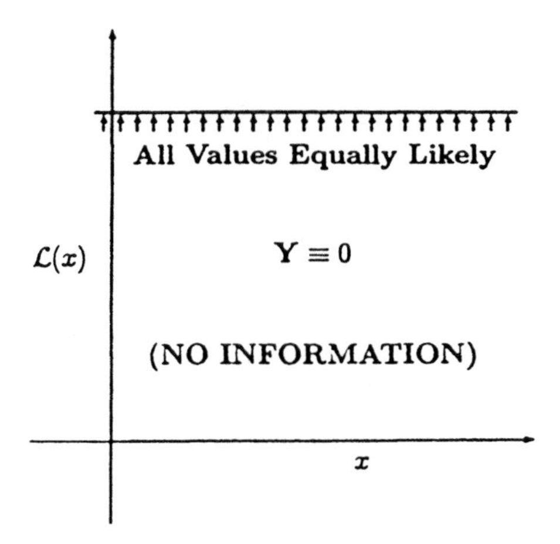

Fig. 8.4 Likelihood without unique maximum.

information for determining the unknown vector **x**. Thus, the measurements may provide *no information* about some linear combinations of the components of **x**, as illustrated in Fig. 8.4.

8.2.3.2 Scaling of Likelihood Functions Maximum-likelihood estimation is based on the argmax of the likelihood function, but for any positive scalar $c > 0$,

$$\mathrm{argmax}(c\mathcal{L}) = \mathrm{argmax}(\mathcal{L}). \tag{8.10}$$

Thus, positive scaling of likelihood functions will have no effect on the maximum-likelihood estimate. As a consequence, likelihood functions can have arbitrary positive scaling.

8.2.3.3 Independent Likelihood Functions The joint probability $P(A\&B)$ of independent events A and B is the product $P(A\&B) = P(A) \times P(B)$. The analogous effect on independent likelihood functions $\mathcal{L}_A$ and $\mathcal{L}_B$ is the pointwise product; that is, at each "point" **x**

$$\mathcal{L}_{A\&B}(\mathbf{x}) = \mathcal{L}_A(\mathbf{x}) \times \mathcal{L}_B(\mathbf{x}). \tag{8.11}$$

8.2.3.4 Pointwise Products of Likelihood Functions One of the remarkable attributes of Gaussian likelihood functions is that their pointwise products are also Gaussian likelihood functions, as illustrated in Fig. 8.5.

A Lemma Given two Gaussian likelihood functions with parameter sets $\{\mathbf{m}_A, \mathbf{Y}_A\}$ and $\{\mathbf{m}_B, \mathbf{Y}_B\}$, their pointwise product is a scaled Gaussian likelihood function with

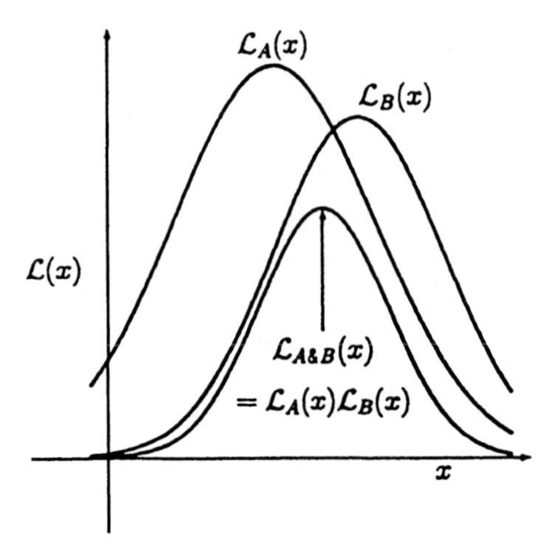

Fig. 8.5 Pointwise products of Gaussian likelihood functions.

parameters $\{\mathbf{m}_{A\&B}, \mathbf{Y}_{A\&B}\}$ such that, for all $\mathbf{x}$, one obtains

$$\exp\left(-\frac{1}{2}[\mathbf{x}-\mathbf{m}_{A\&B}]^T\mathbf{Y}_{A\&B}[\mathbf{x}-\mathbf{m}_{A\&B}]\right)$$
$$= c \times \exp\left(-\frac{1}{2}[\mathbf{x}-\mathbf{m}_{A}]^T\mathbf{Y}_{A}[\mathbf{x}-\mathbf{m}_{A}]\right)$$
$$\times \exp\left(-\frac{1}{2}[\mathbf{x}-\mathbf{m}_{B}]^T\mathbf{Y}_{B}[\mathbf{x}-\mathbf{m}_{B}]\right) \tag{8.12}$$

for some constant $c > 0$.

This is the fundamental lemma about Gaussian likelihood functions, and proving it by construction will give us the Kalman gain matrix as a corollary.

8.2.4 Solving for Combined Information Matrix

One can solve Eq. 8.12 for the parameters $\mathbf{m}_{A\&B}$ and $\mathbf{Y}_{A\&B}$ as functions of the parameters $\mathbf{m}_A$, $\mathbf{Y}_A$, $\mathbf{m}_B$, $\mathbf{Y}_B$.

Taking logarithms of both sides of Eq. 8.12 will produce the equation

$$-\frac{1}{2}[\mathbf{x}-\mathbf{m}_{A\&B}]^T\mathbf{Y}_{A\&B}[\mathbf{x}-\mathbf{m}_{A\&B}] = \log(c) - \frac{1}{2}[\mathbf{x}-\mathbf{m}_{A}]^T\mathbf{Y}_{A}[\mathbf{x}-\mathbf{m}_{A}]$$
$$-\frac{1}{2}[\mathbf{x}-\mathbf{m}_{B}]^T\mathbf{Y}_{B}[\mathbf{x}-\mathbf{m}_{B}]. \tag{8.13}$$

Next, taking the first and second derivatives with respect to the independent variable $\mathbf{x}$ will produce the equations

$$\mathbf{Y}_{A\&B}(\mathbf{x} - \mathbf{m}_{A\&B}) = \mathbf{Y}_A(\mathbf{x} - \mathbf{m}_A) + \mathbf{Y}_B(\mathbf{x} - \mathbf{m}_B), \tag{8.14}$$

$$\mathbf{Y}_{A\&B} = \mathbf{Y}_A + \mathbf{Y}_B, \tag{8.15}$$

respectively.

8.2.4.1 Information is Additive The information matrix of the combined likelihood function ($\mathbf{Y}_{A\&B}$ in Eq. 8.15) equals the sum of the individual information matrices of the component likelihood functions ($\mathbf{Y}_A$ and $\mathbf{Y}_B$ in Eq. 8.15).

8.2.5 Solving for Combined Argmax

Equation 8.14 evaluated at $\mathbf{x} = 0$ is

$$\mathbf{Y}_{A\&B}\mathbf{m}_{A\&B} = \mathbf{Y}_A\mathbf{m}_A + \mathbf{Y}_B\boldsymbol{\mu}_B, \tag{8.16}$$

which can be solved for

$$\mathbf{m}_{A\&B} = \mathbf{Y}^{\dagger}_{A\&B}(\mathbf{Y}_A\mathbf{m}_A + \mathbf{Y}_B\mathbf{m}_B) \tag{8.17}$$

$$= (\mathbf{Y}_A + \mathbf{Y}_B)^{\dagger}(\mathbf{Y}_A\mathbf{m}_A + \mathbf{Y}_B\boldsymbol{\mu}_B), \tag{8.18}$$

where $\dagger$ denotes the Moore–Penrose inverse of a matrix (defined in Section B.1.4.7).

8.2.5.1 Combined Maximum-Likelihood Estimate is a Weighted Average Equations 8.15 and 8.18 are key results for deriving the formula for Kalman gain. All that remains is to define likelihood function parameters for noisy sensors.

8.2.6 Noisy Measurement Likelihoods

The term *measurements* refers to outputs of sensors that are to be used in estimating the argument vector $\mathbf{x}$ of a likelihood function. Measurement models represent how these measurements are related to $\mathbf{x}$, including those errors called *measurement errors* or *sensor noise*. These models can be expressed in terms of likelihood functions with $\mathbf{x}$ as the argument.

8.2.6.1 Measurement Vector The collective output values from a multitude ℓ of sensors can be represented as the components of a vector

$$\mathbf{z} \stackrel{\text{def}}{=} \begin{bmatrix} z_1 \\ z_2 \\ z_3 \\ \vdots \\ z_\ell \end{bmatrix}, \tag{8.19}$$

called the *measurement vector*, a column vector with ℓ rows.

8.2.6.2 Measurement Sensitivity Matrix We suppose that the measured values z_i are linearly[2] related to the unknown vector $\mathbf{x}$ we wish to estimate, namely

$$\mathbf{z} = \mathbf{H}\mathbf{x}, \tag{8.20}$$

where $\mathbf{H}$ is the measurement sensitivity matrix.

8.2.6.3 Measurement Noise Measurement noise is the unpredictable error at the output of the sensors. It is assumed to be additive:

$$\mathbf{z} = \mathbf{H}\mathbf{x} + \mathbf{v} \tag{8.21}$$

or

$$\mathbf{z} = \mathbf{H}\mathbf{x} + \mathbf{J}\mathbf{v}, \tag{8.22}$$

where the measurement noise vector $\mathbf{v}$ is assumed to be zero-mean Gaussian noise with known covariance $\mathbf{R}$:

$$E\langle \mathbf{v} \rangle \stackrel{\text{def}}{=} 0, \tag{8.23}$$

$$\mathbf{R} \stackrel{\text{def}}{=} E\langle \mathbf{v}\mathbf{v}^{\mathrm{T}} \rangle. \tag{8.24}$$

8.2.6.4 Sensor Noise Distribution Matrix The matrix $\mathbf{J}$ in Eq. 8.22 is called a "sensor noise distribution matrix." It models "common mode" sensor noise, in which a lower-dimensional noise source (e.g., power supply noise, electromagnetic interference, or temperature variations) corrupts multiple sensor outputs.

8.2.6.5 Measurement Likelihood A measurement vector $\mathbf{z}$ and its associated covariance matrix of measurement noise $\mathbf{R}$ define a likelihood function for the "true" value of the measurement (i.e., without noise). This likelihood function will have its argmax at

$$\mathbf{m}_z = \mathbf{z} \tag{8.25}$$

and information matrix

$$\mathbf{Y}_z = \mathbf{R}^{-1}, \tag{8.26}$$

assuming that $\mathbf{R}$ is nonsingular.

8.2.6.6 Unknown Vector Likelihoods The same parameters defining measurement likelihoods also define an inferred likelihood function for the true value of

[2]The Kalman filter is defined in terms of the *measurement sensitivity matrix* $\mathbf{H}$, but the *extended Kalman filter* (described in Section 8.6.4) can be defined in terms of a suitably differentiable vector-valued function $\mathbf{h}(\mathbf{x})$.

the unknown vector, with argmax

$$\mathbf{m}_x = \mathbf{H}^{\dagger}\mathbf{m}_z \tag{8.27}$$

$$= \mathbf{H}^{\dagger}\mathbf{z} \tag{8.28}$$

and information matrix

$$\mathbf{Y}_x = \mathbf{H}^{\mathrm{T}}\mathbf{Y}_z\mathbf{H} \tag{8.29}$$

$$= \mathbf{H}^{\mathrm{T}}\mathbf{R}^{-1}\mathbf{H}, \tag{8.30}$$

where the $n\times \ell$ matrix $\mathbf{H}^{\dagger}$ is defined as the Moore–Penrose generalized inverse (defined in Appendix B) of the $\ell \times n$ matrix $\mathbf{H}$. This information matrix will be singular if $\ell < n$ (i.e., if there are fewer sensor outputs than unknown variables), which is not unusual for GPS/INS integration.

8.2.7 Gaussian Maximum-Likelihood Estimate

8.2.7.1 ***Variables*** Gaussian MLE uses the following variables:

$\hat{\mathbf{x}}$, the maximum-likelihood estimate of $\mathbf{x}$. It will always equal the argmax of an associated Gaussian likelihood function, but it can have different values:

- $\hat{\mathbf{x}}(-)$, the predicted value, representing the likelihood function prior to using the measurement results.
- $\hat{\mathbf{x}}(+)$, the corrected value, representing the likelihood function after using the measurement results.

$\mathbf{P}$, the covariance matrix of estimation uncertainty. It will always equal the inverse of the information matrix $\mathbf{Y}$ of the associated likelihood function. It also can have two values:

- $\mathbf{P}(-)$, representing the likelihood function prior to using the measurements.
- $\mathbf{P}(+)$, representing the likelihood function after using the measurements.

$\mathbf{z}$, the vector of measurements.

$\mathbf{H}$, the measurement sensitivity matrix.

$\mathbf{R}$, the covariance matrix of sensor noise.

8.2.7.2 ***Maximum-Likelihood Correction Equations*** The MLE formula for correcting the variables $\hat{\mathbf{x}}$ and $\mathbf{P}$ to reflect the effect of measurements can be derived from Eqs. 8.15 and 8.18 with initial likelihood parameters

$$\mathbf{m}_A = \hat{\mathbf{x}}(-), \tag{8.31}$$

the MLE before measurements, and

$$\mathbf{Y}_A = \mathbf{P}(-)^{-1}, \tag{8.32}$$

the inverse of the covariance matrix of MLE uncertainty before measurements. The likelihood function of $\mathbf{x}$ inferred from the measurements alone (i.e., without taking into account the prior estimate) is represented by the likelihood function parameters

$$\mathbf{Y}_B = \mathbf{H}^T\mathbf{R}^{-1}\mathbf{H}, \tag{8.33}$$

the information matrix of the measurements, and

$$\mathbf{m}_B = \mathbf{H}^{\dagger}\mathbf{z}, \tag{8.34}$$

where $\mathbf{z}$ is the measurement vector and $\dagger$ represents the Moore–Penrose generalized matrix inverse.

8.2.7.3 Covariance Update The Gaussian likelihood function with parameters $\mathbf{m}_{A\&B}$, $\mathbf{Y}_{A\&B}$ of Eqs. 8.15 and 8.18 then represents the state of knowledge about the unknown vector $\mathbf{x}$ combining both sources (i.e., the prior likelihood and the effect of the measurements). That is, the covariance of MLE uncertainty after using the measurements will be

$$\mathbf{P}(+) = \mathbf{Y}_{A\&B}^{-1}, \tag{8.35}$$

and the MLE of $\mathbf{x}$ after using the measurements will then be

$$\hat{\mathbf{x}}(+) = \mathbf{m}_{A\&B}. \tag{8.36}$$

Equation 8.15 can be simplified by applying the following general matrix formula:[3]

$$(\mathbf{A}^{-1} + \mathbf{B}\mathbf{C}^{-1}\mathbf{D})^{-1} = \mathbf{A} - \mathbf{A}\mathbf{B}(\mathbf{C} + \mathbf{D}\mathbf{A}\mathbf{B})^{-1}\mathbf{D}\mathbf{A}, \tag{8.37}$$

where

$$\left.\begin{array}{lcl} \mathbf{A}^{-1} & = & \mathbf{Y}_A, \text{ the prior information matrix for } \hat{\mathbf{x}} \\ \mathbf{A} & = & \mathbf{P}(-), \text{ the prior covariance matrix for } \hat{\mathbf{x}} \\ \mathbf{B} & = & \mathbf{H}^T, \text{ the transpose of the measurement sensitivity matrix} \\ \mathbf{C} & = & \mathbf{R}, \\ \mathbf{D} & = & \mathbf{H}, \text{ the measurement sensitivity matrix,,} \end{array}\right\}, \tag{8.38}$$

so that Eq. 8.35 becomes

$$\left.\begin{array}{lcll} \mathbf{P}(+) & = & \mathbf{Y}_{A\&B}^{-1} & \\ & = & (\mathbf{Y}_A + \mathbf{H}^{\mathrm{T}}\mathbf{R}^{-1}\mathbf{H})^{-1} & \text{(Eq. 8.15)} \\ & = & \mathbf{Y}_A^{-1} - \mathbf{Y}_A^{-1}\mathbf{H}^{\mathrm{T}}(\mathbf{H}\mathbf{Y}_A^{-1}\mathbf{H}^{\mathrm{T}} + \mathbf{R})^{-1}\mathbf{H}\mathbf{Y}_A^{-1} & \text{(Eq. 8.37)} \\ & = & \mathbf{P}(-) - \mathbf{P}(-)\mathbf{H}^{\mathrm{T}}(\mathbf{H}\mathbf{P}(-)\mathbf{H}^{\mathrm{T}} + \mathbf{R})^{-1}\mathbf{H}\mathbf{P}(-), & \end{array}\right\}, \tag{8.39}$$

a form better conditioned for computation.

[3] A formula with many discoverers. Henderson and Searle [81] list some earlier ones.

8.2.7.4 Estimate Correction Equation 8.18 with substitutions from Eqs. 8.31–8.34 will have the form shown in Fig. 8.1

$$
\left.
\begin{aligned}
\hat{\mathbf{x}}(+) &= \mathbf{m}_{A\&B} && \text{(Eq. 8.36)} \\
&= (\mathbf{Y}_A + \mathbf{Y}_B)^{\dagger}(\mathbf{Y}_A \mathbf{m}_A + \mathbf{Y}_B \mathbf{m}_B) && \text{(Eq. 8.18)} \\
&= \underbrace{\mathbf{P}(+)}_{(8.35)}[\underbrace{\mathbf{P}(-)^{-1}}_{(8.32)}\ \underbrace{\hat{\mathbf{x}}(\text{-})}_{(8.31)} + \underbrace{\mathbf{H}^{\mathrm{T}}\mathbf{R}^{-1}\mathbf{H}}_{(8.33)}\ \underbrace{\mathbf{H}^{\dagger} z}_{8.34}] \\
&= [\mathbf{P}(-) - \mathbf{P}(-)\mathbf{H}^{\mathrm{T}}(\mathbf{H}P(-)\mathbf{H}^{\mathrm{T}} + \mathbf{R})^{-1}\mathbf{H}P(-)] \\
&\quad \times [\mathbf{P}(-)^{-1}\hat{\mathbf{x}}(-) + \mathbf{H}^{\mathrm{T}}\mathbf{R}^{-1}\mathbf{H}\mathbf{H}^{\dagger}\mathbf{z}] && \text{(Eq. 8.39)} \\
&= [\mathbf{I} - \mathbf{P}(-)\mathbf{H}^{\mathrm{T}}(\mathbf{H}\mathbf{P}(-)\mathbf{H}^{\mathrm{T}} + \mathbf{R})^{-1}\mathbf{H}] \\
&\quad \times [\hat{\mathbf{x}}(-) + \mathbf{P}(-)\mathbf{H}^{\mathrm{T}}\mathbf{R}^{-1}\mathbf{H}\mathbf{H}^{\dagger}\mathbf{z}] \\
&= \hat{\mathbf{x}}(-) + \mathbf{P}(-)\mathbf{H}^{\mathrm{T}}(\mathbf{H}\mathbf{P}(-)\mathbf{H}^{\mathrm{T}} + \mathbf{R})^{-1} \\
&\quad \times \{[(\mathbf{H}\mathbf{P}(-)\mathbf{H}^{\mathrm{T}} + \mathbf{R})\mathbf{R}^{-1} \\
&\quad - \mathbf{H}\mathbf{P}(-)\mathbf{H}^{\mathrm{T}}\mathbf{R}^{-1}]\mathbf{z} - \mathbf{H}\hat{\mathbf{x}}(-)\} \\
&= \hat{\mathbf{x}}(-) + \mathbf{P}(-)\mathbf{H}^{\mathrm{T}}(\mathbf{H}\mathbf{P}(-)\mathbf{H}^{\mathrm{T}} + \mathbf{R})^{-1} \\
&\quad \times \{[\mathbf{H}\mathbf{P}(-)\mathbf{H}^{\mathrm{T}}\mathbf{R}^{-1} + \mathbf{I} - \mathbf{H}\mathbf{P}(-)\mathbf{H}^{\mathrm{T}}\mathbf{R}^{-1}]\mathbf{z} \\
&\quad - \mathbf{H}\hat{\mathbf{x}}(-)\} \\
&= \hat{\mathbf{x}}(-) + \underbrace{\mathbf{P}(-)H^{\mathrm{T}}(\mathbf{H}\mathbf{P}(-)\mathbf{H}^{\mathrm{T}} + \mathbf{R})^{-1}}_{\overline{\mathbf{K}}} \\
&\quad \times \{\mathbf{z} - \mathbf{H}\hat{\mathbf{x}}(-)\},
\end{aligned}
\right\}, \tag{8.40}
$$

where the matrix $\overline{\mathbf{K}}$ has a special meaning in Kalman filtering.

8.2.8 Kalman Gain Matrix for Maximum-Likelihood Estimation

The last line in Eq. 8.73 has the form of the equation in Fig. 8.1 with Kalman gain matrix

$$\overline{\mathbf{K}} = \mathbf{P}(-)\mathbf{H}^{T}[\mathbf{H}\mathbf{P}(-)\mathbf{H}^{T} + \mathbf{R}]^{-1}, \tag{8.41}$$

which completes the derivation of the Kalman gain matrix based on Gaussian MLE.

8.2.9 Estimate Correction Using Kalman Gain

The Kalman gain expression from Eq. 8.41 can be substituted into Eq. 8.73 to yield

$$\hat{\mathbf{x}}(-) = \hat{\mathbf{x}}(-) + \overline{\mathbf{K}}\left[\mathbf{z} - \mathbf{H}\hat{\mathbf{x}}(-)\right], \tag{8.42}$$

the estimate correction equation to account for the effects of measurements.

8.2.10 Covariance Correction for Measurements

The act of making a measurement and correcting the estimate on the basis of the information obtained reduces the uncertainty about the estimate. The effect this has on the covariance of estimation uncertainty $\mathbf{P}$ can be found by substituting Eq. 8.41 into Eq. 8.39. The result is a simplified equation for the covariance matrix update to correct for the effects of using the measurements:

$$\mathbf{P}(+) = \mathbf{P}(-) - \overline{\mathbf{K}}\mathbf{H}\mathbf{P}(-). \tag{8.43}$$

8.3 PREDICTION

The rest of the Kalman filter is the prediction step, in which the estimate $\hat{\mathbf{x}}$ and its associated covariance matrix of estimation uncertainty $\mathbf{P}$ are propagated from one time epoch to another. This is the part where the dynamics of the underlying physical processes come into play. The "state" of a dynamic process is a vector of variables that completely specify enough of the initial boundary value conditions for propagating the trajectory of the dynamic process forward in time, and the procedure for propagating that solution forward in time is called "state prediction." The model for propagating the covariance matrix of estimation uncertainty is derived from the model used for propagating the state vector.

8.3.1 Stochastic Systems in Continuous Time

The word *stochastic* derives from the Greek expression for *aiming at a target,* indicating some degree of uncertainty in the dynamics of the projectile between launch and impact. That idea has been formalized mathematically as *stochastic systems theory*, in which a *stochastic process* is a model for the evolution over time of a *probability distribution.*

8.3.1.1 White-Noise Processes A white noise process in continuous time is a function whose value at any time is a sample from a zero-mean Gaussian distribution, statistically independent of the values sampled at other times. White noise processes are not integrable functions in the ordinary (Riemann) calculus. A special calculus is required to render them integrable. It is called the *stochastic calculus.* See Ref. 99 for more details on this.

8.3.1.2 Stochastic Differential Equations Ever since the differential calculus was introduced (more or less simultaneously) by Sir Isaac Newton (1643–1727) and Gottfried Wilhelm Leibnitz (1646–1716), we have been using ordinary differential equations as models for the dynamical behavior of systems of all sorts.

In 1827, botanist Robert Brown (1773–1858) described the apparently random motions of small particles in fluids, and the phenomenon came to be called *Brownian motion.* In 1908, French physicist Paul Langevin[4] (1872–1946) published a mathematical model for Brownian motion as a differential equation. It included a random function of time that was eventually characterized as a *white-*

[4]Langevin also invented and developed sonar.

noise process. When the dependent variables in a differential equation include white-noise processes $\mathbf{w}(t)$, it is called a *stochastic differential equation.*

Uncertain dynamical systems are modeled by linear stochastic differential equations of the sort

$$\frac{d}{dt}\mathbf{x}(t) = \mathbf{F}(t)\mathbf{x}(t) + \mathbf{w}(t) \tag{8.44}$$

or

$$\frac{d}{dt}\mathbf{x}(t) = \mathbf{F}(t)\,\mathbf{x}(t) + \mathbf{G}(t)\,\mathbf{w}(t), \tag{8.45}$$

where $\mathbf{x}(t)$ is the *system state vector,* a column vector with n rows; $\mathbf{F}(t)$ is the *dynamic coefficient matrix,* an $n \times n$ matrix; $\mathbf{G}(t)$ is a *dynamic noise distribution matrix,* which can be an identity matrix; and $\mathbf{w}(t)$ is a zero-mean white-noise vector representing *dynamic disturbance noise,* also called *process noise.*

Example 8.1: Stochastic Differential Equation Model for Harmonic Resonator. Dynamical behavior of the one-dimensional damped mass–spring system shown schematically in Fig. 8.6 is modeled by the equations

$$m\frac{d^2\xi}{dt^2} = ma = F = \underbrace{-C_{\text{damping}}\frac{d\xi}{dt}}_{\text{damping force}} - \underbrace{C_{\text{spring}}\xi}_{\text{spring force}} + \underbrace{w(t)}_{\text{disturbance}}$$

or

$$\frac{d^2\xi}{dt^2} + \frac{C_{\text{damping}}}{m}\frac{d\xi}{dt} + \frac{C_{\text{spring}}}{m}\xi = \frac{w(t)}{m}, \tag{8.46}$$

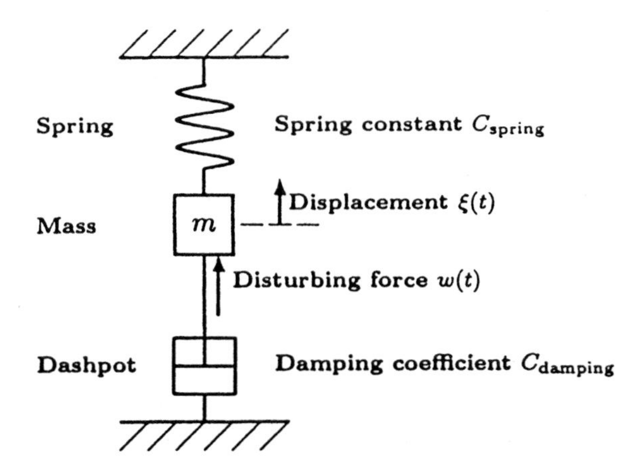

Fig. 8.6 Schematic model for dynamic system of Example 8.1.

where m is the mass attached to spring and damper, ξ is the upward displacement of the mass from its rest position, C_{spring} is the spring constant, C_{damping} is the damping coefficient of the dashpot, and $w(t)$ is the random disturbing force acting on the mass.

8.3.1.3 *Systems of First-Order Linear Differential Equations*

The so-called *state space models* for dynamic systems replace higher-order differential equations with systems of first-order differential equations. This can be done by defining the first $n-1$ derivatives of an nth-order differential equation as state variables.

Example 8.2: State Space Model for Harmonic Resonator. Equation 8.46 is a linear second-order ($n = 2$) differential equation. It can be transformed into a system of two linear first-order differential equations with state variables

$$x_1 \stackrel{\text{def}}{=} \xi \text{ (mass displacement)},$$

$$x_2 \stackrel{\text{def}}{=} \frac{d\xi}{dt} \text{ (mass velocity)},$$

for which

$$\frac{dx_1}{dt} = x_2 \tag{8.47}$$

$$\frac{dx_2}{dt} = \frac{-C_{\text{spring}}}{m} x_1 + \frac{-C_{\text{damping}}}{m} x_2 + \frac{w(t)}{m}. \tag{8.48}$$

8.3.1.4 *Representation in Terms of Vectors and Matrices*

State space models using systems of linear first-order differential equations can be represented more compactly in terms of a *state vector, dynamic coefficient matrix*, and *dynamic disturbance vector*.

Systems of linear first-order differential equations represented in "longhand" form as

$$\begin{array}{ccccccccccccccc}
\frac{dx_1}{dt} & = & f_{11}x_1 & + & f_{12}x_2 & + & f_{13}x_3 & + & \cdots & + & f_{1n}x_n & + & w_1, \\
\frac{dx_2}{dt} & = & f_{21}x_1 & + & f_{22}x_2 & + & f_{23}x_3 & + & \cdots & + & f_{2n}x_n & + & w_2, \\
\frac{dx_3}{dt} & = & f_{31}x_1 & + & f_{32}x_2 & + & f_{33}x_3 & + & \cdots & + & f_{3n}x_n & + & w_3, \\
\vdots & & \vdots & & \vdots & & \vdots & & \ddots & & \vdots & & \vdots \\
\frac{dx_n}{dt} & = & f_{n1}x_1 & + & f_{n2}x_2 & + & f_{n3}x_3 & + & \cdots & + & f_{nn}x_n & + & w_n
\end{array}$$

can be represented more compactly in matrix form as

$$\frac{d}{dt}\mathbf{x} = \mathbf{F}\mathbf{x} + \mathbf{w}, \tag{8.49}$$

where the *state vector* $\mathbf{x}$, *dynamic coefficient matrix* $\mathbf{F}$, and *dynamic disturbance vector* $\mathbf{w}$ are given as

$$\mathbf{x} = \begin{bmatrix} x_1 \\ x_2 \\ x_3 \\ \vdots \\ x_n \end{bmatrix}, \quad \mathbf{F} = \begin{bmatrix} f_{11} & f_{12} & f_{13} & \cdots & f_{1n} \\ f_{21} & f_{22} & f_{23} & \cdots & f_{2n} \\ f_{13} & f_{23} & f_{33} & \cdots & f_{3n} \\ \vdots & \vdots & \vdots & \ddots & \vdots \\ f_{n1} & f_{n2} & f_{n3} & \cdots & f_{nn} \end{bmatrix}, \quad \mathbf{w} = \begin{bmatrix} w_1 \\ w_2 \\ w_3 \\ \vdots \\ w_n \end{bmatrix},$$

respectively.

Example 8.3: Harmonic Resonator Model in Matrix Form. For the system of linear differential equations 8.47 and 8.48, we obtain

$$\mathbf{x} = \begin{bmatrix} x_1 \\ x_2 \end{bmatrix},$$
$$\mathbf{F} = \begin{bmatrix} 0 & 1 \\ -C_{\text{spring}}/m & -C_{\text{damping}}/m \end{bmatrix}$$
$$\mathbf{w}(t) = \begin{bmatrix} 0 \\ w(t)/m \end{bmatrix}.$$

8.3.1.5 Eigenvalues of Dynamic Coefficient Matrices The coefficient matrix $\mathbf{F}$ of a system of linear differential equations $\dot{\mathbf{x}} = \mathbf{F}\mathbf{x} + \mathbf{w}$ has effective units of reciprocal time, or frequency (in units of radians per second). It is perhaps then not surprising that the characteristic values (eigenvalues) of $\mathbf{F}$ are the characteristic frequencies of the dynamic system represented by the differential equations.

The eigenvalues of an $n \times n$ matrix $\mathbf{F}$ are the roots $\{\lambda_i\}$ of its *characteristic polynomial*

$$\det(\lambda\mathbf{I} - \mathbf{F}) = \sum_{k=0}^{n} a_n \lambda^n = 0. \tag{8.50}$$

The eigenvalues of $\mathbf{F}$ have the same interpretation as the poles of the related system transfer function, in that the dynamic system $\dot{\mathbf{x}} = \mathbf{F}\mathbf{x} + \mathbf{w}$ is stable if and only if the solutions of the characteristic equation $\det(\lambda\mathbf{I} - \mathbf{F}) = 0$ lie in the left half-plane.

Example 8.4: Damping and Resonant Frequency for Underdamped Harmonic Resonator. For the dynamic coefficient matrix

$$\mathbf{F} = \begin{bmatrix} 0 & 1 \\ -C_{\text{spring}}/m & -C_{\text{damping}}/m \end{bmatrix} \tag{8.51}$$

in Example 8.3, the eigenvalues of $\mathbf{F}$ are the roots of its characteristic polynomial

$$\det(\lambda\mathbf{I} - \mathbf{F}) = \det\begin{bmatrix} \lambda & -1 \\ C_{\text{spring}}/m & \lambda + C_{\text{damping}}/m \end{bmatrix}$$
$$= \lambda^2 + \frac{C_{\text{damping}}}{m}\lambda + \frac{C_{\text{spring}}}{m},$$

which are

$$\lambda = -\frac{C_{\text{damping}}}{2m} \pm \frac{1}{2m}\sqrt{C^2_{\text{damping}} - 4mC_{\text{spring}}}.$$

If the discriminant

$$C^2_{\text{damping}} - 4mC_{\text{spring}} < 0,$$

then the mass–spring system is called *underdamped*, and its eigenvalues are a complex conjugate pair

$$\lambda = -\frac{1}{\tau_{\text{damping}}} \pm \omega_{\text{resonant}}\mathbf{i}$$

with real part

$$-\frac{1}{\tau_{\text{damping}}} = -\frac{C_{\text{damping}}}{2m}$$

and imaginary part

$$\omega_{\text{resonant}} = \frac{1}{2m}\sqrt{4mC_{\text{spring}} - C^2_{\text{damping}}}. \tag{8.52}$$

The alternative parameter

$$\tau_{\text{damping}} = \frac{2m}{C_{\text{damping}}}$$

is called the *damping time constant* of the system, and the other parameter ω_{resonant} is the resonant frequency in units of radians per second.

The dynamic coefficient matrix for the damped harmonic resonator model can also be expressed in terms of the resonant frequency and damping time constant as

$$\mathbf{F}_{\text{harmonic resonator}} = \begin{bmatrix} 0 & 1 \\ -\omega^2 - 1/\tau^2 & -2/\tau \end{bmatrix}. \tag{8.53}$$

As long as the damping coefficient $C_{\text{damping}} > 0$, the eigenvalues of this system will lie in the left half-plane. In that case, the damped mass–spring system is guaranteed to be stable.

8.3.1.6 Matrix Exponential Function The matrix exponential function is defined in Section B.6.4 of Appendix B (on the CD-ROM) as

$$\exp(\mathbf{M}) \stackrel{\text{def}}{=} \sum_{k=0}^{\infty} \frac{1}{k!} \mathbf{M}^k \tag{8.54}$$

for square matrices **M**. The result is a square matrix of the same dimension as **M**.
This function has some useful properties:

1. The matrix $\mathbf{N} = \exp(\mathbf{M})$ is always invertible and $\mathbf{N}^{-1} = \exp(-\mathbf{M})$.
2. If **M** is antisymmetric (i.e., its matrix transpose $\mathbf{M}^T = -\mathbf{M}$), then $\mathbf{N} = \exp(\mathbf{M})$ is an *orthogonal* matrix (i.e., its matrix transpose $\mathbf{N}^T = \mathbf{N}^{-1}$).
3. The eigenvalues of $\mathbf{N} = \exp(\mathbf{M})$ are the (scalar) exponential functions of the eigenvalues of **M**.
4. If $\mathbf{M}(s)$ is an integrable function of a scalar s, then the derivative

$$\frac{d}{dt}\left(\int^{t} \mathbf{M}(s)\,ds\right) = \mathbf{M}(t)\left(\int^{t} \mathbf{M}(s)\,ds\right). \tag{8.55}$$

8.3.1.7 Forward Solution The forward solution of a differential equation is a solution in terms of initial conditions. The property of the matrix exponential function shown in Eq. 8.55 can be used to define the forward solution of Eq. 8.49 as

$$\mathbf{x}(t) = \exp\left(\int_{t_0}^{t} \mathbf{F}(s)\,ds\right)\left[\mathbf{x}(t_0) + \int_{t_0}^{t} \exp\left(-\int_{t_0}^{s} \mathbf{F}(r)\,dr\right)\mathbf{w}(s)\,ds\right], \tag{8.56}$$

where $\mathbf{x}(t_0)$ is the *initial value* of the state vector **x** for $t \geq t_0$.

8.3.1.8 Time-Invariant Systems If the dynamic coefficient matrix **F** of Eq. 8.49 does not depend on t (time), then the problem is called *time invariant*. In that case

$$\int_{t_0}^{T} \mathbf{F} ds = (t - t_0)\,\mathbf{F}, \tag{8.57}$$

and the forward solution

$$\mathbf{x}(t) = \exp\left[(t - t_0)\,\mathbf{F}\right]\left\{\mathbf{x}(t_0) + \int_{t_0}^{t} \exp\left[-(s - t_0)\,\mathbf{F}\right]\mathbf{w}(s)\,ds\right\}. \tag{8.58}$$

8.3.2 Stochastic Systems in Discrete Time

8.3.2.1 Zero-Mean White Gaussian Noise Sequences *A zero-mean white Gaussian* noise process in discrete time is a sequence of *independent* samples

$\dots, \mathbf{w}_{k-1}, \mathbf{w}_k, \mathbf{w}_{k+1}, \dots$ from a normal probability distribution $N(0, \mathbf{Q}_k)$ with zero-mean and known finite covariances $\mathbf{Q}_k$. In Kalman filtering, it is not necessary (but not unusual) that the covariance of all samples be the same.

Sampling is called independent if the expected values of outer products

$$E\left\langle \mathbf{w}_i \mathbf{w}_j^T \right\rangle = \begin{cases} 0, & i \neq j, \\ \mathbf{Q}_i, & i = j, \end{cases} \tag{8.59}$$

for all integer indices i and j of the random process.

Zero-mean white Gaussian noise sequences are the fundamental random processes used in Kalman filtering. However, it is *not* necessary that all noise sources in the modeled sensors and dynamic systems be zero-mean white Gaussian noise sequences. It is only necessary that they can be modeled in terms of such processes.

8.3.2.2 Gaussian Linear Stochastic Processes in Discrete Time A linear stochastic processes model in discrete time has the form

$$\mathbf{x}_k = \Phi_{k-1}\mathbf{x}_{k-1} + \mathbf{w}_{k-1}, \tag{8.60}$$

where $\mathbf{w}_k$ is a zero-mean white Gaussian noise process with known covariances $\mathbf{Q}_k$ and the vector $\mathbf{x}$ represents the state of a dynamic system.

This model for "marginally random" dynamics is quite useful for representing physical systems (e.g., land vehicles, seacraft, aircraft) with zero-mean random disturbances (e.g., winds or currents). The state transition matrix Φ_k represents the known dynamic behavior of the system, and the covariance matrices $\mathbf{Q}_k$ represent the unknown random disturbances. Together, they model the propagation of the necessary statistical properties of the state variable $\mathbf{x}$.

Example 8.5: Harmonic Resonator with White Acceleration Disturbance Noise. If the disturbance acting on the harmonic resonator of Examples 8.1–8.6 were zero-mean white acceleration noise with variance $\sigma^2_{\text{disturbance}}$, then its disturbance noise covariance matrix would have the form

$$\mathbf{Q} = \begin{bmatrix} 0 & 0 \\ 0 & \sigma^2_{\text{disturbance}} \end{bmatrix}. \tag{8.61}$$

8.3.3 State Space Models for Discrete Time

Measurements are the outputs of sensors sampled at *discrete times* $\cdots < t_{k-1} < t_k < t_{k+1} < \cdots$. The Kalman filter uses these values to estimate the state of the associated dynamic systems at those discrete times.

If we let $\dots, \mathbf{x}_{k-1}, \mathbf{x}_k, \mathbf{x}_{k+1}, \dots$ be the corresponding state vector values of a linear dynamic system at those discrete times, then each of these values can be determined from the previous value by using Eq. 8.58 in the form

$$\mathbf{x}_k = \Phi_{k-1}\mathbf{x}_{k-1} + \mathbf{w}_{k-1}, \tag{8.62}$$

$$\Phi_{k-1} \stackrel{\text{def}}{=} \exp\left(\int_{t_{k-1}}^{t_k} \mathbf{F}(s)\,ds\right), \tag{8.63}$$

$$\mathbf{w}_{k-1} \stackrel{\text{def}}{=} \Phi_k \int_{t_{k-1}}^{t_k} \exp\left(-\int_{t_{k-1}}^{t_k} \mathbf{F}(s)\,ds\right) \mathbf{w}(t)\,dt. \tag{8.64}$$

Equation 8.62 is the discrete-time dynamic system model corresponding to the continuous-time dynamic system model of Eq. 8.49.

The matrix Φ_{k-1} (defined in Eq. 8.63) in the discrete-time model (Eq. 8.62) is called a *state transition matrix* for the dynamic system defined by $\mathbf{F}$. Note that Φ depends only on $\mathbf{F}$, and not on the dynamic disturbance function $\mathbf{w}(t)$.

The noise vectors $\mathbf{w}_k$ are the discrete-time analog of the dynamic disturbance function $\mathbf{w}(t)$. They depend on their continuous-time counterparts $\mathbf{F}$ and $\mathbf{w}$.

Example 8.6: State Transition Matrix for Harmonic Resonator Model. The underdamped harmonic resonator model of Example 8.4 has no time-dependent terms in its coefficient matrix (Eq. 8.51), making it a time-invariant model with state transition matrix

$$\Phi = \exp(\Delta t\,\mathbf{F}) \tag{8.65}$$

$$= e^{-\Delta t/\tau} \begin{bmatrix} \cos(\omega\,\Delta t) + \sin(\omega\,\Delta t)/\omega\tau & \sin(\omega\,\Delta t)/\omega \\ -\left[\sin(\omega\,\Delta t)/\omega\tau^2\right]\left[1+\omega^2\tau^2\right] & \cos(\omega\,\Delta t) - \sin(\omega\,\Delta t)/\omega\tau \end{bmatrix}, \tag{8.66}$$

where $\omega = \omega_{\text{resonant}}$, the resonant frequency; $\tau = \tau_{\text{damping}}$, the damping time constant; and Δt is the discrete timestep.

The eigenvalues of $\mathbf{F}$ were shown to be $-1/\tau_{\text{damping}} \pm \mathbf{i}\omega_{\text{resonant}}$, so the eigenvalues of $\mathbf{F}\,\Delta t$ will be $-\Delta t\,\tau_{\text{damping}} \pm \mathbf{i}\,\Delta t\,\omega_{\text{resonant}}$ and the eigenvalues of Φ will be

$$\exp\left(-\frac{\Delta t}{\tau_{\text{damping}}} \pm \mathbf{i}\,\omega_{\text{resonant}}\,\Delta t\right) = e^{-\Delta t/\tau}\left[\cos(\omega\,\Delta t) \pm \mathbf{i}\,\sin(\omega\,\Delta t)\right].$$

A discrete-time dynamic system will be stable only if the eigenvalues of Φ lie inside the unit circle (i.e., $|\lambda_\ell| < 1$).

8.3.4 Dynamic Disturbance Noise Distribution Matrices

A common noise source can disturb more than one independent component of the state vector representing a dynamic system. Forces applied to a rigid body, for example, can affect rotational dynamics as well as translational dynamics. This sort of coupling of common disturbance noise sources into different components of the state dynamics can be represented by using a *noise distribution matrix* **G** in the form

$$\frac{d}{dt}\mathbf{x} = \mathbf{F}\mathbf{x} + \mathbf{G}\mathbf{w}(t), \tag{8.67}$$

where the components of $\mathbf{w}(t)$ are the common disturbance noise sources and the matrix **G** represents how these disturbances are distributed among the state vector components.

The covariance of state vector disturbance noise will then have the form $\mathbf{G}\mathbf{Q}_w\mathbf{G}^T$, where $\mathbf{Q}_w$ is the covariance matrix for the white-noise process $\mathbf{w}(t)$.

The analogous model in discrete time has the form

$$\mathbf{x}_k = \Phi_{k-1}\mathbf{x}_{k-1} + \mathbf{G}_{k-1}\mathbf{w}_{k-1}, \tag{8.68}$$

where $\{\mathbf{w}_k\}$ is a zero-mean white-noise process in discrete time.

In either case (i.e., continuous or discrete time), it is possible to use the noise distribution matrix for noise scaling, as well, so that the components of $\mathbf{w}_k$ can be independent, uncorrelated unit normal variates and the noise covariance matrix $\mathbf{Q}_w = \mathbf{I}$, the identity matrix.

8.3.5 Predictor Equations

The linear stochastic process model parameters Φ and **Q** can be used to calculate how the discrete-time process variables $\boldsymbol{\mu}$ (mean) and **P** (covariance) evolve over time.

Using Eq. 8.60 and taking expected values, we obtain

$$\begin{aligned}
\hat{\mathbf{x}}_k &= \boldsymbol{\mu}_k \\
&\stackrel{\text{def}}{=} E\langle \mathbf{x}_k \rangle \\
&= E\langle \Phi_{k-1}\mathbf{x}_{k-1} + \mathbf{w}_{k-1} \rangle \\
&= \Phi_{k-1}E\langle \mathbf{x}_{k-1} \rangle + E\langle \mathbf{w}_{k-1} \rangle \\
&= \Phi_{k-1}\hat{\mathbf{x}}_{k-1} + 0 \\
&= \Phi_{k-1}\hat{\mathbf{x}}_{k-1},
\end{aligned} \tag{8.69}$$

$$\begin{aligned}
\mathbf{P}_k &\stackrel{\text{def}}{=} E\langle (\hat{\mathbf{x}}_k - \mathbf{x}_k)(\hat{\mathbf{x}}_k - \mathbf{x}_k)^T \rangle \\
&= E\langle (\Phi_{k-1}\hat{\mathbf{x}}_{k-1} - \Phi_{k-1}\mathbf{x}_{k-1} - \mathbf{w}_{k-1})(\Phi_{k-1}\hat{\mathbf{x}}_{k-1} - \Phi_{k-1}\mathbf{x}_{k-1} - \mathbf{w}_{k-1})^T \rangle \\
&= \Phi_{k-1}\underbrace{E\langle (\hat{\mathbf{x}}_{k-1} - \mathbf{x}_{k-1})(\hat{\mathbf{x}}_{k-1} - \mathbf{x}_{k-1})^T \rangle}_{\mathbf{P}_{k-1}}\Phi_{k-1}^T + \underbrace{E\langle \mathbf{w}_{k-1}\,\mathbf{w}_{k-1}^T \rangle}_{\mathbf{Q}_{k-1}} \\
&\quad + \text{terms with zero expected value} \\
&= \Phi_{k-1}\mathbf{P}_{k-1}\Phi_{k-1}^T + \mathbf{Q}_{k-1}.
\end{aligned} \tag{8.70}$$

Equations 8.69 and 8.70 are the essential predictor equations for Kalman filtering.

8.4 SUMMARY OF KALMAN FILTER EQUATIONS

8.4.1 Essential Equations

The complete equations for the Kalman filter are summarized in Table 8.1.

8.4.2 Common Terminology

The symbols used in Table 8.1 for the variables and parameters of the Kalman filter are essentially those used in the original paper by Kalman [104], and this notation is fairly common in the literature.

The following are some terms commonly used for the symbols in Table 8.1:

$\mathbf{H}$ is the *measurement sensitivity matrix* or *observation matrix.*

$\mathbf{H}\hat{\mathbf{x}}_k(-)$ is the *predicted measurement.*

$\mathbf{z} - H\hat{\mathbf{x}}_k(-)$, the difference between the measurement vector and the predicted measurement, is the *innovations vector.*

$\overline{\mathbf{K}}$ is the *Kalman gain.*

$\mathbf{P}_k(-)$ is the *predicted* or *a priori* value of estimation covariance.

$\mathbf{P}_k(+)$ is the *corrected* or *a posteriori* value of estimation covariance.

$\mathbf{Q}_k$ is the *covariance of dynamic disturbance noise.*

$\mathbf{R}$ is the *covariance of sensor noise* or *measurement uncertainty.*

$\hat{\mathbf{x}}_k(-)$ is the *predicted* or *a priori* value of the estimated state vector.

TABLE 8.1. Essential Kalman Filter Equations

Predictor (Time or Temporal Updates)			
Predicted state vector:			
$\hat{\mathbf{x}}_k(-)$	$=$	$\Phi_k\hat{\mathbf{x}}_{k-1}(+)$	(Eq. 8.69)
Predicted covariance matrix:			
$\mathbf{P}_k(-)$	$=$	$\Phi_k\mathbf{P}_{k-1}(+)\Phi_k^{\mathrm{T}} + \mathbf{Q}_{k-1}$	(Eq. 8.70)
Corrector (Measurement or Observational Updates)			
Kalman gain:			
$\overline{\mathbf{K}}_k$	$=$	$\mathbf{P}_k(-)\mathbf{H}_k^{\mathrm{T}}(\mathbf{H}_k\mathbf{P}_k(-)\mathbf{H}_k^{\mathrm{T}} + \mathbf{R}_k)^{-1}$	(Eq. 8.41)
Corrected state estimate:			
$\hat{\mathbf{x}}_k(+)$	$=$	$\hat{\mathbf{x}}_k(-) + \overline{\mathbf{K}}_k(\mathbf{z}_k - \mathbf{H}_k\hat{\mathbf{x}}_k(-))$	(Eq. 8.42)
Corrected covariance matrix:			
$\mathbf{P}_k(+)$	$=$	$\mathbf{P}_k(-) - \overline{\mathbf{K}}_k\mathbf{H}_k\mathbf{P}_k(-)$	(Eq. 8.43)

$\hat{\mathbf{x}}_k(+)$ is the *corrected* or *a posteriori* value of the estimated state vector.

$\mathbf{z}$ is the *measurement vector* or *observation vector.*

8.4.3 Data Flow Diagrams

The matrix-level data flow of the Kalman filter implementation for a time-varying problem is diagrammed in Fig. 8.7, with the inputs shown on the left, the outputs

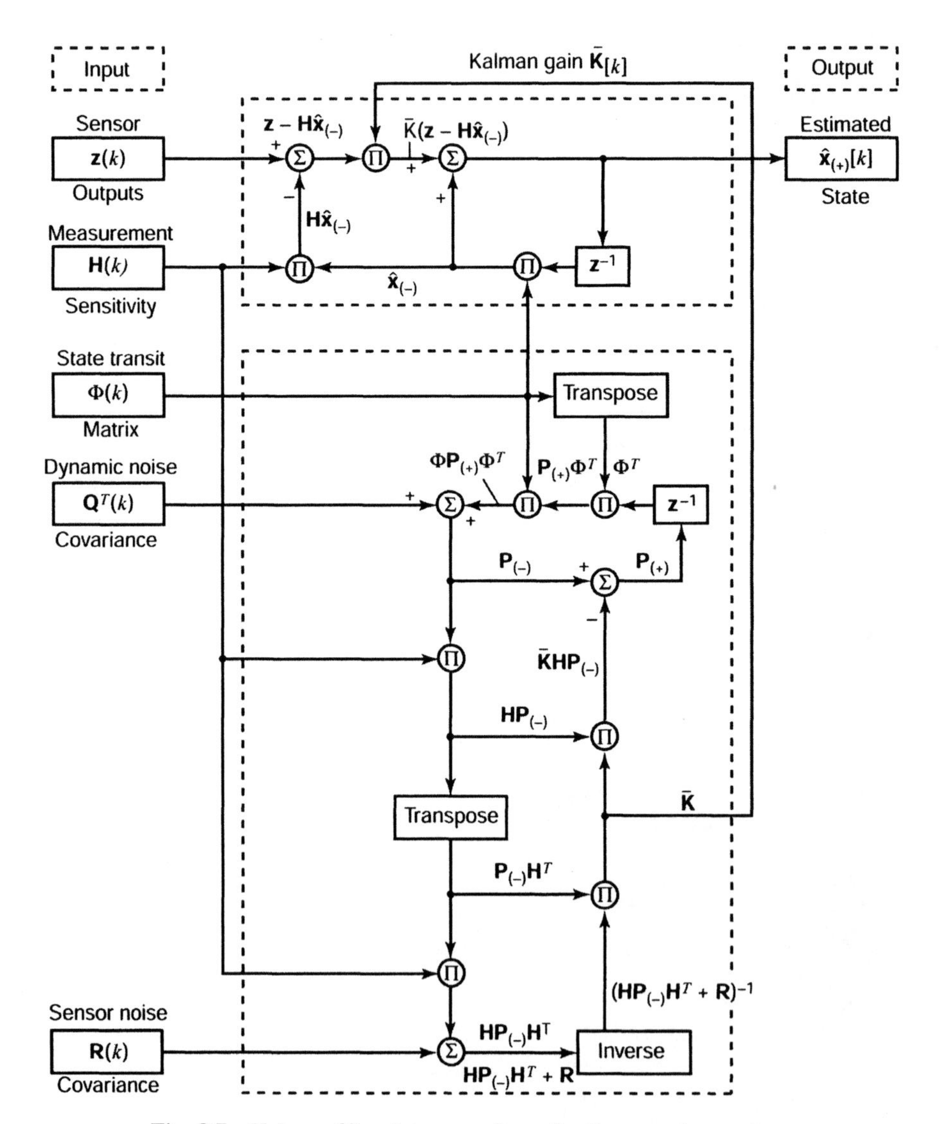

Fig. 8.7 Kalman filter data array flows for time-varying system.

(corrected estimates) on the right, and the symbol z^{-1} representing the unit delay operator.

The dashed lines in the figure enclose two computation loops. The top loop is the estimation loop, with the feedback gain (Kalman gain) coming from the bottom loop. The bottom loop implements the Riccati equation solution used to calculate the Kalman gain. This bottom loop runs "open loop," in that there is no feedback mechanism to stabilize it in the presence of roundoff errors. Numerical instability problems with the Riccati equation propagation loop were discovered soon after the introduction of the Kalman filter.

8.5 ACCOMMODATING TIME-CORRELATED NOISE

The fundamental noise processes in the basic Kalman filter model are zero-mean white Gaussian noise processes $\{\mathbf{w}_k\}$, called *dynamic disturbance, plant noise,* or *process noise* and $\{\mathbf{v}_k\}$, called *sensor noise, measurement noise,* or *observation noise.*

GPS signal propagation errors and INS position errors are not white noise processes, but are correlated over time. Fortunately, time-correlated noise processes can easily be accommodated in Kalman filtering by adding state variables to the Kalman filter model. A correlated noise process $\boldsymbol{\xi}_k$ can be modeled by a linear stochastic system model of the sort

$$\boldsymbol{\xi}_k = \Phi_{k-1}\boldsymbol{\xi}_{k-1} + \mathbf{w}_{k-1}, \tag{8.71}$$

where $\{\mathbf{w}_k\}$ is a zero-mean white Gaussian noise process, and then augment the state vector by appending the new variable $\boldsymbol{\xi}_k$

$$\mathbf{x}_{\text{augmented}} = \begin{bmatrix} \mathbf{x}_{\text{original}} \\ \boldsymbol{\xi} \end{bmatrix} \tag{8.72}$$

and modify the parameter matrices Φ, $\mathbf{Q}$, and $\mathbf{H}$ accordingly.

8.5.1 Correlated Noise Models

8.5.1.1 Autocovariance Functions Correlation of a random sequence $\{\xi\}$ is characterized by its discrete-time *autocovariance function* $\mathbf{P}_{\boldsymbol{\xi}}[\Delta k]$, a function of the delay index Δk defined as

$$\mathbf{P}_{\boldsymbol{\xi}}[\Delta k] \stackrel{\text{def}}{=} \mathrm{E}_k\langle(\boldsymbol{\xi}_k - \boldsymbol{\mu}_\xi)(\boldsymbol{\xi}_{k+\Delta k} - \boldsymbol{\mu}_\xi)^T\rangle, \tag{8.73}$$

where $\boldsymbol{\mu}_\xi$ is the mean value of the random sequence $\{\boldsymbol{\xi}_k\}$.

For *white*-noise processes

$$\mathbf{P}[\Delta k] = \begin{cases} 0, & \Delta k \neq 0, \\ \mathbf{Q}, & \Delta k = 0, \end{cases} \tag{8.74}$$

where $\mathbf{Q}$ is the covariance of the white-noise process.

8.5.1.2 Random Walks Random walks, also called *Wiener processes*, are cumulative sums of white-noise processes $\{\mathbf{w}_k\}$

$$\xi_k = \xi_{k-1} + \mathbf{w}_{k-1}, \tag{8.75}$$

a stochastic process model with state transition matrix $\Phi = \mathbf{I}$, an identity matrix.

Random walks are notoriously unstable, in the sense that the covariance of the variate ξ_k grows linearly with k and without bound as $k \to \infty$. In general, if any of the eigenvalues of a state transition matrix fall on or outside the unit circle in the complex plane (as they all do for identity matrices), the variate of the stochastic process can fail to have a finite steady-state covariance matrix. However, as was demonstrated by R. E. Kalman in 1960, the covariance of *uncertainty* in the *estimated* system state vector can still converge to a finite steady-state value, even if the process itself is unstable.

8.5.1.3 Exponentially Correlated Noise Exponentially correlated random processes have finite, constant steady-state covariances. A scalar exponentially random process $\{\xi_k\}$ has a model of the sort

$$\xi_k = e^{-\Delta t/\tau}\xi_{k-1} + w_{k-1}, \tag{8.76}$$

where Δt is the time period between samples and τ is the exponential decay time constant of the process. The steady-state variance σ^2 of such a process is the solution to its steady-state variance equation

$$\sigma_\infty^2 = e^{-2\,\Delta t/\tau}\sigma_\infty^2 + Q \tag{8.77}$$

$$= \frac{Q}{1 - e^{-2\Delta t/\tau}}, \tag{8.78}$$

where Q is the variance of the scalar zero-mean white-noise process $\{w_k\}$.

The autocovariance sequence of an exponentially correlated random process in discrete time has the general form

$$P[\Delta k] = \sigma_\infty^2 \exp\frac{-\Delta k}{N_c}, \tag{8.79}$$

which falls off exponentially on either side of its peak value σ_∞^2 (the process variance) at $\Delta k = 0$. The parameter N_c is called the *correlation number* of the process, where $N_c = \tau/\Delta t$ for correlation time τ and sample interval Δt.

8.5.1.4 Harmonic Noise Harmonic noise includes identifiable frequency components, such as those from AC power or from mechanical or electrical resonances. A stochastic process model for such sources has already been developed in the examples of this chapter.

8.5.1.5 SA Noise Autocorrelation A pseudorandom clock dithering algorithm is described in U.S. Patent 4,646,032 [212] including a parametric model of the autocorrelation function (autocovariance function divided by variance) of the resulting timing errors. Knowledge of the dithering algorithm does not necessarily give the user any advantage, but there is at least a suspicion that this may be the algorithm used for SA dithering of the individual GPS satellite time references. Its theoretical autocorrelation function is plotted in Fig. 8.8 along with an exponential correlation curve. The two are scaled to coincide at the autocorrelation coefficient value of $1/e \approx 0.36787944\ldots$, the argument at which correlation time is defined. Unlike exponentially correlated noise, this source has greater short-term correlation and less long-term correlation.

The correlation time of SA errors determined from GPS signal analysis is on the order of $10^2 - 10^3$ s. It is possible that the actual correlation time is variable, which might explain the range of values reported in the literature.

Although this is not an exponential autocorrelation function, it could perhaps be modeled as such.

8.5.1.6 Slow Variables SA timing errors (if present) are only one of a number of slowly varying error sources in GPS/INS integration. Slow variables may also include many of the calibration parameters of the inertial sensors, which can be responding to temperature variations or other unknown but slowly changing influences. Like SA errors, these other slow variations of these variables can often be tracked and compensated by combining the INS navigation estimates with the

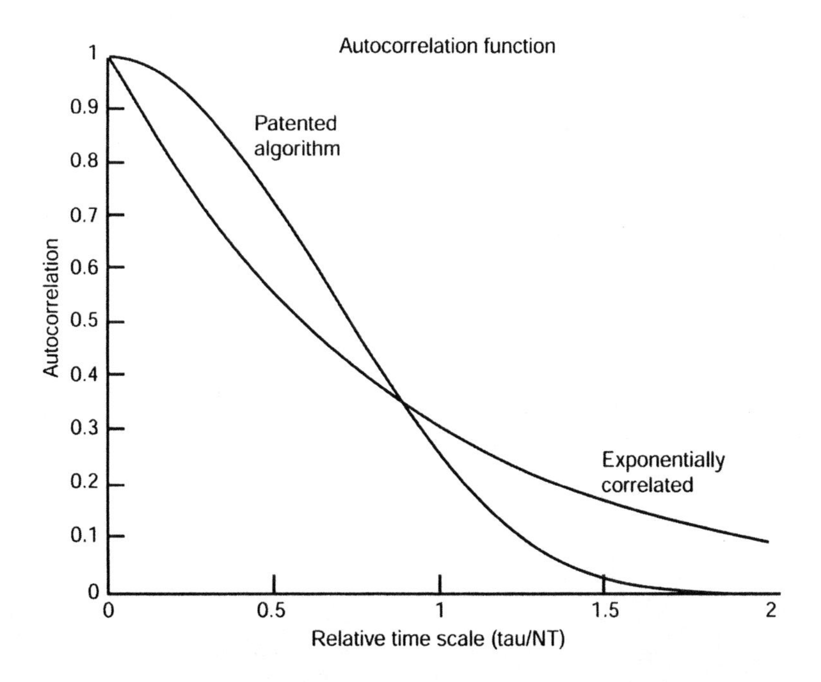

Fig. 8.8 Autocorrelation function for pseudonoise algorithm.

GPS-derived estimates. What is different about the calibration parameters is that they are involved nonlinearly in the INS system model.

8.5.2 Empirical Sensor Noise Modeling

Noise models used in Kalman filtering should be reasonably faithful representations of the true noise sources. Sensor noise can often be measured directly and used in the design of an appropriate noise model. Dynamic process noise is not always so accessible, and its models must often be inferred from indirect measurements.

8.5.2.1 Spectral Characterization Spectrum analyzers and spectrum analysis software make it relatively easy to calculate the power spectral density of sampled noise data, and the results are useful for characterizing the type of noise and identifying likely noise models.

The resulting noise models can then be simulated using pseudorandom sequences, and the power spectral densities of the simulated noise can be compared to that of the sampled noise to verify the model.

The power spectral density of white noise is constant across the spectrum, and each successive integral changes its slope by -20 dB/decade of frequency, as illustrated in Fig. 8.9.

8.5.2.2 Shaping Filters The spectrum of white noise is flat, and the amplitude spectrum of the output of a filter with white-noise input will have the shape of the

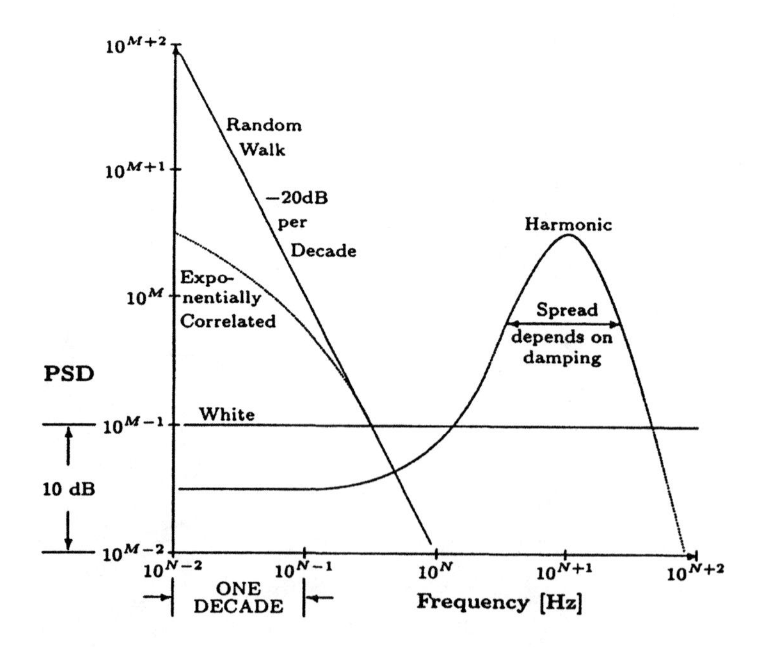

Fig. 8.9 Spectral properties of some common noise types.

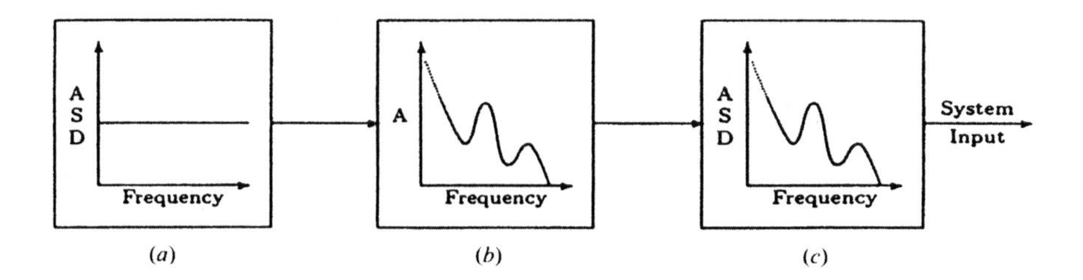

Fig. 8.10 Putting white noise through shaping filters: (*a*) white-noise source; (*b*) linear shaping filter; (*c*) shaped noise source.

amplitude transfer function of the filter, as illustrated in Fig. 8.10. Therefore, any noise spectrum can be approximated by white noise passed through a *shaping filter* to yield the desired shape. All correlated noise models for Kalman filters can be implemented by shaping filters.

8.5.3 State Vector Augmentation

8.5.3.1 Correlated Dynamic Disturbance Noise A model for a linear stochastic process model in discrete time with uncorrelated and correlated disturbance noise has the form

$$\mathbf{x}_k = \Phi_{x,\,k-1}\mathbf{x}_{k-1} + \mathbf{G}_{w_x,k-1}\mathbf{w}_{k-1} + \mathbf{D}_{\xi,k-1}\boldsymbol{\xi}_{k-1}, \tag{8.80}$$

where $\mathbf{w}_{k-1}$ is zero-mean white (i.e., uncorrelated) disturbance noise, $\mathbf{G}_{w_x,k-1}$ is white-noise distribution matrix, $\boldsymbol{\xi}_{k-1}$ is zero-mean correlated disturbance noise, and $\mathbf{D}_{\xi\,x,\,k-1}$ is correlated noise distribution matrix.

If the correlated dynamic disturbance noise can be modeled as yet another linear stochastic process

$$\boldsymbol{\xi}_k = \Phi_{\xi,k-1}\boldsymbol{\xi}_{k-1} + \mathbf{G}_{w_\xi,\,k-1}\mathbf{w}_{\xi,\,k-1} \tag{8.81}$$

with only zero-mean white-noise inputs $\{\mathbf{w}_{u,k}\}$, then the augmented state vector

$$\mathbf{x}_{\text{aug},k} \stackrel{\text{def}}{=} \begin{bmatrix} \mathbf{x}_k \\ \boldsymbol{\xi}_k \end{bmatrix} \tag{8.82}$$

has a stochastic process model

$$\begin{aligned} \mathbf{x}_{\text{aug},\,k} &= \begin{bmatrix} \Phi_{x,\,k-1} & \mathbf{D}_{\xi\,x,\,k-1} \\ 0 & \Phi_{\xi,k-1} \end{bmatrix} x_{\text{aug},\,k-1} \\ &\quad + \begin{bmatrix} \mathbf{G}_{w_x,\,k-1} & 0 \\ 0 & \mathbf{G}_{w_\xi,\,k-1} \end{bmatrix} \begin{bmatrix} \mathbf{w}_{x,\,k-1} \\ \mathbf{w}_{\xi,\,k-1} \end{bmatrix} \end{aligned} \tag{8.83}$$

having only uncorrelated disturbance noise with covariance

$$\mathbf{Q}_{\text{aug},\,k-1} = \begin{bmatrix} \mathbf{Q}_{wx,k-1} & 0 \\ 0 & \mathbf{Q}_{w\xi,k-1} \end{bmatrix}. \tag{8.84}$$

The new measurement sensitivity matrix for this augmented state vector will have the block form

$$\mathbf{H}_{\text{aug},\,k} = \begin{bmatrix} \mathbf{H}_k & 0 \end{bmatrix}. \tag{8.85}$$

The augmenting block is zero in this case because the uncorrelated noise source is dynamic disturbance noise, not sensor noise.

*8.5.3.2 **Correlated Sensor Noise*** The same sort of state augmentation can be done for correlated sensor noise $\{\xi_k\}$,

$$\mathbf{z}_k = \mathbf{H}_k \mathbf{x}_k + \mathbf{A}_k \mathbf{v}_k + \mathbf{B}_k \xi_k, \tag{8.86}$$

with the same type of model for the correlated noise (Eq. 8.81) and using the same augmented state vector (Eq. 8.82), but now with a different augmented state transition matrix

$$\mathbf{\Phi}_{\text{aug},\,k-1} = \begin{bmatrix} \mathbf{\Phi}_{x,k-1} & 0 \\ 0 & \mathbf{\Phi}_{\xi,k-1} \end{bmatrix} \tag{8.87}$$

and augmented measurement sensitivity matrix

$$\mathbf{H}_{\text{aug},k} = [\mathbf{H}_k \;\; \mathbf{B}_k]. \tag{8.88}$$

*8.5.3.3 **Correlated Noise in Continuous Time*** There is an analogous procedure for state augmentation using continuous-time models. If $\boldsymbol{\xi}(t)$ is a correlated noise source defined by a model of the sort

$$\frac{d}{dt}\boldsymbol{\xi} = \mathbf{F}_\xi \boldsymbol{\xi} + \mathbf{w}_\xi \tag{8.89}$$

for $\mathbf{w}_\xi(t)$ a white-noise source, then any stochastic process model of the sort

$$\frac{d}{dt}\mathbf{x} = \mathbf{F}_x \mathbf{x} + \mathbf{w}_x(t) + \boldsymbol{\xi}(t) \tag{8.90}$$

with this correlated noise source can also be modeled by the augmented state vector

$$\mathbf{x}_{\text{aug}} \stackrel{\text{def}}{=} \begin{bmatrix} \mathbf{x} \\ \boldsymbol{\xi} \end{bmatrix} \tag{8.91}$$

as

$$\frac{d}{dt}\mathbf{x}_{\text{aug}} = \begin{bmatrix} \mathbf{F}_x & \mathbf{I} \\ 0 & \mathbf{F}_\xi \end{bmatrix} \mathbf{x}_{\text{aug}} + \begin{bmatrix} \mathbf{w}_x \\ \mathbf{w}_\xi \end{bmatrix} \tag{8.92}$$

with only uncorrelated disturbance noise.

8.6 NONLINEAR AND ADAPTIVE IMPLEMENTATIONS

Although the Kalman filter is defined for linear dynamic systems with linear sensors, it has been applied more often than not to real-world applications without truly linear dynamics or sensors—and usually with remarkably great success.

The following subsections show how this is done.

8.6.1 Nonlinear Dynamics

State dynamics for nonlinear systems can be expressed in the functional form

$$\frac{d}{dt}\mathbf{x}(t) = \mathbf{f}(\mathbf{x}, t) + \mathbf{w}(t). \tag{8.93}$$

For this to be linearized, the function $\mathbf{f}$ must be differentiable, with Jacobian matrix

$$\mathbf{F}(\mathbf{x}, t) \stackrel{\text{def}}{=} \underbrace{\left.\frac{\partial \mathbf{f}}{\partial \mathbf{x}}\right|_{\hat{\mathbf{x}}(t)}}_{\text{extended}} \quad \text{or} \quad \underbrace{\left.\frac{\partial \mathbf{f}}{\partial \mathbf{x}}\right|_{\mathbf{x}_{\text{nominal}}(t)}}_{\text{linearized}}, \tag{8.94}$$

where the extended Kalman filter uses the estimated trajectory for evaluating the Jacobian, and linearized Kalman filtering uses a nominal trajectory $\mathbf{x}_{\text{nominal}}(t)$, which may come from a simulation.

8.6.1.1 Nonlinear Dynamics with Control In applications with control variables $\mathbf{u}(t)$, Eq. 8.93 can also be expressed in the form

$$\frac{d}{dt}\mathbf{x} = \mathbf{f}[\mathbf{x}, \mathbf{u}(t), t] + \mathbf{w}(t), \tag{8.95}$$

in which case the control vector $\mathbf{u}$ may also appear in the Jacobian matrix $\mathbf{F}$.

8.6.1.2 Propagating Estimates The estimate $\hat{\mathbf{x}}$ is propagated by solving the differential equation

$$\frac{d}{dt}\hat{\mathbf{x}} = \mathbf{f}(\hat{\mathbf{x}}, t), \tag{8.96}$$

using whatever means necessary (e.g., Runge–Kutta integration). The solution is called the *trajectory* of the estimate.

8.6.1.3 Propagating Covariances The covariance matrix for nonlinear systems is also propagated over time as the solution to the matrix differential equation

$$\frac{d}{dt}\mathbf{P}(t) = \mathbf{F}(\mathbf{x}(t), t)\,\mathbf{P}(t) + \mathbf{P}(t)\,\mathbf{F}^{\mathrm{T}}(\mathbf{x}(t), t) + \mathbf{Q}(t), \tag{8.97}$$

where the values of $\mathbf{F}(\mathbf{x}, t)$ from Eq. 8.63 must be calculated along a trajectory $\mathbf{x}(t)$. This trajectory can be the solution for the estimated value $\hat{\mathbf{x}}$ calculated using the Kalman filter and Eq. 8.94 (for the extended Kalman filter) or along any "nominal" trajectory (for the "linearized" Kalman filter).

8.6.2 Nonlinear Sensors

Nonlinear Kalman filtering can accommodate sensors that are not truly linear but can at least be represented in the functional form

$$\mathbf{z}_k = \mathbf{h}_k(\mathbf{x}_k) + \mathbf{v}_k, \tag{8.98}$$

where $\mathbf{h}$ is a smoothly differentiable function of $\mathbf{x}$. For example, even linear sensors with nonzero biases (offsets) $\mathbf{b}_{\text{sensor}}$ will have sensor models of the sort

$$\mathbf{h}(\mathbf{x}) = \mathbf{H}\mathbf{x} + \mathbf{b}_{\text{sensor}}, \tag{8.99}$$

in which case the Jacobian matrix

$$\frac{\partial \mathbf{h}}{\partial \mathbf{x}} = \mathbf{H}. \tag{8.100}$$

8.6.2.1 Predicted Sensor Outputs The predicted value of nonlinear sensor outputs uses the full nonlinear function applied to the estimated state vector:

$$\hat{\mathbf{z}}_k = \mathbf{h}_k(\hat{\mathbf{x}}_k). \tag{8.101}$$

8.6.2.2 Calculating Kalman Gains The value of the measurement sensitivity matrix $\mathbf{H}$ used in calculating Kalman gain is evaluated as a Jacobian matrix

$$\mathbf{H}_k = \underbrace{\left.\frac{\partial \mathbf{h}}{\partial \mathbf{x}}\right|_{\mathbf{x}=\hat{\mathbf{x}}}}_{\text{extended}} \quad \text{or} \quad \underbrace{\left.\frac{\partial \mathbf{h}}{\partial \mathbf{x}}\right|_{\mathbf{x}=\mathbf{x}_{\text{nominal}}}}_{\text{linearized}}, \tag{8.102}$$

where the first value (used for extended Kalman filtering) uses the estimated trajectory for evaluation of partial derivatives and the second value uses a nominal trajectory (used for the linearized Kalman filtering).

8.6.3 Linearized Kalman Filter

Perhaps the simplest approach to Kalman filtering for nonlinear systems uses linearization of the system model about a nominal trajectory. This approach is necessary for preliminary analysis of systems during the system design phase, when there may be several potential trajectories defined by different mission scenarios. The essential implementation equations for this case are summarized in Table 8.2.

8.6.4 Extended Kalman Filtering

This approach is due to Stanley F. Schmidt, and it has been used successfully in an enormous number of nonlinear applications. It is a form of nonlinear Kalman filtering with all Jacobian matrices (i.e., **H** and/or **F**) evaluated at $\hat{\mathbf{x}}$, the estimated state. The essential extended Kalman filter equations are summarized in Table 8.3; the major differences from the conventional Kalman filter equations of Table 8.1 are

1. Integration of the nonlinear integrand $\dot{\mathbf{x}} = \mathbf{f}(\mathbf{x})$ to predict $\hat{\mathbf{x}}_k(-)$
2. Use of the nonlinear function $\mathbf{h}_k(\hat{\mathbf{x}}_k(-))$ in measurement prediction
3. Use of the Jacobian matrix of the dynamic model function **f** as the dynamic coefficient matrix **F** in the propagation of the covariance matrix
4. Use of the Jacobian matrix of the measurement function **h** as the measurement sensitivity matrix **H** in the covariance correction and Kalman gain equations

TABLE 8.2. Linearized Kalman Filter Equations

Predictor (Time Updates)	
Predicted state vector:	
$\hat{\mathbf{x}}_k(-) = \hat{\mathbf{x}}_{k-1}(+) + \int_{t_{k-1}}^{t_k} \mathbf{f}(\hat{\mathbf{x}}, t)\, dt$	(Eq. 8.93)
Predicted covariance matrix:	
$\dot{\mathbf{P}} = \mathbf{F}\mathbf{P} + \mathbf{P}\mathbf{F}^{\mathrm{T}} + \mathbf{Q}(t)$	(Eq. 8.97)
$\mathbf{F} = \frac{\partial \mathbf{f}}{\partial \mathbf{x}}\Big\vert_{\mathbf{x}=\mathbf{x}_{\mathrm{nom}}(t)}$	(Eq. 8.94)
or	
$\mathbf{P}_k(-) = \Phi_k \mathbf{P}_{k-1}(+) \Phi_k^T + \mathbf{Q}_{k-1}$	(Eq. 8.70)
Corrector (Measurement Updates)	
Kalman gain:	
$\overline{\mathbf{K}}_k = \mathbf{P}_k(-)\mathbf{H}_k^T [\mathbf{H}_k \mathbf{P}_k(-) \mathbf{H}_k^T + \mathbf{R}_k]^{-1}$	(Eq. 8.41)
$\mathbf{H}_k = \frac{\partial \mathbf{h}}{\partial \mathbf{x}}\Big\vert_{\mathbf{x}=\mathbf{x}_{\mathrm{nom}}}$	(Eq. 8.102)
Corrected state estimate:	
$\hat{\mathbf{x}}_k(+) = \hat{\mathbf{x}}_k(-) + \overline{\mathbf{K}}_k[\mathbf{z}_k - \mathbf{h}_k(\hat{\mathbf{x}}_k(-))]$	(Eqs. 8.42, 8.101)
Corrected covariance matrix:	
$\mathbf{P}_k(+) = \mathbf{P}_k(-) - \overline{\mathbf{K}}_k \mathbf{H}_k \mathbf{P}_k(-)$	(Eq. 8.43)

TABLE 8.3. Extended Kalman Filter Equations

Predictor (Time Updates)	
Predicted state vector:	
$\hat{\mathbf{x}}_k(-) = \hat{\mathbf{x}}_{k-1}(+) + \int_{t_{k-1}}^{t_k} \mathbf{f}(\hat{\mathbf{x}}, t)\, dt$	(Eq. 8.93)
Predicted covariance matrix:	
$\dot{\mathbf{P}} = \mathbf{F}\mathbf{P} + \mathbf{P}\mathbf{F}^{\mathrm{T}} + \mathbf{Q}(t)$	(Eq. 8.97)
$\mathbf{F} = \frac{\partial \mathbf{f}}{\partial \mathbf{x}}\Big\vert_{\mathbf{x}=\hat{\mathbf{x}}(t)}$	(Eq. 8.94)
or	
$\mathbf{P}_k(-) = \Phi_k \mathbf{P}_{k-1}(+)\Phi_k^T + \mathbf{Q}_{k-1}$	(Eq. 8.70)
Corrector (Measurement Updates)	
Kalman gain:	
$\overline{\mathbf{K}}_k = \mathbf{P}_k(-)\mathbf{H}_k^T[\mathbf{H}_k \mathbf{P}_k(-)\mathbf{H}_k^T + \mathbf{R}_k]^{-1}$	(Eq. 8.41)
$\mathbf{H}_k = \frac{\partial \mathbf{h}}{\partial \mathbf{x}}\Big\vert_{\mathbf{x}=\hat{\mathbf{x}}}$	(Eq. 8.102)
Corrected state estimate:	
$\hat{\mathbf{x}}_k(+) = \hat{\mathbf{x}}_k(-) + \overline{\mathbf{K}}_k[\mathbf{z}_k - \mathbf{h}_k(\hat{\mathbf{x}}_k(-))]$	(Eqs. 8.42, 8.101)
Corrected covariance matrix:	
$\mathbf{P}_k(+) = \mathbf{P}_k(-) - \overline{\mathbf{K}}_k \mathbf{H}_k \mathbf{P}_k(-)$	(Eq. 8.43)

8.6.5 Adaptive Kalman Filtering

In adaptive Kalman filtering, nonlinearities in the model arise from making parameters of the model into functions of state variables. For example, the time constant τ of a scalar exponentially correlated process

$$x_k = \exp \frac{-\Delta t}{\tau} x_{k-1} + w_k$$

may be unknown or slowly time-varying, in which case it can be made part of the augmented state vector

$$\hat{\mathbf{x}}_{\mathrm{aug}} = \begin{bmatrix} \hat{x} \\ \hat{\tau} \end{bmatrix}$$

with state transition matrix

$$\Phi = \begin{bmatrix} \exp\left(-\Delta t/\hat{\tau}\right) & \Delta t \exp(-\Delta t/\hat{\tau})\hat{x}/\hat{\tau}^2 \\ 0 & \exp(-\Delta t/\tau^*) \end{bmatrix},$$

where $\tau^* >> \tau$ is the correlation time constant of the variations in $\hat{\tau}$.

Example 8.7: Tracking Time-Varying Frequency and Damping. Consider the problem of tracking the phase components of a damped harmonic oscillator with slowly time-varying resonant frequency and damping time constant. The state variables for this nonlinear dynamic system are x_1, the in-phase component of the oscillator output signal (i.e, the only observable component); x_2, the quadrature-phase component of the signal; x_3, the damping time constant of the oscillator (nominally 5 s); and x_4, the frequency of oscillator (nominally 2 π rad/s, or 1 Hz).

The dynamic coefficient matrix will be

$$\mathbf{F} = \begin{bmatrix} -1/x_3 & x_4 & x_1/x_3^2 & x_2 \\ -x_4 & -1/x_3 & x_2/x_3^2 & -x_1 \\ 0 & 0 & -1/\tau_\tau & 0 \\ 0 & 0 & 0 & -1/\tau_\omega \end{bmatrix},$$

where τ_τ is the correlation time for the time-varying oscillator damping time constant and τ_ω is the correlation time for the time-varying resonant frequency of the oscillator.

If only the in-phase component or the oscillator output can be sensed, then the measurement sensitivity matrix will have the form

$$\mathbf{H} = \begin{bmatrix} 1 & 0 & 0 & 0 \end{bmatrix}$$

Figure 8.11 is a sample output of the MATLAB m-file `osc_ekf.m` on the accompanying CD-ROM, which implements this extended Kalman filter. Note that it tracks the phase, amplitude, frequency, and damping of the oscillator.

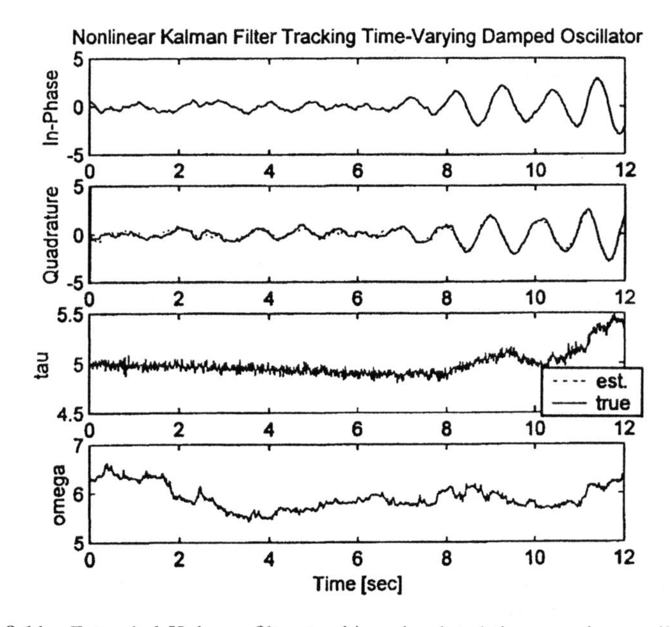

Fig. 8.11 Extended Kalman filter tracking simulated time-varying oscillator.

The unknown or time-varying parameters can also be in the measurement model. For example, a sensor output with time-varying scale factor S and bias b can be modeled by the nonlinear equation $z = Sx + b$ and linearized using augmented state vector

$$\mathbf{x}_{\text{aug}} = \begin{bmatrix} x \\ S \\ b \end{bmatrix}$$

and measurement sensitivity matrix

$$\mathbf{H} = \begin{bmatrix} \hat{S} & \hat{x} & 1 \end{bmatrix}.$$

8.7 KALMAN–BUCY FILTER

The discrete-time form of the Kalman filter is well suited for computer implementation, but is not particularly natural for engineers, who find it more natural think about dynamic systems in terms of differential equations.

The analog of the Kalman filter in continuous time is the Kalman–Bucy filter, developed jointly by Richard Bucy[5] and Rudolf Kalman [105].

8.7.1 Implementation Equations

The fundamental equations of the Kalman–Bucy filter are shown in Table 8.4.

People already familiar with differential equations may find the Kalman–Bucy filter more intuitive and easier to work with than the Kalman filter—despite complications of the stochastic calculus. To its credit, the Kalman–Bucy filter requires only one equation each for propagation of the estimate and its covariance, whereas the Kalman filter requires two (for prediction and correction).

However, if the result must eventually be implemented in a digital processor, then it will have to be put into discrete-time form. Formulas for this transformation are given below. Those who prefer to "think in continuous time"

TABLE 8.4. Kalman–Bucy Filter Equations

State equation (unified predictor/corrector):
$\frac{d}{dt}\hat{\mathbf{x}}(t) = \mathbf{F}(t)\,\hat{\mathbf{x}}(t) + \underbrace{\mathbf{P}(t)\mathbf{H}^T(t)\mathbf{R}^{-1}(t)}_{\overline{\mathbf{K}}_{\text{KB}}(t)}[\mathbf{z}(t) - \mathbf{H}(t)\mathbf{x}(t)]$
Covariance equation (unified predictor/corrector):
$\dot{\mathbf{P}}(t) = \mathbf{F}(t)\,\mathbf{P}(t) + \mathbf{P}(t)\,\mathbf{F}^T(t) + \mathbf{Q}(t) - \overline{\mathbf{K}}_{\text{KB}}(t)\mathbf{R}(t)\overline{\mathbf{K}}_{\text{KB}}^T$

[5]Bucy recognized the covariance equation as a form of the nonlinear differential equation studied by Jacopo Francesco Riccati [162] (1676–1754), and that the equation was equivalent to spectral factorization in the Wiener filter.

can develop the problem solution first in continuous time as a Kalman–Bucy filter, then transform the result to Kalman filter form for implementation.

8.7.2 Kalman–Bucy Filter Parameters

Formulas for the Kalman filter parameters $\mathbf{Q}_k$ and $\mathbf{R}_k$ as functions of the Kalman–Bucy filter parameters $\mathbf{Q}(t)$ and $\mathbf{R}(t)$ can be derived from the process models.

8.7.2.1 **$\mathbf{Q}(t)$ *and* $\mathbf{Q}_k$** The relationship between these two distinct matrix parameters depends on the coefficient matrix $\mathbf{F}(t)$ in the stochastic system model:

$$\mathbf{Q}_k = \int_{t_{k-1}}^{t_k} \exp\left(\int_t^{t_k} \mathbf{F}(s)\, ds\right) \mathbf{Q}(t) \exp\left(\int_t^{t_k} \mathbf{F}(s)\, ds\right)^T dt.$$

8.7.2.2 **$\mathbf{R}(t)$ *and* $\mathbf{R}_k$** This relationship will depend on how the sensor outputs in continuous time are filtered before sampling for the Kalman filter. If the sensor outputs were simply sampled without filtering, then

$$\mathbf{R}_k = \mathbf{R}(t_k). \tag{8.103}$$

However, it is common practice to use antialias filtering of the sensor outputs before sampling for Kalman filtering. Filtering of this sort can also alter the parameter $\mathbf{H}$ between the two implementations. For an integrate-and-hold filter (an effective antialiasing filter), this relationship has the form

$$\mathbf{R}_k = \int_{t_{k-1}}^{t_k} \mathbf{R}(t)\, dt, \tag{8.104}$$

in which case the measurement sensitivity matrix for the Kalman filter will be $\mathbf{H}_K = \Delta t \mathbf{H}_{\text{KB}}$, where $\mathbf{H}_{\text{KB}}$ is the measurement sensitivity matrix for the Kalman–Bucy filter.

8.8 GPS RECEIVER EXAMPLES

The following is a simplified example of the expected performance of a GPS receiver using (1) DOP calculations and (2) covariance analysis using the Riccati equations of a Kalman filter[6] for given sets of GPS satellites. These examples are implemented in the MATLAB m-file `GPS_perf.m` on the accompanying CD.

8.8.1 Satellite Models

This example demonstrates how the Kalman filter converges to its minimum error bound and how well the GPS system performs as a function of the different phasings of the four available satellites. In the simulations, the available satellites

[6]There are more Kalman filter models for GNSS in Section 10.2.

and their respective initial phasings include the following:

Satellite No.	Ω_0 (deg)	θ_0 (deg)
1	326	68
2	26	340
3	146	198
4	86	271
5	206	90

The simulation runs two cases to demonstrate the criticality of picking the correctly phased satellites. Case 1 chooses satellites 1, 2, 3, and 4 as an example of an optimum set of satellites. Case 2 utilizes satellites 1, 2, 3, and 5 as an example of a nonoptimal set of satellites that will result in the dreaded "GDOP chimney" measure of performance.

Here, the GPS satellites are assumed to be in a circular orbital trajectory at a 55° inclination angle. The angle Ω_0 is the right ascension of the satellite and θ_0 is the angular location of the satellite in its circular orbit. It is assumed that the satellites orbit the earth at a constant rate $\dot{\theta}$ with a period of approximately 43,082 s or slightly less than half of a day. The equations of motion that describe the angular phasing of the satellites are given, as in the simulation

$$\Omega(t) = \Omega_0 - \dot{\Omega}t,$$

$$\theta(t) = \theta_0 + \dot{\theta}t,$$

where the angular rates are given as

$$\dot{\Omega} = 2\frac{\pi}{86164},$$

$$\dot{\theta} = 2\frac{\pi}{43082},$$

where t is in seconds. The projects simulate the GPS system from t = 0 s to 3600 s as an example of the available satellite visibility window.

8.8.2 Measurement Model

In both the GDOP and Kalman filter models, the common observation matrix equations for discrete points is

$$z_k = H_k x_k + v_k,$$

where z, H, and v are the vectors and matrices for the kth observation point in time k. This equation is usually linearized when calculating the pseudorange by defining $z = \rho - \rho_0 = H^{[1]}x + v$.

Measurement noise v is usually assumed to be $N(0,\ R)$ (normally distributed with zero mean and variance R). The covariance of receiver error R is usually assumed to be the same error for all measurements as long as all the same conditions exist for all time intervals (0–3600 s) of interest. By defining the measurement Z as the difference in position, the measurement sensitivity matrix H can be linearized and approximated as $H^{[1]}$ (i.e., first-order linear approximation) by defining

$$H^{[1]} = \frac{\partial \rho_r^i}{\partial x_i},$$

where i refers to the n different states of the Kalman filter and ρ_r is the reference pseudorange.

8.8.3 Coordinates

The orbital frame coordinates used in this simulation simplify the mathematics by using a linear transformation between the ECEF coordinate system to a locally level coordinate frame as the observer's local reference frame. Then, the satellite positions become

$$x' = y, \quad y' = z, \quad z' = x - R_e,$$

where (x', y', z') are the locally level coordinates of the satellites and (x, y, z) are the original ECEF coordinates. Here, R_e is the earth's radius. This assumes a user position at (0,0,0) in locally level coordinates, which makes the math simpler because now the pseudorange can be written as

$$\rho_1(t) = \sqrt{(x_1(t) - 0)^2 + (y_1(t) - 0)^2 + (z_1(t) - 0)^2},$$
$$h_x^{[1]}(t) = -(x_1(t) - 0)/\rho_1(t),$$

where $h^{[1]}$ represents the partial of the pseudorange with respect to x (component of the $H^{[1]}$ matrix). Therefore, the default earth and orbit constants are defined as

$$R_{\text{sat}} = 26560000.0, \quad R_e = 6380000.0, \quad \alpha = 55^\circ.$$

8.8.4 Measurement Sensitivity Matrix

The definition of the different elements of the $H^{[1]}$ matrix are

$$x_1(t) = R_{\text{sat}}\{\cos[\theta(t)]\ \sin[\Omega(t)] + \sin[\theta(t)]\cos[\Omega(t)]\cos\alpha\},$$
$$y_1(t) = R_{\text{sat}}\{\sin[\theta(t)]\}\ \sin\alpha,$$
$$z_1(t) = R_{\text{sat}}\{\cos[\theta(t)]\cos[\Omega(t)] - \sin[\theta(t)]\sin[\Omega(t)]\cos\alpha\} - R_e$$
$$\rho_1(t) = \sqrt{[x_1(t)]^2 + [y_1(t)]^2 + [z_1(t)]^2},$$
$$h_x(t) = -x_1(t)/\rho_1(t),$$
$$h_y(t) = -y_1(t)/\rho_1(t),$$
$$h_z(t) = -z_1(t)/\rho_1(t),$$

and likewise for each of the other four satellites.

The complete $H^{[1]}$ matrix can then be defined as

$$H(t) = \begin{bmatrix} h1_x(t) & h1_y(f) & h1_z(t) & 1 & 0 \\ h2_x(t) & h2_y(t) & h2_z(t) & 1 & 0 \\ h3_x(t) & h3_y(t) & h3_z(t) & 1 & 0 \\ h4_x(t) & h4_y(t) & h4_z(t) & 1 & 0 \end{bmatrix},$$

where the last two columns refer to the clock bias and clock drift.

In calculating GDOP, only the clock bias is used in the equations, so the H matrix becomes

$$H(t) = \begin{bmatrix} h1_x(t) & h1_y(t) & h1_z(t) & 1 \\ h2_x(t) & h2_y(t) & h2_z(t) & 1 \\ h3_x(t) & h3_y(t) & h3_z(t) & 1 \\ h4_x(t) & h4_y(t) & h4_z(t) & 1 \end{bmatrix}.$$

The calculation of the GDOP and various other DOPs are then defined in terms of this $H(t)$ matrix as a function of time t:

$$\begin{aligned} A(t) &= [H(t)^T H(t)]^{-1}, \\ \text{GDOP}(t) &= \sqrt{\text{tr}[A(t)]}, \\ \text{PDOP}(t) &= \sqrt{A(t)_{1,1} + A(t)_{2,2} + A(t)_{3,3}}, \\ \text{HDOP}(t) &= \sqrt{A(t)_{1,1} + A(t)_{2,2,}}, \\ \text{VDOP}(t) &= \sqrt{A(t)_{3,3}}, \\ \text{TDOP}(t) &= \sqrt{A(t)_{4,4}}. \end{aligned}$$

8.8.5 Implementation Results

8.8.5.1 DOP Calculations In the MATLAB implementation, the GDOP, PDOP, HDOP, VDOP, and TDOP, are defined and plotted for the two different cases of satellite phasings:

Case 1: Good Geometry. The results from case 1 (satellites 1, 2, 3, and 4) show an excellent GDOP ranging to less 3.2 as a function of time. Figure 8.12 shows the variation of GDOP in meters as a function of time. This is a reasonable GDOP. Figure 8.13 shows all of the DOPs in meters as a function of time.

Case 2: Bad Geometry. Case 2 satellite phasing results in the infamous GDOP "chimney peak" during that time when satellite geometry fails to provide observability of user position. Figure 8.14 shows the resulting GDOP plots.

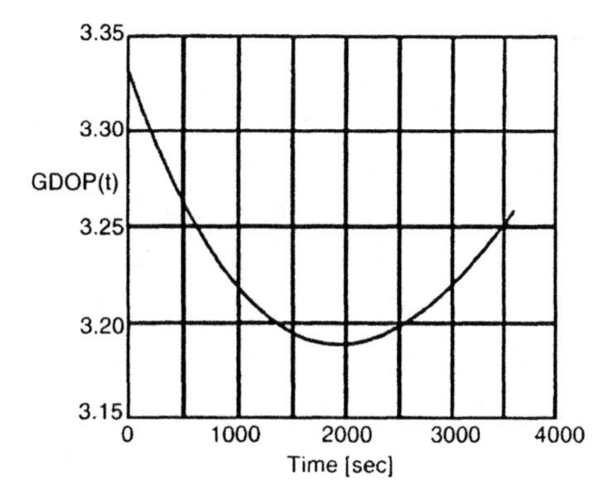

Fig. 8.12 Case 1 GDOP.

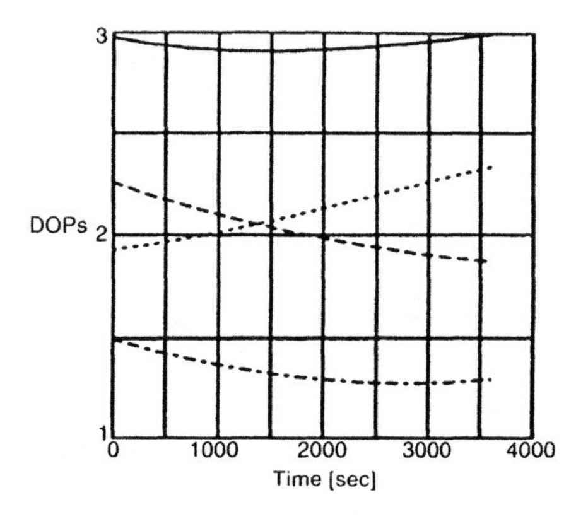

Fig. 8.13 Case 1 DOPs.

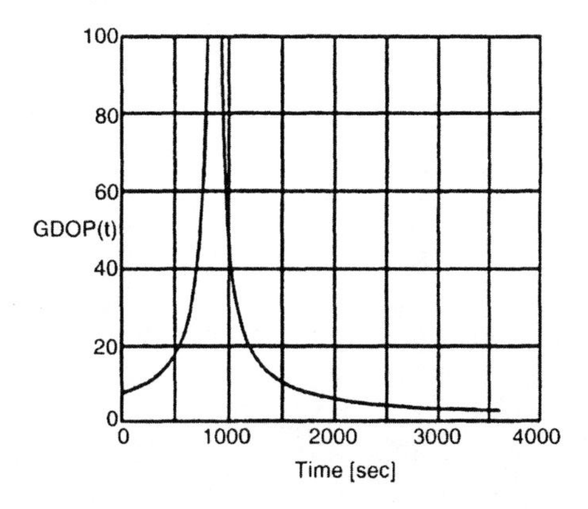

Fig. 8.14 Case 2 GDOP.

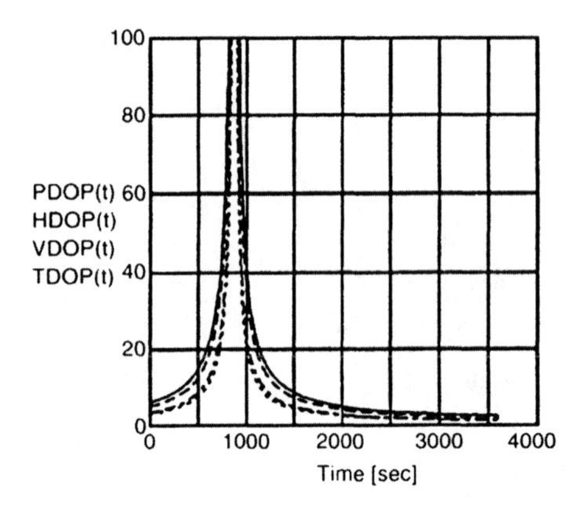

Fig. 8.15 Case 2 DOPs.

It shows that two satellites out of four are close to each other and thereby do not provide linearly independent equations. This combination of satellites cannot be used to find the user position, clock drift, and biases. Figure 8.15 is a multiplot of all the DOPs.

8.8.5.2 Kalman Filter Implementation For the second part of the example, a covariance analysis of the GPS/Kalman filter system is used to evaluate the performance of the system, given initial position estimates and estimates of receiver R and system dynamic Q noise. This type of analysis is done if actual measurement data is not available and can serve as a predictor of how well the system will converge to a residual error estimate in the position and time. The masking error in the Q matrix is

$$\mathbf{Q} = \begin{bmatrix} 0.333 & 0 & 0 & 0 & 0 \\ 0 & 0.333 & 0 & 0 & 0 \\ 0 & 0 & 0.333 & 0 & 0 \\ 0 & 0 & 0 & 0.0833 & 0 \\ 0 & 0 & 0 & 0 & 0.142 \end{bmatrix}, \tag{8.105}$$

where dimensions are in meters squared and meters squared per second squared.

The receiver noise R matrix in meters squared is

$$\mathbf{R} = \begin{bmatrix} 225 & 0 & 0 & 0 \\ 0 & 225 & 0 & 0 \\ 0 & 0 & 225 & 0 \\ 0 & 0 & 0 & 225 \end{bmatrix}.$$

The initial transformation matrix between the first and next measurement is the matrix

$$\Phi = \begin{bmatrix} 1 & 0 & 0 & 0 & 0 \\ 0 & 1 & 0 & 0 & 0 \\ 0 & 0 & 1 & 0 & 0 \\ 0 & 0 & 0 & 1 & 1 \\ 0 & 0 & 0 & 0 & 1 \end{bmatrix}.$$

The assumed initial error estimate was 100 m and is represented by the $\mathbf{P}_{0}(+)$ matrix and is an estimate of how far off the initial measurements are from the actual points:

$$\mathbf{P}_{0}(+) = \begin{bmatrix} 10,000 & 0 & 0 & 0 & 0 \\ 0 & 10,000 & 0 & 0 & 0 \\ 0 & 0 & 10,000 & 0 & 0 \\ 0 & 0 & 0 & 90,000 & 0 \\ 0 & 0 & 0 & 0 & 900 \end{bmatrix}. \tag{8.106}$$

These assume a clock bias error of 300 m and a drift of 30 m/s. The discrete extended Kalman filtering equations, as listed in Table 8.3, are the a priori *covariance matrix*

$$\mathbf{P}_{k}(-) = \Phi \mathbf{P}_{k-1}(+)\Phi^{T} + \mathbf{Q}_{k-1},$$

the *Kalman gain equation*

$$\overline{\mathbf{K}}_{k} = \mathbf{P}_{k}(-)\mathbf{H}_{k}^{[1]T}[\mathbf{H}_{k}^{[1]}\mathbf{P}_{k}(-)\mathbf{H}_{k}^{[1]T} + \mathbf{R}_{k}]^{-1},$$

and the a posteriori *covariance matrix*

$$\mathbf{P}_{k}(+) = \{\mathbf{I} - \overline{\mathbf{K}}_{k}\mathbf{H}_{k}^{[1]}\}\mathbf{P}_{k}(-).$$

The diagonal elements of the covariance matrices $\mathbf{P}_{k}(-)$ (predicted) and $\mathbf{P}_{k}(+)$ (corrected) are plotted as an estimate of how well the individual x, y, z and clock drift errors converge as a function of time for $t = 1$ s to $t = 150$ s.

In a real system, the $\mathbf{Q}$, $\mathbf{R}$, and Φ matrices and Kalman gain estimates are under control of the designer and need to be varied individually to obtain an acceptable residual covariance error. This example only analyzes the covariance estimates for the given $\mathbf{Q}$, $\mathbf{R}$, and Φ matrices, which turned out to be a satisfactory set of inputs.

Simulation Procedure Start simulation for $t = 0, \ldots, 3600$. Case 1: Satellites 1, 2, 3, and 4:

$$\Omega_0 1 = 326\frac{\pi}{180}, \quad \Omega_0 2 = 26\frac{\pi}{180},$$

$$\Omega_0 3 = 146\frac{\pi}{180}, \quad \Omega_0 4 = 86\frac{\pi}{180},$$

$$\theta_0 1 = 68\frac{\pi}{180}, \quad \theta_0 2 = 340\frac{\pi}{180},$$
$$\theta_0 3 = 198\frac{\pi}{180}, \quad \theta_0 4 = 271\frac{\pi}{180}.$$

Define rate variables:

$$\dot{\Omega}_r = 2\frac{\pi}{86,164}, \quad \dot{\theta}_r = 2\frac{\pi}{43,082}.$$

The angular rate equations are

$$\Omega 1(t) = \Omega_0 1 - \Omega_r t, \quad \theta 1(t) = \theta_0 1 + \theta_r t,$$
$$\Omega 2(t) = \Omega_0 2 - \Omega_r t, \quad \theta 2(t) = \theta_0 2 + \theta_r t,$$
$$\Omega 3(t) = \Omega_0 3 - \Omega_r t, \quad \theta 3(t) = \theta_0 3 + \theta_r t,$$
$$\Omega 4(t) = \Omega_0 4 - \Omega_r t, \quad \theta 4(t) = \theta_0 4 + \theta_r t$$

The default earth constants are

$$\mathrm{R_{sat}} = 26560000.0, \quad \mathrm{R_e} = 6380000.0, \quad \cos\alpha = \cos 55^\circ, \quad \sin\alpha = \sin 55^\circ,$$

For satellite 1:

$$x_1(t) = \mathrm{R_{sat}}\{\cos[\theta_1(t)]\sin[\Omega_1(t)] + \sin[\theta_1(t)]\cos[\Omega_1(t)]\cos[\alpha]\}$$
$$y_1(t) = \mathrm{R_{sat}}\{\sin[\theta_1(t)]\sin[\alpha]$$
$$z_1(t) = \mathrm{R_{sat}}\{\cos[\theta_1(t)]\cos[\Omega_1(t)] - \sin[\theta(t)]\sin[\Omega_1(t)]\cos[\alpha]\}] - \mathrm{R_e},$$
$$\rho_1(t) = \sqrt{[(x_1(t)]^2 + [y_1(t)]^2 + [z_1(t)]^2}$$

and the H matrix elements are

$$h1_x(t) = \frac{-x_1(t)}{\rho_1(t)}, \quad h1_y(t) = \frac{-y_1(t)}{\rho_1(t)}, \quad h1_z(t) = \frac{-z_1(t)}{\rho_1(t)}.$$

For satellite 2

$$x_2(t) = \mathrm{R_{sat}}\{\cos[\theta_2(t)]\sin[\Omega_2(t)] + \sin[\theta_2(t)]\cos[\Omega_2(t)]\cos[\alpha]\}$$
$$y_2(t) = \mathrm{R_{sat}}\sin[\theta_2(t)]\sin[\alpha]$$
$$z_2(t) = \mathrm{R_{sat}}\{\cos[\theta_2(t)]\cos[\Omega_2(t)] - \sin[\theta_2(t)]\sin[\Omega_2(t)]\cos[\alpha]\} - \mathrm{R_e}$$
$$\rho_2(t) = \sqrt{[x_2(t)]^2 + [y_2(t)]^2 + [z_2(t)]^2},$$

and the H matrix elements are

$$h2_x(t) = \frac{-x_2(t)}{\rho_2(t)}, \quad h2_y(t) = \frac{-y_2(t)}{\rho_2(t)}, \quad h2_z(t) = \frac{-z_2(t)}{\rho_2(t)}$$

For satellite 3

$$\begin{aligned}
x_3(t) &= \mathrm{R_{sat}}\{\cos[\theta_3(t)]\sin[\Omega_3(t)] + \sin[\theta_3(t)]\cos[\Omega_3(t)]\cos[\alpha]\} \\
y_3(t) &= \mathrm{R_{sat}}\sin[\theta_3(t)]\sin[\alpha] \\
z_3(t) &= \mathrm{R_{sat}}\{\cos[\theta_3(t)]\cos[\Omega_3(t)] - \sin[\theta_3(t)]\sin[\Omega_3(t)]\cos[\alpha]\} - \mathrm{R_e} \\
\rho_3(t) &= \sqrt{[x_3(t)]^2 + [y_3(t)]^2 + z_3(t)]^2},
\end{aligned}$$

and the **H** matrix elements are

$$h3_x(t) = \frac{-x_3(t)}{\rho_3(t)}, \quad h3_y(t) = \frac{-y_3(t)}{\rho_3(t)}, \quad h3_z(t) = \frac{-z_3(t)}{\rho_3(t)}.$$

For satellite 4

$$\begin{aligned}
x_4(t) &= \mathrm{R_{sat}}\{\cos[\theta_4(t)]\sin[\Omega_4(t)] + \sin[\theta_4(t)]\cos[\Omega_4(t)]\cos[\alpha]\} \\
y_4(t) &= \mathrm{R_{sat}}\sin[\theta_4(t)]\sin[\alpha] \\
z_4(t) &= \mathrm{R_{sat}}\{\cos[\theta_4(t)]\cos[\Omega_4(t)] - \sin[\theta_4(t)]\sin[\Omega_4(t)]\cos[\alpha]\} - \mathrm{R_e} \\
\rho_4(t) &= \sqrt{[x_4(t)]^2 + [y_4(t)]^2 + [z_4(t)]^2},
\end{aligned}$$

and the **H** matrix elements are

$$h4_x(t) = \frac{-x_4(t)}{\rho_4(t)}, \quad h4_y(t) = \frac{-y_4(t)}{\rho_4(t)}, \quad h4_z(t) = \frac{-z_4(t)}{\rho_4(t)}.$$

Complete $\mathbf{H}^{[1]}$ matrix:

$$\mathbf{H}^{[1]}(t) = \begin{bmatrix} h1_x(t) & h1_y(t) & h1_z(t) & 1 & 0 \\ h2_x(t) & h2_y(t) & h2_z(t) & 1 & 0 \\ h3_x(t) & h3_y(t) & h3_z(t) & 1 & 0 \\ h4_x(t) & h4_y(t) & h4_z(t) & 1 & 0 \end{bmatrix}.$$

The H matrix used in the GDOP calculation is

$$H^{[1]}(t) = \begin{bmatrix} h1_x(t) & h1_y(t) & h1_z(t) & 1 \\ h2_x(t) & h2_y(t) & h2_z(t) & 1 \\ h3_x(t) & h3_y(t) & h3_z(t) & 1 \\ h4_x(t) & h4_y(t) & h4_z(t) & 1 \end{bmatrix}.$$

The noise matrix is

$$Q = \begin{bmatrix} 0.333 & 0 & 0 & 0 & 0 \\ 0 & 0.333 & 0 & 0 & 0 \\ 0 & 0 & 0.333 & 0 & 0 \\ 0 & 0 & 0 & 0.0833 & 0 \\ 0 & 0 & 0 & 0 & 0.142 \end{bmatrix}.$$

The initial guess of the $\mathbf{P}_0(+)$ matrix is

$$\mathbf{P}_0(+) = \begin{bmatrix} 10,000 & 0 & 0 & 0 & 0 \\ 0 & 10,000 & 0 & 0 & 0 \\ 0 & 0 & 10,000 & 0 & 0 \\ 0 & 0 & 0 & 90,000 & 0 \\ 0 & 0 & 0 & 0 & 900 \end{bmatrix}$$

and the $\mathbf{R}$ matrix is

$$\mathbf{R} = \begin{bmatrix} 225 & 0 & 0 & 0 \\ 0 & 225 & 0 & 0 \\ 0 & 0 & 225 & 0 \\ 0 & 0 & 0 & 225 \end{bmatrix},$$

$$A(t) = [H^{[1]T}(t)H^{[1]}(t)]^{-1},$$

$$\text{GDOP}(t) = \sqrt{\text{tr}[A(t)]}.$$

Kalman Filter Simulation Results Figure 8.16 shows the square roots of the covariance terms P_{11} (RMS east position uncertainty), both predicted (dashed line) and corrected (solid line). After a few iterations, the RMS error in the x position is less than 5 m. Figure 8.17 shows the corresponding RMS north position uncertainty in meters, and Figure 8.18 shows the corresponding RMS vertical position uncertainty in meters.

Figures 8.19 and 8.20 show the square roots of the error covariances in clock bias and clock drift rate in meters.

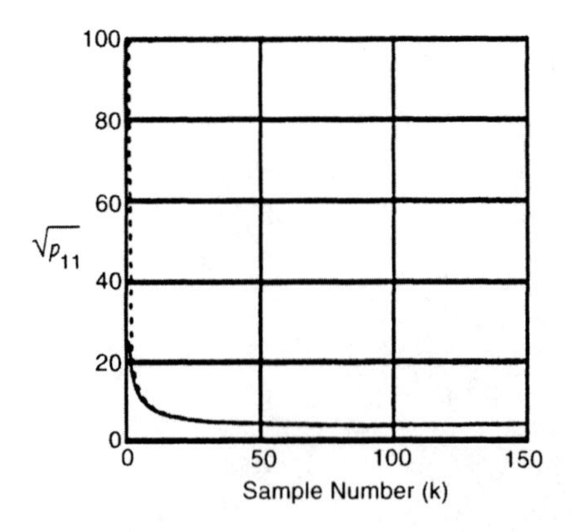

Fig. 8.16 RMS east position error.

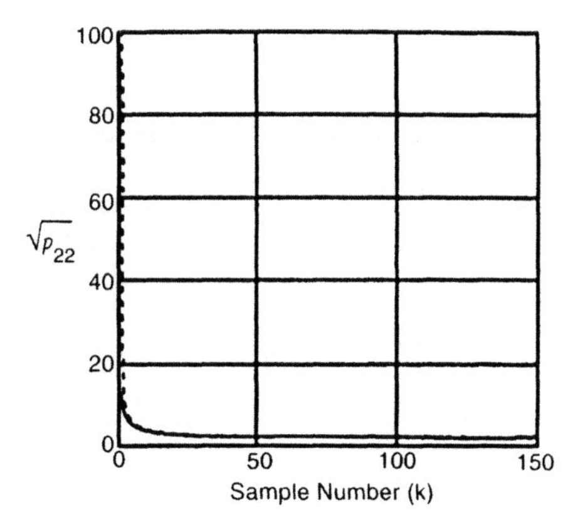

Fig. 8.17 RMS north position error.

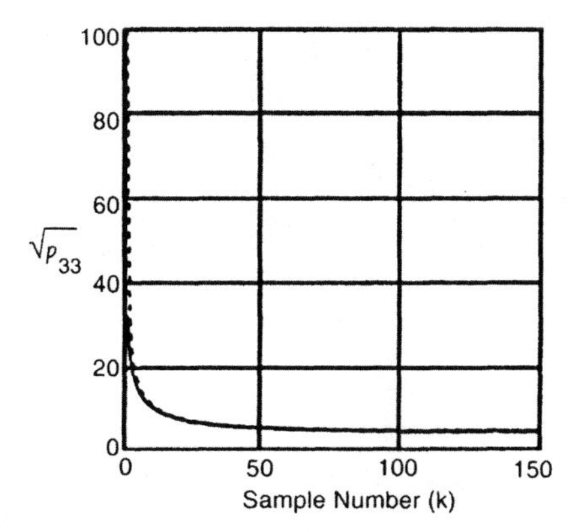

Fig. 8.18 RMS vertical position error.

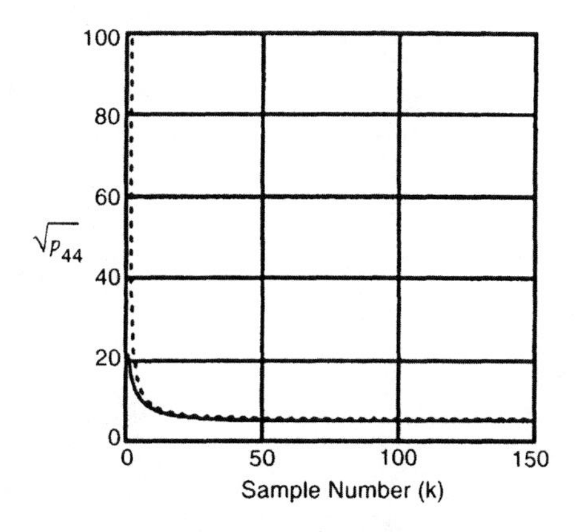

Fig. 8.19 RMS clock error.

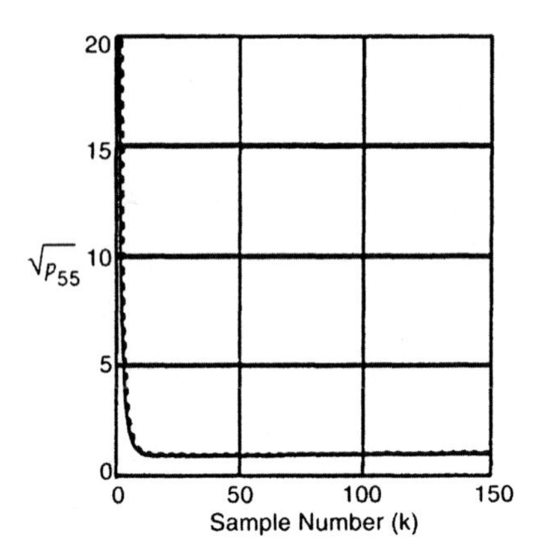

Fig. 8.20 RMS drift error.

8.9 OTHER KALMAN FILTER IMPROVEMENTS

There have been many "improvements" in the Kalman filter since 1960. Some are changes in the methods of computation, some use the Kalman filter model to solve related nonfiltering problems, and some make use of Kalman filtering variables for addressing other applications-related problems. We present here some that have been found useful in GPS/INS integration. More extensive coverage of the underlying issues and solution methods is provided in Ref. 66.

8.9.1 Schmidt–Kalman Suboptimal Filtering

This is a method proposed by Stanley F. Schmidt [173] for reducing the processing and memory requirements for Kalman filtering, with predictable performance degradation. It has been used in GPS navigation as a means of eliminating additional variables (one per GPS satellite) required for Kalman filtering with time-correlated pseudorange errors (originally for SA errors, also useful for uncompensated ionospheric propagation delays).

8.9.1.1 State Vector Partitioning Schmidt–Kalman filtering partitions the state vector into "essential" variables (designated by the subscript e) and "unessential" variables (designated by the subscript u)

$$\mathbf{x} = \begin{bmatrix} \mathbf{x}_e \\ \mathbf{x}_u \end{bmatrix}, \tag{8.107}$$

where $\mathbf{x}_e$ is the $n_e \times 1$ subvector of essential variables to be estimated, $\mathbf{x}_u$ is the $n_u \times 1$ subvector that will not be estimated, and

$$n_e + n_u = n, \text{ the total number of state variables.} \tag{8.108}$$

Even though the subvector $\mathbf{x}_u$ of nuisance variables is not estimated, the effects of not doing so must be reflected in the covariance matrix $\mathbf{P}_{ee}$ of uncertainty in the estimated variables. For that purpose, the Schmidt–Kalman filter calculates the covariance matrix $\mathbf{P}_{uu}$ of uncertainty in the unestimated state variables and the cross-covariance matrix $\mathbf{P}_{ue}$ between the two types. These other covariance matrices are used in the calculation of the Schmidt–Kalman gain.

8.9.1.2 Implementation Equations The essential implementation equations for the Schmidt–Kalman (SK) filter are listed in Table 8.5. These equations have been arranged for reusing intermediate results to reduce computational requirements.

8.9.1.3 Simulated Performance in GPS Position Estimation Figure 8.21 is the output of the MATLAB m-file `SchmidtKalmanTest.m` on the accompanying CD. This is a simulation using the vehicle dynamic model Damp2, described in Section 10.2.2.5, with 29 GPS satellites (almanac of March 8, 2006), 9–11 of which were in view (15° above the horizon) at any one time, and

$n_e = 9$, the number of essential state variables (3 each of position, velocity, acceleration)

$n_u = 29$, the number of unessential state variables (propagation delays)

TABLE 8.5. Summary Implementation of Schmidt–Kalman Filter

Corrector (Observational Update)		
$\mathbf{C}$	=	$\mathbf{H}_{e,k}\left[\mathbf{P}_{ee,k}(-)\mathbf{H}_{e,k}^T + \mathbf{P}_{eu,k}(-)\mathbf{H}_{u,k}^T\right]$
		$+\mathbf{H}_{u,k}\left[\mathbf{P}_{ue,k}(-)\mathbf{H}^T + \mathbf{P}_{uu,k}(-)\,\mathbf{H}_{u,k}^T\right]$
$\overline{\mathbf{K}}_{\mathrm{SK},k}$	=	$\left[\mathbf{P}_{ee,k}(-)\mathbf{H}_{e,k}^T + \mathbf{P}_{eu,k}(-)\,\mathbf{H}_{u,k}^T\right]\mathbf{C}$
$\hat{\mathbf{x}}_{e,k}(+)$	=	$\hat{\mathbf{x}}_{e,k}(-) + \overline{\mathbf{K}}_{\mathrm{SK},k}\left[\mathbf{z}_k - \mathbf{H}_{ek}\mathbf{x}_{e,k}(-)\right]$
$\mathbf{A}$	=	$\mathbf{I}_{n_e} - \overline{\mathbf{K}}_{\mathrm{SK},k}\mathbf{H}_{e,k}$
$\mathbf{B}$	=	$\mathbf{A}\mathbf{P}_{eu,k}(-)\,\mathbf{H}_{u,k}^T\overline{\mathbf{K}}_{\mathrm{SK},k}^T$
$\mathbf{P}_{ee,k}(+)$	=	$\mathbf{A}\mathbf{P}_{ee,k} - \mathbf{A}^{\mathrm{T}} - \mathbf{B} - \mathbf{B}^T$
		$+\overline{\mathbf{K}}_{\mathrm{SK},k}(\mathbf{H}_{u,k}\mathbf{P}_{uu,k} - \mathbf{H}_{u,k}^T + \mathbf{R}_k)\overline{\mathbf{K}}_{\mathrm{SK},k}^T$
$\mathbf{P}_{eu,k}(+)$	=	$\mathbf{A}\mathbf{P}_{eu,k}(-) - \overline{\mathbf{K}}_{\mathrm{SK},k}\mathbf{H}_{u,k}\mathbf{P}_{uu,k}(-)$
$\mathbf{P}_{ue,k}(+)$	=	$\mathbf{P}_{eu,k}(+)^{\mathrm{T}}$
$\mathbf{P}_{uu,k}(+)$	=	$\mathbf{P}_{uu,k}(-)$
Predictor (Time Update)		
$\hat{\mathbf{x}}_{e,k+1}(-)$	=	$\Phi_{e,k}\hat{\mathbf{x}}_{e,k}(+)$
$\mathbf{P}_{ee,k+1-}$	=	$\Phi_{e,k}\mathbf{P}_{ee,k}(+)\,\Phi_{e,k}^T + \mathbf{Q}_{ee}$
$\mathbf{P}_{eu,k+1}(-)$	=	$\Phi_{e,k}\mathbf{P}_{eu,k}(+)\Phi_{u,k}^T$
$\mathbf{P}_{ue,k+1}(-)$	=	$\mathbf{P}_{eu,k+1}(-)^T$
$\mathbf{P}_{uu,k+1-}$	=	$\Phi_{u,k}\mathbf{P}_{uu,k+}\Phi_{u,k}^T + \mathbf{Q}_{uu}$

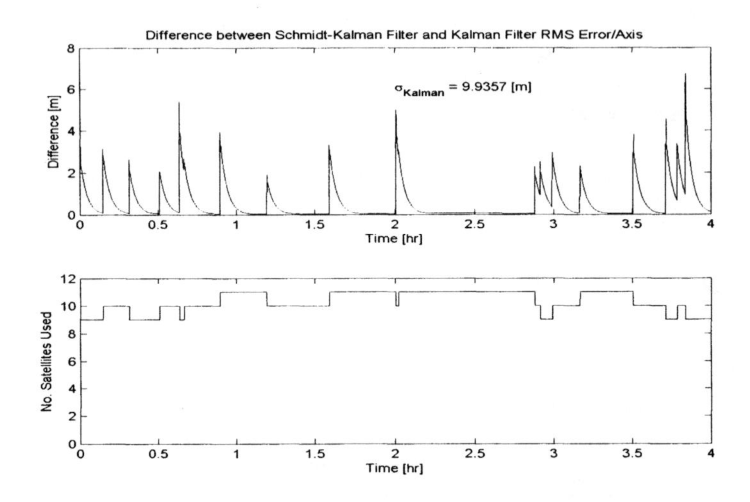

Fig. 8.21 Simulation comparing Schmidt–Kalman and Kalman filters.

$$\Delta t = 1 \text{ s, time interval between filter updates}$$
$$\sigma_{\text{pos}}(0) = 20 \text{ m, initial position uncertainty, RMS/axis}$$
$$\sigma_{\text{vel}} = 200 \text{ m/s, RMS of random vehicle speed}(\sim 447 \text{ mi/h})$$
$$\sigma_{\text{acc}} = 0.5 \text{ g, RMS of random vehicle acceleration}$$
$$\sigma_{\text{prop}} = 10 \text{ m, RMS propagation delay uncertainty (steady-state)}$$
$$\sigma_{\rho} = 10 \text{ m, RMS pseudorange measurement noise (white, zero-mean)}$$
$$\tau_{\text{prop}} = 150 \text{ s, correlation time constant of propagation delay}$$
$$\tau_{\text{acc}} = 120 \text{ s, correlation time constant of random vehicle acceleration}$$

The variables plotted in the top graph are the "un-RSS" differences

$$\sigma_{\text{difference}} = \sqrt{\sigma_{\text{SK}}^2 - \sigma_{\text{K}}^2}$$

between the mean-squared position uncertainties of the Schmidt–Kalman filter (σ_{SK}^2) and the Kalman filter ($\sigma_{\text{K}}^2 \approx 10^2 \text{ m}^2$). The peak errors introduced by the Schmidt–Kalman filter are a few meters to several meters, and transient. The error spikes generally coincide with the changes in the number of satellites used (plotted in the bottom graph). This would indicated that, for this GPS application anyway, the Schmidt–Kalman filter performance comes very close to that of the Kalman filter, except when a new satellite with unknown propagation delay is first used. Even then, the errors introduced by using a new satellite generally die down after a few correlation time constants of the propagation delay errors.

8.9.2 Serial Measurement Processing

It has shown been [106] that it is more efficient to process the components of a measurement vector serially, one component at a time, than to process them as a vector. This may seem counterintuitive, but it is true even if its implementation requires a transformation of measurement variables to make the associated measurement noise covariance $\mathbf{R}$ a diagonal matrix (i.e., with noise uncorrelated from one component to another).

8.9.2.1 ***Measurement Decorrelation*** If the covariance matrix $\mathbf{R}$ of measurement noise is *not* a diagonal matrix, then it can be made so by $\mathbf{UDU}^T$ decomposition (Eq. B.22) and changing the measurement variables

$$\mathbf{R}_{\text{correlated}} = \mathbf{U}_R \mathbf{D}_R \mathbf{U}_R^T, \tag{8.109}$$

$$\mathbf{R}_{\text{decorrelated}} \stackrel{\text{def}}{=} \mathbf{D}_R \text{ (a diagonal matrix)}, \tag{8.110}$$

$$\mathbf{z}_{\text{decorrelated}} \stackrel{\text{def}}{=} \mathbf{U}_R \backslash \mathbf{z}_{\text{correlated}}, \tag{8.111}$$

$$\mathbf{H}_{\text{decorrelated}} \stackrel{\text{def}}{=} \mathbf{U}_R \backslash \mathbf{H}_{\text{correlated}}, \tag{8.112}$$

where $\mathbf{R}_{\text{correlated}}$ is the nondiagonal (i.e., correlated component-to-component) measurement noise covariance matrix and the new *decorrelated* measurement vector $\mathbf{z}_{\text{decorrelated}}$ has a diagonal measurement noise covariance matrix $\mathbf{R}_{\text{decorrelated}}$ and measurement sensitivity matrix $\mathbf{H}_{\text{decorrelated}}$.

8.9.2.2 ***Serial Processing of Decorrelated Measurements*** The components of $\mathbf{z}_{\text{decorrelated}}$ can now be processed one component at a time using the corresponding row of $\mathbf{H}_{\text{decorrelated}}$ as its measurement sensitivity matrix and the corresponding diagonal element of $\mathbf{R}_{\text{decorrelated}}$ as its measurement noise variance.

A "pidgin-MATLAB" implementation for this procedure is listed in Table 8.6, where the final line is a "symmetrizing" procedure designed to improve robustness.

8.9.3 Improving Numerical Stability

8.9.3.1 ***Effects of Finite Precision*** Computer roundoff limits the precision of numerical representation in the implementation of Kalman filters. It has been known to cause severe degradation of filter performance in many applications, and alternative implementations of the Kalman filter equations (the Riccati equations, in particular) have been shown to improve robustness against roundoff errors.

Computer roundoff for floating-point arithmetic is often characterized by a single parameter $\varepsilon_{\text{roundoff}}$, which is the smallest number such that

$$1 + \varepsilon_{\text{roundoff}} > 1 \text{ in machine precision.} \tag{8.113}$$

TABLE 8.6. Implementation Equations for Serial Measurement Update

$\mathbf{x} = \hat{\mathbf{x}}_k(-)$;
$\mathbf{P} = \mathbf{P}_k(-)$;
`for j = 1:`ℓ,
  $z = \mathbf{z}_k(\mathrm{j})$;
  $\mathbf{H} = \mathbf{H}_k(\mathrm{j}, :)$;
  $R = \mathbf{R}_{\text{decorrelated}}(\mathrm{j}, \mathrm{j})$;
  $\overline{\mathbf{K}} = \mathbf{PH}'/(\mathbf{HPH}' + R)$;
  $\hat{\mathbf{x}} = \overline{\mathbf{K}}(z - \mathbf{Hx})$;
  $\mathbf{P} = \mathbf{P} - \overline{\mathbf{K}}\mathbf{HP}$;
end;
$\hat{\mathbf{x}}_k(+) = \mathbf{x}$;
$\mathbf{P}_k(+) = (\mathbf{P} + \mathbf{P}')/2$; (symmetrize)

It is the value assigned to the parameter `eps` in MATLAB. In 64-bit ANSI/IEEE Standard floating-point arithmetic (MATLAB precision on PCs) `eps` $= 2^{-52}$.

The following example, due to Dyer and McReynolds [49] , shows how a problem that is well conditioned, as posed, can be made ill-conditioned by the filter implementation.

Example 8.8: Ill-Conditioned Measurement Sensitivity. Consider the filtering problem with measurement sensitivity matrix

$$\mathbf{H} = \begin{bmatrix} 1 & 1 & 1 \\ 1 & 1 & 1+\delta \end{bmatrix}$$

and covariance matrices

$$\mathbf{P}_0 = \mathbf{I}_3, \text{ and } \mathbf{R} = \delta^2 \mathbf{I}_2,$$

where $\mathbf{I}_n$ denotes the $n \times n$ identity matrix and the parameter δ satisfies the constraints

$$\delta^2 < \varepsilon_{\text{roundoff}} \text{ but } \delta > \varepsilon_{\text{roundoff}}.$$

In this case, although $\mathbf{H}$ clearly has rank 2 in machine precision, the product $\mathbf{HP}_0\mathbf{H}^T$ with *roundoff* will equal

$$\begin{bmatrix} 3 & 3+\delta \\ 3+\delta & 3+2\delta \end{bmatrix},$$

which is singular. The result is unchanged when $\mathbf{R}$ is added to $\mathbf{HP}_0\mathbf{H}^T$. In this case, then, the filter observational update fails because the matrix $\mathbf{HP}_0\mathbf{H}^T + \mathbf{R}$ is not invertible.

8.9.3.2 Alternative Implementations The covariance correction process (observational update) in the solution of the Riccati equation was found to be the dominant source of numerical instability in the Kalman filter implementation, and the more common symptoms of failure were asymmetry of the covariance matrix (easily fixed) or (worse by far) negative terms on its diagonal. These implementation problems could be avoided for some problems by using more precision, but they were eventually solved for most applications by using alternatives to the covariance matrix **P** as the dependent variable in the covariance correction equation. However, each of these methods required a compatible method for covariance prediction. Table 8.7 lists several of these compatible implementation methods for improving the numerical stability of Kalman filters.

Figure 8.22 illustrates how these methods perform on the ill-conditioned problem of Example 8.8 as the conditioning parameter $\delta \to 0$. For this particular test case, using 64-bit floating-point precision (52-bit mantissa), the accuracy of the Carlson [31] and Bierman [17] implementations degrade more gracefully than do the others as $\delta \to \varepsilon$, the machine precision limit. The Carlson and Bierman solutions still maintain about nine digits ($\approx$30 bits) of accuracy at $\delta \approx \sqrt{\varepsilon}$, when the other methods have essentially no bits of accuracy in the computed solution.

These results, by themselves, do not prove the general superiority of the Carlson and Bierman solutions for the Riccati equation. Relative performance of alternative implementation methods may depend on details of the specific application, and for many applications, the standard Kalman filter implementation will suffice. For many other applications, it has been found sufficient to constrain the covariance matrix to remain symmetric.

The MATLAB m-file `shootout.m` on the accompanying CD generates Fig. 8.22, using m-files with the same names as those of the solution methods in Fig. 8.22. For detailed derivations of these methods, see Ref. 66.

Conditioning and Scaling Considerations The data formatting differences between Cholesky factors (Carlson implementation) and modified Cholesky factors (Bierman–Thornton implementation) are not always insignificant, as illustrated by the following example.

TABLE 8.7. Compatible Methods for Solving the Riccati Equation

Covariance Matrix Format	Riccati Equation Implementation Methods	
	Corrector	Predictor
Symmetric nonnegative definite	Kalman [104]	Kalman [104]
	Joseph [30]	Kalman [104]
Square Cholesky factor **C**	Potter [12, 155]	$\mathbf{C}_{k+1}(-) = \Phi_k \Phi \mathbf{C}_k(+)$
Triangular Cholesky factor **C**	Carlson [31]	Kailath–Schmidt[a]
Triangular Cholesky factor **C**	Morf–Kailath combined [143]	
Modified Cholesky factors **U**, **D**	Bierman [17]	Thornton [187]

[a] From unpublished sources.

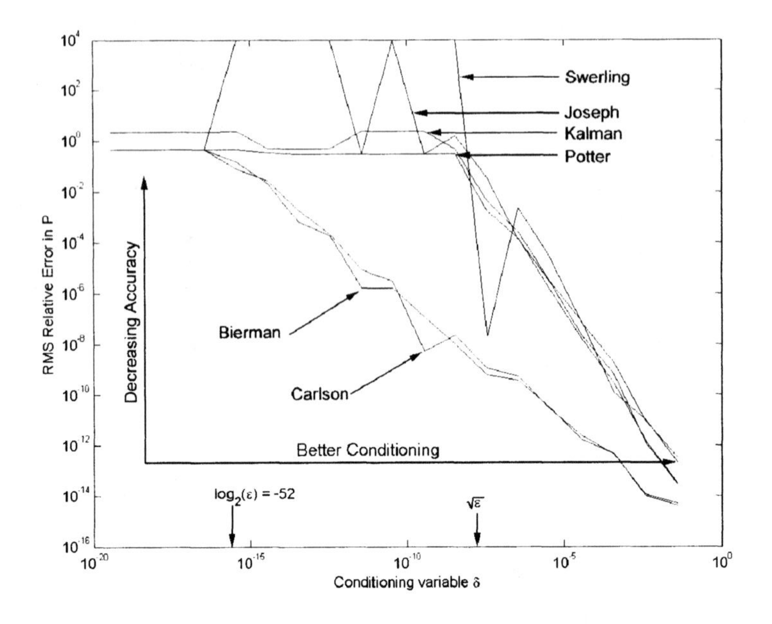

Fig. 8.22 Degradation of numerical solutions with problem conditioning.

Example 8.9: Cholesky Factor Scaling and Conditioning. The $n \times n$ covariance matrix

$$\mathbf{P} = \begin{bmatrix} 10^0 & 0 & 0 & \cdots & 0 \\ 0 & 10^{-2} & 0 & \cdots & 0 \\ 0 & 0 & 10^{-4} & \cdots & 0 \\ \vdots & \vdots & \vdots & \ddots & \vdots \\ 0 & 0 & 0 & \cdots & 10^{2-2n} \end{bmatrix} \tag{8.114}$$

has condition number 10^{2n-2}. Its Cholesky factor

$$\mathbf{C} = \begin{bmatrix} 10^0 & 0 & 0 & \cdots & 0 \\ 0 & 10^{-1} & 0 & \cdots & 0 \\ 0 & 0 & 10^{-2} & \cdots & 0 \\ \vdots & \vdots & \vdots & \ddots & \vdots \\ 0 & 0 & 0 & \cdots & 10^{1-n} \end{bmatrix} \tag{8.115}$$

has condition number 10^{n-1}. However, its modified Cholesky factors are $\mathbf{U} = \mathbf{I}_n$ (condition number $= 1$) and $\mathbf{D} = \mathbf{P}$ (condition number $= 10^{2n-2}$).

The condition numbers of the different factors are plotted versus matrix dimension n in Fig. 8.23. As a rule, one would like matrix condition numbers to be $\ll 1/\epsilon$, where ϵ is the machine precision limit (the smallest number such that

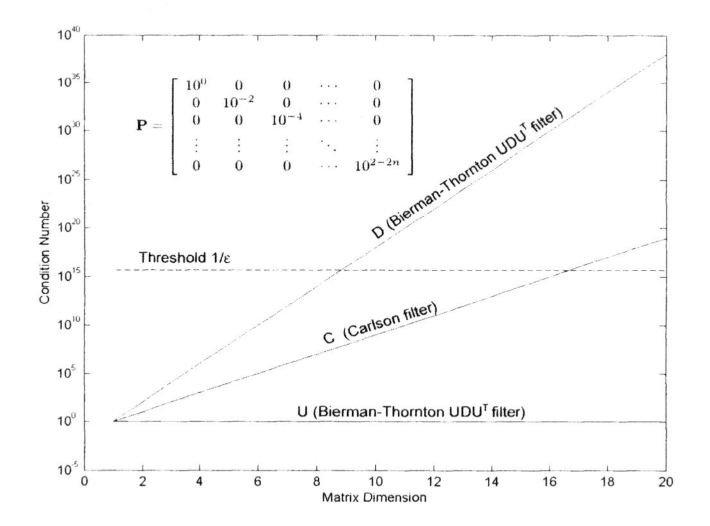

Fig. 8.23 Example 8.9: Conditioning of Cholesky factors.

$1+\epsilon > 1$ in machine precision, equal to 2^{-52} in the IEEE 64-bit precision used by MATLAB on most PCs). This threshold is labeled $1/\epsilon$ on the plot.

If the implementation is to be done in fixed-point arithmetic, scaling also becomes important. For this example, the nonzero elements of **D** and **C** will have the same relative dynamic ranges as the condition numbers.

8.9.4 Kalman Filter Monitoring

8.9.4.1 Rejecting Anomalous Sensor Data Anomalous sensor data can result from sensor failures or from corruption of the signals from sensors, and it is important to detect these events before the anomalous data corrupt the estimate. The filter is not designed to accept errors due to sensor failures or signal corruption, and they can seriously degrade the accuracy of estimates. The Kalman filter has infinite impulse response, so errors of this sort can persist for some time.

Detecting Anomalous Sensor Data Fortunately, the Kalman filter implementation includes parameters that can be used to detect anomalous data. The Kalman gain matrix

$$\overline{\mathbf{K}}_k = \mathbf{P}_k(-)\mathbf{H}_k^T \underbrace{(\mathbf{H}_k\mathbf{P}_k(-)\mathbf{H}_k^T + \mathbf{R}_k)^{-1}}_{\mathbf{Y}_{\nu k}} \tag{8.116}$$

includes the factor

$$\mathbf{Y}_{\nu k} = (\mathbf{H}_k\mathbf{P}_k(-)\mathbf{H}_k^T + \mathbf{R}_k)^{-1}, \tag{8.117}$$

the information matrix of innovations. The innovations are the measurement residuals

$$\boldsymbol{v}_k \stackrel{\text{def}}{=} \mathbf{z}_k - \mathbf{H}_k \hat{\mathbf{x}}_k(-), \tag{8.118}$$

the differences between the apparent sensor outputs and the predicted sensor outputs. The associated likelihood function for innovations is

$$\mathcal{L}(\boldsymbol{v}_k) = \exp(-\tfrac{1}{2}\boldsymbol{v}_k^T \mathbf{Y}_{\upsilon k} \boldsymbol{v}_k), \tag{8.119}$$

and the log-likelihood is

$$\log[\mathcal{L}(\boldsymbol{v}_k)] = -\boldsymbol{v}_k^T \mathbf{Y}_{\upsilon k} \boldsymbol{v}_k, \tag{8.120}$$

which can easily be calculated. The equivalent statistic

$$\chi^2 = \frac{\boldsymbol{v}_k^T \mathbf{Y}_{\upsilon k} \boldsymbol{v}_k}{\ell} \tag{8.121}$$

(i.e., without the sign change and division by 2, but divided by the dimension of $\boldsymbol{v}_k$) is nonnegative with a minimum value of zero. If the Kalman filter were perfectly modeled and all white-noise sources were Gaussian, this would be a chi-squared statistic with distribution as plotted in Fig. 8.24. An upper limit threshold value on χ^2 can be used to detect anomalous sensor data, but a practical value of that threshold should be determined by the operational values of χ^2, not the theoretical values. Thus, first its range of values should be determined by monitoring the system in operation, then a threshold value χ^2_{max} chosen such that the fraction of good data rejected when $\chi^2 > \chi^2_{\text{max}}$ will be acceptable.

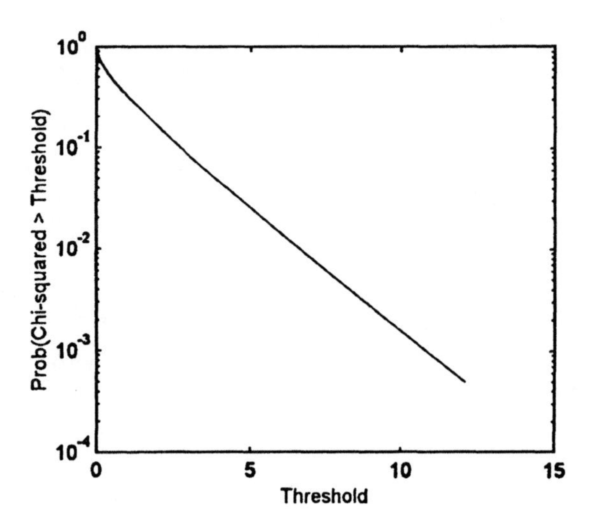

Fig. 8.24 Chi-squared distribution.

Exception Handling for Anomalous Sensor Data The log-likelihood test can be used to detect and reject anomalous data, but it can also be important to use the measurement innovations in other ways:

1. As a minimum, to raise an alarm whenever something anomalous has been detected
2. To tally the relative frequency of sensor data anomalies, so that trending or incipient failure may be detectable
3. To aid in identifying the source, such as which sensor or system may have failed.

8.9.4.2 Monitoring Filter Health Filter health monitoring methods are useful for detecting disparities between the physical system and the model of the system used in Kalman filtering (useful in filter development), for detecting numerical instabilities in the solution of the Riccati equation, and for detecting the onset of poor observability conditions. We have discussed in Section 8.9.4.1 the monitoring methods for detecting when sensors fail, or for detecting gradual degradation of sensors.

Covariance Analysis Covariance analysis in this context means monitoring selected diagonal terms of the covariance matrix $\mathbf{P}$ of estimation uncertainty. These are the variances of state estimation uncertainty. System requirements are often specified in terms of the variance or RMS uncertainties of key state variables, and this is a way of checking whether these requirements are being met. It is not always possible to cover all operational trajectories in the design of the sensor system. It is possible that situations can occur when these requirements are not being met in operation, and it can be useful to know that.

Checking Covariance Symmetry The so-called "square root" filtering methods presented in Section 8.9.3 are designed to ensure that the covariance matrix of estimation uncertainty (the dependent variable of the matrix Riccati equation) remains symmetric and positive definite. Otherwise, the fidelity of the solution of the Riccati equation can degrade to the point that it corrupts the Kalman gain, and that can corrupt the estimate. If you should choose not to use square-root filtering, then you may need some assurance that the decision was justified.

Verhaegen and Van Dooren [198] have shown that asymmetry of $\mathbf{P}$ is one of the factors contributing to numerical instability of the Riccati equation. If square-root filtering is not used, then the covariance matrix can be "symmetrized" occasionally by adding it to its transpose and rescaling:

$$\mathbf{P} := \tfrac{1}{2}(\mathbf{P} + \mathbf{P}^T). \tag{8.122}$$

This trick has been used for many years to head off numerical instabilities.

Checking Innovations Means and Autocorrelations Innovations are the differences between what comes out of the sensors and what was expected, based on the estimated system state. If the system were perfectly modeled in the Kalman filter, the innovations would be a zero-mean white-noise process and its autocorrelation function would be zero except at zero delay. The departure of the empirical autocorrelation of innovations from this model is a useful tool for analysis of mismodeling in real-world applications.

Calculation of Autocovariance and Autocorrelation Functions The mean of the innovations should be zero. If not, the mean must be subtracted from the innovations before calculating the autocovariance and autocorrelation functions of the innovations.

For vector-valued innovations, the autocovariance function is a matrix-valued function, defined as

$$\mathbf{A}_{\text{covar},k} \stackrel{\text{def}}{=} E_i \langle \Delta \mathbf{z}_{\upsilon,i} \Delta \mathbf{z}_{\upsilon,i+k}^T \rangle, \tag{8.123}$$

$$\Delta \mathbf{z}_{\upsilon,i} \stackrel{\text{def}}{=} \mathbf{z}_i - \mathbf{H}_i \hat{\mathbf{x}}_i(-) \text{ (innovations)}, \tag{8.124}$$

and the autocorrelation function is defined by

$$\mathbf{A}_{\text{correl},k} \stackrel{\text{def}}{=} \mathbf{D}_\sigma^{-1} \mathbf{A}_{\text{covar},k} \mathbf{D}_\sigma^{-1}, \tag{8.125}$$

$$\mathbf{D}_\sigma \stackrel{\text{def}}{=} \begin{bmatrix} \sigma_1 & 0 & 0 & \cdots & 0 \\ 0 & \sigma_2 & 0 & \cdots & 0 \\ 0 & 0 & \sigma_3 & \cdots & 0 \\ \vdots & \vdots & \vdots & \ddots & \vdots \\ 0 & 0 & 0 & \cdots & \sigma_\ell \end{bmatrix}, \tag{8.126}$$

$$\sigma_j \stackrel{\text{def}}{=} \sqrt{\left\{ A_{\text{covar},0} \right\}_{jj}}, \tag{8.127}$$

where the jth diagonal element $\left\{ A_{\text{covar},0} \right\}_{jj}$ of $\mathbf{A}_{\text{covar},0}$ is the variance of the jth component of the innovations vector.

Calculation of Spectra and Cross-Spectra The Fourier transforms of the diagonal elements of the autocovariance function $\mathbf{A}_{\text{covar},k}$ (i.e., as functions of k) are the power spectral densities (spectra) of the corresponding components of the innovations, and the Fourier transforms of the off-diagonal elements are the cross-spectra between the respective components.

Simple patterns to look for in interpretation of the results include the following:

1. Nonzero means of innovations may indicate the presence of uncompensated sensor output biases, or mismodeled output biases. The modeled variance of the bias may be seriously underestimated, for example.
2. Short-term means increasing or varying with time may indicate output noise that is a random walk or an exponentially correlated process.

3. Exponential decay of the autocorrelation functions is a reasonable indication of unmodeled (or mismodeled) random walk or exponentially correlated noise.
4. Spectral peaks may indicate unmodeled harmonic noise, but it could also indicate that there is an unmodeled harmonic term in the state dynamic model.
5. The autocovariance function at zero delay, $\mathbf{A}_{\text{covar},0}$, should equal $\mathbf{HPH}^T + \mathbf{R}$ for time-invariant or very slowly time-varying systems. If $\mathbf{A}_{\text{covar},0}$ is much bigger than $\mathbf{HPH}^T + \mathbf{R}$, it could indicate that $\mathbf{R}$ is too small or that the process noise $\mathbf{Q}$ is too small, either of which may cause $\mathbf{P}$ to be too small. If $\mathbf{A}_{\text{covar},0}$ is much smaller than $\mathbf{HPH}^T + \mathbf{R}$, than $\mathbf{R}$ and/or $\mathbf{Q}$ may be too large.
6. If the off-diagonal elements of $\mathbf{A}_{\text{correl},0}$ are much bigger than those of $\mathbf{D}_\sigma^{-1}\left(\mathbf{HPH}^T + \mathbf{R}\right)\mathbf{D}_\sigma^{-1}$, then there may be unmodeled correlations between sensor outputs. These correlations could be caused by mechanical vibration or power supply noise, for example.

PROBLEMS

8.1 Given the scalar plant and observation equations

$$x_k = x_{k-1}, \qquad z_k = x_k + v_k \sim N(0, \sigma_v^2)$$

and white noise with $\sigma_v^2 = 1$

$$E(x_0) = 1, \qquad E(x_0^2) = \mathbf{P}_0,$$

find the estimate of x_k and the steady-state covariance.

8.2 Given the vector plant and scalar observation equations

$$x_k = \begin{bmatrix} 1 & 1 \\ 0 & 1 \end{bmatrix} x_{k-1} + w_{k-1} \quad \text{(normal and white)},$$

$$z_k = [1 \;\; 0]x_k + v_k, \text{(normal and white)},$$

$$Ew_k = 0, \qquad Q_k = \begin{bmatrix} 0 & 0 \\ 0 & 1 \end{bmatrix},$$

$$Ev_k = 0, \qquad R_k = 1 + (-1)^k,$$

find the covariances and Kalman gains for $k = 10$, $\quad \mathbf{P}_0 = \begin{bmatrix} 10 & 0 \\ 0 & 10 \end{bmatrix}$.

8.3 Given

$$x_k = \begin{bmatrix} 1 & 1 \\ 0 & 1 \end{bmatrix} x_{k-1} + \begin{bmatrix} 1 \\ \frac{1}{2} \end{bmatrix} (-g),$$

$$z_k = [1\ 0]x_k + v_k \sim \text{normal and white},$$

where g is gravity, find $\hat{x}_1$, $\mathbf{P}_k(+)$ for $k = 6$:

$$\hat{x}_0 = \begin{bmatrix} 90 \\ 1 \end{bmatrix}, \quad \mathbf{P}_0 = \begin{bmatrix} 10 & 0 \\ 0 & 2 \end{bmatrix},$$

$$E(v_k) = 0, \quad E(v_k^2) = 2$$

$$Z_1 = 1$$

8.4 Given

$$x_k = -2x_{k-1} + w_{k-1},$$

$$z_k = x_k + v_k \sim \text{normal and white},$$

$$E(v_k) = 0, \quad E(v_k^2) = 1,$$

$$E(w_k) = 0, \quad E(w_k w_j) = e^{-|k-j|},$$

find the covariances and Kalman gains for $k = 3$, $\mathbf{P}_0 = 10$.

8.5 Given

$$E[(w_k - 1)(w_j - 1)] = e^{-|k-j|},$$

find the discrete time equation model for the sequence $\{w_k\}$.

8.6 Given

$$E\langle[w(t_1) - 1][w(t_2) - 1]\rangle = e^{-|t_1 - t_2|},$$

find the stochastic differential equation model.

8.7 Based on the 24-satellite GPS constellation, five satellite trajectories are selected, and their parameters tabulated accordingly:

$\alpha = 55°$		
Satellite ID	Ω_0 (deg)	Θ_0 (deg)
6	272.847	268.126
7	332.847	80.956
8	32.847	111.876
9	92.847	135.226
10	152.847	197.046

(a) Choose correctly phased satellites of four.

(b) Calculate DOPs to show their selection by plots for 5 satellites.

(c) Use Kalman filter equations for $\mathbf{P}_k(-)$, $\overline{K}_k$, and $\mathbf{P}_k(+)$ to show the errors. Draw the plots. This should be done with good GDOP for 4 and 5 satellites.

Choose user positions at (0, 0, 0) for simplicity. (see section 8.8.)

9

INERTIAL NAVIGATION SYSTEMS

A descriptive overview of the fundamental concepts of inertial navigation is presented in Chapter 2. The focus here is on the necessary details for INS implementation and GNSS/INS integration.

We begin with an overview of some of the technologies used for INS, and how these devices and systems are modeled mathematically. Next, we cover the mathematical modeling necessary for system implementation. Finally, we will delve into mathematical models characterizing how INS errors propagate over time. These last models will be essential for GNSS/INS integration.

9.1 INERTIAL SENSOR TECHNOLOGIES

A sampling of inertial sensor technologies used in inertial navigation is presented in Table 9.1. There are many more, but these will serve to illustrate the great diversity of technologies applied to inertial navigation. How these devices function will be explained briefly. A more thorough treatment of inertial sensor designs is given in [190].

The histories by Draper [47], Mackenzie [129], Mueller [144], and Wrigley [215] contain much more information on the history of inertial systems and sensor technologies and the individuals involved. See also [8, 63, 110, 168] and the cited references below for more historical and technical details.

9.1.1 Early Gyroscopes

9.1.1.1 Momentum Wheel Gyroscopes (MWGs) The earth itself is a giant momentum wheel gyroscope, the spin axis of which remains pointing at the pole

Global Positioning Systems, Inertial Navigation, and Integration, Second Edition, by M. S. Grewal, L. R. Weill, and A. P. Andrews

TABLE 9.1. Some Basic Inertial Sensor Technologies

Sensor	Gyroscope				Accelerometer	
Physical effect used	Conservation of angular momentum	Coriolis effect	Sagnac effect	Gyroscopic precession	Electromagnetic force	Strain under load
Sensor implementation methods	Angular displacement	Vibration	Ring laser	Angular displacement	Drag cup	Piezoelectric
	Torque rebalance	Rotation	Fiberoptic	Torque rebalance	Electromagnetic	Piezoresistive

star Polaris. The toy top is essentially a MWG that is probably older than recorded history. These are devices with stored angular momentum, a vector quantity that tends to remain constant and pointing in a fixed inertial direction unless disturbed by torques.

Jean Bernard Léon Foucault (1819–1868) had used a momentum wheel gyroscope to measure the rotation of the earth in 1852, and it was he who coined the term "gyroscope" from the Greek words for "turn" (γύρος) and "view" (σκοπός).

9.1.1.2 Gyrocompass Technology Gyroscope technology advanced significantly in the early 1900s, when gyrocompasses were developed to replace magnetic compasses, which would not work on iron ships (see Section 2.2.3.3). The first patent for a gyrocompass was issued to M. G. van den Bos in 1885, but the first known practical device was designed by Hermann Anschütz-Kaempfe[1] in 1903. It performed well in the laboratory but did not do well in sea trials—especially when the host ship was heading northeast-southwest or northwest–southeast in heavy seas. Anschütz-Kaempfe's cousin Maximilian Schuler analyzed its dynamics and determined that lateral accelerations due to rolling of the ship were the cause of the observed errors. Schuler also found that the gyrocompass suspension could be tuned to eliminate this error sensitivity [174]. This has come to be called *Schuler tuning*. It essentially tunes the pendulum period of the suspended mass to mimic a gravity pendulum with an effective arm length equal to the radius of curvature of the earth (called a *Schuler pendulum*). The period of this pendulum (called the *Schuler period*) equals the orbital period of a satellite at the same altitude (about 84.4 min at sea level).

Inertial navigators also experience oscillatory errors at the Schuler period, as described in Section 9.5.2.1.

9.1.1.3 Bearing Technologies The earliest gyroscopes used bearing technology from wheeled vehicles, industrial rotating machinery and clocks—including sleeve bearings, thrust bearings, jewel bearings and (later on) ball bearings. Bearing technologies developed explicitly for gyroscopes include the following:

[1]When the American inventor Elmer Sperry attempted to patent his gyrocompass in Europe in 1914, he was sued by Anschütz-Kaempfe. Sperry hired the former Swiss patent examiner Albert Einstein as an expert witness, but Einstein's testimony tended to support the claims of Anschütz-Kaempfe, which prevailed.

Dry-tuned gyroscopes use flexible coupling between the momentum wheel and its shaft, as illustrated in Fig. 9.1, with spring constants "tuned" so that the momentum wheel is effectively decoupled from bearing torques at the wheel rotation rate. This is not so much a bearing technology as a momentum wheel isolation technology. The very popular AN/ASN-141 military INS (USAF Standard Navigator) used dry-tuned gyroscopes.

Gas bearing gyroscopes support the momentum wheel on a spherical bearing, as illustrated in Fig. 9.2, with a thin layer of gas between the moving parts. Once operating, this type of bearing has essentially no wear. It was used in the gyroscopes of the USAF Minuteman I–III ICBMs, which were operated for years without being turned off.

Electrostatic gyroscopes (ESGs) have spherical beryllium rotors suspended by electrostatic forces inside a spherical cavity lined with suspension electrodes, as illustrated in Fig. 9.3. The electrostatic gyroscope in the USAF B-52 INS in the 1980s used optical readouts from a pattern etched on the hollow rotor to determine the direction of the rotor spin axis. The Electrically Supported Gyro Navigation (ESGN) system in USN Trident-class submarines in the late twentieth century used a solid rotor with deliberate radial mass unbalance, and determined the direction of the rotor spin axis from the effect this has on the suspension servo signals. Neither of these

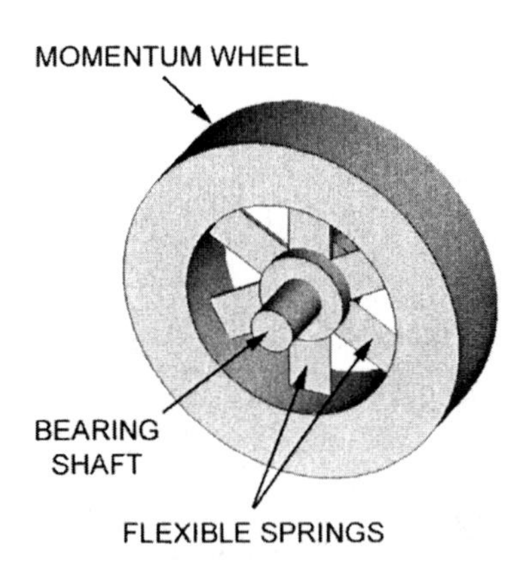

Fig. 9.1 Dry-tuned gyroscope.

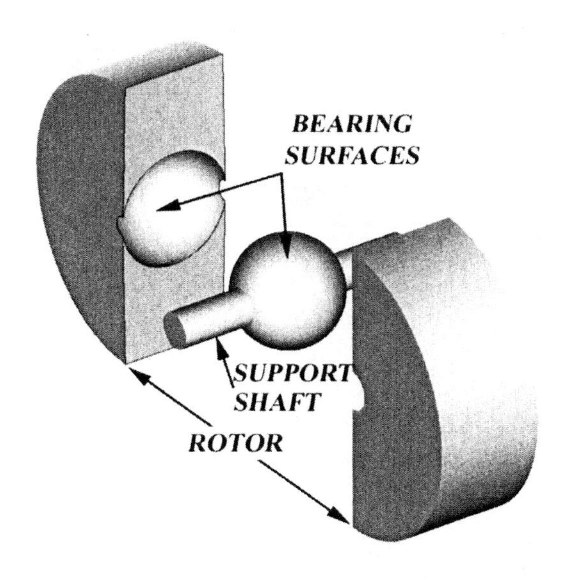

Fig. 9.2 Gas bearing gyroscope, with momentum wheel split to show bearing.

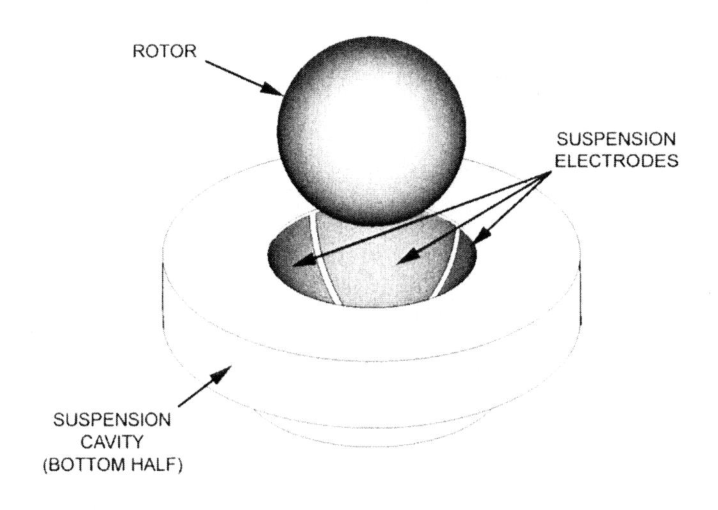

Fig. 9.3 Electrostatic gyroscope.

TABLE 9.2. Performance Grades for Gyroscopes

Performance Parameter	Performance Units	Performance Grades		
		Inertial	Intermediate	Moderate
Maximum input	deg/h	10^2–10^6	10^2–10^6	10^2–10^6
	deg/s	10^{-2}–10^2	10^{-2}–10^2	10^{-2}–10^2
Scale factor	part/part	10^{-6}–10^{-4}	10^{-4}–10^{-3}	10^{-3}–10^{-2}
Bias stability	deg/h	10^{-4}–10^{-2}	10^{-2}–10	10–10^2
	deg/s	10^{-8}–10^{-6}	10^{-6}–10^{-3}	10^{-3}–10^{-2}
Bias drift	deg/$\sqrt{h}$	10^{-4}–10^{-3}	10^{-2}–10^{-1}	1–10
	deg/$\sqrt{s}$	10^{-6}–10^{-5}	10^{-5}–10^{-4}	10^{-4}–10^{-3}

gyroscopes was torqued, except for induction torquing during spinup and spindown.

9.1.1.4 Gyroscope Performance Grades Gyroscopes used in inertial navigation are called "inertial grade," which generally refers to a range of sensor performance, depending on INS performance requirements. Table 9.2 lists some generally accepted performance grades used for gyroscopes, based on their intended applications but not necessarily including integrated GNSS/INS applications.

These are only rough order-of-magnitude ranges for the different error characteristics. Sensor requirements are determined largely by the application. For example, gyroscopes for gimbaled systems can generally use much smaller input ranges than can those for strapdown applications.

9.1.2 Early Accelerometers

Pendulum clocks were used in the eighteen century for measuring the acceleration due to gravity, but these devices were not usable on moving platforms.

9.1.2.1 Drag Cup Accelerometer An early "integrating" accelerometer design is illustrated in Fig. 9.4. It has two independent moving parts, both able to rotate on a common shaft axis:

1. A bar magnetic, the rotation rate of which is controlled by a dc motor.
2. A nonferrous metal "drag cup" that will be dragged along by the currents induced in the metal by the moving magnetic field, producing a torque on the drag cup that is proportional to the magnet rotation rate. The drag cup also has a proof mass attached to one point, so that acceleration along the "input axis" direction shown in the illustration will also create a torque on the drag cup.

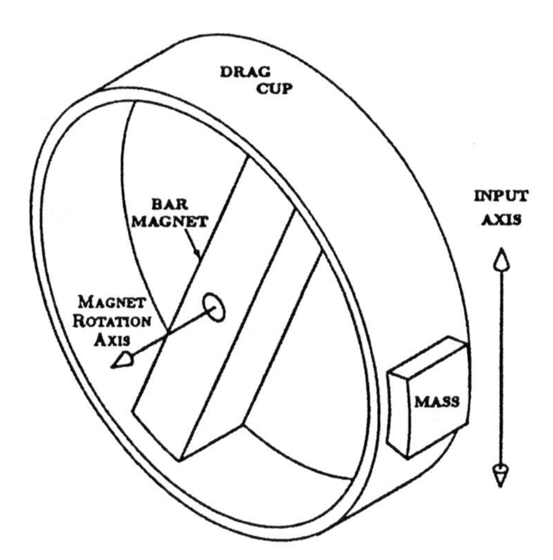

Fig. 9.4 Drag cup accelerometer.

The DC current to the motor is servoed to keep the drag cup from rotating, so that the magnet rotation rate will be proportional to acceleration and each rotation of the magnet will be proportional to the resulting velocity increment over that time period. At very low input accelerations (e.g., during gimbaled IMU leveling), inhomogeneities in the drag cup material can introduce harmonic noise in the output.

This same sort of drag cup, without the proof mass and with a torsion spring restraining the drag cup, has been used for decades for automobile speedometers. A flexible shaft from the drive wheels drove the magnet, so that the angular deflection of the drag cup would be proportional to speed.

9.1.2.2 Vibrating-Wire Accelerometer This is another early digital accelerometer design, with the output a frequency difference proportional to input acceleration.

The resonant frequencies of vibrating wires (or strings) depend upon the length, density, and elastic modulus of the wire and on the square of the tension in the wire. The motions of the wires must be sensed (e.g., by capacitance pickoffs) and forced (e.g., electrostatically or electromagnetically) to be kept in resonance. The wires can then be used as digitizing force sensors, as illustrated in Fig. 9.5. The configuration shown is for a single-axis accelerometer, but the concept can be expanded to a three-axis accelerometer by attaching pairs of opposing wires in three orthogonal directions.

In the "push–pull" configuration shown, any lateral acceleration of the proof mass will cause one wire frequency to increase and the other to decrease. Furthermore, if the preload tensions in the wires are servoed to keep the sum of their

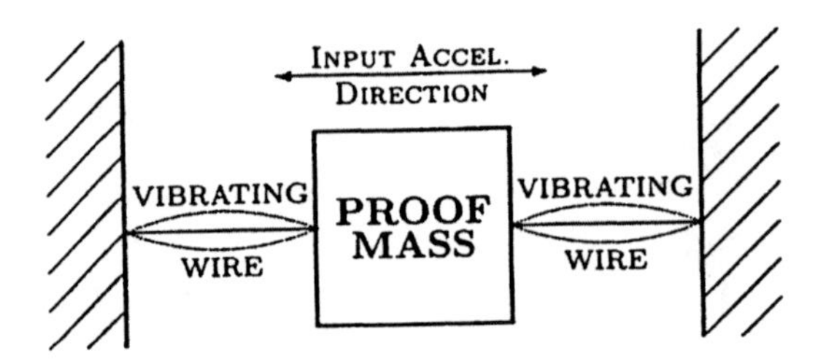

Fig. 9.5 Single-axis vibrating-wire accelerometer.

frequencies constant, then the difference frequency

$$\omega_{\text{left}}^2 - \omega_{\text{right}}^2 \propto ma, \tag{9.1}$$

$$\omega_{\text{left}} - \omega_{\text{right}} \propto \frac{ma}{\omega_{\text{left}} + \omega_{\text{right}}}, \tag{9.2}$$

$$\propto a. \tag{9.3}$$

Both the difference frequency $\omega_{\text{left}} - \omega_{\text{right}}$ and the sum frequency $\omega_{\text{left}} + \omega_{\text{right}}$ (used for preload tension control) can be obtained by mixing and filtering the two wire position signals from the resonance forcing servo loop. Each cycle of the difference frequency then corresponds to a constant delta velocity, making the sensor inherently digital.

9.1.2.3 Gyroscopic Accelerometers Some of the earlier designs for accelerometers for inertial navigation used the acceleration-sensitive precession of momentum wheel gyroscopes, as illustrated in Fig. 9.6.

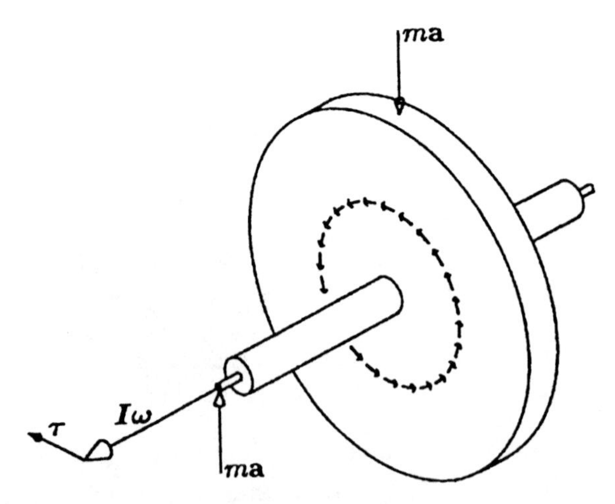

Fig. 9.6 Acceleration-sensitive precession of momentum wheel gyroscope.

This has the center of support offset from the center of mass of the momentum wheel, a condition known as "mass unbalance." For a mass-unbalanced design like the one shown in the figure, precession rate will be proportional to acceleration. If the angular momentum and mass offset of the gyro can be kept constant, this relationship will extremely linear over several orders of magnitude.

Gyroscopic accelerometers are integrating accelerometers. Angular precession rate is proportional to acceleration, so the change in precession angle will be proportional to velocity change along the input axis direction.

Accelerometer designs based on gyroscopic precession are still used in the most accurate floated system [129].

9.1.2.4 Accelerometer Performance Ranges Table 9.3 lists accelerometer and gyroscope performance ranges compatible with the INS performance ranges listed in Chapter 2, Section 2.2.4.3.

9.1.3 Feedback Control Technology

9.1.3.1 Feedback Control Practical inertial navigation began to evolve in the 1950s, using technologies that had evolved in the early twentieth century, or were coevolving throughout the mid-twentieth century. These technologies include classical control theory and feedback[2] control technology. Its mathematical underpinnings included analytic function theory, Laplace transforms and Fourier transforms, and its physical implementations were dominated by analog electronics and electromagnetic transducers.

Feedback Control Servomechanisms Servomechanisms are electronic and/or electromechanical devices for implementing feedback control. The term "servo" is often used as a noun (short for *servomechanism*), adjective (e.g., "servo control") and verb (meaning to control with a servomechanism).

TABLE 9.3. INS and Inertial Sensor Performance Ranges

System or Sensor	Performance Ranges: High	Medium	Low	Units
INS	$\leq 10^{-1}$	≈ 1	≥ 10	nmi/h[a]
Gyroscopes	$\leq 10^{-3}$	$\approx 10^{-2}$	$\geq 10^{-1}$	deg/h
Accelerometers	$\leq 10^{-7}$	$\approx 10^{-6}$	$\geq 10^{-5}$	g (9.8 m/s^2)

[a]Nautical miles per hour CEP rate (defined in Section 2.2.4.2).

[2]A concept discovered by Harold S. Black (1898–1983) in 1927 and applied to the operational amplifier [20]. According to his own account, the idea of negative feedback occurred to Black when he was commuting to his job at the West Street Laboratories of Western Electric in New York (later part of Bell Labs) on the Hudson River Ferry. He wrote it down on the only paper available to him (a copy of the *New York Times*), dated it and signed it, and had it witnessed and signed by a colleague when he arrived at work. His initial patent application was refused by the U.S. Patent Office on the grounds that it was a "perpetual motion" device.

Transducers These are devices which convert measurable physical quantities to electrical signals, and vice versa. Early servo transducers for INS included analog shaft angle encoders (angle to signal) and torquers (signal to torque) for gimbal bearings, and torquers for controlling the direction of angular momentum in momentum wheel gyroscopes.

9.1.3.2 Gimbal Control In the mid-1930s, Robert H. Goddard used momentum wheel gyroscopes for feedback attitude control of rockets, and gyros and accelerometers were used for missile guidance in Germany during World War II [8]. These technologies (along with many of their developers) were transferred to the United States and the Soviet Union immediately after the war [47, 215].

All feedback control loops are used to null something, usually the difference between some reference signal and a measured signal. Servos are used in gimbaled systems for controlling the gimbals to keep the gyro outputs at specified values (e.g., earthrate), which keeps the ISA in a specified orientation relative to navigation coordinates, independent of host vehicle dynamics.

9.1.3.3 Torque Feedback Gyroscopes These use a servo loop to apply just enough torque on the momentum wheel to keep the spin axis from moving relative to its enclosure, and use the applied torque (or the motor current required to generate it) as a measure of rotation rate. If the feedback torque is delivered in precisely repeatable pulses, then each pulse represents a fixed angular rotation $\delta\theta$, and the pulse count in a fixed time interval Δt will be proportional to the net angle change $\Delta\theta$ over that time period (plus quantization error). The result is a digital integrating gyroscope.

Pulse Quantization Quantization pulse size determines quantization error, and smaller quantization levels are preferred. The feedback pulse quantization size also has an effect on outer control loops, such as those used for nulling the east gyro output to align a gimbaled IMU in heading. When the east gyro output is close to being nulled and its pulse rate approaches zero, quantization pulse size will determine how long one has to wait for an LSB of the gyro output.

9.1.3.4 Torque Feedback Gyroscopic Accelerometers These use a torque feedback loop to keep the momentum wheel rotation axis in a gyroscopic accelerometer from moving relative to the instrument housing. In the *pulse integrating gyroscopic accelerometer* (PIGA), the feedback torque is delivered in repeatable pulses. Each pulse then represents a fixed velocity change δv along the input acceleration axis, and the net velocity change during a fixed time interval Δt will be proportional to the pulse count in that period.

Nulling the outputs of the north and east accelerometers of a gimbaled IMU during leveling is affected by pulse quantization the same way that nulling the east gyro outputs is influenced by quantization (see "Pulse Quantization" above).

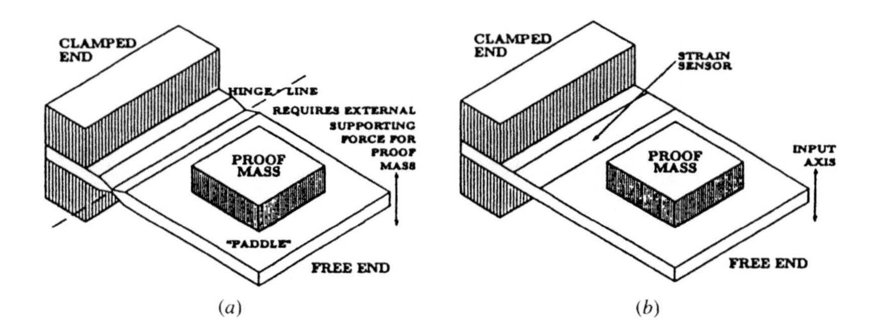

Fig. 9.7 Pendulous (*a*) and beam (*b*) accelerometers.

9.1.3.5 Force Feedback Accelerometers Except for gyroscopic accelerometers, all other practical accelerometers measure (in various ways) the specific force required to make a proof mass follow the motions of the host vehicle.[3]

Pendulous Accelerometers One of the design challenges for accelerometers is how to support a proof mass rigidly in two dimensions and allow it to be completely free in the third dimension. Pendulous accelerometers use a hinge to support the proof mass in two dimensions, as illustrated in Fig. 9.7*a*, so that it is free to move only in the input axis direction, normal to the "paddle" surface. This design requires an external supporting force to keep the proof mass from moving in that direction, and the force required to do it will be proportional to the acceleration that would otherwise be disturbing the proof mass.

Electromagnetic Accelerometer (EMA) Electromagnetic accelerometers (EMAs) are pendulous accelerometers using electromagnetic force to keep the paddle from moving. A common design uses a voice coil attached to the paddle, as illustrated in Fig. 9.8. Current through the voice coil provides the force on the proof mass to keep the paddle centered in the instrument enclosure. This is similar to the speaker cone drive in permanent magnet speakers, with the magnetic flux through the coils provided by permanent magnets. The coil current is controlled through a feedback servo loop including a paddle position sensor such as a capacitance pickoff. The current in this feedback loop through the voice coil will be proportional to the disturbing acceleration.

Integrating Accelerometers For *pulse-integrating accelerometers*, the feedback current is supplied in discrete pulses with very repeatable shapes, so that each pulse is proportional to a fixed change in velocity. An up/down counter keeps

[3]This approach was turned inside-out around 1960, when satellites designed to measure low levels of atmospheric drag at the outer edges of the atmosphere used a free-floating proof mass inside the satellite, protected from drag forces, and measured the thrust required to make the satellite follow the drag-free proof mass.

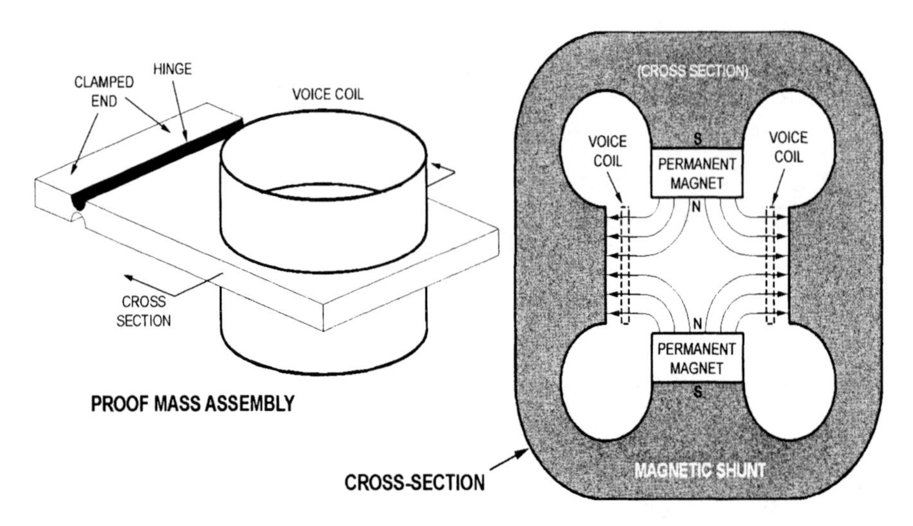

Fig. 9.8 Electromagnetic accelerometer (EMA).

track of the net pulse count between samples of the digitized accelerometer output. The pulse feedback electromagnetic accelerometer is an integrating accelerometer, in that each pulse output corresponds to a constant increment in velocity δv. The electromagnetic accelerometer (EMA) illustrated in Fig. 9.8 is another type of integrating accelerometer, similar to the PIGA, as is the beam accelerometer of Fig. 9.7(*b*) with SAW strain sensor.

9.1.4 Rotating Coriolis Multisensors

9.1.4.1 Coriolis Effect Gustav Gaspard de Coriolis (1792–1843) published a report in 1835 [39] describing the effects of coordinate rotation on Newton's laws of motion. Bodies with no applied acceleration maintain constant velocity in nonrotating coordinates, but appear to experience additional apparent accelerations in rotating frames. The "Coriolis effect" is an apparent acceleration of the form

$$\mathbf{a}_{\text{coriolis}} = -2\boldsymbol{\omega} \otimes \mathbf{v}_{\text{rotating}}, \tag{9.4}$$

where $\boldsymbol{\omega}$ is the coordinate rotation rate vector, $\otimes$ represents the vector cross-product, and $\mathbf{v}_{\text{rotating}}$ is the velocity of the body measured in rotating coordinates.

9.1.4.2 Rotating Coriolis Gyroscope These are gyroscopes that measure the coriolis acceleration on a rotating wheel. An example of such a two-axis gyroscope is illustrated in Fig. 9.9. For sensing rotation, it uses an accelerometer mounted off axis on the rotating member, with its acceleration input axis parallel to the rotation axis of the base. When the entire assembly is rotated about any axis normal to its own rotation axis, the accelerometer mounted on the rotating base senses a sinusoidal coriolis acceleration.

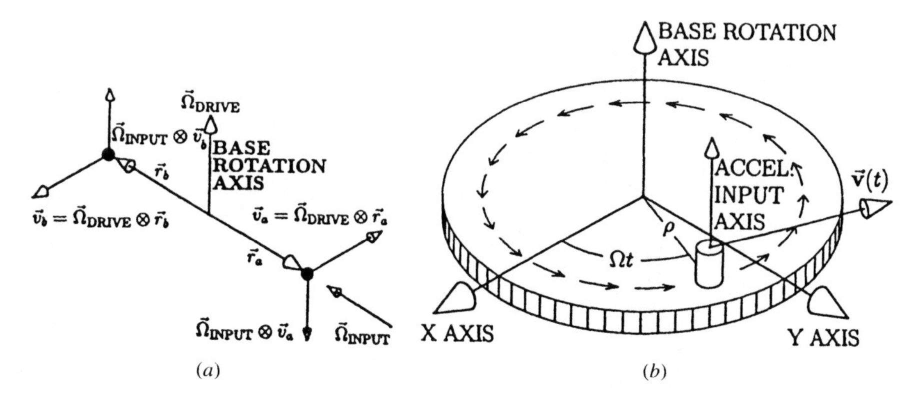

Fig. 9.9 Rotating coriolis gyroscope: (*a*) Function; (*b*) sensing.

The position **x** and velocity **v** of the rotated accelerometer with respect to inertial coordinates will be

$$\mathbf{x}(t) = \rho \begin{bmatrix} \cos(\Omega_{\text{drive}} t) \\ \sin(\Omega_{\text{drive}} t) \\ 0 \end{bmatrix}, \tag{9.5}$$

$$\mathbf{v}(t) = \frac{d}{dt}\mathbf{x}(t), \tag{9.6}$$

$$= \rho\Omega_{\text{drive}} \begin{bmatrix} -\sin(\Omega_{\text{drive}} t) \\ \cos(\Omega_{\text{drive}} t) \\ 0 \end{bmatrix}, \tag{9.7}$$

where Ω_{drive} is the drive rotation rate and ρ is the offset distance of the accelerometer from the base rotation axis.

The input axis of the accelerometer is parallel to the rotation axis of the base, so it is insensitive to rotations about the base rotation axis (z axis). However, if this apparatus is rotated with components $\Omega_{x,\text{ input}}$ and $\Omega_{y,\text{ input}}$ orthogonal to the z axis, then the coriolis acceleration of the accelerometer will be the vector cross-product

$$\mathbf{a}_{\text{coriolis}}(t) = -2 \begin{bmatrix} \Omega_{x,\text{ input}} \\ \Omega_{y,\text{ input}} \\ 0 \end{bmatrix} \otimes \mathbf{v}(t) \tag{9.8}$$

$$= -2\rho\Omega_{\text{drive}} \begin{bmatrix} \Omega_{x,\text{ input}} \\ \Omega_{y,\text{ input}} \\ 0 \end{bmatrix} \otimes \begin{bmatrix} -\sin(\Omega_{\text{drive}} t) \\ \cos(\Omega_{\text{drive}} t) \\ 0 \end{bmatrix} \tag{9.9}$$

$$= 2\rho\, \Omega_{\text{drive}} \begin{bmatrix} 0 \\ 0 \\ -\Omega_{x,\text{ input}} \cos(\Omega_{\text{drive}} t) + \Omega_{y,\text{ input}} \sin(\Omega_{drive} t) \end{bmatrix} \tag{9.10}$$

The rotating z-axis accelerometer will then sense the z-component of coriolis acceleration,

$$a_{z,\,\text{input}}(t) = \rho\Omega_{\text{drive}}[\Omega_{x,\,\text{input}}\cos(\Omega_{drive}t) - \Omega_{y,\,\text{input}}\sin(\Omega_{\text{drive}}t)], \tag{9.11}$$

which can be demodulated to recover the phase components $\rho\,\Omega_{\text{drive}}\Omega_x$ (in phase) and $\rho\,\Omega_{\text{drive}}\Omega_{y,\,\text{input}}$ (in quadrature), each of which is proportional to a component of the input rotation rate. Demodulation of the accelerometer output removes the DC bias, so this implementation is insensitive to accelerometer bias errors.

9.1.4.3 Rotating Multisensor Another accelerometer can be mounted on the moving base of the rotating coriolis gyroscope, but with its input axis tangential to its direction of motion. Its outputs can be demodulated in similar fashion to implement a two-axis accelerometer with zero effective bias error. The resulting multisensor is a two-axis gyroscope and two-axis accelerometer.

9.1.5 Laser Technology and Lightwave Gyroscopes

Lasers are phase-coherent light sources. Phase-coherent light traveling around a closed planar path will experience a slight phase shift each lap that is proportional to the inertial rotation rate of its planar path (the Sagnac effect). Lightwave gyroscopes compare the phases of two phase-coherent beams traveling in opposite directions around the same path. All require mechanical stability to optical tolerances, and all exhibit some level of angle random walk.

The two common types of laser gyroscopes are illustrated in Fig. 9.10 and described below.

9.1.5.1 Ring Laser Gyroscopes (RLGs) Ring laser gyroscopes use a lasing segment within a closed polygonal light path with mirrors at the corners. These are effectively digital rate integrating gyroscopes, with the phase rate between the counterrotating beams proportional to inertial rotation rate.

The first ring laser gyroscopes were developed in the 1960s, soon after the first practical lasers had been developed. It would take about a decade to make them

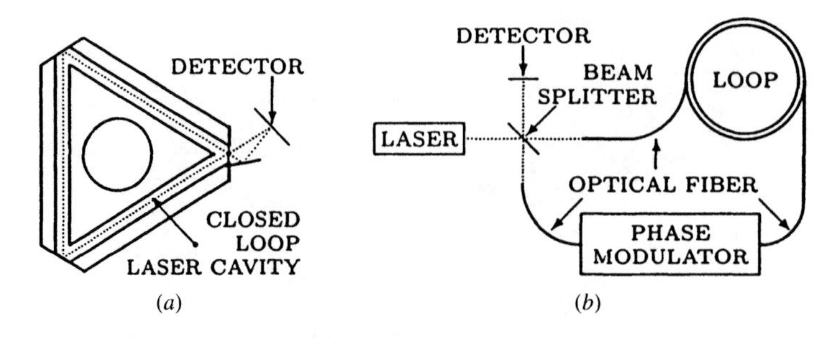

Fig. 9.10 Laser gyroscopes: (*a*) ring laser gyro (RLG); (*b*) fiberoptic gyro.

practical, however. The problem was scattering of the counterrotating beams off the RLG mirrors that causes the frequencies of the two beams to "lock in," creating a serious dead-zone near zero input rate. Most practical RLG designs employ some sort of dithering to avoid the dead zone. A *Zero Lock Gyroscope*[4] (ZLG) uses a combination of laser frequency splitting and out-of-plane path segments to eliminate lockin.

9.1.5.2 Fiberoptic Gyroscopes (FOGs) Fiberoptic gyroscopes use thousands of turns of optical fiber to increase the phase sensitivity, multiplying the number of turns by the phase shift per turn. A common external laser source can be used for both beams. There are two basic strategies for sensing rotation rates:

Open-Loop FOGs Open-loop designs compare the phases of the two counterrotating beams. They are effectively rate gyroscopes, with the relative phase change between the counterrotating light beams proportional to the inertial rotation rate normal the plane of the lightpath. Those used in inertial navigation typically have dynamic ranges in the order of 10^3, sensitivities (i.e., minimum detectable inputs) $\geq 10^{-2}$ degrees per hour and bias stabilities in the order of 1 degree per hour (or more) [196].

Closed-Loop Integrating FOGs (IFOGs) Closed-loop designs use feedback of the output phase to a light modulator in the loop to null the output. These are effectively rate integrating gyroscopes. They can have dynamic ranges in the order of 10^6, nonlinearity errors $\leq 10^{-5}$, and bias stability in the order of 10^{-2} degrees per hour (or better) [196].

9.1.6 Vibratory Coriolis Gyroscopes (VCGs)

9.1.6.1 VCG Principles The first functional vibrational coriolis gyroscope was probably the 1851 Foucault pendulum, in which the coriolis effect from the rotation of the earth causes the plane of pendulum mass oscillation to rotate. Modern designs have replaced the gravity pendulum (which does not travel well) with relatively high-frequency mechanical resonances, but the principle of operation remains essentially the same. A mass particle resonating with velocity $\mathbf{v}_0 \cos(\Omega t)$ fixed to a body rotating at rate $\boldsymbol{\omega}_{\text{input}}$ would experience a time-varying coriolis acceleration

$$\mathbf{a}_{\text{coriolis}}(t) = \left[-2\boldsymbol{\omega}_{\text{input}} \otimes \mathbf{v}_0\right] \cos(\Omega t), \tag{9.12}$$

which is at the same frequency as the driving acceleration, but at right angles to the particle velocity.

Equation 9.12 only describes the mass particle acceleration due to rotation. There are dozens of electromechanical gyro designs using this mass particle dynamic model to provide an output signal proportional to rotation rate [88].

[4]"Zero Lock Gyro" and "ZLG" are trademarks of Northrop Grumman Corp.

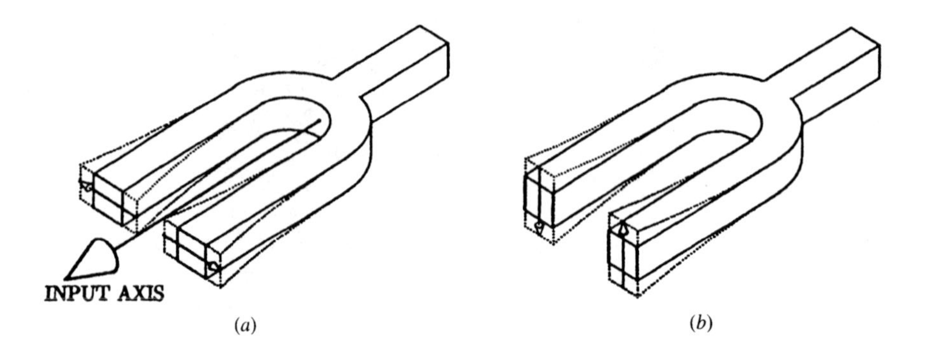

Fig. 9.11 Vibration modes of tuning fork gyroscope: (*a*) input mode; (*b*) output mode.

Transmission of vibrational energy to the supporting structure is a significant error mechanism for most vibratory sensors. Two designs which do provide good vibration isolation are the tuning fork gyro and the hemispherical resonator gyro.

9.1.6.2 Tuning Fork Gyroscope A tuning fork gyro illustrated in Fig. 9.11 is driven in a balanced vibration mode with its tines coming together and apart in unison (Fig. 9.11*a*), creating no vibrational stress in the handle. Its sensitive axis is parallel to the handle. Rotation about this axis is orthogonal to the direction of tine velocity, and the resulting coriolis acceleration will be in the direction of $\omega \otimes v$, which excites the output vibration mode shown in Fig. 9.11*b*. This unbalanced "twisting" mode will create a torque couple through the handle, and some designs use a double-ended fork to transfer this mode to a second set of output tines.

9.1.6.3 Hemispherical Resonator Gyroscope (HRG) In 1890, physicist G. H. Bryan observed that the nodes of the resonant modes of wine glasses (produced by rubbing a wet finger around the rim) precessed when the wine glass was rotated. This became the basis for the hemispherical resonator gyroscope (HRG) (also called the "wine glass" gyroscope), which uses resonant modes of bowl-shaped structures on stems (similar to a wine glass) with vibratory displacements normal to the edges of the bowl. When the device is rotated about its stem axis (its input axis), the nodes of the vibration modes rotate around the stem at a rate proportional to input rotation rate. Like many gyroscopes, they exhibit angle random walk ($< 10^{-3}$ degree per root hour in some HRG designs). They are very rugged. They can be made to operate through the radiation bursts from nuclear events, because mechanical resonances will persist and the phase change will continue to accumulate during periods of high neutron fluence when the drive and sensing electronics are turned off, then recover the accumulated angular displacements after turnon. (Most momentum wheel gyroscopes can do this, too.)

9.1.7 MEMS Technology

Microelectromechanical systems (MEMS) evolved from silicon semiconductor manufacturing in the late 1970s as an inexpensive mass manufacturing technology for sensors at sub-millimeter scales. At these scales, the ratio of surface area to volume becomes enormous, and electrostatic forces are significant. Vibration frequencies also scale up as size shrinks, and this makes vibratory coriolis gyroscopes very effective at MEMS scales. Electrostatic or piezoelectric forcing is used in most MEMS vibratory coriolis gyroscopes.

9.1.7.1 Open-Loop MEMS Accelerometers Many MEMS accelerometers are "open loop," in the sense that no force-feedback control loop is used on the proof mass. The cantilever beam accelerometer design illustrated in Fig. 9.7*b* senses the strain at the root of the beam resulting from support of the proof mass under acceleration load. The surface strain near the root of the beam will be proportional to the applied acceleration. This type of accelerometer can be manufactured relatively inexpensively using MEMS technologies, with a surface strain sensor (e.g., piezoelectric capacitor or ion implanted piezoresistor) to measure surface strain.

9.1.7.2 Rotational Vibratory Coriolis Gyroscope (RVCG) Many vibratory coriolis gyroscopes are MEMS devices. The rotational vibratory coriolis gyroscope is a MEMS device first developed at C. S. Draper Laboratory in the 1980s, then jointly with industry. It uses a momentum wheel coupled to a torsion spring and driven by a rotational electrostatic "comb drive" at resonance to create sinusoidal angular momentum in the wheel. If the device is turned about any axis in the plane of the wheel, the coriolis effect will introduce sinusoidal tilting about the orthogonal axis in the plane of the wheel, as illustrated in Fig. 9.12*a*. This sinusoidal tilting is sensed by four capacitor sensors in close proximity to the wheel underside, as illustrated in Fig. 9.12*b*. All the supporting electronics for controlling the momentum wheel and extracting the angular rate measurements fits on an application-specific integrate circuit (ASIC) that is only slightly bigger than the active device.

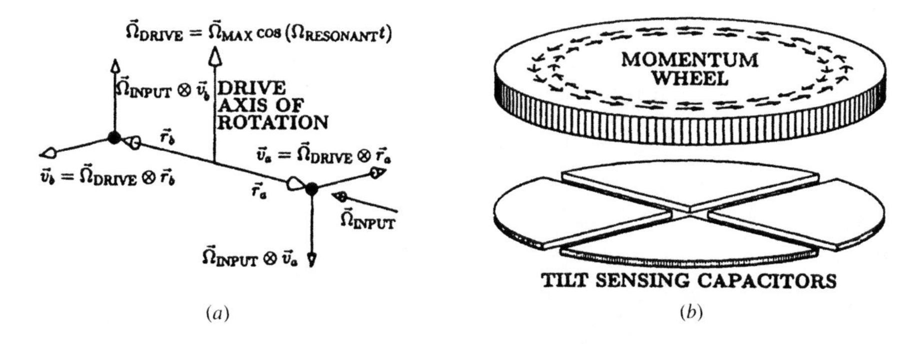

Fig. 9.12 Rotational vibratory coriolis gyroscope: (*a*) Function; (*b*) sensing.

9.2 INERTIAL SYSTEMS TECHNOLOGIES

9.2.1 Early Requirements

The development of inertial navigation in the United States started around 1950, during the Cold War with the Soviet Union. Cold War weapons projects in the United States that would need inertial navigation included the following:

1. *Long-range bombers* could not rely on radionavigation technologies of World War II for missions into the Soviet Union, because they could easily be jammed. Efforts started around 1950 and led by Charles Stark Draper in the Servomechanisms Laboratory at MIT (now the C. S. Draper Laboratory) were focused on developing airborne inertial navigation systems. The first successful flights with INS[5] across the United States had terminal errors in the order of a few kilometers.
2. Hymnan Rickover began studying and promoting nuclear propulsion for the U.S. Navy immediately after World War II, leading to the development of the first nuclear submarine, *Nautilus,* launched in 1954. Nuclear submarines would be able to remain submerged for months, and needed an accurate navigation method that did not require exposing the submarine to airborne radar detection.
3. The *Navaho Project* started in the early 1950s to develop a long-range air-breathing supersonic cruise missile to carry a 15,000-lb payload (the atomic bomb of that period) 5500 miles with a terminal navigation accuracy of about one nautical mile (1.85 km). The prime contractor, North American Aviation, developed an INS for this system. The project was canceled in 1957, when nuclear weaponry and rocketry had improved to the point that thermonuclear devices could be carried on rockets. However, a derivative of the Navaho INS survived, and the nuclear submarine *Nautilus* used it to successfully navigate under the arctic ice cap in 1958.
4. The intercontinental ballistic missiles (ICBMs) that replaced Navaho would not have been practical without an INS to guide them. The INS made each missile self-contained and able to control itself from liftoff without any external aiding. Their accuracy requirements were not far from those for Navaho.

The accuracy requirements of many of these systems was determined by the radius of destruction of nuclear weapons.

9.2.2 Computer Technology

9.2.2.1 Early Computers Computers in 1950 used vacuum-tube technology, and they tended to be expensive, large, heavy, and power-hungry. Computers

[5]One of these systems did not use gimbal rings, but had the INS stabilized platform mounted on a spherical bearing [129].

of that era occupied large rooms or whole buildings [160], they were extremely slow by today's standards, and the cost of each bit of memory was in the order of a U.S. dollar in today's money.

9.2.2.2 The Silicon Revolution Silicon transistors and integrated circuits began to revolutionize computer technology in the 1950s. Some early digital inertial navigation systems of that era used specialized digital differential analyzer circuitry to doubly integrate acceleration. Later in the 1950s, INS computers used magnetic core or magnetic drum memories. The Apollo moon missions (1969–1972) used onboard computers with magnetic core memories, and magnetic memory technology would dominate until the 1970s. The cost per bit would drop as low as a few cents before semiconductor memories came along. Memory prices fell by several orders of magnitude in the next few decades.

9.2.2.3 Impact on INS Technology Faster, cheaper computers enabled the development of strapdown inertial technology. Many vehicles (e.g., torpedos) had been using strapdown gyroscopes for steering control since the early twentieth century, but now they could be integrated with accelerometers to make a strapdown INS. This eliminated the need for expensive gimbals, but it also required considerable progress in attitude estimation algorithms [23]. Computers also enabled "modern" estimation and control, based on state space models. This would have a profound effect on sensor integration capabilities for INS.

9.2.3 Early Strapdown Systems

A gimbaled INS was carried on each of nine Apollo command modules from the earth to the moon and back between December 1968 and December 1972, but a strapdown INS was carried on each of the six[6] Lunar Excursion Modules (LEMs) that shuttled two astronauts from lunar orbit to the lunar surface and back.

By the mid-1970s, strapdown systems could demonstrate "medium" accuracy (1 nmi/h CEP). A strapdown contender for the U.S. Air Force Standard Navigator contract in 1979 was the Autonetics[7] N73 navigator, using electrostatic gyroscopes, electromagnetic accelerometers and a navigation computer with microprogrammed instructions and nonvolatile"magnetic wire" memory. In that same time period, the first INSs with ring laser gyros appeared in commercial aircraft.

A few years later, GPS appeared.

[6]Two additional LEMs were carried to the moon but did not land there. The Apollo 13 LEM did not make its intended lunar landing, but played a far more vital role in crew survival.

[7]Then a division of Rockwell International, now part of the Boeing Company.

9.2.4 INS and GNSS

9.2.4.1 Advantages of INS The main advantages of inertial navigation over other forms of navigation are as follows:

1. It is *autonomous* and does not rely on any external aids or on visibility conditions. It can operate in tunnels or underwater as well as anywhere else.
2. It is inherently well suited for integrated navigation, guidance, and control of the host vehicle. Its IMU measures the derivatives of the variables to be controlled (e.g., position, velocity, attitude).
3. It is immune to jamming and inherently stealthy. It neither receives nor emits detectable radiation and requires no external antenna that might be detectable by radar.

9.2.4.2 Disadvantages of INS These include the following:

1. Mean-squared navigation errors increase with time.
2. Cost, including
   (a) Acquisition cost, which can be an order of magnitude (or more) higher than that of GNSS receivers.
   (b) Operations cost, including the crew actions and time required for initializing position and attitude. Time required for initializing INS attitude by gyrocompass alignment is measured in minutes. Time to first fix for GNSS receivers is measured in seconds.
   (c) Maintenance cost. Electromechanical avionics systems (e.g., INS) tend to have higher failure rates and repair costs than do purely electronic avionics systems (e.g., GNSS).
3. Size and weight, which have been shrinking:
   (a) Earlier INS systems weighed tens to hundreds of kilograms.
   (b) Later "mesoscale" INSs for integration with GPS weighed 1–10 kgms.
   (c) Developing MEMS sensors are targeted for gram-size systems.

   INS weight has a multiplying effect on vehicle system design, because it requires increased structure and propulsion weight as well.
4. Power requirements, which have been shrinking along with size and weight but are still higher than those for GPS receivers.
5. Temperature control and heat dissipation, which is proportional to (and shrinking with) power consumption.

9.2.4.3 Competition from GPS Since the 1970s, U.S. commercial air carriers have been required by FAA regulations to carry two INS systems on all flights over water. The cost of these two systems is on the order of 10^5 U.S. dollars. The relatively high cost of INS was one of the factors leading to the development of

GPS. After deployment of GPS in the 1980s, the few remaining applications for "standalone" (i.e., unaided) INS include submarines, which cannot receive GPS signals while submerged, and intercontinental ballistic missiles, which cannot rely on GPS availability in time of war.

9.2.4.4 Synergism with GNSS GNSS integration has not only made inertial navigation perform better, it has made it cost less. Sensor errors that were unacceptable for stand-alone INS operation became acceptable for integrated operation, and the manufacturing and calibration costs for removing these errors could be eliminated. Also, new low-cost MEMS manufacturing methods could be applied to meet the less stringent sensor requirements for integrated operation.

The use of integrated GNSS/INS for mapping the gravitational field near the earth's surface has also enhanced INS performance by providing more detailed and accurate gravitational models.

Inertial navigation also benefits GNSS performance by carrying the navigation solution during loss of GNSS signals and allowing rapid reacquisition when signals become available.

Integrated systems have found applications that neither GNSS nor INS could perform alone. These include low-cost systems for precise autonomous control of vehicles operating at the surface of the earth, including automatic landing systems for aircraft and autonomous control of surface mining equipment, surface grading equipment, and farm equipment.

9.3 INERTIAL SENSOR MODELS

Mathematical models for how inertial sensors perform are used throughout the INS development cycle. They include the following:

1. Models used in designing the sensors to meet specified performance metrics.
2. Models used to calibrate and compensate for fixed errors, such as scale factor and bias variations. The extreme performance requirements for inertial sensors cannot be met within manufacturing tolerances. Fortunately, the last few orders of magnitude improvement in performance can be achieved through calibration. These models are generally of two types:
   (a) Models based on engineering data and the principles of physics, such as the models carried over from the design trade offs. These models generally have a known cause for each observed effect.
   (b) Abstract, general-purpose mathematical models such as polynomials, used to fit observed error data in such a way that the sensor output errors can be effectively corrected.

3. Error models used in GNSS/INS integration for determining the optimal weighting (Kalman gain) in combining GNSS and INS navigation data.
4. Sensor models used in GNSS/INS integration for recalibrating the INS continuously while GNSS data are available. This approach allows the INS to operate more accurately during periods of GNSS signal outage.

9.3.1 Zero-Mean Random Errors

These are the standard types of error models used in Kalman filtering, described in the previous chapter.

9.3.1.1 White Sensor Noise This is usually lumped together under "electronic noise," which may come from power supplies, intrinsic noise in semiconductor devices, or from quantization errors in digitization.

9.3.1.2 Exponentially Correlated Noise Temperature sensitivity of sensor bias will often look like a time-varying additive noise source, driven by external ambient temperature variations or by internal heat distribution variations.

9.3.1.3 Random-Walk Sensor Errors Random-walk errors are characterized by variances that grow linearly with time and power spectral densities that fall off as $1/\text{frequency}^2$ (i.e., 20 dB per decade; see Section 8.5.1.2).

There are specifications for random walk noise in inertial sensors, but mostly for the integrals of their outputs, and not in the outputs themselves. For example, the "angle random walk" from a rate gyroscope is equivalent to white noise in the angular rate outputs. In similar fashion, the integral of white noise in accelerometer outputs would be equivalent to "velocity random walk."

The random-walk error model has the form

$$\begin{aligned}
\epsilon_k &= \epsilon_{k-1} + w_{k-1} \\
\sigma_k^2 &\stackrel{\text{def}}{=} \mathrm{E}\left\langle \epsilon_k^2 \right\rangle \\
&= \sigma_{k-1}^2 + \mathrm{E}\left\langle w_{k-1}^2 \right\rangle \\
&= \sigma_0^2 + k\, Q_w \quad \text{for time-invariant systems,} \\
Q_w &\stackrel{\text{def}}{=} \mathrm{E}_k \left\langle w_k^2 \right\rangle .
\end{aligned}$$

The value of Q_w will be in units of squared-error per discrete time step Δt. Random-walk error sources are usually specified in terms of standard deviations, that is, error units per square-root of time unit. Gyroscope angle random walk errors, for example, might be specified in $\text{deg}/\sqrt{\text{h}}$. Most navigation-grade gyroscopes (including RLG, HRG, IFOG) have angle random-walk errors in the order of 10^{-3} $\text{deg}/\sqrt{\text{h}}$ or less.

9.3.1.4 Harmonic Noise Temperature control schemes (including building HVAC systems) often introduce cyclical errors due to thermal transport lags, and these can cause harmonic errors in sensor outputs, with harmonic periods that scale with device dimensions. Also, suspension and structural resonances of host vehicles introduce harmonic accelerations, which can excite acceleration-sensitive error sources in sensors.

9.3.1.5 "1/f" Noise This noise is characterized by power spectral densities that fall off as $1/f$, where f is frequency. It is present in most electronic devices, its causes are not well understood, and it is usually modeled as some combination of white noise and random walk.

9.3.2 Systematic Errors

These are errors that can be calibrated and compensated.

9.3.2.1 Sensor-Level Models These are sensor output errors in addition to additive zero-mean white noise and time-correlated noise considered above. The same models apply to accelerometers and gyroscopes. Some of the more common types of sensor errors are illustrated in Fig. 9.13:

(a) Bias, which is any nonzero sensor output when the input is zero
(b) Scale factor error, often resulting from aging or manufacturing tolerances

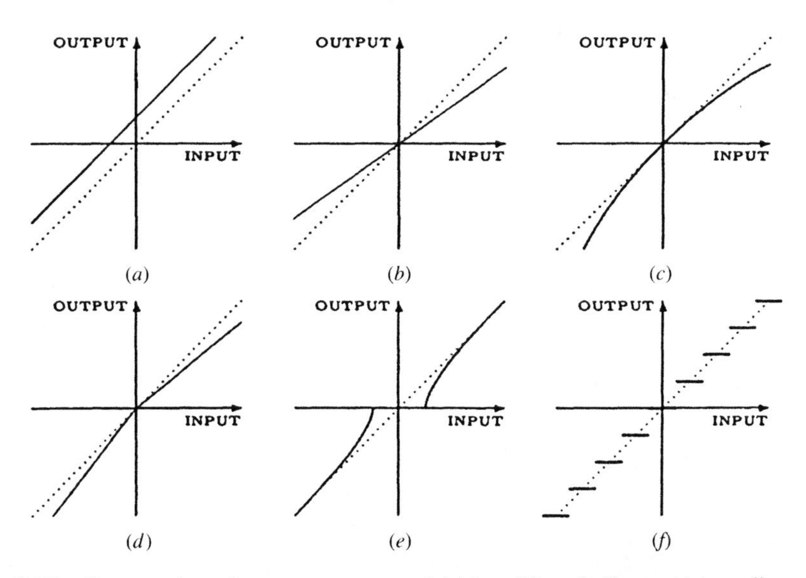

Fig. 9.13 Common input/output error types: (a) bias; (b) scale Factor; (c) nonlinearity; (d) $\pm$ asymmetry; (e) dead zone; (f) quantization.

(*c*) Nonlinearity, which is present in most sensors to some degree
(*d*) Scale factor sign asymmetry (often from mismatched push–pull amplifiers)
(*e*) A dead zone, usually due to mechanical stiction or lockin [for a ring laser gyroscope (RLG)]
(*f*) Quantization error, inherent in all digitized systems. It may not be zero-mean when the input is held constant, as it could be under calibration conditions.

We can recover the sensor input from the sensor output so long as the input/output relationship is known and invertible. Dead-zone errors and quantization errors are the only ones shown with this problem. The cumulative effects of both types (dead zone and quantization) often benefit from zero-mean input noise or dithering. Also, not all digitization methods have equal cumulative effects. Cumulative quantization errors for sensors with frequency outputs are bounded by ± 0.5 least significant bit (LSB) of the digitized output, but the variance of cumulative errors from independent sample-to-sample A/D conversion errors can grow linearly with time.

9.3.2.2 ISA-Level Models For a cluster of $N \geq 3$ gyroscopes or accelerometers, the effects of individual *biases, scale factors*, and *input axis misalignments* can be modeled by an equation of the form

$$\underbrace{\mathbf{z}_{\text{input}}}_{3\times 1} = \underbrace{\mathbf{M}_{\text{scale factor \& misalignment}}}_{3\times N} \underbrace{\mathbf{z}_{\text{output}}}_{N\times 1} + \underbrace{\mathbf{b}_z}_{3\times 1}, \tag{9.13}$$

where the components of the vector $\mathbf{b}_z$ are three aggregate biases, the components of the $\mathbf{z}_{\text{input}}$ and $\mathbf{z}_{\text{output}}$ vectors are the sensed values (accelerations or angular rates) and output values from the sensors, respectively, and the elements m_{ij} of the "scale factor and misalignment matrix" $\mathbf{M}$ represent the individual scale factor deviations and input axis misalignments as illustrated in Fig. 9.14 for $N = 3$ orthogonal input axes. The larger arrows in the figure represent the nominal input axis directions (labeled #1, #2, and #3) and the smaller arrows (labeled m_{ij}) represent the directions of scale factor deviations ($i = j$) and input axis misalignments ($i \neq j$).

Equation 9.13 is in "compensation form"; that is, it represents the inputs as functions of the outputs. The corresponding "error form" is

$$\mathbf{z}_{\text{output}} = \mathbf{M}^{\dagger} \left(\mathbf{z}_{\text{input}} - \mathbf{b}_z \right) \tag{9.14}$$

where $\dagger$ represents the Moore–Penrose matrix inverse.

The compensation model of Eq. 9.13 is the one used in system implementation for compensating sensor outputs using a single constant matrix $\mathbf{M}$ and vector $\mathbf{b}_z$.

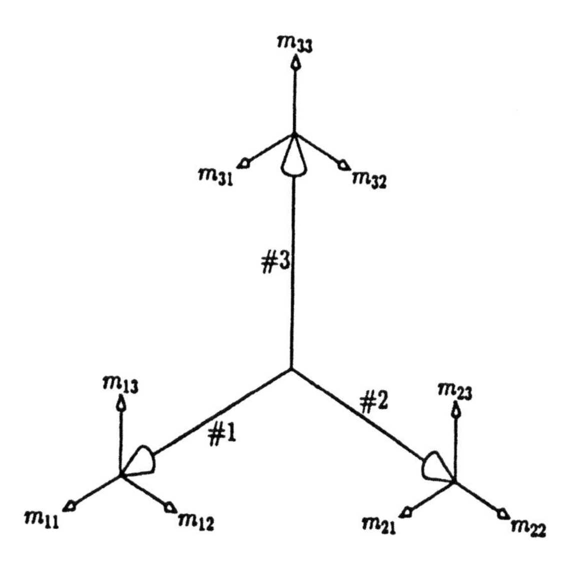

Fig. 9.14 Directions of modeled sensor cluster errors.

9.3.2.3 Calibrating Sensor Biases, Scale Factors, and Misalignments *Calibration* is the process of taking sensor data to characterize sensor inputs as functions of sensor outputs. It amounts to estimating the values of $\mathbf{M}$ and $\mathbf{b}_z$, given input-output pairs $[\mathbf{z}_{\text{input},k}, \mathbf{z}_{\text{output},k}]$, where $\mathbf{z}_{\text{input},k}$ is known from controlled calibration conditions (e.g., for accelerometers, from the direction and magnitude of gravity, or from conditions on a shake table or centrifuge) and $\mathbf{z}_{\text{output},k}$ is measured under these conditions.

Calibration Data Processing The full set of data under K sets of calibration conditions yields a system of $3K$ linear equations

$$\underbrace{\begin{bmatrix} z_{1,\text{input},1} \\ z_{2,\text{input},1} \\ z_{3,\text{input},1} \\ \vdots \\ z_{3,\text{input},K} \end{bmatrix}}_{3K \text{ knowns}} = \underbrace{\begin{bmatrix} z_{1,\text{output},1} & z_{2,\text{output},1} & z_{3,\text{output},1} & \cdots & 0 \\ 0 & 0 & 0 & \cdots & 0 \\ 0 & 0 & 0 & \cdots & 1 \\ \vdots & \vdots & \vdots & \ddots & \vdots \\ 0 & 0 & 0 & \cdots & 1 \end{bmatrix}}_{\mathbf{Z},\text{ a } 3K\times(3N+3) \text{ matrix of knowns}} \underbrace{\begin{bmatrix} m_{1,1} \\ m_{1,2} \\ m_{1,3} \\ \vdots \\ b_{3,z} \end{bmatrix}}_{3N+3 \text{ unknowns}} \tag{9.15}$$

in the $3N$ unknown parameters $m_{i,j}$ (the elements of the matrix $\mathbf{M}$) and 3 unknown parameters $b_{i,z}$ (rows of the 3-vector $\mathbf{b}_z$), which will be overdetermined for $K > N + 1$. In that case, the system of linear equations may be solvable for the $3(N+1)$ calibration parameters by using the method of least-squares

$$\begin{bmatrix} m_{1,1} \\ m_{1,2} \\ m_{1,3} \\ \vdots \\ b_{3,z} \end{bmatrix} = \left[\mathbf{Z}^T\mathbf{Z}\right]^{-1} \mathbf{Z}^{\mathrm{T}} \begin{bmatrix} z_{1,\text{input},1} \\ z_{2,\text{input},1} \\ z_{3,\text{input},1} \\ \vdots \\ z_{3,\text{input},K} \end{bmatrix}, \tag{9.16}$$

provided that the matrix $\mathbf{Z}^T\mathbf{Z}$ is nonsingular.

Calibration Parameters The values of $\mathbf{M}$ and $\mathbf{b}_z$ determined in this way are called *calibration parameters.*

Estimation of the calibration parameters can also be done using Kalman filtering, a by product of which would be the covariance matrix of calibration parameter uncertainty. This matrix is also useful in modeling system-level performance.

9.3.3 Other Calibration Parameters

9.3.3.1 Nonlinearities Sensor input/output nonlinearities are generally modeled by polynomials:

$$z_{\text{input}} = \sum_{i=0}^{N} a_i \, z_{\text{output}}^i, \tag{9.17}$$

where the first two parameters a_0 = bias and a_1 = scale factor. The polynomial input output model of Eq. 9.17 is linear in the calibration parameters, so they can still be calibrated using a system of linear equations—as was used for scale factor and bias.

The generalization of Eq. 9.17 to vector-valued inputs and outputs includes all the cross-power terms between different sensors, but it also includes multidimensional data structures in place of the scalar parameters a_i. Such a model would, for example, include the acceleration sensitivities of gyroscopes and the rotation rate sensitivities of accelerometers.

9.3.3.2 Sensitivities to Other Measurable Conditions Most inertial sensors are also thermometers, and part of the art of sensor design is to minimize their temperature sensitivities. Other bothersome sensitivities include acceleration sensitivity of gyroscopes and rotation rate sensitivities of accelerometers (already mentioned above).

Compensating for temperature sensitivity requires adding one or more thermometers to the sensors and taking calibration data over the expected operational temperature range, but the other sensitivities can be "cross-compensated" by using the outputs of the other inertial sensors. The accelerometer outputs can be used in compensating for acceleration sensitivities of gyroscopes, and the gyro outputs can be used in compensating for angular rate sensitivities of accelerometers.

9.3.3.3 Other Accelerometer Models

Centrifugal Acceleration Effects Accelerometers have input axes defining the component(s) of acceleration that they measure. There is a not-uncommon superstition that these axes must intersect at a point to avoid some unspecified error source. That is seldom the case, but there can be some differential sensitivity to centrifugal accelerations due to high rotation rates and relative displacements between accelerometers. The effect is rather weak, but not always negligible. It is modeled by the equation

$$a_{i,\,\text{centrifugal}} = \omega^2 r_i, \tag{9.18}$$

where ω is the rotation rate and r_i is the displacement component along the input axis from the axis of rotation to the effective center of the accelerometer. Even manned vehicles can rotate at $\omega \approx 3$ rad/s, which creates centrifugal accelerations of about 1 g at $r_i = 1$ m and 0.001 g at 1 mm. The problem is less significant, if not insignificant, for MEMS-scale accelerometers that can be mounted within millimeters of one another.

Center of Percussion Because ω can be measured, sensed centrifugal accelerations can be compensated, if necessary. This requires designating some reference point within the instrument cluster and measuring the radial distances and directions to the accelerometers from that reference point. The point within the accelerometer required for this calculation is sometimes called its "center of percussion." It is effectively the point such that rotations about all axes through the point produce no sensible centrifugal accelerations, and that point can be located by testing the accelerometer at differential reference locations on a rate table.

Angular Acceleration Sensitivities Pendulous accelerometers are sensitive to angular acceleration about their hinge lines, with errors equal to $\dot{\omega}\Delta_{\text{hinge}}$, where $\dot{\omega}$ is the angular acceleration in radians per second squared and Δ_{hinge} is the displacement of the accelerometer proof mass (at its center of mass) from the hinge line. This effect can reach the 1-g level for $\Delta_{\text{hinge}} \approx 1$ cm and $\dot{\omega} \approx 10^3$ rad/s^2, but these extreme conditions are rarely persistent enough to matter in most applications.

9.3.4 Calibration Parameter Instability

INS calibration parameters are not always exactly constant. Their values can change over the operational life of the INS. Specifications for calibration stability generally divide these calibration parameter variations into two categories: (1) changes from one system turnon to the next and (2) slow "parameter drift" during operating periods.

9.3.4.1 Calibration Parameter Changes Between Turn-ons These are changes that occur between a system shutdown and the next startup. They may be caused by temperature transients during shutdowns and turnons, or by what is termed "aging." They are generally considered to be independent from turn-on to turnon, so the model for the covariance of calibration errors for the kth turnon would be of the form

$$\mathbf{P}_{\text{calib},\,k} = \mathbf{P}_{\text{calib},\,k-1} + \mathbf{\Delta} P_{\text{calib}}, \tag{9.19}$$

where $\mathbf{\Delta} P_{\text{calib}}$ is the covariance of turnon-to-turnon parameter changes. The initial value $\mathbf{P}_{\text{calib},\,0}$ at the end of calibration is usually determinable from error covariance analysis of the calibration process. Note that this is the covariance model for a random walk, the covariance of which grows without bound.

9.3.4.2 Calibration Parameter Drift This term applies to changes that occur in the operational periods between startups and shutdowns. The calibration parameter uncertainty covariance equation has the same form as Eq. 9.19, but with $\mathbf{\Delta} P_{\text{calib.}}$ now representing the calibration parameter drift in the time interval $\Delta t = t_k - t_{k-1}$ between successive discrete times within an operational period.

Detecting Error Trends Incipient sensor failures can sometimes be predicted by observing the variations over time of the sensor calibration parameters. One of the advantages of tightly coupled GNSS/INS integration is that INS sensors can be continuously calibrated all the time that GNSS data are available. System health monitoring can then include tests for the trends of sensor calibration parameters, setting threshold conditions for failing the INS system, and isolating a likely set of causes for the observed trends.

9.3.5 Auxilliary Sensors

9.3.5.1 Attitude Sensors Nongyroscopic attitude sensors can also be used as aids in inertial navigation. These include the following:

Magnetic sensors are used primarily for coarse heading initialization, to speed up INS alignment.

Star trackers are used primarily for space-based or near-space applications. The U-2 spy plane, for example, used an inertial-platform-mounted star tracker to maintain INS alignment on long flights.

Optical alignment systems used on some space launch systems. Some use Porro prisms mounted on the inertial platform to maintain optical line-of-sight reference through ground-based theodolites to reference directions at the launch complex.

GNSS receiver systems using antenna arrays and carrier phase interferometry have been used for heading initialization in artillery fire control systems, for example, but the same technology could be used for INS aiding. The systems generally have baselines in the order of several meters, which could limit their utility for some host vehicles.

9.3.5.2 Altitude Sensors These include barometric altimeters and radar altimeters. Without GNSS inputs, some sort of altitude sensor is required to stabilize INS vertical channel errors.

9.4 SYSTEM IMPLEMENTATION MODELS

9.4.1 One-Dimensional Example

This example is intended as an introduction to INS technology for the uninitiated. It illustrates some of the key properties of inertial sensors and inertial system implementations.

If we all lived in one-dimensional "line land," then there could be no rotation and no need for gyroscopes. In that case, an INS would need only one accelerometer and navigation computer (all one-dimensional, of course), and its implementation would be as illustrated in Fig. 9.15 (in two dimensions), where the dependent variable x denotes position in one dimension and the independent variable t is time.

This implementation for one dimension has many features common to implementations for three dimensions:

1. *Accelerometers cannot measure gravitational acceleration.* An accelerometer effectively measures the force acting on its proof mass to make it follow its mounting base, which includes only nongravitational accelerations applied through physical forces acting on the INS through its host vehicle. Satellites, which are effectively in free fall, experience no sensible accelerations.
2. Accelerometers have *scale factors*, which are the ratios of input acceleration units to output signal magnitude units (e.g., meters per second squared per volt). The signal must be rescaled in the navigation computer by multiplying by this scale factor.

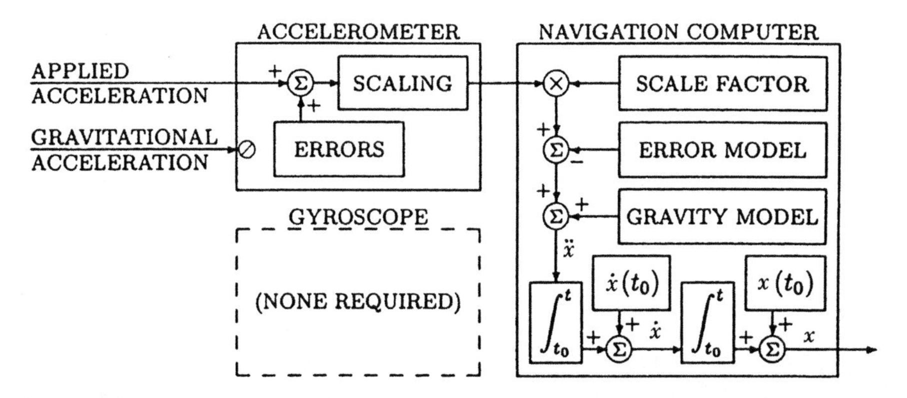

Fig. 9.15 INS functional implementation for a one-dimensional world.

3. Accelerometers have *output errors*, including
   (a) Unknown constant *offsets*, also called *biases*.
   (b) Unknown constant *scale factor errors*.
   (c) Unknown *sensor input axis misalignments*.
   (d) Unknown *nonconstant variations* in bias and scale factor.
   (e) Unknown zero-mean additive *noise* on the sensor outputs, including quantization noise and electronic noise. The noise itself is not predictable, but its statistical properties may be used in Kalman filtering to estimate drifting scale factor and biases.
4. *Gravitational accelerations must be modeled* and calculated in the navigational computer, then added to the sensed acceleration (after error and scale compensation) to obtain the net acceleration $\ddot{x}$ of the INS.
5. The navigation computer must integrate acceleration to obtain velocity. *This is a definite integral and it requires an initial value*, $\dot{x}(t_0)$; that is, the INS implementation in the navigation computer must start with a known initial velocity.
6. The navigation computer must also integrate velocity ($\dot{x}$) to obtain position (x). *This is also a definite integral and it also requires an initial value*, $x(t_0)$. The INS implementation in the navigation computer must start with a known initial location, too.

Inertial navigation in three dimensions requires more sensors and more signal processing than in one dimension, and it also introduces more possibilities for implementation (e.g., gimbaled or strapdown)

9.4.2 Initialization and Alignment

9.4.2.1 Navigation Initialization INS initialization is the process of determining initial values for system position, velocity, and attitude in navigation coordinates. INS position initialization ordinarily relies on external sources such as GNSS or manual entry by crew members. INS velocity initialization can be accomplished by starting when it is zero (i.e., the host vehicle is not moving) or (for vehicles carried in or on other vehicles) by reference to the carrier velocity. (See alignment method 3 below.) INS attitude initialization is called *alignment*.

9.4.2.2 Sensor Alignment INS alignment is the process of aligning the stable platform axes parallel to navigation coordinates (for gimbaled systems) or that of determining the initial values of the coordinate transformation from sensor coordinates to navigation coordinates (for strapdown systems).

Alignment Methods Four basic methods for INS alignment are as follows:

1. *Optical alignment,* using either of the following:
   (a) Optical line-of-sight reference to a ground-based direction (e.g., using a ground-based theodolite and a mirror on the platform). Some space

boosters have used this type of optical alignment, which is much faster and more accurate than gyrocompass alignment. Because it requires a stable platform for mounting the mirror, it is applicable only to gimbaled systems.

(b) An onboard star tracker, used primarily for alignment of gimbaled or strapdown systems in space or near-space (e.g., above all clouds).

2. *Gyrocompass alignment* of stationary vehicles, using the sensed direction of acceleration to determine the local vertical and the sensed direction of rotation to determine north. Latitude can be determined by the angle between the earth rotation vector and the horizontal, but longitude must be determined by other means and entered manually or electronically. This method is inexpensive but the most time-consuming (several minutes, typically).
3. *Transfer alignment* in a moving host vehicle, using velocity matching with an aligned and operating INS. This method is typically several times faster than gyrocompass alignment, but it requires another INS on the host vehicle and it may require special maneuvering of the host vehicle to attain observability of the alignment variables. It is commonly used for in-air INS alignment for missiles launched from aircraft and for on-deck INS alignment for aircraft launched from carriers. Alignment of carrier-launched aircraft may also use the direction of the velocity impulse imparted by the steam catapult.
4. *GNSS-aided alignment* using position matching with GNSS to estimate the alignment variables. It is an integral part of integrated GNSS/INS implementations. It does not require the host vehicle to remain stationary during alignment, but there will be some period of time after turnon (a few minutes, typically) before system navigation errors settle to acceptable levels.

Gyrocompass alignment is the only one of these that requires no external aiding. Gyrocompass alignment is not necessary for integrated GNSS/INS, although many INSs may already be configured for it.

INS Gyrocompass Alignment Accuracy A rough rule of thumb for gyrocompass alignment accuracy is

$$\sigma^2_{\text{gyrocompass}} > \sigma^2_{\text{acc}} + \frac{\sigma^2_{\text{gyro}}}{15^2 \cos^2(\phi_{\text{geodetic}})}, \tag{9.20}$$

where $\sigma_{\text{gyrocompass}}$ is the minimum achievable RMS alignment error in radians, σ_{acc} is the RMS accelerometer accuracy in g values, σ_{gyro} is the RMS gyroscope accuracy in degrees per hour, 15° per hour is the rotation rate of the earth, and ϕ_{geodetic} is the latitude at which gyrocompassing is performed.

Alignment accuracy is also a function of the time allotted for it, and the time required to achieve a specified accuracy is generally a function of sensor error magnitudes (including noise) and the degree to which the vehicle remains stationary.

Gimbaled INS Gyrocompass Alignment Gyrocompass alignment for gimbaled systems is a process for aligning the inertial platform axes with the navigation coordinates using only the sensor outputs while the host vehicle is essentially stationary. For systems using ENU navigation coordinates, for example, the platform can be tilted until two of its accelerometer inputs are zero, at which time both input axes will be horizontal. In this locally leveled orientation, the sensed rotation axis will be in the north–up plane, and the platform can be slewed about the vertical axis to null the input of one of its horizontal gyroscopes, at which time that gyroscope input axis will point east–west. That is the basic concept used for gyrocompass alignment, but practical implementation requires filtering[8] to reduce the effects of sensor noise and unpredictable zero-mean vehicle disturbances due to loading activities and/or wind gusts.

Strapdown INS Gyrocompass Alignment Gyrocompass alignment for strapdown systems (see Fig. 9.16) is a process for "virtual alignment" by determining the sensor cluster attitude with respect to navigation coordinates using only the sensor outputs while the system is essentially stationary.

If the sensor cluster could be firmly affixed to the earth and there were no sensor errors, then the sensed acceleration vector $\mathbf{a}_{\text{output}}$ in sensor coordinates would be in the direction of the local vertical, the sensed rotation vector ω_{output} would be in the direction of the earth rotation axis, and the unit column vectors

$$\mathbf{1}_U = \frac{\mathbf{a}_{\text{output}}}{|\mathbf{a}_{\text{output}}|}, \tag{9.21}$$

$$\mathbf{1}_N = \frac{w_{\text{output}} - (\mathbf{1}_U^T w_{\text{output}})\mathbf{1}_U}{|w_{\text{output}} - (\mathbf{1}_U^T w_{\text{output}})\mathbf{1}_U|}, \tag{9.22}$$

$$\mathbf{1}_E = \mathbf{1}_N \otimes \mathbf{1}_U \tag{9.23}$$

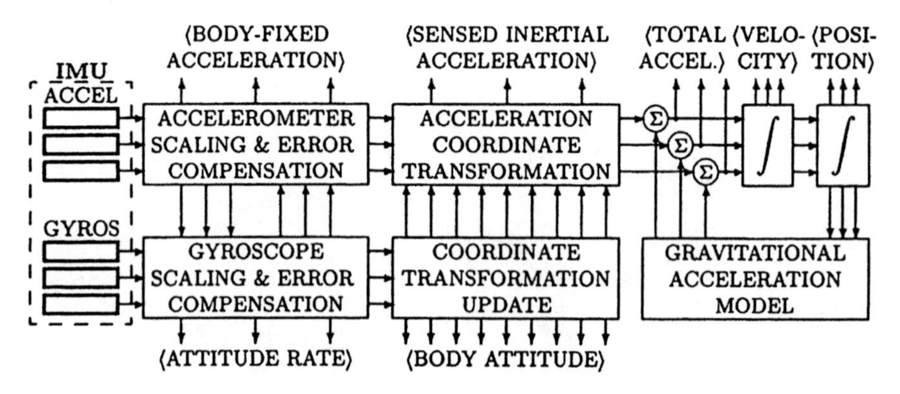

Fig. 9.16 Outputs (in angular brackets) of simple strapdown INS.

[8]The vehicle dynamic model used for gyrocompass alignment filtering can be "tuned" to include the major resonance modes of the vehicle suspension.

would define the initial value of the coordinate transformation matrix from sensor-fixed coordinates to ENU coordinates:

$$\mathbf{C}_{\mathrm{ENU}}^{\mathrm{sensor}} = [\mathbf{1}_E|\mathbf{1}_N|\mathbf{1}_U]^T . \tag{9.24}$$

In practice, the sensor cluster is usually mounted in a vehicle that is not moving over the surface of the earth but may be buffeted by wind gusts and/or disturbed by fueling and payload handling. Gyrocompassing then requires some amount of filtering to reduce the effects of vehicle buffeting and sensor noise. The gyrocompass filtering period is typically on the order of several minutes for a medium-accuracy INS but may continue for hours or days for high-accuracy systems.

9.4.3 Earth Models

Inertial navigation and satellite navigation require models for the shape, gravity, and rotation of the earth.

9.4.3.1 Navigation Coordinates Descriptions of the major coordinates used in inertial navigation and GNSS/INS integration are described in Appendix C (on the CD). These include coordinate systems used for representing the trajectories of GNSS satellites and user vehicles in the near-earth environment and for representing the attitudes of host vehicles relative to locally level coordinates, including the following:

1. Inertial coordinates:
   (a) Earth-centered inertial (ECI), with origin at the center of mass of the earth and principal axes in the directions of the vernal equinox (defined in Section C.2.1) and the rotation axis of the earth.
   (b) Satellite orbital coordinates, as illustrated in Fig. C.4 and used in GPS ephemerides.
2. Earth-fixed coordinates:
   (a) Earth-centered, earth-fixed (ECEF), with origin at the center of mass of the earth and principal axes in the directions of the prime meridian (defined in Section C.3.5) at the equator and the rotation axis of the earth.
   (b) Geodetic coordinates, based on an ellipsoid model for the shape of the earth. Longitude in geodetic coordinates is the same as in ECEF coordinates, and geodetic latitude as defined as the angle between the equatorial plane and the normal to the reference ellipsoid surface. Geodetic latitude can differ from geocentric latitude by as much as 12 arc minutes, equivalent to about 20 km of northing distance.
   (c) Local tangent plane (LTP) coordinates, also called "locally level coordinates," essentially representing the earth as being locally flat. These coordinates are particularly useful from a human factors standpoint for

representing the attitude of the host vehicle and for representing local directions. They include

i. East–north–up (ENU), shown in Fig. C.7
ii. North–east–down (NED), which can be simpler to relate to vehicle coordinates and
iii. Alpha wander, rotated from ENU coordinates through an angle α about the local vertical (see Fig. C.8)

3. Vehicle-fixed coordinates:
   (a) Roll–pitch–yaw (RPY) (axes shown in Fig. C.9).

Transformations between these different coordinate systems are important for representing vehicle attitudes, for resolving inertial sensor outputs into inertial navigation coordinates, and for GNSS/INS integration. Methods used for representing and implementing coordinate transformations are also presented in Appendix C, Section C.4.

9.4.3.2 Earth Rotation Our earth is the mother of all clocks. It has given us the time units of days, hours, minutes, and seconds we use to manage our lives. Not until the discovery of atomic clocks based on hyperfine transitions were we able to observe the imperfections in our earth clock. Despite these, we continue to use earth rotation as our primary time reference, adding or subtracting leap seconds to atomic clocks to keep them synchronized to the rotation of the earth. These time variations are significant for GNSS navigation, but not for inertial navigation.

WGS84 Earthrate Model The value of earthrate in the World Geodetic System 1984 (WGS84) earth model used by GPS is $7,292,115,167 \times 10^{-14}$ radians per second, or about 15.04109°/h. This is its *sidereal* rotation rate with respect to distant stars. Its mean rotation rate with respect to the nearest star (our sun), as viewed from the rotating earth, is 15°/h, averaged over one year.

9.4.3.3 GPS Gravity Models Accurate gravity modeling is important for maintaining ephemerides for GPS satellites, and models developed for GPS have been a boon to inertial navigation as well. However, spatial resolution of the earth gravitational field required for GPS operation may be a bit coarse compared to that for precision inertial navigation, because the GPS satellites are not near the surface and the mass concentration anomalies that create surface gravity anomalies. GPS orbits have very little sensitivity to surface-level undulations of the gravitational field on the order of 100 km or less, but these can be important for high-precision inertial systems.

9.4.3.4 INS Gravity Models Because an INS operates in a world with gravitational accelerations it is unable to sense and unable to ignore, it must use a reasonably faithful model of gravity.

Gravity models for the earth include centrifugal acceleration due to rotation of the earth as well as true gravitational accelerations due to the mass distribution of the earth, but they do not generally include oscillatory effects such as tidal variations.

Gravitational Potential Gravitational potential is defined to be zero at a point infinitely distant from all massive bodies and to decrease toward massive bodies such as the earth. That is, a point at infinity is the reference point for gravitational potential.

In effect, the gravitational potential at a point in or near the earth is defined by the potential energy lost by a unit of mass falling to that point from infinite altitude. In falling from infinity, potential energy is converted to kinetic energy, $mv_{\text{escape}}^2/2$, where v_{escape} is the *escape velocity*. Escape velocity at the surface of the earth is about 11 km/s.

Gravitational Acceleration Gravitational acceleration is the negative gradient of gravitational potential. Potential is a scalar function, and its gradient is a vector. Because gravitational potential increases with altitude, its gradient points upward and the negative gradient points downward.

Equipotential Surfaces An equipotential surface is a surface of constant gravitational potential. If the ocean and atmosphere were not moving, then the surface of the ocean at static equilibrium would be an equipotential surface. *Mean sea level* is a theoretical equipotential surface obtained by time averaging the dynamic effects.

Ellipsoid Models for Earth Geodesy is the process of determining the shape of the earth, often using ellipsoids as approximations of an equipotential surface (e.g., mean sea level), as illustrated in Fig. 9.17. The most common ones are ellipsoids of revolution, but there are many reference ellipsoids based on different survey data. Some are global approximations and some are local approximations. The global approximations deviate from a spherical surface by about ± 10 km, and locations on the earth referenced to different ellipsoidal approximations can differ from one another by $10^2 - 10^3$ m.

Geodetic latitude on a reference ellipsoid is measured in terms of the angle between the equator and the normal to the ellipsoid surface, as illustrated in Fig. 9.17.

Orthometric height is measured along the (curved) plumbline.

WGS84 Ellipsoid The World Geodetic System (WGS) is an international standard for navigation coordinates. WGS84 is a reference earth model released in 1984. It approximates mean sea level by an ellipsoid of revolution with its rotation axis coincident with the rotation axis of the earth, its center at the center of mass of the earth, and its prime meridian through Greenwich. Its semimajor axis (equatorial radius) is defined to be 6,378,137 m, and its semiminor axis (polar radius) is defined to be 6,356,752.3142 m.

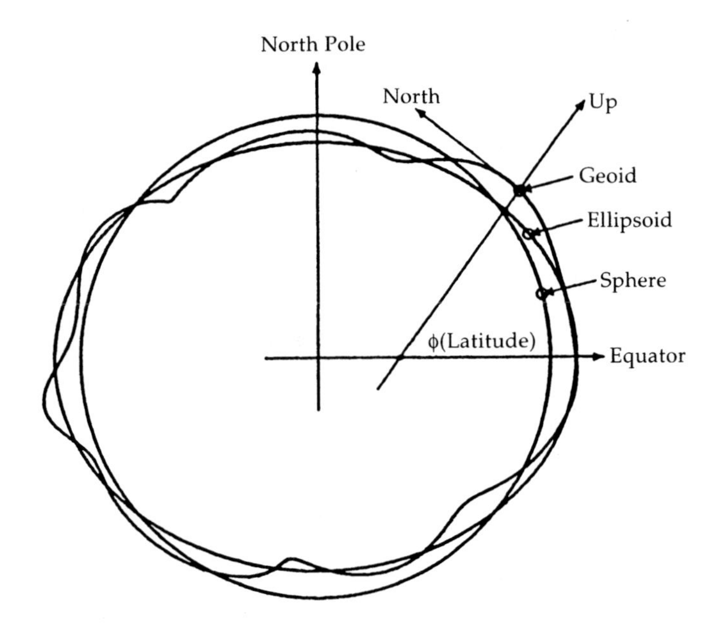

Fig. 9.17 Equipotential surface models for earth.

Geoid Models Geoids are approximations of mean sea-level orthometric height with respect to a reference ellipsoid. Geoids are defined by additional higher order shapes, such as spherical harmonics of height deviations from an ellipsoid, as illustrated in Fig. 9.17. There are many geoid models based on different data, but the more recent, most accurate models depend heavily on GPS data. Geoid heights deviate from reference ellipsoids by tens of meters, typically.

The WGS84 geoid heights vary about ± 100 m from the reference ellipsoid. As a rule, oceans tend to have lower geoid heights and continents tend to have higher geoid heights. Coarse 20-m contour intervals are plotted versus longitude and latitude in Fig. 9.18, with geoid regions above the ellipsoid shaded gray.

9.4.3.5 Longitude and Latitude Rates The second integral of acceleration in locally level coordinates should result in the estimated vehicle position. This integral is somewhat less than straightforward when longitude and latitude are the preferred horizontal location variables.

The rate of change of vehicle altitude equals its vertical velocity, which is the first integral of net (i.e., including gravity) vertical acceleration. The rates of change of vehicle longitude and latitude depend on the horizontal components of vehicle velocity, but in a less direct manner. The relationship between longitude and latitude rates and east and north velocities is further complicated by the oblate shape of the earth.

The rates at which these angular coordinates change as the vehicle moves tangent to the surface will depend on the radius of curvature of the reference

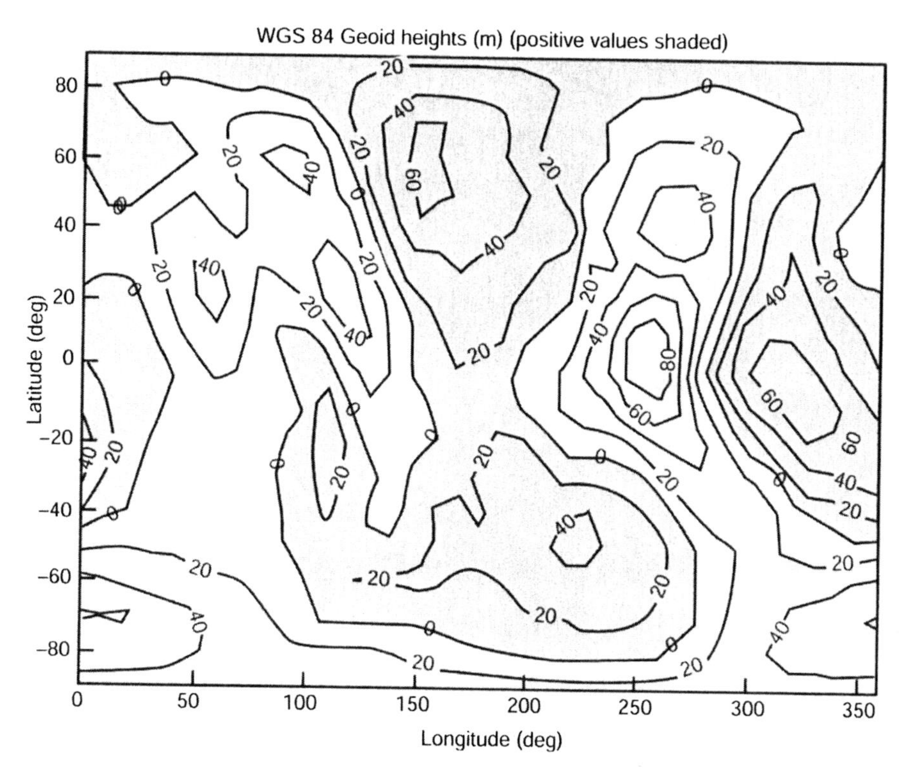

Fig. 9.18 WGS84 geoid heights.

surface model, which is an ellipsoid of revolution for the WGS84 model. Radius of curvature can depend on the direction of travel, and for an ellipsoidal model there is one radius of curvature for north–south motion and another radius of curvature for east–west motion.

Meridional Radius of Curvature The radius of curvature for north–south motion is called the "meridional" radius of curvature, because north–south travel is along a meridian (i.e., line of constant longitude). For an ellipsoid of revolution (the WGS84 model), all meridians have the same shape, which is that of the ellipse that was rotated to produce the ellipsoidal surface model. The tangent circle with the same radius of curvature as the ellipse is called the "*osculating*" *circle* (*osculating* means "kissing"). As illustrated in Fig. 9.19 for an oblate earth model, the radius of the meridional osculating circle is smallest where the geocentric radius is largest (at the equator), and the radius of the osculating circle is largest where the geocentric radius is smallest (at the poles). The osculating circle lies inside or on the ellipsoid at the equator and outside or on the ellipsoid at the poles and passes through the ellipsoid surface for latitudes in between.

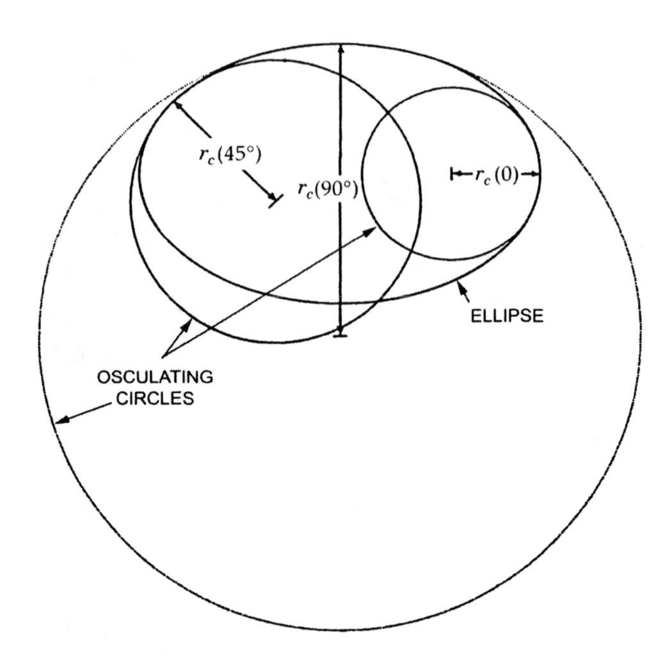

Fig. 9.19 Ellipse and osculating circles.

The formula for meridional radius of curvature as a function of geodetic latitude (ϕ_{geodetic}) is

$$r_M = \frac{b^2}{a[1 - e^2 \sin^2(\phi_{\text{geodetic}})]^{3/2}} \tag{9.25}$$

$$= \frac{a(1 - e^2)}{\left[1 - e^2 \sin^2(\phi_{\text{geodetic}})\right]^{3/2}}, \tag{9.26}$$

where a is the semimajor axis of the ellipse, b is the semiminor axis, and $e^2 = (a^2 - b^2)/a^2$ is the eccentricity squared.

Geodetic Latitude Rate The rate of change of geodetic latitude as a function of north velocity is then

$$\frac{d\phi_{\text{geodetic}}}{dt} = \frac{v_N}{r_M + h}, \tag{9.27}$$

and geodetic latitude can be maintained as the integral

$$\phi_{\text{geodetic}}(t_{\text{now}}) = \phi_{\text{geodetic}}(t_{\text{start}}) + \int_{t_{\text{start}}}^{t_{\text{now}}} \frac{v_{N(t)}\, dt}{a(1 - e^2)/\{1 - e^2 \sin^2[\phi_{\text{geodetic}}(t)]\}^{3/2} + h(t)}, \tag{9.28}$$

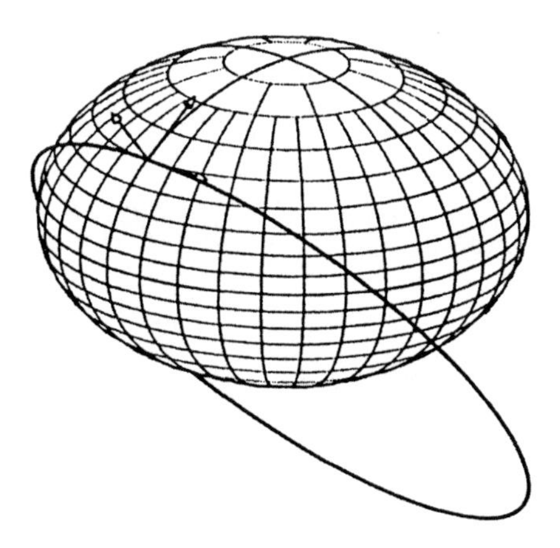

Fig. 9.20 Transverse osculating circle.

where $h(t)$ is height above (+) or below (−) the ellipsoid surface and $\phi_{geodetic}(t)$ will be in radians if $v_N(t)$ is in meters per second and $r_M(t)$ and $h(t)$ are in meters.

Transverse Radius of Curvature The radius of curvature of the reference ellipsoid surface in the east–west direction (i.e., orthogonal to the direction in which the meridional radius of curvature is measured) is called the *transverse radius of curvature*. It is the radius of the osculating circle in the local east–up plane, as illustrated in Fig. 9.20, where the arrows at the point of tangency of the transverse osculating circle are in the local ENU coordinate directions. As this figure illustrates, on an oblate earth, the plane of a transverse osculating circle does not pass through the center of the earth, except when the point of osculation is at the equator. (All osculating circles at the poles are in meridional planes.) Also, unlike meridional osculating circles, transverse osculating circles generally lie outside the ellipsoidal surface, except at the point of tangency and at the equator, where the transverse osculating circle *is* the equator.

The formula for the transverse radius of curvature on an ellipsoid of revolution is

$$r_T = \frac{a}{\sqrt{1 - e^2 \sin^2(\phi_{\text{geodetic}})}}, \tag{9.29}$$

where a is the semimajor axis of the generating ellipse and e is its eccentricity.

Longitude Rate The rate of change of longitude as a function of east velocity is then

$$\frac{d\theta}{dt} = \frac{v_E}{\cos(\phi_{\text{geodetic}})\,(r_T + h)} \tag{9.30}$$

and longitude can be maintained by the integral

$$\theta\,(t_{\text{now}}) = \theta\,(t_{\text{start}}) + \int_{t_{\text{start}}}^{t_{\text{now}}} \frac{v_E(t)\;dt}{\cos[\phi_{\text{geodetic}}(t)]\,\left(a/\sqrt{1-e^2\sin^2(\phi_{\text{geodetic}}(t))}+h(t)\right)}, \tag{9.31}$$

where $h(t)$ is height above (+) or below (−) the ellipsoid surface and θ will be in radians if $v_E(t)$ is in meters per second and $r_T(t)$ and $h(t)$ are in meters. Note that this formula has a singularity at the poles, where $\cos(\phi_{\text{geodetic}}) = 0$, a consequence of using latitude and longitude as location variables.

WGS84 Reference Surface Curvatures The apparent variations in meridional radius of curvature in Fig. 9.19 are rather large because the ellipse used in generating Fig. 9.19 has an eccentricity of about 0.75. The WGS84 ellipse has an eccentricity of about 0.08, with geocentric, meridional, and transverse radius of curvature as plotted in Fig. 9.21 versus geodetic latitude. For the WGS84 model:

- Mean geocentric radius is about 6371 km, from which it varies by −14.3 km (−0.22%) to +7.1 km (+0.11%).
- Mean meridional radius of curvature is about 6357 km, from which it varies by −21.3 km (−0.33%) to 42.8 km (+0.67%).
- Mean transverse radius of curvature is about 6385 km, from which it varies by −7.1 km (−0.11%) to +14.3 km (+0.22%).

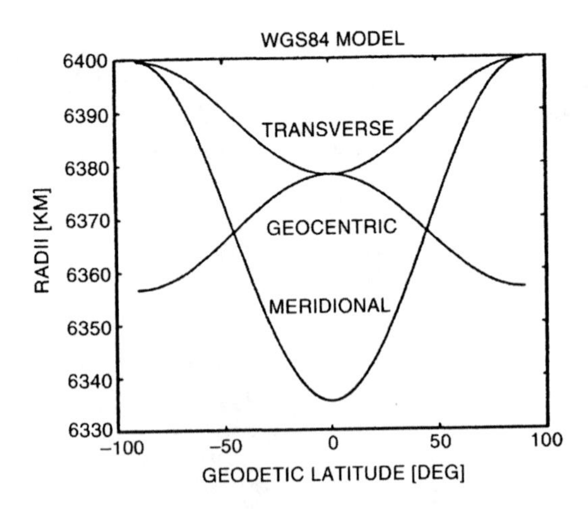

Fig. 9.21 Radii of WGS84 reference ellipsoid.

Because these vary by several parts per thousand, one must take radius of curvature into account when integrating horizontal velocity increments to obtain longitude and latitude.

9.4.4 Gimbal Attitude Implementations

The primary function of gimbals is to isolate the ISA from vehicle rotations, but they are also used for other INS functions.

9.4.4.1 Accelerometer Recalibration Navigation accuracy is very sensitive to accelerometer biases, which can shift as a result of thermal transients in turnon/turnoff cycles, and can also drift randomly over time. Fortunately, the gimbals can be used to calibrate accelerometer biases in a stationary 1-g environment. In fact, both bias and scale factor can be determined by using the gimbals to point the accelerometer input axis straight up and straight down, and recording the respective accelerometer outputs a_{up} and a_{down}. Then the bias $a_{\text{bias}} = \left(a_{\text{up}} + a_{\text{down}}\right)/2$ and scale factor $s = \left(a_{\text{up}} - a_{\text{down}}\right)/2\,g_{\text{local}}$, where g_{local} is the local gravitational acceleration.

9.4.4.2 Gyrocompass Alignment This is the process of determining the orientation of the ISA with respect to locally level coordinates (e.g., NED or ENU). Gyrocompassing allows the ISA to be oriented with its sensor axes aligned parallel to the north, east, and vertical directions. It is accomplished using three servo loops. The two "leveling" loops slew the ISA until the outputs of the nominally "north" and "east" accelerometer outputs are zeroed, and the "heading" loop slews the ISA around the third orthogonal axis (i.e., the vertical one) until the output of the nominally "east-pointing" gyro is zeroed.

9.4.4.3 Vehicle Attitude Determination The gimbal angles determine the vehicle attitude with respect to the ISA, which has a controlled orientation with respect to locally level coordinates. Each gimbal angle encoder output determines the relative rotation of the structure outside gimbal axis relative to the structure inside the gimbal axis, the effect of each rotation can be represented by a 3×3 rotation matrix, and the coordinate transformation matrix representing the attitude of vehicle with respect to the ISA will be the ordered product of these matrices.

For example, in the gimbal structure shown in Fig. 2.6, each gimbal angle represents an Euler angle for vehicle rotations about the vehicle roll, pitch and yaw axes. Then the transformation matrix from vehicle roll–pitch–yaw coordinates to locally level east–north–up coordinates will be

$$C_{\text{ENU}}^{\text{RPY}} = \begin{bmatrix} S_Y C_P & C_R C_Y + S_R S_Y S_P & -S_R C_Y + C_R S_Y S_P \\ C_Y C_P & -C_R S_Y + S_R C_Y S_P & S_R S_Y + C_R C_Y S_P \\ S_P & -S_R C_P & -C_R C_P \end{bmatrix}, \quad (9.32)$$

where

$$
\begin{aligned}
S_R &= \sin\,(\text{roll angle})\,,\\
C_R &= \cos\,(\text{roll angle})\,,\\
S_P &= \sin\,(\text{pitch angle})\,,\\
C_P &= \cos\,(\text{pitch angle})\,,\\
S_Y &= \sin\,(\text{yaw angle})\,,\\
C_Y &= \cos\,(\text{yaw angle})\,.
\end{aligned}
$$

9.4.4.4 ISA Attitude Control The primary purpose of gimbals is to stabilize the ISA in its intended orientation. This is a 3-degree-of-freedom problem, and the solution is unique for three gimbals. That is, there are three attitude-control loops with (at least) three sensors (the gyroscopes) and three torquers. Each control loop can use a PID controller, with the commanded torque distributed to the three torquers according to the direction of the torquer/gimbal axis with respect to the gyro input axis, somewhat as illustrated in Fig. 9.22, where

DISTURBANCES includes the sum of all torque disturbances on the individual gimbals and the ISA, including those due to mass unbalance and acceleration, air currents, torque motor errors, etc.

GIMBAL DYNAMICS is actually quite a bit more complicated than the rigid-body torque equation

$$\boldsymbol{\tau} = \mathbf{M}_{\text{inertia}}\dot{\boldsymbol{\omega}},$$

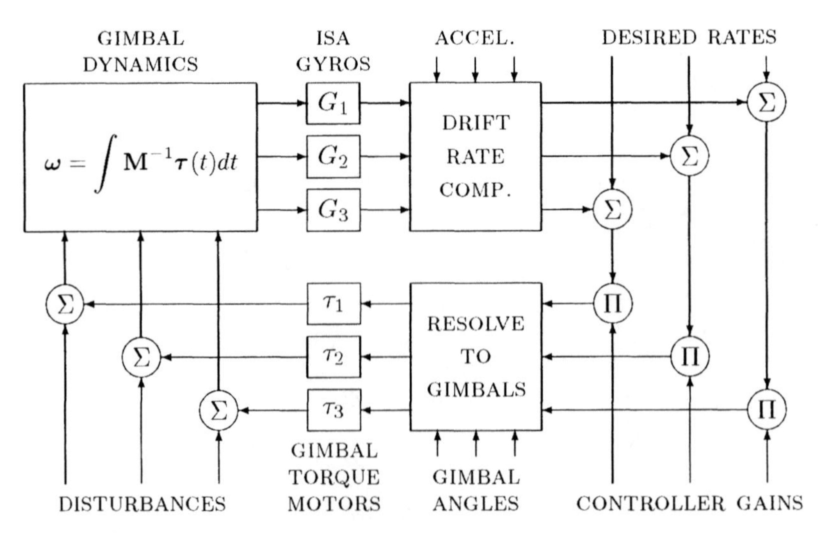

Fig. 9.22 Simplified control flow diagram for three gimbals.

which is the torque analog of $\mathbf{F} = m\mathbf{a}$, where $\mathbf{M}_{\text{inertia}}$ is the moment of inertia matrix. The IMU is not a rigid body, and the gimbal torque motors apply torques *between* the gimbal elements (i.e., ISA, gimbal rings and host vehicle).

DESIRED RATES refers to the rates required to keep the ISA aligned to a moving coordinate frame (e.g., locally level).

RESOLVE TO GIMBALS is where the required torques are apportioned among the individual torquer motors on the gimbal axes.

The actual control loop is more complicated than that shown in the figure, but it does illustrate in general terms how the sensors and actuators are used.

For systems using four gimbals to avoid gimbal lock, the added gimbal adds another degree of freedom to be controlled. In this case, the control law usually adds a fourth constraint (e.g., maximize the minimum angle between gimble axes) to avoid gimbal lock.

9.4.5 Strapdown Attitude Implementations

9.4.5.1 Strapdown Attitude Problem Early on, strapdown systems technology had an "attitude problem," which was the problem of representing attitude rate in a format amenable to accurate computer integration. The eventual solution was to represent attitude in different mathematical formats as it is processed from raw gyro outputs to the matrices used for transforming sensed acceleration to inertial coordinates for integration.

Figure 9.23 illustrates the resulting major gyro signal processing operations, and the formats of the data used for representing attitude information. The processing starts with gyro outputs and ends with a coordinate transformation matrix from sensor coordinates to the coordinates used for integrating the sensed accelerations.

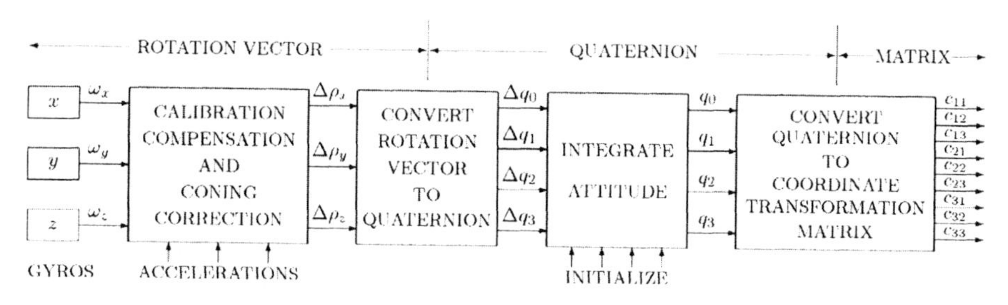

Fig. 9.23 Strapdown attitude representations.

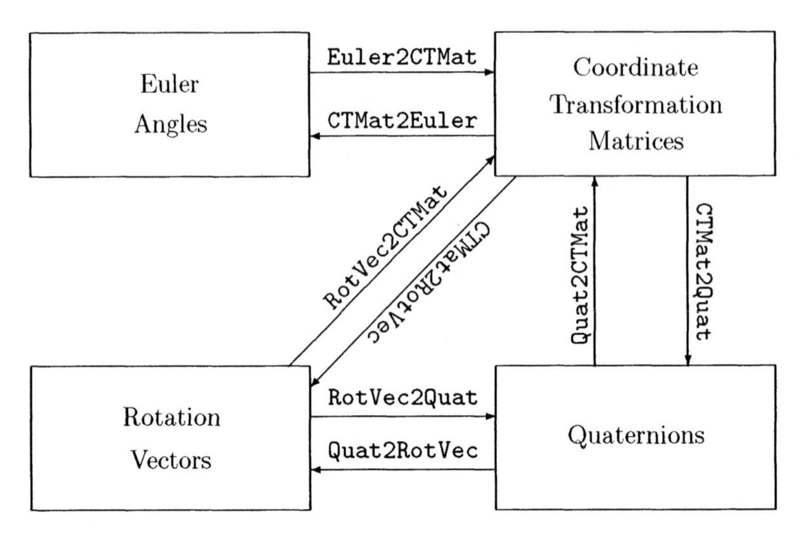

Fig. 9.24 Attitude representation formats and MATLAB transformations.

9.4.5.2 MATLAB Implementations The diagram in Fig. 9.24 shows four different representations used for relative attitudes, and the names of the MATLAB script m-files (i.e., with the added ending `.m`) on the accompanying CD-ROM for transforming from one representation to another.

9.4.5.3 Coning Motion This type of motion is a problem for attitude integration when the frequency of motion is near or above the sampling frequency. It is usually a consequence of host vehicle frame vibration modes where the INS is mounted, and INS shock and vibration isolation is often designed to eliminate or substantially reduce this type of rotational vibration.

Coning motion is an example of an attitude trajectory (i.e., attitude as a function of time) for which the integral of attitude rates does *not* equal the attitude change. An example trajectory would be

$$\boldsymbol{\rho}(t) = \theta_{\text{cone}} \begin{bmatrix} \cos\left(\Omega_{\text{coning}} t\right) \\ \sin\left(\Omega_{\text{coning}} t\right) \\ 0 \end{bmatrix}, \tag{9.33}$$

$$\dot{\boldsymbol{\rho}}(t) = \theta_{\text{cone}} \Omega_{\text{coning}} \begin{bmatrix} -\sin\left(\Omega_{\text{coning}} t\right) \\ \cos\left(\Omega_{\text{coning}} t\right) \\ 0 \end{bmatrix}, \tag{9.34}$$

where θ_{cone} is the *cone angle* of the motion and Ω_{coning} is the *coning frequency* of the motion, as illustrated in Fig. 9.25.

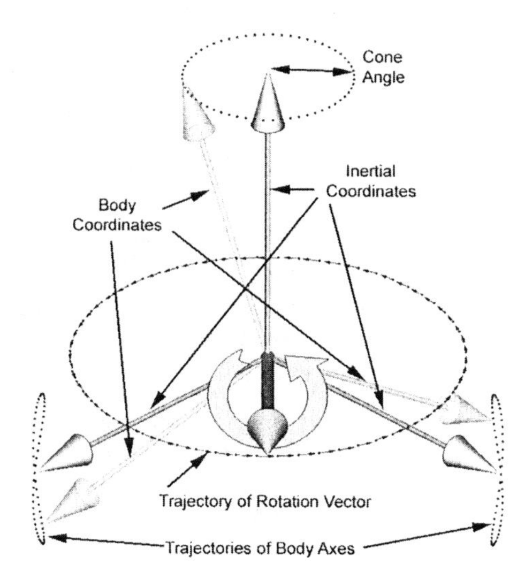

Fig. 9.25 Coning motion.

The coordinate transformation matrix from body coordinates to inertial coordinates (Eq. C.112) will be

$$
\begin{aligned}
\mathbf{C}_{\text{inertial}}^{\text{body}}(\boldsymbol{\rho}) &= \cos\theta \mathbf{I} + (1-\cos\theta) \\
&\times \begin{bmatrix} \cos(\Omega_{\text{coning}}t)^2 & \sin(\Omega_{\text{coning}}t)\cos(\Omega_{\text{coning}}t) & 0 \\ \sin(\Omega_{\text{coning}}t)\cos(\Omega_{\text{coning}}t) & \sin(\Omega_{\text{coning}}t)^2 & 0 \\ 0 & 0 & 0 \end{bmatrix} \\
&+ \sin\theta \begin{bmatrix} 0 & 0 & \sin(\Omega_{\text{coning}}t) \\ 0 & 0 & -\cos(\Omega_{\text{coning}}t) \\ -\sin(\Omega_{\text{coning}}t) & \cos(\Omega_{\text{coning}}t) & 0 \end{bmatrix},
\end{aligned}
\tag{9.35}
$$

and the measured inertial rotation rates in body coordinates will be

$$
\boldsymbol{\omega}_{\text{body}} = \mathbf{C}_{\text{body}}^{\text{inertial}} \dot{\boldsymbol{\rho}}_{\text{inertial}} \tag{9.36}
$$

$$
= \theta_{\text{cone}}\Omega_{\text{coning}} \left[\mathbf{C}_{\text{inertial}}^{\text{body}}\right]^T \begin{bmatrix} -\sin(\Omega_{\text{coning}}t) \\ \cos(\Omega_{\text{coning}}t) \\ 0 \end{bmatrix}, \tag{9.37}
$$

$$
= \begin{bmatrix} -\theta_{\text{cone}}\,\Omega_{\text{coning}}\,\sin(\Omega_{\text{coning}}\,t)\cos(\theta_{\text{cone}}) \\ \theta_{\text{cone}}\,\Omega_{\text{coning}}\,\cos(\Omega_{\text{coning}}\,t)\cos(\theta_{\text{cone}}) \\ -\sin(\theta_{\text{cone}})\,\theta_{\text{cone}}\,\Omega_{\text{coning}} \end{bmatrix}. \tag{9.38}
$$

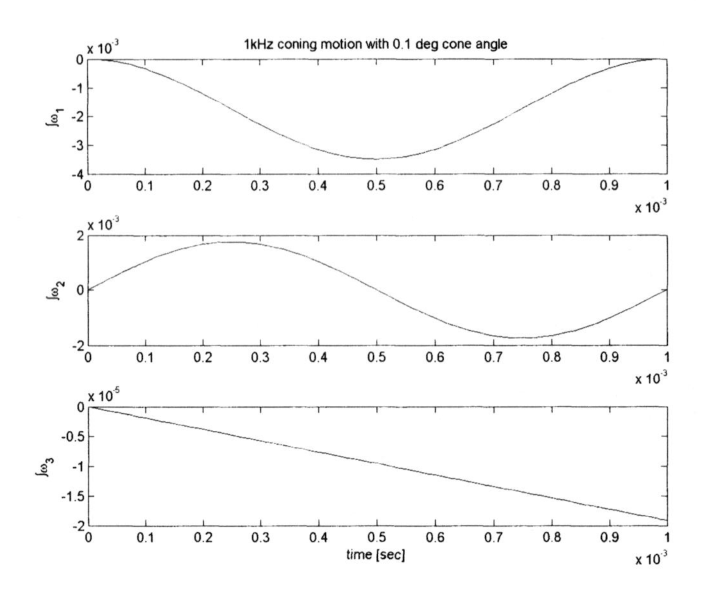

Fig. 9.26 Coning error for 1 deg cone angle, 1kHz coning rate

The integral of $\boldsymbol{\omega}_{\text{body}}$

$$\int_{s=0}^{t} \boldsymbol{\omega}_{\text{body}}(s)ds = \begin{bmatrix} -\theta_{\text{cone}} \cos(\theta_{\text{cone}}) \left[1 - \cos\left(\Omega_{\text{coning}}\, t\right)\right] \\ \theta_{\text{cone}} \cos(\theta_{\text{cone}}) \sin\left(\Omega_{\text{coning}}\, t\right) \\ -\sin(\theta_{\text{cone}})\, \theta_{\text{cone}}\, \Omega_{\text{coning}}\, t \end{bmatrix}, \quad (9.39)$$

which is what a rate integrating gyroscope would measure.

The solutions for $\theta_{\text{cone}} = 0.1^\circ$ and $\Omega_{\text{coning}} = 1$ k Hz are plotted over one cycle (1 ms) in Fig. 9.26. The first two components are cyclical, but the third component accumulates linearly over time at about -1.9×10^{-5} radians in 10^{-3} second, which is a bit more than -1°/s. *This is why coning error compensation is important.*

9.4.5.4 Rotation Vector Implementation This implementation is primarily used at a faster sampling rate than the nominal sampling rate (i.e., that required for resolving measured accelerations into navigation coordinates). It is used to remove the nonlinear effects of coning and skulling motion that would otherwise corrupt the accumulated angle rates over the nominal intersample period. This implementation is also called a "coning correction."

Bortz Model for Attitude Dynamics This exact model for attitude integration based on measured rotation rates and rotation vectors was developed by John Bortz [23]. It represents ISA attitude with respect to the reference inertial coordinate frame in terms of the rotation vector $\boldsymbol{\rho}$ required to rotate the reference

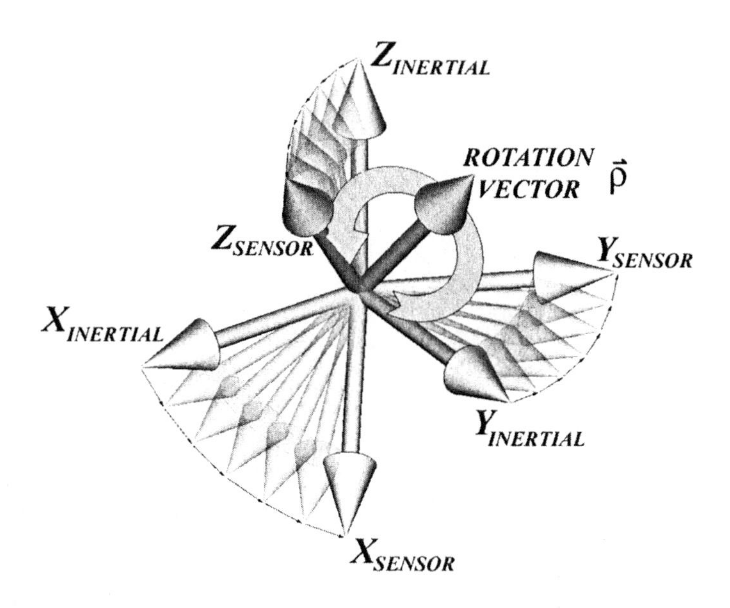

Fig. 9.27 Rotation vector representing coordinate transformation.

inertial coordinate frame into coincidence with the sensor-fixed coordinate frame, as illustrated in Fig. 9.27.

The Bortz dynamic model for attitude then has the form

$$\dot{\boldsymbol{\rho}} = \boldsymbol{\omega} + \mathbf{f}_{\text{Bortz}}(\boldsymbol{\omega}, \boldsymbol{\rho}), \tag{9.40}$$

where $\boldsymbol{\omega}$ is the vector of measured rotation rates. The Bortz "noncommutative rate vector"

$$\mathbf{f}_{\text{Bortz}}(\boldsymbol{\omega}, \boldsymbol{\rho}) = \frac{1}{2}\boldsymbol{\rho} \otimes \boldsymbol{\omega} + \frac{1}{\|\boldsymbol{\rho}\|^2}\left\{1 - \frac{\|\boldsymbol{\rho}\| \sin(\|\boldsymbol{\rho}\|)}{2\left[1 - \cos(\|\boldsymbol{\rho}\|)\right]}\right\} \boldsymbol{\rho} \otimes (\boldsymbol{\rho} \otimes \boldsymbol{\omega}), \tag{9.41}$$

$$|\boldsymbol{\rho}| < \frac{\pi}{2}. \tag{9.42}$$

Equation 9.40 represents the rate of change of attitude as a nonlinear differential equation that is linear in the measured instantaneous body rates $\boldsymbol{\omega}$. Therefore, by integrating this equation over the nominal intersample period $[0, \Delta t]$ with initial value $\boldsymbol{\rho}(0) = 0$, an exact solution of the body attitude change over that period can be obtained in terms of the net rotation vector

$$\Delta\boldsymbol{\rho}(\Delta t) = \int_0^{\Delta t} \dot{\boldsymbol{\rho}}(\boldsymbol{\rho}(s), \boldsymbol{\omega}(s))\, ds \tag{9.43}$$

that avoids all the noncommutativity errors, and satisfies the constraint of Eq. 9.42 as long as the body cannot turn 180° in one sample interval Δt. In practice, the integral is done numerically with the gyro outputs ω_1, ω_2, ω_3 sampled at intervals $\delta t \ll \Delta t$. The choice of δt is usually made by analyzing the gyro outputs under operating conditions (including vibration isolation), and selecting a sampling frequency $1/\delta t$ above the Nyquist frequency for the observed attitude rate spectrum. The frequency response of the gyros also enters into this design analysis.

The MATLAB function `fBrotz.m` on the CD-ROM calculates $\mathbf{f}_{\text{Bortz}}(\boldsymbol{\omega})$ defined by Eq. 9.41.

9.4.5.5 Quaternion Implementation The quaternion representation of vehicle attitude is the most reliable, and it is used as the "holy point" of attitude representation. Its value is maintained using the incremental rotations $\Delta\boldsymbol{\rho}$ from the rotation vector representation, and the resulting values are used to generate the coordinate transformation matrix for accumulating velocity changes in inertial coordinates.

Converting Incremental Rotations to Incremental Quaternions An incremental rotation vector $\Delta\boldsymbol{\rho}$ from the Bortz coning correction implementation of Eq. 9.43 can be converted to an equivalent incremental quaternion $\Delta\mathbf{q}$ by the operations

$$\Delta\theta = |\Delta\boldsymbol{\rho}| \text{ (rotation angle in radians)}, \tag{9.44}$$

$$\mathbf{u} = \frac{1}{\theta}\Delta\boldsymbol{\rho} \tag{9.45}$$

$$= \begin{bmatrix} u_1 \\ u_2 \\ u_3 \end{bmatrix} \text{ (unit vector)}, \tag{9.46}$$

$$\Delta\mathbf{q} = \begin{bmatrix} \cos\left(\frac{\theta}{2}\right) \\ u_1 \sin\left(\frac{\theta}{2}\right) \\ u_2 \sin\left(\frac{\theta}{2}\right) \\ u_3 \sin\left(\frac{\theta}{2}\right) \end{bmatrix}, \tag{9.47}$$

$$= \begin{bmatrix} \Delta q_0 \\ \Delta q_1 \\ \Delta q_2 \\ \Delta q_3 \end{bmatrix} \text{ (unit quaternion)}. \tag{9.48}$$

Quaternion Implementation of Attitude Integration If $\mathbf{q}_{k-1}$ is the quaternion representing the prior value of attitude, $\Delta\mathbf{q}$ is the quaternion representing the change in attitude, and $\mathbf{q}_k$ is the quaternion representing the updated value of attitude, then the update equation for quaternion representation of attitude is

$$\mathbf{q}_k = \Delta\mathbf{q} \times \mathbf{q}_{k-1} \times \Delta\mathbf{q}^\star, \tag{9.49}$$

where the postsuperscript * represents the conjugate of a quaternion.

9.4.5.6 Direction Cosines Implementation The coordinate transformation matrix $\mathbf{C}_{\text{inertial}}^{\text{body}}$ from body-fixed coordinates to inertial coordinates is needed for transforming discretized velocity changes measured by accelerometers into inertial coordinates for integration. The quaternion representation of attitude is used for computing $\mathbf{C}_{\text{inertial}}^{\text{body}}$.

Quaternions to Direction Cosines Matrices The direction cosines matrix $\mathbf{C}_{\text{inertial}}^{\text{body}}$ from body-fixed coordinates to inertial coordinates can be computed from its equivalent unit quaternion representation

$$\mathbf{q}_{\text{inertial}}^{\text{body}} = \begin{bmatrix} q_0 \\ q_1 \\ q_2 \\ q_3 \end{bmatrix} \tag{9.50}$$

as

$$\mathbf{C}_{\text{inertial}}^{\text{body}} = \left(2\,q_0^2 - 1\right)\mathbf{I}_3 + 2\begin{bmatrix} q_1 \\ q_2 \\ q_3 \end{bmatrix} \times \begin{bmatrix} q_1 \\ q_2 \\ q_3 \end{bmatrix}^T - 2\,q_0 \begin{bmatrix} q_1 \\ q_2 \\ q_3 \end{bmatrix} \otimes \tag{9.51}$$

$$= \begin{bmatrix} \left(2\,q_0^2 - 1 + 2\,q_1^2\right) & (2\,q_1q_2 + 2\,q_0q_3) & (2\,q_1q_3 - 2\,q_0q_2) \\ (2\,q_1q_2 - 2\,q_0q_3) & \left(2\,q_0^2 - 1 + 2\,q_2^2\right) & \left(2\,q_2^2 + 2\,q_0q_1\right) \\ (2\,q_1q_3 + 2\,q_0q_2) & \left(2\,q_2^2 - 2\,q_0q_1\right) & \left(2\,q_0^2 - 1 + 2\,q_3^2\right) \end{bmatrix}. \tag{9.52}$$

9.4.6 Navigation Computer and Software Requirements

Inertial navigation systems operate under conditions of acceleration, shock, and vibration that may preclude the use of hard disks or standard mounting and interconnect methods. As a consequence, INS computers tend to be somewhat specialized. The following sections list some of the requirements placed on navigation computers and software that tend to set them apart.

9.4.6.1 Physical and Operational Requirements These include

1. Size, weight, form factor, and available input power.
2. Environmental conditions such as shock/vibration, temperature, and electromagnetic interference (EMI).
3. Memory (how much and how fast), throughput (operations per second), wordlength/precision.
4. Time required between power-on and full operation, and minimum time between turnoff and turnon. (Some vehicles shut down all power during fueling, for example.)
5. Reliability, shelf-life, and storage requirements.

6. Operating life. Some applications (e.g., missiles) have operating lifetimes of minutes or seconds, others (e.g., military and commercial aircraft) may operate nearly continuously for decades.
7. Additional application-specific requirements such as radiation hardening.

9.4.6.2 Operating Systems Inertial navigation is a *real-time* process. The tasks of sampling the sensor outputs, and of integrating attitude rates, accelerations and velocities must be scheduled at precise time intervals, and the results must be available after limited delay times. The top-level operating system which prioritizes and schedules these and other tasks must be a *real time operating system* (RTOS). It may also be required to communicate with other computers in various ways.

9.4.6.3 Interface Requirements These not only include the operational interfaces to sensors and displays but may also include communications interfaces and specialized computer interfaces to support navigation software development and verification.

9.4.6.4 Software Development Because INS failures could put host vehicle crews and passengers at risk, it is very important during system development to demonstrate high reliability of the software. INS software is usually developed offline on a general-purpose computer interfaced to the navigation computer. Software development environments for INS typically include code editors, cross-compilers, navigation computer emulators, hardware simulators, hardware-in-the-loop interfaces, and specialized source–code–online interfaces to the navigation computer for monitoring, debugging and verifying the navigation software on the navigation computer. Software developed for manned missions must be acceptably reliable, which requires metrics for demonstrating reliability. In addition, real-time programmers for INS do tend to be a different breed of cat from general-purpose programmers, and software development cost can be a significant fraction of overall system development cost [28].

9.5 SYSTEM-LEVEL ERROR MODELS

The system-level implementation models discussed in the previous section are for the internal implementation of the inertial navigation system, itself. These are models for the peculiarities of the sensors and software that contribute to navigation errors. They are used in INS design analysis for predicting performance as a function of component characteristics. They are also used within the implementation for compensating the corrupting influence of sensor and software error tolerances on the measured and inferred vehicle dynamic variables.

The system-level error models in this section are for implementing GNSS/INS integration. These models represent how the resulting navigation errors will propagate over time, as functions of the error parameters of the inertial sensors. They are used in two ways:

1. In so-called "loosely coupled" approaches for keeping track of the uncertainty in the INS navigation solution to use in a Kalman gain for combining the GNSS and INS navigation solutions to maintain an integrated navigation solution. When GNSS signals are not available, the model can still be used to propagate the estimated INS errors and subtract them from the uncompensated INS navigation solution. The resulting compensated INS navigation solution can then be used to speed up detection and reacquisition of GNSS signals—if and when they become available again.
2. In more "tightly coupled" approaches using GNSS measurements to estimate and compensate for random variations in the calibration parameters of individual sensors in the INS. These approaches continually re-calibrate the INS when GNSS signals are available. They are functionally similar to the loosely coupled approaches, in that they still carry forward the calibration-compensated INS navigation solution when GNSS signals are unavailable. Unlike the loosely coupled approaches, however, they are feedback-based and not as sensitive to modeling errors.

Model Diversity There is no universal INS error model for GNSS/INS integration, because there is no standard design for an INS. There may be differences between different GNSS systems, and between generations of GPS satellites, but GNSS error models are all quite similar. Differences between error models for INSs, on the other hand, can be anything but minor. There are some broad INS design types (e.g., gimbaled vs. strapdown), but there are literally thousands of different inertial sensor designs that can be used for each INS type.

Methodology We present here a variety of inertial system error models, which will be sufficient for many of the sensors in common use, but not for every conceivable inertial sensor. For applications with sensor characteristics different from those used here, the use of these error models in GNSS/INS integration will serve to illustrate the general integration *methodology,* so that users can apply the same methodology to GNSS/INS integration with other sensor error models, as well.

9.5.1 Error Sources

9.5.1.1 Initialization Errors Inertial navigators can only integrate sensed accelerations to propagate initial estimates of position and velocity. Systems without GNSS aiding require other sources for their initial estimates of position and velocity. Initialization errors are the errors in these initial values.

9.5.1.2 Alignment Errors Most standalone INS implementations include an initial period for alignment of the gimbals (for gimbaled systems) or attitude direction cosines (for strapdown systems) with respect to the navigation axes. Errors remaining at the end of this period are the alignment errors. These include *tilts* (rotations about horizontal axes) and *heading errors* (rotations about the vertical axis).

Tilt errors introduce acceleration errors through the miscalculation of gravitational acceleration, and these propagate primarily as Schuler oscillations (see Section 9.5.2.1) plus a non-zero-mean position error approximately equal to the tilt error in radians times the radius from the earth center. Initial azimuth errors primarily rotate the system trajectory about the starting point, and there are secondary effects due to coriolis accelerations and excitation of Schuler oscillations.

9.5.1.3 Sensor Compensation Errors Sensor calibration is a procedure for estimating the parameters of models used in sensor error compensation. It is not uncommon for these modeled parameters to change over time and between turnons, and designing sensors to make the parameters sufficiently constant can also make the sensors relatively expensive. Costs resulting from stringent requirements for parameter stability can be reduced significantly for sensors that will be used in integrated GNSS/INS applications, because Kalman filter–based GNSS/INS integration can use the differences between INS-derived position and GNSS-derived position to make corrections to the calibration parameters.

These nonconstant sensor compensation parameters are not true parameters (i.e., constants), but "slow variables," which change slowly relative to the other dynamic variables. Other slow variables in the integrated system model include the satellite clock offsets for selective availability (SA).

The GNSS/INS integration filter implementation requires models for how variations in the compensation parameters propagate into navigation errors. These models are derived in Section 9.3 for the more common types of sensors and their compensation parameters.

9.5.1.4 Gravity Model Errors The influence of unknown gravity modeling errors on vehicle dynamics is usually modeled as a zero-mean exponentially correlated acceleration process (see Section 8.5)

$$\delta \mathbf{a}_k = e^{-\Delta t/\tau_{\text{correlation}}} \delta \mathbf{a}_{k-1} + \mathbf{w}_k, \tag{9.53}$$

where Δt is the filter period, the correlation time

$$\tau_{\text{correlation}} \approx \frac{d_{\text{correlation}}}{|v_{\text{horizontal}}|}, \tag{9.54}$$

$v_{\text{horizontal}}$ is horizontal velocity, $d_{\text{correlation}}$ is the horizontal correlation distance of gravity anomalies (usually on the order of 10^4–10^5 m), w_k is a zero-mean white-noise process with covariance matrix

$$\mathbf{Q}_{\text{gravity model}} \stackrel{\text{def}}{=} \mathrm{E}\left\langle \mathbf{w}_k \mathbf{w}_k^T \right\rangle \tag{9.55}$$

$$\approx a_{\text{RMS}}^2 (1 - e^{-2\Delta t/\tau})\, \mathbf{I}, \tag{9.56}$$

a_{RMS}^2 is the variance of acceleration error, and $\mathbf{I}$ is an identity matrix. The correlation distance $d_{\text{correlation}}$ and RMS acceleration disturbance a_{RMS} will generally depend upon the local terrain. Here, $d_{\text{correlation}}$ tends to be larger and a_{RMS} smaller as terrain becomes more gentle or (for aircraft) as altitude increases.

The effects of gravity modeling errors in the vertical direction will be mediated by vertical channel stabilization.

9.5.2 Navigation Error Propagation

The dynamics of INS error propagation are strongly influenced by the fact that gravitational accelerations point toward the center of the earth and decrease in magnitude with altitude and is somewhat less influenced by the fact that the earth rotates.

9.5.2.1 Schuler Oscillations of INS Errors The dominant effect of alignment errors on free-inertial navigation is from *tilts*, also called *leveling errors*. These are errors in the estimated direction of the local vertical in sensor-fixed coordinates. The way in which tilts translate into navigation errors is through a process called *Schuler oscillation*. These are oscillations at the same period that Schuler had identified for gyrocompassing errors (see Section 9.1.1.) The analogy between these Schuler oscillations of INS errors and those of a simple gravity pendulum is illustrated in Figure 9.28. The physical force acting on the mass of a gravity pendulum is the vector sum of gravitational acceleration (mg) and the tension T in the support are, as shown in Figure 9.28*a*. The analogous acceleration driving inertial navigation errors is the difference between the modeled gravitational acceleration (which changes direction with *estimated* location) and the actual gravitational acceleration, as shown in Figure 9.28*b*. In the case of the gravity pendulum, the physical mass of the pendulum is oscillating. In the case of INS errors, only the estimated position, velocity, and acceleration *errors* oscillate. The gravity pendulum is a physical device, but the Schuler pendulum is a theoretical model to illustrate how INS errors behave.

In either case, the restoring acceleration is approximately related to displacement δ from the equilibrium position by the harmonic equation

$$\ddot{\delta} \approx -\frac{g}{L \text{ or } R}\,\delta, \tag{9.57}$$

$$= -\omega^2\,\delta \tag{9.58}$$

$$\omega = \sqrt{\frac{g}{L \text{ or } R}}, \tag{9.59}$$

the solution for which is

$$\delta(t) = |\delta|_{\max} \cos\left(\omega t + \phi\right), \tag{9.60}$$

where g is the acceleration due to gravity at the surface of the earth, L is the length of the support arm of the gravity pendulum, R is the radius to the center

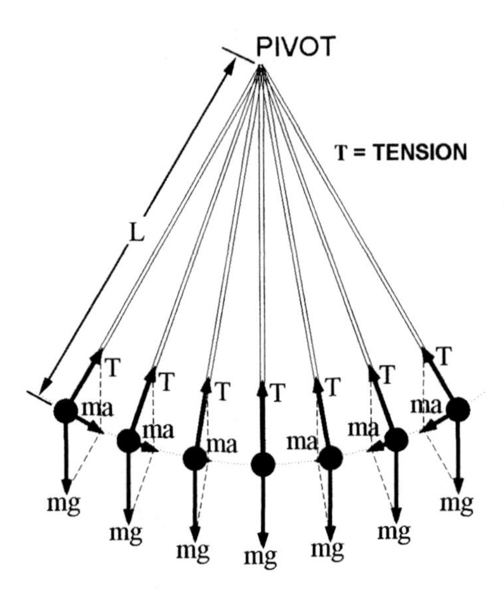

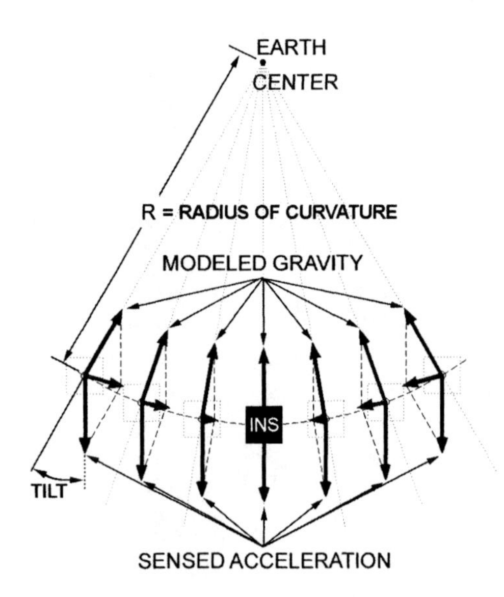

Fig. 9.28 Pendulum model for Schuler oscillations of INS errors: (*a*) gravity pendulum model; (*b*) Schuler oscillation model.

of the earth, $|\delta|_{max}$ is the peak displacement, ω is the oscillation frequency (in radians per second), and ϕ is an arbitrary oscillation phase angle.

In the case of the gravity pendulum, the period of oscillation

$$T_{\text{gravity pendulum}} = \frac{2\pi}{\omega}, \tag{9.61}$$

$$= \frac{2\pi\sqrt{L}}{\sqrt{g}}, \tag{9.62}$$

and for the Schuler pendulum

$$T_{\text{Schuler}} = \frac{2\pi\sqrt{R}}{\sqrt{g}}, \tag{9.63}$$

$$\approx \frac{2 \times 3.1416 \times \sqrt{6378170\ (\text{m})},}{\sqrt{9.8\ (\text{m/s}^2)}} \tag{9.64}$$

$$\approx 5069\ (\text{s}) \tag{9.65}$$

$$\approx 84.4\ (\text{min}) \tag{9.66}$$

at the surface of the earth.

This $\approx$ 84.4-min period is called the Schuler period. It is also the orbital period of a satellite at that radius (neglecting atmospheric drag), and the exponential time constant of altitude error instability in pure inertial navigation (see Section 9.5.2.2).

The corresponding angular frequency (Schuler frequency)

$$\Omega_{\text{Schuler}} = \frac{2\pi}{T_{\text{Schuler}}} \tag{9.67}$$

$$\approx 0.00124\ (\text{rad/s}) \tag{9.68}$$

$$\approx 0.0002\ (\text{Hz}). \tag{9.69}$$

Dependence on Position and Direction. A spherical earth model was used to illustrate the Schuler pendulum. The period of Schuler oscillation actually depends on the radius of curvature of the equipotential surface of the gravity model, which can be different in different directions and vary with longitude, latitude, and altitude. However, the variations in Schuler period due to these effects are generally in the order of parts per thousand, and are usually ignored.

Impact on INS Performance Schuler oscillations include variations in the INS errors in position, velocity, and acceleration (or tilt), which are all related harmonically. Thus, if the peak position displacement from the median location is

$|\delta|_{\max}$, then

$$\delta(t) = |\delta|_{\max} \cos(\Omega_{\text{Schuler}}\, t + \phi) \text{ (position error)}, \tag{9.70}$$

$$\dot{\delta}(t) = -|\delta|_{\max}\, \Omega_{\text{Schuler}} \sin(\Omega_{\text{Schuler}}\, t + \phi) \text{ (velocity error)}, \tag{9.71}$$

$$\ddot{\delta}(t) = -|\delta|_{\max}\, \Omega^2_{\text{Schuler}} \cos(\Omega_{\text{Schuler}}\, t + \phi) \text{ (acceleration error)}, \tag{9.72}$$

$$\ddot{\delta}(t) = \frac{-|\delta|_{\max}\, \Omega^2_{\text{Schuler}}}{g} \cos(\Omega_{\text{Schuler}}\, t + \phi) \text{ (acceleration error in } g \text{ values)}, \tag{9.73}$$

$$= \tau(t) \text{ (tilt error in radians)}. \tag{9.74}$$

Note, however, that when the initial INS error is a pure tilt error (i.e., no position error), the peak position error will be $2\,|\delta|_{\max}$ after ≈ 42.2 min from the starting location, and the RMS position error will be $\left(1 + 1/\sqrt{2}\right)|\delta|_{\max}$. If the initial INS error is a pure tilt error, then the true INS position would at one end of the Schuler pendulum swing—not the middle—and the peak and RMS position errors

$$|\delta|_{\text{peak}} = \frac{2\,\tau_{\text{initial}}\, g}{\Omega^2_{\text{Schuler}}}, \tag{9.75}$$

$$|\delta|_{\text{RMS}} = \frac{\left(1 + \frac{1}{\sqrt{2}}\right)\tau_{\text{initial}}\, g}{\Omega^2_{\text{Schuler}}}, \tag{9.76}$$

as plotted in Fig. 9.29. This shows why alignment tilt errors are so important in free inertial navigation. Tilts as small as one milliradian can cause peak position excursions as big as 10 km after 42 min.

Figure 9.30 is a plot generated using the MATLAB INS Toolbox from GPSoft, showing how an initial northward velocity error of 0.1 ms excites Schuler oscillations in the INS navigation errors, and how coriolis accelerations rotate the direction of oscillation—just like a Foucault pendulum with Schuler period. The total simulated time is about/14 h, enough time for 10 Schuler oscillation periods.

For a maximum velocity error $|\dot{\delta}|_{\max} = 0.1$ m/s, the maximum expected position error would be

$$\begin{aligned}|\delta|_{\max} &= \frac{|\dot{\delta}|_{\max}}{\Omega_{\text{Schuler}}} \\ &\approx \frac{0.1 \text{ m/s}}{0.00124 \text{ rad/s}} \\ &\approx 80 \text{ meter},\end{aligned}$$

which is just about the maximum excursion seen in Fig. 9.30.

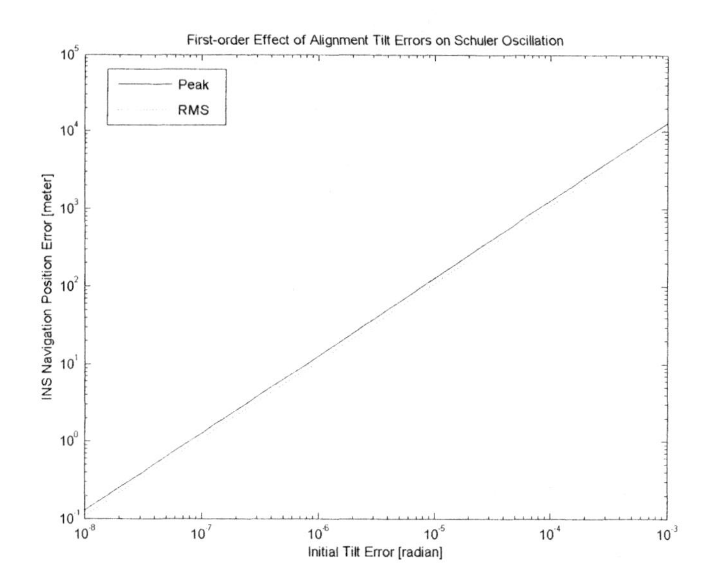

Fig. 9.29 Effect of tilt errors on Schuler oscillations.

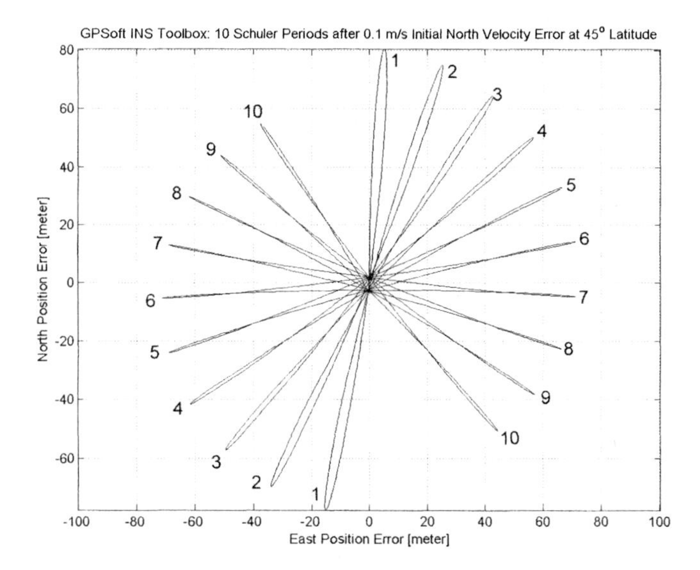

Fig. 9.30 GPSoft INS Toolkit simulation of Schuler oscillations.

9.5.2.2 Vertical Channel Instability The modeled vertical acceleration due to gravity in the downward direction, as a function of height h above the reference geoid, is dominated by the first term

$$a_{\text{gravity}} = \frac{\text{GM}}{(R+h)^2} + \text{ less significant terms}, \tag{9.77}$$

where R is the reference radius and GM is the model gravitational constant, $\approx 398{,}600\ \text{km}^3/\text{s}^2$ for the WGS84 gravity model. Consequently, the vertical gradient of downward gravity in the downward direction will be

$$-\frac{\partial a_{\text{gravity}}}{\partial h} \approx \frac{a_{\text{gravity}}}{(R+h)} \tag{9.78}$$

$$\approx \frac{a_{\text{gravity}}}{R}, \tag{9.79}$$

which is positive. Therefore, if δh is the altitude estimation error, it satisfies a differential equation of the form

$$\ddot{\delta}_h \approx \frac{a_{\text{gravity}}}{R} \delta_h, \tag{9.80}$$

which is exponentially unstable with solution

$$\delta_h(t) \approx \delta_h(t_0) \exp\left[\sqrt{\frac{a_{\text{gravity}}}{R}}\,(t - t_0)\right]. \tag{9.81}$$

Not surprisingly, the exponential time constant of this vertical channel instability

$$T_{\text{vertical channel}} = \sqrt{\frac{R}{a_{\text{gravity}}}} \tag{9.82}$$

$$\approx 84.4 \text{ min}, \tag{9.83}$$

the Schuler period (see Section 9.5.2.1).

9.5.2.3 Coriolis Coupling The coriolis effect couples position error rates (velocities) into into their second derivative (accelerations) as

$$\frac{d}{dt}\begin{bmatrix} \delta v_E \\ \delta v_N \\ \delta v_U \end{bmatrix} = -2\,\Omega_{\text{earth}} \begin{bmatrix} 0 & -\sin\phi & \cos\phi \\ \sin\phi & 0 & 0 \\ -\cos\phi & 0 & 0 \end{bmatrix} \begin{bmatrix} \delta v_E \\ \delta v_N \\ \delta v_U \end{bmatrix}, \tag{9.84}$$

$$\Omega_{\text{earth}} \approx 7.3 \times 10^{-5}\ (\text{rad/s})\ (\text{earthrate}), \tag{9.85}$$

where ϕ is geodetic latitude. Adding Schuler oscillations yields the model

$$\frac{d}{dt}\delta v_E = -\Omega^2_{\text{Schuler}}\delta E + 2\Omega_{\text{earth}}\sin\phi\,\delta v_N - 2\Omega_{\text{earth}}\cos\phi\,\delta v_U, \tag{9.86}$$

$$\frac{d}{dt}\delta v_N = -\Omega^2_{\text{Schuler}}\delta N - 2\Omega_{\text{earth}}\sin\phi\,\delta v_E, \tag{9.87}$$

$$\frac{d}{dt}\delta v_U = 2\Omega_{\text{Earth}}\cos\phi\,\delta v_E, \tag{9.88}$$

where δE is east position error and δN is north position error. This effect can be seen in Fig. 9.30, in which the coriolis acceleration causes the error trajectory to swerve to the right. The sign of $\sin\phi$ in Eq. 9.86 is negative in the southern hemisphere, so it would swerve to the left there.

Note that, through coriolis coupling, vertical channel errors couple into east channel errors.

9.5.3 Sensor Error Propagation

Errors made in compensating for inertial sensor errors will cause navigation errors. Here, we derive some approximating formulas for how errors in individual compensation parameters propagate into velocity and position errors.

9.5.3.1 Scale Factors, Biases, and Misalignments The models for bias, scale factor and input axis misalignment compensation are the same for gyroscopes and accelerometers. The compensated sensor output

$$\mathbf{z}_{\text{comp}}, = \mathbf{M}\mathbf{z}_{\text{output}} + \mathbf{b}_z \tag{9.89}$$

for $\mathbf{z} = \boldsymbol{\omega}$ (for gyroscopes) or $\mathbf{z} = \mathbf{a}$ (for accelerometers). The sensitivity of the compensated output to bias is then

$$\frac{\partial \mathbf{z}_{\text{comp}}}{\partial \mathbf{b}_z} = \mathbf{I}_3, \tag{9.90}$$

the 3×3 identity matrix, and the sensitivity to the elements $m_{j,k}$ of $\mathbf{M}$ are

$$\frac{\partial \mathbf{z}_{\text{comp},i}}{\partial m_{j,k}} = \begin{cases} 0, & i \neq j \\ z_{\text{output},k} & i = j \end{cases} \tag{9.91}$$

If we put these 12 calibration parameters in vector form as

$$\mathbf{p}_{\text{comp}} = [\, b_{z,1} \quad b_{z,2} \quad b_{z,3} \quad m_{1,1} \quad m_{1,2} \quad m_{1,3} \quad m_{2,1} \quad m_{2,2} \quad \dots$$
$$m_{2,3} \quad m_{3,1} \quad m_{3,2} \quad m_{3,3} \,]^T, \tag{9.92}$$

then the matrix of partial derivatives

$$\frac{\partial \mathbf{z}_{\text{comp},\, i}}{\partial \mathbf{p}_{\text{comp}}} = \begin{bmatrix} 1 & 0 & 0 \\ 0 & 1 & 0 \\ 0 & 0 & 1 \\ z_{\text{output},\,1} & 0 & 0 \\ z_{\text{output},\,2} & 0 & 0 \\ z_{\text{output},\,3} & 0 & 0 \\ 0 & z_{\text{output},\,1} & 0 \\ 0 & z_{\text{output},\,2} & 0 \\ 0 & z_{\text{output},\,3} & \\ 0 & 0 & z_{\text{output},\,1} \\ 0 & 0 & z_{\text{output},\,2} \\ 0 & 0 & z_{\text{output},\,3} \end{bmatrix}. \tag{9.93}$$

For analytical purposes, this matrix of partial derivatives would be evaluated under "nominal" conditions, which could be for $\mathbf{M} = \mathbf{I}$ and $\mathbf{b}_z = 0$. In that case, $\mathbf{z}_{\text{output}} = \mathbf{z}_{\text{input}}$, the bias sensitivities will be unitless (e.g., g/g), the scale factor sensitivities will be in units of parts-per-part and the misalignment sensitivities will be in output units per radian.

This matrix can be augmented with additional calibration parameters, such as acceleration-sensitivities for gyroscopes or temperature sensitivities. It only requires taking the necessary partial derivatives.

9.5.3.2 Accelerometer Compensation Error Propagation Acceleration errors due to accelerometer compensation errors in sensor-fixed coordinates and navigation coordinates will then be

$$\delta \mathbf{a}_{\text{sensor}} \approx \frac{\partial \mathbf{a}_{\text{comp}}}{\partial \mathbf{p}_{\text{acc.comp}}} \delta \mathbf{p}_{\text{acc.comp}}, \tag{9.94}$$

$$\delta \mathbf{a}_{\text{nav}} = \mathbf{C}_{\text{nav}}^{\text{sensor}} \delta \mathbf{a}_{\text{sensor}} \tag{9.95}$$

$$\approx \mathbf{C}_{\text{nav}}^{\text{sensor}} \frac{\partial \mathbf{a}_{\text{comp}}}{\partial \mathbf{p}_{\text{acc.comp}}} \delta \mathbf{p}_{\text{acc.comp}}, \tag{9.96}$$

where $\delta \mathbf{p}_{\text{acc.comp}}$ is the vector of accelerometer compensation parameter errors, the partial derivative matrix is the one in Eq. 9.93 with $\mathbf{z} = \mathbf{a}$, and $\mathbf{C}_{\text{nav}}^{\text{sensor}}$ is the coordinate transformation matrix from sensor-fixed coordinates to navigation coordinates. For example

- For gimbaled systems aligned to locally level coordinates with sensor axes pointing north and east

$$\mathbf{C}_{\text{nav}}^{\text{sensor}} = \mathbf{I}, \text{ the identity matrix.} \tag{9.97}$$

- For a carouseled gimbal system aligned to locally level coordinates

$$\mathbf{C}_{\text{nav}}^{\text{sensor}} = \begin{bmatrix} \cos\alpha & \sin\alpha & 0 \\ -\sin\alpha & \cos\alpha & 0 \\ 0 & 0 & 1 \end{bmatrix}, \tag{9.98}$$

where α is the carousel angle.

- For a strapdown system aligned to body coordinates of the host vehicle

$$\mathbf{C}_{\text{nav}}^{\text{sensor}} = \mathbf{C}_{\text{nav}}^{\text{body}}. \tag{9.99}$$

- For a carouseled strapdown system rotating about the body-fixed yaw axis

$$\mathbf{C}_{\text{nav}}^{\text{sensor}} = \mathbf{C}_{\text{nav}}^{\text{body}} \begin{bmatrix} \cos\alpha & \sin\alpha & 0 \\ -\sin\alpha & \cos\alpha & 0 \\ 0 & 0 & 1 \end{bmatrix}. \tag{9.100}$$

Velocity and Position Errors. The velocity error sensitivities to each of the compensation parameters will be the integral over time of the acceleration sensitivities, and the position error sensitivities to each of the compensation parameters will be the integral over time of the velocity sensitivities. However, the accelerations must be transformed into navigation coordinates before integration:

$$\delta\mathbf{v}_{\text{nav}}(t) = \delta\mathbf{v}_{\text{nav}}(t_0) + \int_0^t \delta\mathbf{a}_{\text{nav}}(s)\, ds \tag{9.101}$$

$$= \delta\mathbf{v}_{\text{nav}}(t_0) + \int_{t_0}^t \mathbf{C}_{\text{nav}}^{\text{sensor}}(s)\ \delta\mathbf{a}_{\text{sensor}}(s)\ ds \tag{9.102}$$

$$\approx \delta\mathbf{v}_{\text{nav}}(t_0) + \int_{t_0}^t \mathbf{C}_{\text{nav}}^{\text{sensor}}(s) \frac{\partial\mathbf{a}_{\text{comp}}}{\partial\mathbf{p}_{\text{acc.comp}}}(s)\ \delta\mathbf{p}_{\text{acc.comp}}, \tag{9.103}$$

$$\delta\mathbf{x}_{\text{nav}}(t) \approx \delta\mathbf{x}_{\text{nav}}(t_0) + (t - t_0)\,\delta\mathbf{v}_{\text{nav}}(t_0) \tag{9.104}$$

$$+ \int\int_{t_0}^t \mathbf{C}_{\text{nav}}^{\text{sensor}}(s) \frac{\partial\mathbf{a}_{\text{comp}}}{\partial\mathbf{p}_{\text{acc.comp}}}(s)\ \delta\mathbf{p}_{\text{acc.comp}} ds, \tag{9.105}$$

where $\delta\mathbf{x}_{\text{nav}}$ is the navigation position error due to compensation parameter errors. The GNSS navigation solution will not include $\delta\mathbf{x}_{\text{nav}}$, and it is the difference between the INS and GNSS solutions that can be used to estimate the compensation parameter errors.

9.5.3.3 Gyroscope Compensation Error Propagation The principal effect of gyroscope compensation errors on inertial navigation position errors is from the

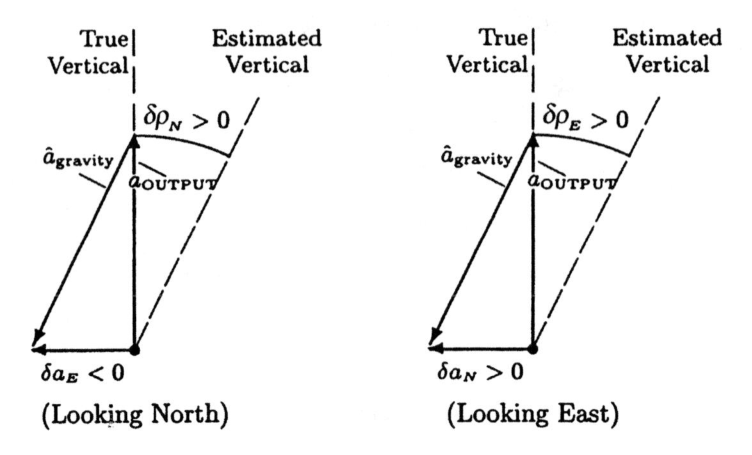

Fig. 9.31 Acceleration errors due to tilts.

miscalculation of gravitational acceleration due to the resulting tilt errors, as illustrated in Fig. 9.31, where

$$\delta a_E \approx -g\delta\rho_N, \tag{9.106}$$

$$\delta a_N \approx g\delta\rho_E, \tag{9.107}$$

and $\delta\rho_N$ is a small rotation about the north-pointing coordinate axis and $\delta\rho_E$ is the corresponding east tilt component.

Tilt Errors Small tilt errors due to calibration errors can be approximated as the horizontal north and east components of a rotation vector $\delta\boldsymbol{\rho}$:

$$\frac{d}{dt}\delta\boldsymbol{\rho}_{\text{nav}}(t) \approx \delta\boldsymbol{\omega}_{\text{nav}}(t) \tag{9.108}$$

$$= \mathbf{C}_{\text{nav}}^{\text{sensor}}(t)\delta\,\boldsymbol{\omega}_{\text{sensor}}(t) \tag{9.109}$$

$$\approx \mathbf{C}_{\text{nav}}^{\text{sensor}}(t)\frac{\partial\boldsymbol{\omega}_{\text{comp}}}{\partial\mathbf{p}_{\text{gyro.comp}}}(t)\,\delta\mathbf{p}_{\text{gyro.comp}} \tag{9.110}$$

The east and north tilt components can then be substituted into Eqs. 9.106 and 9.107 to obtain the equations for position error due to tilts. Schuler oscillations are excited when these position errors, in turn, cause tilts.

Velocity Errors For the tilt error angles $\delta\rho_E$, $\delta\rho_N$ in radians and $g \approx 9.8$ m/s^2, the corresponding velocity errors will propagate as

$$\frac{d}{dt}\delta v_E(t) \approx g\delta\rho_N(t), \tag{9.111}$$

$$\frac{d}{dt}\delta v_N(t) \approx -g\delta\rho_E(t). \tag{9.112}$$

Effect of Heading Errors Navigation error sensitivity to rotational error ρ_U about the local vertical (i.e., heading error) is usually smaller than the sensitivities to tilt-related errors. The time rate-of-change of position errors due to heading error are

$$\frac{d}{dt}\delta x_E(t) \approx -\delta\rho_U v_N, \tag{9.113}$$

$$\frac{d}{dt}\delta x_N(t) \approx \delta\rho_U v_E, \tag{9.114}$$

where δx_E and δx_N are the navigation error components due to heading error $\delta\rho_U$ (measured counterclockwise in radians) and v_E and v_N are the vehicle velocity components in the east and north directions, respectively.

9.5.4 Examples

9.5.4.1 Damping Vertical Channel Errors The propagation of altitude error δh over time is governed by Eq. 9.80:

$$\begin{aligned}\frac{d^2}{dt^2}\delta h &\approx \frac{a_{\text{gravity}}}{R+h}\delta h \\ &\approx \frac{\delta h}{T^2_{\text{Schuler}}},\end{aligned} \tag{9.115}$$

where the Schuler period $T_{\text{Schuler}} \approx 5046$ s at sea level. The INS initialization procedure can make the initial value $\delta h(t_0)$ very small (e.g., in the order of a meter). It can also make the vertical component of the initial gravity model error very small (e.g., by making it match the sensed vertical acceleration during alignment). Thereafter, the vertical channel navigation error due to zero-mean white accelerometer noise $w(t)$ will propagate according to the model

$$\frac{d}{dt}\begin{bmatrix}\delta h(t) \\ \delta v_U(t)\end{bmatrix} = \underbrace{\begin{bmatrix}0 & 1 \\ \frac{1}{T^2_{\text{Schuler}}} & 0\end{bmatrix}}_{\mathbf{F}_{\text{vert.chan}}}\begin{bmatrix}\delta h(t) \\ \delta v_U(t)\end{bmatrix} + \begin{bmatrix}0 \\ w(t)\end{bmatrix}, \tag{9.116}$$

where $\delta v_U(t) = \frac{d}{dt}\delta h(t)$ is vertical velocity error.

The equivalent state transition matrix in discrete time with timestep Δt is

$$\begin{aligned}\Phi_{\text{vert.chan}} &= \exp(\mathbf{F}\,\Delta t) \\ &= \frac{1}{2}e^{\frac{\Delta t}{T_{\text{Schuler}}}}\begin{bmatrix}1 & T_{\text{Schuler}} \\ T^{-1}_{\text{Schuler}} & 1\end{bmatrix} \\ &\quad + \frac{1}{2}e^{-\frac{\Delta t}{T_{\text{Schuler}}}}\begin{bmatrix}1 & -T_{\text{Schuler}} \\ -T^{-1}_{\text{Schuler}} & 1\end{bmatrix},\end{aligned} \tag{9.117}$$

and the corresponding Riccati equation for propagation of the covariance matrix $\mathbf{P}_{\text{vert.chan}}$ of vertical channel navigation errors has the form

$$\mathbf{P}_{\text{vert.chan},\,k} = \Phi_{\text{vert.chan}}\mathbf{P}_{\text{vert.chan},\,k-1}\Phi^{T}_{\text{vert.chan}} + \begin{bmatrix} 0 & 0 \\ 0 & q_{\text{accelerometer}} \end{bmatrix}, \quad (9.118)$$

where $q_{\text{accelerometer}}$ is the incremental variance of velocity uncertainty per timestep Δt due to vertical altimeter noise. For example, for velocity random walk errors specified as having VRW meter per second per root hour, we obtain

$$q_{\text{accelerometer}} = \frac{\text{VRW}^2\,\Delta t}{3600}. \quad (9.119)$$

Figure 9.32 is a plot of altitude uncertainty versus time after INS initialization for a range of accelerometer white noise levels, from 10^{-2} to 10^2 m/s/$\sqrt{\text{hr}}$. All the solid-line plots will increase over time without bound.

Barometric Altimeter for Vertical Channel Damping. If a barometric altimeter is to be used for vertical channel stabilization, then the altimeter error $\delta h_{\text{altimeter}}$ will not be zero-mean white noise, but something more like a zero-mean exponentially correlated error. This sort of error has a discrete-time model of the form

$$\delta h_{\text{altimeter},\,k} = \exp\frac{\Delta t}{\tau_{\text{altimeter}}}\,\delta h_{\text{altimeter},\,k} + w_k, \quad (9.120)$$

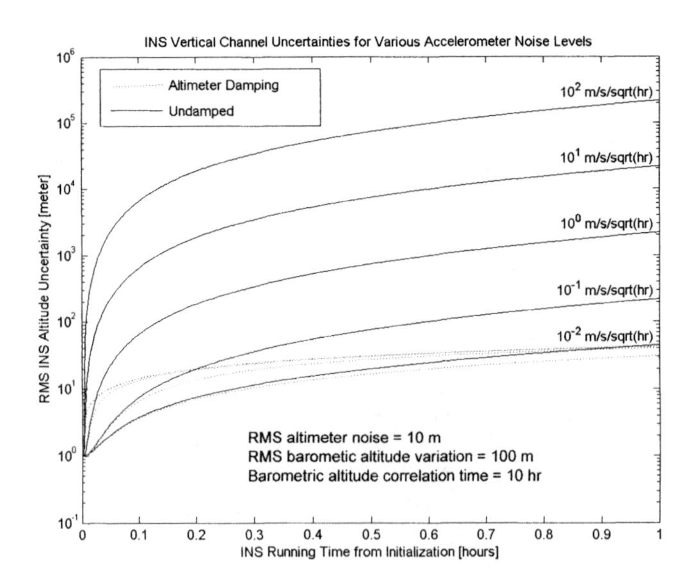

Fig. 9.32 INS vertical instability, with and without altimeter aiding.

where $\tau_{\text{altimeter}}$ is the correlation time and the white-noise sequence $\{w_k\}$ has variance

$$q_{\text{altimeter}} = \left(1 - \exp\frac{-2\Delta t}{\tau_{\text{altimeter}}}\right)\sigma^2_{\text{altitude}}, \tag{9.121}$$

for steady-state altitude variance $\sigma^2_{\text{altitude}}$.

In this case, the augmented vertical channel state vector

$$\mathbf{x}_{\text{aug.vert.chan}} = \begin{bmatrix} \delta h(t) \\ \delta v_U(t) \\ \delta h_{\text{altimeter}} \end{bmatrix}, \tag{9.122}$$

and the resulting Riccati equations for state uncertainty will be

$$\mathbf{P}_{\text{aug.vert.chan},\,k(-)} = \mathbf{\Phi}_{\text{aug.vert.chan}}\mathbf{P}_{\text{aug.vert.chan},\,k-1(+)}\mathbf{\Phi}^T_{\text{aug.vert.chan}} + \begin{bmatrix} 0 & 0 & 0 \\ 0 & q_{\text{accelerometer}} & 0 \\ 0 & 0 & q_{\text{altimeter}} \end{bmatrix} \text{ (a priori)}, \tag{9.123}$$

$$\mathbf{\Phi}_{\text{aug.vert.chan}} = \begin{bmatrix} \mathbf{\Phi}_{\text{vert.chan}} & 0 \\ 0 & \exp(-\Delta t \,/\, \tau_{\text{altimeter}}) \end{bmatrix} \text{ (state transition)}, \tag{9.124}$$

$$\mathbf{P}_{\text{aug.vert.chan},\,k(+)} = \mathbf{P}_{\text{aug.vert.chan},\,k(-)} - \overline{\mathbf{K}}_k\mathbf{H}\mathbf{P}_{\text{aug.vert.chan},\,k(-)} \text{ (a posteriori)}, \tag{9.125}$$

$$\overline{\mathbf{K}}_k = \frac{\mathbf{P}_{\text{aug.vert.chan},\,k(-)}\mathbf{H}^{\mathrm{T}}}{\mathbf{H}\mathbf{P}_{\text{aug.vert.chan},\,k(-)}\mathbf{H}^{\mathrm{T}} + R_{\text{altimeter}}} \text{ (Kalman gain)}, \tag{9.126}$$

$$\mathbf{H} = \begin{bmatrix} 1 & 0 & 1 \end{bmatrix} \text{ (measurement sensitivity)}, \tag{9.127}$$

where $R_{\text{altimeter}}$ is mean-squared altimeter noise, exclusive of the correlated component $\delta h_{\text{altimeter}}$.

The dotted lines in Fig. 9.32 are plots of altitude uncertainty with vertical channel damping, using a barometric altimeter. The assumed atmospheric and altimeter model parameters are written on the plot. These show much better performance than does the undamped case, over the same range of accelerometer noise levels. In all cases, the damped results do not continue to grow without bound.

Figure 9.32 was generated by the m-file `VertChanErr.m` on the CD-ROM.

9.5.4.2 Carouseling Accelerometer bias errors $\delta\mathbf{b}_{\mathrm{acc}}$ couple into horizontal navigation errors, as modeled in Eqs. 9.86–9.87 and 9.96:

$$\frac{d}{dt}\begin{bmatrix}\delta E\\ \delta N\\ \delta v_E\\ \delta v_N\end{bmatrix} = \begin{bmatrix} 0 & 0 & 1 & 0\\ 0 & 0 & 0 & 1\\ -\Omega_{\mathrm{Schuler}}{}^2 & 0 & 0 & 2\,\Omega_{\mathrm{earth}}\sin(\phi)\\ 0 & -\Omega_{\mathrm{Schuler}}{}^2 & -2\,\Omega_{\mathrm{earth}}\sin(\phi) & 0 \end{bmatrix} \times \begin{bmatrix}\delta E\\ \delta N\\ \delta v_E\\ \delta v_N\end{bmatrix} + \begin{bmatrix} 0_{2\times 1}\\ \begin{bmatrix}1 & 0 & 0\\ 0 & 1 & 0\end{bmatrix}\mathbf{C}_{\mathrm{nav}}^{\mathrm{sensor}}\,\delta\mathbf{b}_{\mathrm{acc}}\end{bmatrix}, \tag{9.128}$$

where $\delta\mathbf{b}_{\mathrm{acc}}$ is the vector of accelerometer biases and $\mathbf{C}_{\mathrm{nav}}^{\mathrm{sensor}}$ can have any of the values given in Eqs. 9.97–9.100.

Figure 9.33 is a plot of fourteen hours of simulated INS position errors resulting from 10-μg north accelerometer bias on a gimbaled INS, with and without carouseling. The simulated carousel rotation period is 5 min, and the resulting navigation errors are reduced by more than an order of magnitude. *This shows why carouseling (and indexing) is a popular implementation scheme.*

The plot in Figure 9.33 was generated by the MATLAB m-file `AccBias-Carousel.m` on the CD-ROM. Note that it exhibits the same coriolis-coupled

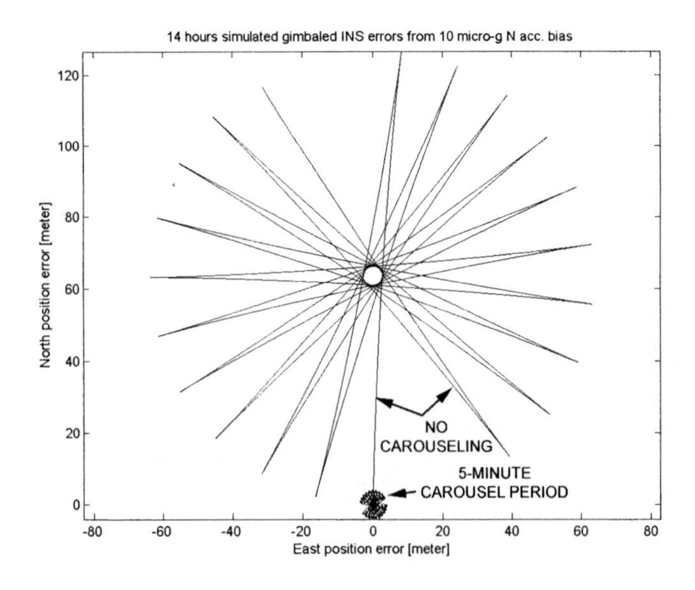

Fig. 9.33 Fourteen hours of simulated gimbaled INS navigation errors, with and without carouseling.

Schuler oscillations as in Fig. 9.30, which was the result of an initial north velocity error. In this case, it is a north acceleration error, the result of which is that the Schuler oscillation center is offset from the starting point.

PROBLEMS

9.1 In the one-dimensional "line land" world of Section 9.4.1, "an INS requires no gyroscopes. How many gyroscopes would be required for two-dimensional navigation in "flat land?"

9.2 Derive the equivalent formulas in terms of Y (yaw angle), P (pitch angle), and R (roll angle) for unit vectors $\mathbf{1}_R, \mathbf{1}_P, \mathbf{1}_Y$ in NED coordinates and $\mathbf{1}_N, \mathbf{1}_E, \mathbf{1}_D$ in RPY coordinates, corresponding to Eqs. C.86–C.91 of Appendix C.

9.3 Explain why accelerometers cannot sense gravitational accelerations.

9.4 Show that the matrix $\mathbf{C}_{\text{inertial}}^{\text{body}}$ defined in Eq. 9.52 is orthogonal by showing that $\mathbf{C}_{\text{inertial}}^{\text{body}} \times \mathbf{C}_{\text{inertial}}^{\text{body}\,T} = \mathbf{I}$, the identity matrix. (*Hint*: Use $q_0^2 + q_1^2 + q_2^2 + q_3^2 = 1$.)

9.5 Calculate the numbers of computer multiplies and adds required for

(a) Gyroscope scale factor/misalignment/bias compensation (Eq. 9.13 with $N = 3$)

(b) Accelerometer scale factor/misalignment/bias compensation (Eq. 9.13 with $N = 3$)

(c) Transformation of accelerations to navigation coordinates (Fig. 9.16) using quaternion rotations (Eq. C.243) requiring two quaternion products (Eq. C.234).

If the INS performs these 100 times per second, how many operations per second will be required?

9.6 Calculate the maximum tilt error for the Schuler oscillations shown in Fig. 9.30. Does this agree with the sensitivities plotted in Fig. 9.29?

10

GNSS/INS INTEGRATION

10.1 BACKGROUND

10.1.1 Sensor Integration

GNSS/INS integration is a form of *sensor integration* or *sensor fusion,* which involves combining the outputs of different sensor systems to obtain a better estimate of what they are sensing.

A GNSS receiver is a position sensor. It may use velocity estimates to reduce filter lags, but its primary output is the position of its antenna relative to an earth-centered coordinate system. GNSS position errors will depend on the availability and geometric distribution of GNSS satellites it can track, and other error sources described in Chapter 5. The resulting RMS position errors will be bounded, except for those times when there are not enough satellite signals available for a position solution.

An INS uses acceleration and attitude (or attitude rate) sensors, but its primary output as a sensor system is also position—the position of its ISA relative to an earth-centered coordinate system. INS position errors depend on the quality of its inertial sensors and earth models, described in Chapter 9. Although their short-term position errors are very smooth, RMS position errors are **not** bounded. They do tend to grow over time, and without bound.

This chapter is about practical methods for combining GNSS and INS outputs to improve overall system *performance metrics*, including

Global Positioning Systems, Inertial Navigation, and Integration, Second Edition, by M. S. Grewal, L. R. Weill, and A. P. Andrews

- RMS position estimation error under nominal GNSS conditions, when the receiver can track enough satellites to obtain a good position estimate.
- RMS position estimation error when the GNSS receiver cannot track enough satellites to obtain a good position estimate. This can happen as a result of
  - Background noise or signal jamming,
  - Blocking by the leaf canopy, buildings, tunnels, etc.
  - GNSS system failures
  - Vehicle attitudes pointing the GNSS antenna pattern downward, or
  - Violent vehicle dynamics causing loss of signal phase lock.
- RMS velocity estimation error, which is important for
  - Aircraft autonomous landing systems (ALS),
  - Attitude transfer alignment to auxiliary INS
  - Guided weapon delivery
  - Rendezvous and docking maneuvers
- RMS attitude estimation error, which is important for
  - Sensor pointing
  - Host vehicle guidance and control
- Maintaining GNSS satellite signal lock, which can be difficult under severe dynamic conditions. In effect, INS accelerometers measure the derivative of signal Doppler shift, which can be used to improve GNSS receiver phase-lock control margins.

Evaluating these metrics for a proposed system application requires statistical models for how the component error sources contribute to overall system performance under conditions of the intended mission trajectories.

The purpose of this chapter is to present these models, and the techniques for applying them to evaluate the expected integrated performance for a specific INS design, GNSS configuration and ensemble of mission trajectories.

10.1.2 The Influence of Host Vehicle Trajectories on Performance

Host vehicle dynamics have a strong influence on GNSS navigation performance and INS performance. We will demonstrate in Section 10.2 how host vehicle dynamic uncertainty affects the achievable navigation accuracy of GNSS receivers. Trajectory profiles of acceleration and attitude rate also affect INS performance, because the errors due to inertial sensor scale factors tend to grow linearly with the input accelerations and attitude rates.

The combined influence of host vehicle dynamics on integrated GNSS/INS performance can usually be characterized by a representative set of host vehicle trajectories, from which we can estimate the expected performance of the integrated GNSS/INS system over an ensemble of mission applications.

For example, a representative set of trajectories for a regional passenger jet might include several trajectories from gate to gate, including taxiing, takeoff, climbout, cruise, midcourse heading and altitude changes, approach, landing, and taxiing. Trajectories with different total distances and headings should represent the expected range of applications. These can even be weighted according to the expected frequency of use. With such a set of trajectories, one can assess expected performances with different INS error characteristics and different satellite and pseudolite geometries.

Similarly, for a standoff air-to-ground weapon, an ensemble of trajectories with different approach and launch geometries and different target impact constraints can be used to evaluate RMS miss distances with GNSS jamming at different ranges from the target.

We will here demonstrate integrated GNSS/INS performance using some simple trajectory simulators, just to show the benefits of GNSS/INS integration. In order to quantify the expected performance for a specific application, however, you must use your own representative set of trajectories.

10.1.3 Loosely and Tightly Coupled Integration

The design process for GNSS/INS integration includes the tradeoff between performance and cost, and cost can be strongly influenced by the extent to which GNSS/INS integration requires some modification of the inner workings of the GNSS receiver or INS. The terms "loosely coupled" and "tightly coupled" are used to describe this attribute of the problem.

10.1.3.1 Loosely Coupled Implementations The most loosely coupled implementations use only the standard outputs of the GNSS receiver and INS as inputs to a sensor system integrating filter (often a Kalman filter), the outputs of which are the estimates of "navigation variables" (including position and velocity) based on both subsystem outputs. Although each subsystem (GNSS or INS) may already include its own Kalman filter, the integration architecture does not modify it in any way.

10.1.3.2 More Tightly Coupled Implementations The more tightly coupled implementations use less standard subsystem outputs, such as pseudoranges from GNSS receivers or raw accelerations from inertial navigators. These outputs generally require software changes within "standalone" GNSS receivers or INSs, and may even require hardware changes. The filter model used for system integration may also include variables such as GNSS signal propagation delays or accelerometer scale factor errors, and the estimated values of these variables may used in the internal implementations of the GNSS receiver or INS.

Tightly coupled implementations may impact the internal inputs within the GNSS receiver or INS, as well. The acceleration outputs from the INS can be used to tighten the GNSS Doppler tracking loops, and the position estimates from the INS can be used for faster reacquisition after GNSS outages. Also, the INS accelerometers and gyroscopes can be recalibrated in real time to improve free-inertial performance during GNSS signal outages.

10.1.3.3 Examples There are many possible levels of coupling between the extremes. Figure 10.1 shows some of these intermediate approaches.

10.1.3.4 Incomplete Ordering The loose/tight ordering not "complete," in the sense that it is not always possible to decide whether one implementation is looser or tighter than another.

A loosely coupled implementation designed for a given GNSS receiver, INS and surface ship, for example, may differ significantly from one designed for a highly maneuverable aircraft using the same GNSS receiver and INS. Both may be equally "loose," in the sense that they require no modifications of the GNSS receiver or INS, but the details of the integrating filter will almost certainly differ. The possibilities for equally loose but different implementations only multiplies when we consider different GNSS receivers and INS hardware.

Similarly, there is no unique and well-defined "ultimately tightly coupled" implementation, because there is no unique GNSS receiver or INS. As the technology advances, the possibilities for tighter coupling continue to grow, and the available hardware and software for implementing it will almost certainly continue to improve. Something even better may always come along.

10.1.4 Antenna/ISA Offset Correction

The "holy point" for a GNSS navigation receiver is its antenna. It is where the relative phases of all received signals are determined, and it is the location determined by the navigation solution.

The INS holy point is its inertial sensor assembly (ISA), which is where the accelerations and attitude rates of the host vehicle are measured and integrated.

The distance between the two navigation solutions can be large enough[1] to be accounted for when combining the two navigation solutions. In that case, the displacement of the antenna from the ISA can be specified as a parameter vector

$$\boldsymbol{\delta}_{\text{ant,RPY}} = \begin{bmatrix} \delta_{\text{ant, R}} \\ \delta_{\text{ant, P}} \\ \delta_{\text{ant, Y}} \end{bmatrix} \tag{10.1}$$

in body-fixed roll–pitch–yaw (RPY) coordinates. Then the displacement in north–east–down (NED) coordinates will be

$$\boldsymbol{\delta}_{\text{ant, NED}} = \mathbf{C}_{\text{NED}}^{\text{RPY}} \boldsymbol{\delta}_{\text{ant, RPY}} \tag{10.2}$$

$$= \begin{bmatrix} C_Y P_P & -S_Y C_R + C_Y S_P S_R & S_Y S_R + C_Y S_P C_R \\ S_Y C_P & C_Y C_R + S_Y S_P S_R & -C_Y S_R + S_Y S_P C_R \\ -S_P & C_P S_R & C_P C_R \end{bmatrix} \boldsymbol{\delta}_{\text{ant, RPY}}, \tag{10.3}$$

[1] It can be tens of meters for ships, where the INS may located well below deck and the antenna is mounted high on a mast.

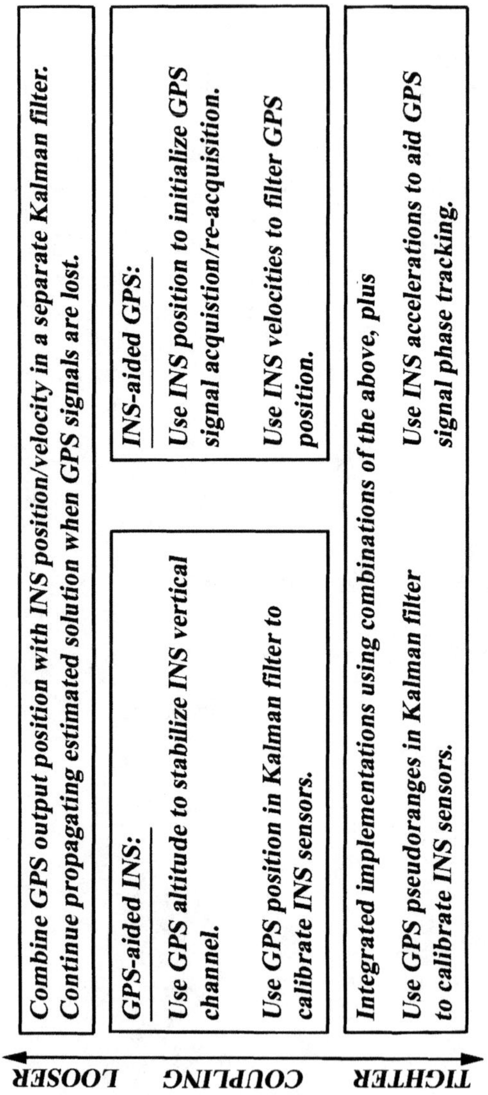

LOOSER ← COUPLING → TIGHTER

Combine GPS output position with INS position/velocity in a separate Kalman filter. Continue propagating estimated solution when GPS signals are lost.

GPS-aided INS:

Use GPS altitude to stabilize INS vertical channel.

Use GPS position in Kalman filter to calibrate INS sensors.

INS-aided GPS:

Use INS position to initialize GPS signal acquistion/re-acquisition.

Use INS velocities to filter GPS position.

Integrated implementations using combinations of the above, plus

Use GPS pseudoranges in Kalman filter to calibrate INS sensors.

Use INS accelerations to aid GPS signal phase tracking.

Fig. 10.1 Loosely and tightly coupled implementations.

where $\mathbf{C}_{\text{NED}}^{\text{RPY}}$ is the coordinate transformation from RPY to NED coordinates, defined in terms of the vehicle attitude Euler angles by

$$\begin{aligned}
S_R &\stackrel{\text{def}}{=} \sin(\text{roll}), \\
C_R &\stackrel{\text{def}}{=} \cos(\text{roll}), \\
S_P &\stackrel{\text{def}}{=} \sin(\text{pitch}), \\
C_P &\stackrel{\text{def}}{=} \cos(\text{pitch}), \\
S_Y &\stackrel{\text{def}}{=} \sin(\text{yaw}), \\
C_Y &\stackrel{\text{def}}{=} \cos(\text{yaw}), \\
\text{roll} &\stackrel{\text{def}}{=} \text{vehicle roll angle}, \\
\text{pitch} &\stackrel{\text{def}}{=} \text{vehicle pitch angle}, \\
\text{yaw} &\stackrel{\text{def}}{=} \text{vehicle yaw/heading angle}.
\end{aligned}$$

Usually, the matrix $\mathbf{C}_{\text{NED}}^{\text{RPY}}$ and/or the roll, pitch, and yaw angles are variables in INS implementation software.

Once $\boldsymbol{\delta}_{\text{ant., NED}}$ is computed, it can be used to relate the two navigation positions in NED coordinates:

$$\mathbf{x}_{\text{ant, NED}} = \mathbf{x}_{\text{ISA, NED}} + \boldsymbol{\delta}_{\text{ant, NED}}, \tag{10.4}$$

$$\mathbf{x}_{\text{ISA, NED}} = \mathbf{x}_{\text{ant, NED}} - \boldsymbol{\delta}_{\text{ant, NED}}, \tag{10.5}$$

eliminating the potential source of error.

This correction is generally included in integrated GNSS/INS systems. It requires a procedure for requesting and entering the components of $\boldsymbol{\delta}_{\text{ant, NED}}$ during system installation.

10.2 EFFECTS OF HOST VEHICLE DYNAMICS

Host vehicle dynamics impact GNSS/INS performance in a number of ways, including the following:

1. A GNSS receiver is primarily a position sensor. To counter unpredictable dynamics, it must use a vehicle dynamic model with higher derivatives of position. This adds state variables that dilute the available (pseudorange) information, the net effect of which is reduction of position accuracy with increased host vehicle dynamic activity. The filtering done within GNSS receivers influences loosely coupled GNSS/INS integration through the impact it has on GNSS position errors. The examples in Section 8.8

introduced Kalman filter models for estimating GPS receiver antenna position and clock bias. sections 10.2.1–10.2.3 expand the modeling to include filters optimized for specific classes of host vehicle trajectories.

2. Host vehicle dynamics translate into phase dynamics of the received satellite signals. This impacts phase-lock capability in three ways:
   (a) It increases RMS phase error, which increases pseudorange error. This contributes to position solution error. It also degrades to clock error correction capability.
   (b) It can cause the phase-lock loop to fail, causing momentary loss of signal information.
   (c) At best, it increases signal acquisition time. At worst, it may cause signal acquisition to fail.
3. High dynamic rates increase the sensitivity of INS performance to inertial sensor errors, especially scale factor errors. For strapdown systems, it greatly increases the influence of gyroscope scale factor errors on navigation performance.

The effects are reduced by GNSS/INS integration. To understand how, we will need to examine how these effects work and how integration can mitigate their effects. Impacts on GNSS host vehicle dynamic filtering and mitigation methods are discussed in this section. Mitigation methods for the other effects are discussed in the next two sections.

10.2.1 Vehicle Tracking Filters

Starting around 1950, radar systems were integrated with computers to detect and track Soviet aircraft that might invade the continental United States [160]. The computer software included filters to identify and track individual aircraft within a formation. These "tracking filters" generated estimates of position and velocity for each aircraft, and they could be tuned to follow the unpredictable maneuvering capabilities of Soviet bombers of that era.

The same sorts of tracking filters are used in GNSS receivers to estimate the position and velocity of GNSS antennas on host vehicles with unpredictable dynamics. Important issues in the design and implementation of these filters include the following:

1. In what ways does vehicle motion affect GNSS navigation performance?
2. Which characteristics of vehicle motions influence the choice of tracking filter models?
3. How do we determine these characteristics for a specified vehicle type?

These issues are addressed in the following subsections.

10.2.1.1 Dynamic Dilution of Information In addition to the "dilution of precision" related to satellite geometry, there is a GNSS receiver "dilution of information" problem related to vehicle dynamics. In essence, if more information (in the measurements) is required to make up for the uncertainty of vehicle movement, then less information is left over for determining the instantaneous antenna position and clock bias.

For example, the location of a receiver antenna at a fixed position on the earth can be specified by three unknown constants (i.e., position coordinates in three dimensions). Over time, as more and more measurements are used, the accuracy of the estimated position should improve. If the vehicle is moving, however, only the more recent measurements relate to the current antenna position.

10.2.1.2 Effect on Position Uncertainty Figure 10.2 is a plot of the contribution vehicle dynamic characteristics make to GPS position estimation uncertainty for a range of host vehicle dynamic capabilities. In order to indicate the contributions that vehicle dynamics make to position uncertainty, this demonstration assumes that other contributory error sources are either negligible or nominal, e.g.,

- No receiver clock bias. (That will come later.)
- 10 m RMS time-correlated pseudorange error due to iono spheric delay, receiver bias, interfrequency biases, etc.
- 60 s pseudorange error correlation time.

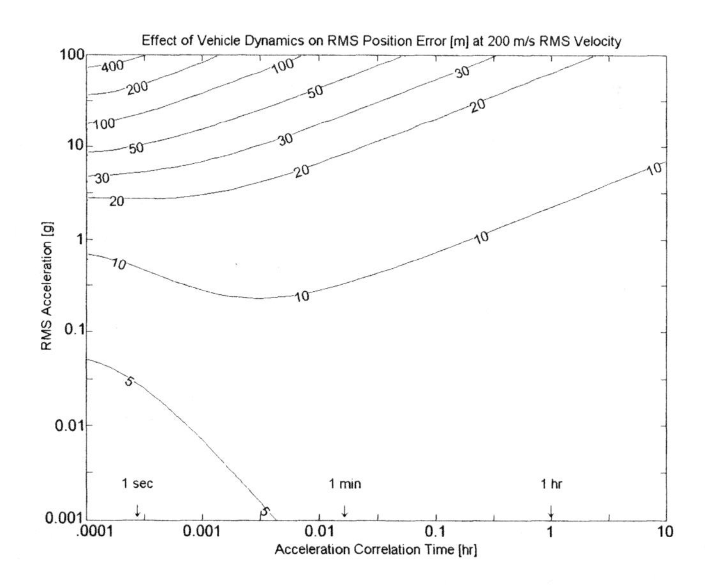

Fig. 10.2 DAMP2 tracker performance versus $\sigma_{\rm acc}$ and $\tau_{\rm acc}$.

- Pseudoranges of each available satellite sampled every second.
- 10 m RMS pseudorange uncorrelated measurement noise.
- 29-satellite GPS configuration of March 8, 2006.
- Only those satellites more than 15° above the horizon were used.
- 200 m/s RMS host vehicle velocity, representing a high-performance aircraft or missile.
- Host vehicle at 40° north latitude.
- Results averaged over 1 h of simulated operation.

Figure 10.2 is output from the MATLAB m-file `Damp2eval.m` on the accompanying CD-ROM. It performs a set of GPS tracking simulations using the "DAMP2" tracking filter described in Table 10.1 and Section 10.2.2.5. This filter allows the designer to specify the RMS velocity, RMS acceleration, and acceleration correlation time of the host vehicle, and the plot shows how these two dynamic characteristics influence position estimation accuracy.

These results would indicate that navigation performance is more sensitive to vehicle acceleration magnitude than to its correlation time. Five orders of magnitude variation in correlation time do not cause one order of magnitude variation in RMS position estimation accuracy. At short correlation times, five orders of magnitude variation in RMS acceleration[2] cause around three orders of magnitude variation in RMS position estimation accuracy. These simulations were run at 40° latitude. Changing simulation conditions may change the results somewhat.

The main conclusion is that unpredictable vehicle motion does, indeed, compromise navigation accuracy.

10.2.2 Specialized Host Vehicle Tracking Filters

In Kalman filtering, dynamic models are completely specified by two matrix parameters:

1. The dynamic coefficient matrix $\mathbf{F}$ (or equivalent state transition matrix $\mathbf{\Phi}$)
2. The dynamic disturbance covariance matrix $\mathbf{Q}$

The values of these matrix parameters for six different vehicle dynamic models are listed in Table 10.1. They are all time-invariant (i.e., constant). As a consequence, the corresponding state transition matrices

$$\Phi = \exp(\mathbf{F}\Delta t)$$

are also constant, and can be computed using the matrix exponential function (`expm` in MATLAB).

[2]The RMS acceleration used here does not include the acceleration required to counter gravity.

Table 10.1 also lists the independent and dependent parameters of the models. The independent parameters can be specified by the filter designer. Because the system model is time-invariant, the finite dependent variables are determinable from the steady-state matrix Riccati differential equation,

$$0 = \mathbf{F}\mathbf{P}_{\infty} + \mathbf{P}_{\infty}\mathbf{F}^{\mathrm{T}} + \mathbf{Q}, \tag{10.6}$$

the solution of which exists only if the eigenvalues of **F** lie in the left-half complex plane. However, even in those cases where the full matrix Riccati differential equation has no finite solution, a reduced equation with a submatrix of $\mathbf{P}_{\infty}$ and corresponding submatrix of **F** may still have a finite steady-state solution. For those with "closed form" solutions that can be expressed as formulas, the solutions are listed below with the model descriptions.

The TYPE2 filter, for example, does not have a steady-state solution for its Riccati equation without measurements. As a consequence, we cannot use mean-squared velocity as a TYPE2 filter parameter for modeling vehicle maneuverability. However, we can still solve the Riccati equation with GNSS measurements (which is not time-invariant) to characterize position uncertainty as a function of mean-squared vehicle acceleration (modeled as a zero-mean white noise process).

10.2.2.1 Unknown Constant Tracking Model This model was used in Section 8.8. There are no parameters for vehicle dynamics, because there are no vehicle dynamics. The Kalman filter state variables are three components of position, shown below as NED coordinates. The only model parameters are three values of $\sigma^2_{\mathrm{pos}}(0)$ for three direction components. These represent the initial position uncertainties before measurements start. The value of $\sigma^2_{\mathrm{pos}}(0)$ can be different in different directions. The necessary Kalman filter parameters for a stationary antenna are then

$$\mathbf{P}_0 = \begin{bmatrix} \sigma^2_{\mathrm{north}} & 0 & 0 \\ 0 & \sigma^2_{\mathrm{east}} & 0 \\ 0 & 0 & \sigma^2_{\mathrm{down}} \end{bmatrix}, \text{ the initial mean-squared position uncertainty,}$$

$$\Phi = \mathbf{I}, \text{ the } 3 \times 3 \text{ identity matrix,}$$

$$\mathbf{Q} = 0, \text{ the } 3 \times 3 \text{ zero matrix.}$$

The initial position uncertainty, as modeled by $\sigma^2_{\mathrm{pos}}(0)$, may also influence GNSS signal acquisition search time. The other necessary Kalman filter parameters (**H** and **R**) come from the pseudorange measurement model, which was addressed in Section 8.8.

10.2.2.2 Damped Harmonic Resonator GNSS antennas can experience harmonic displacements in the order of several centimeters from host vehicle resonant modes, which are typically at frequencies in the order of ≈ 1 Hz (the

TABLE 10.1. Vehicle Dynamic Models for GNSS Receivers

Model Name	Model Parameters (Each Axis) F	Q	Independent Variables	Dependendent Variables
Unknown constant	0	0	$\sigma^2_{\text{pos}}(0)$	None
Harmonic resonator (Example 8.3)	$\begin{bmatrix} 0 & 1 \\ -\omega^2 - \frac{1}{\tau^2} & \frac{-2}{\tau} \end{bmatrix}$	$\begin{bmatrix} 0 & 0 \\ 0 & \sigma^2_{\text{acc}}\Delta t^2 \end{bmatrix}$	σ^2_{pos} ω τ	σ^2_{acc}
TYPE2	$\begin{bmatrix} 0 & 1 \\ 0 & 0 \end{bmatrix}$	$\begin{bmatrix} 0 & 0 \\ 0 & \sigma^2_{\text{acc}}\Delta t^2 \end{bmatrix}$	σ^2_{acc}	$\sigma^2_{\text{pos}} \rightarrow \infty$ $\sigma^2_{\text{vel}} \rightarrow \infty$
DAMP1	$\begin{bmatrix} 0 & 1 \\ 0 & -1/\tau_{\text{vel}} \end{bmatrix}$	$\begin{bmatrix} 0 & 0 \\ 0 & \sigma^2_{\text{acc}}\Delta t^2 \end{bmatrix}$	σ^2_{vel} τ_{vel}	$\sigma^2_{\text{pos}} \rightarrow \infty$ σ^2_{acc}
DAMP2	$\begin{bmatrix} 0 & 1 & 0 \\ 0 & \frac{-1}{\tau_{\text{vel}}} & 1 \\ 0 & 0 & \frac{-1}{\tau_{\text{acc}}} \end{bmatrix}$	$\begin{bmatrix} 0 & 0 & 0 \\ 0 & 0 & 0 \\ 0 & 0 & \sigma^2_{\text{jerk}}\Delta t^2 \end{bmatrix}$	σ^2_{vel} σ^2_{acc} τ_{acc}	$\sigma^2_{\text{pos}} \rightarrow \infty$ τ_{vel} $\rho_{\text{vel,acc}}$ σ^2_{jerk}
DAMP3	$\begin{bmatrix} \frac{-1}{\tau_{\text{pos}}} & 1 & 0 \\ 0 & \frac{-1}{\tau_{\text{vel}}} & 1 \\ 0 & 0 & \frac{-1}{\tau_{\text{acc}}} \end{bmatrix}$	$\begin{bmatrix} 0 & 0 & 0 \\ 0 & 0 & 0 \\ 0 & 0 & \sigma^2_{\text{jerk}}\Delta t^2 \end{bmatrix}$	σ^2_{pos} σ^2_{vel} σ^2_{acc} τ_{acc}	τ_{pos} τ_{vel} $\rho_{\text{pos,vel}}$ $\rho_{\text{pos,acc}}$ $\rho_{\text{vel,acc}}$ σ^2_{jerk}

suspension resonance of most passenger cars) to several Hz—but the effect is small compared to other error sources.

However, a model of this sort (developed in Examples 8.1–8.7) is needed for INS gyrocompass alignment, which is addressed in Chapter 9.

10.2.2.3 TYPE2 Tracking Model The TYPE2 tracker is older than Kalman filtering. Given sufficient measurements, it can estimate position and velocity in three dimensions. (Type 1 trackers do not estimate velocity.) The tracker uses a host vehicle dynamic model with zero-mean white-noise acceleration, unbounded steady-state mean-squared velocity (not particularly reasonable), and unbounded steady-state mean-squared position variation (quite reasonable). When GNSS signals are lost, the velocity uncertainty variance will grow without bound unless something is done about it—such as limiting velocity variance to some maximum value. Trackers based on this model can do an adequate job when GNSS signals are present.

The model parameters shown in Table 10.1 are for a single-direction component, and do not include position. The full tracking model will include three position components and three velocity components. The necessary Kalman filter parameters for a 3D TYPE2 tracking filter include

$$\mathbf{P}_0 = \begin{bmatrix} \sigma^2_{\text{north}} & 0 & 0 & 0 & 0 & 0 \\ 0 & \sigma^2_{\text{east}} & 0 & 0 & 0 & 0 \\ 0 & 0 & \sigma^2_{\text{down}} & 0 & 0 & 0 \\ 0 & 0 & 0 & \sigma^2_{v,\text{north}} & 0 & 0 \\ 0 & 0 & 0 & 0 & \sigma^2_{v,\text{east}} & 0 \\ 0 & 0 & 0 & 0 & 0 & \sigma^2_{v,\text{down}} \end{bmatrix},$$

$$\Phi = \begin{bmatrix} 1 & 0 & 0 & \Delta t & 0 & 0 \\ 0 & 1 & 0 & 0 & \Delta t & 0 \\ 0 & 0 & 1 & 0 & 0 & \Delta t \\ 0 & 0 & 0 & 1 & 0 & 0 \\ 0 & 0 & 0 & 0 & 1 & 0 \\ 0 & 0 & 0 & 0 & 0 & 1 \end{bmatrix},$$

$$\mathbf{Q} = \begin{bmatrix} 0 & 0 & 0 & 0 & 0 & 0 \\ 0 & 0 & 0 & 0 & 0 & 0 \\ 0 & 0 & 0 & 0 & 0 & 0 \\ 0 & 0 & 0 & \sigma^2_{\text{acc}}\Delta t^2 & 0 & 0 \\ 0 & 0 & 0 & 0 & \sigma^2_{\text{acc}}\Delta t^2 & 0 \\ 0 & 0 & 0 & 0 & 0 & \sigma^2_{\text{acc}}\,\Delta t^2 \end{bmatrix},$$

where σ^2_{acc} is the only adjustable parameter value. Adjusting it for a particular application may take some experimenting.

10.2.2.4 DAMP1 Tracking Model This type of tracking filter is based on the Langevin equation[3]

$$\frac{d}{dt}v(t) = \underbrace{-\frac{1}{\tau_{\text{vel}}}}_{F} v(t) + w(t), \tag{10.7}$$

where $v(t)$ is a velocity component and $w(t)$ is a zero-mean white-noise process in continuous time.

It differs from the TYPE2 tracker in that it includes a velocity damping time constant τ_{vel}, which is enough to put an eigenvalue of $\mathbf{F}$ in the left-half complex plane and allow a steady-state variance for velocity. This is more realistic as a model for a vehicle with finite speed capabilities. Also, the parameter τ_{vel} is a measure of persistence of velocity, which would be useful for distinguishing the dynamics of an oil tanker, say, from those of a jet ski.

The values of $\mathbf{P}_0$ and $\mathbf{Q}$ will be the same as for the TYPE2 tracker, and the state transition matrix

$$\Phi = \begin{bmatrix} 1 & 0 & 0 & \Delta t & 0 & 0 \\ 0 & 1 & 0 & 0 & \Delta t & 0 \\ 0 & 0 & 1 & 0 & 0 & \Delta t \\ 0 & 0 & 0 & \varepsilon & 0 & 0 \\ 0 & 0 & 0 & 0 & \varepsilon & 0 \\ 0 & 0 & 0 & 0 & 0 & \varepsilon \end{bmatrix}$$

$$\varepsilon = \exp\left(-\Delta t/\tau_{\text{vel}}\right).$$

The steady-state solution of the Riccati equation can be used to solve for

$$\sigma_{\text{acc}}^2 = \sigma_{\text{vel}}^2 \left[1 - \exp\left(-2\Delta t/\tau_{\text{vel}}\right)\right]/\Delta t^2. \tag{10.8}$$

Thus, one can specify the vehicle maneuver capability in terms of its mean-square velocity σ_{vel}^2 and velocity correlation time τ_{vel}, and use Eq. 10.8 to specify compatible values for the modeled $\mathbf{Q}$ matrix.

10.2.2.5 DAMP2 Tracking Model This is an even more realistic model for a vehicle with finite speed *and acceleration* capabilities. It also includes an acceleration time correlation constant τ_{acc}, which is useful for distinguishing the more lively vehicle types from the more sluggish ones. The corresponding steady-state Riccati equation used for making the model a function of RMS velocity and acceleration is not as easy to solve in closed form, however.

[3]The first known stochastic differential equation, was published by Paul Langevin (1872–1946) in 1908 [116]. Langevin was a prolific scientist with pioneering work in many areas, including para- and diamagnetism, and sonar.

The 2×2 submatrix of the state transition matrix Φ relating velocity and acceleration along a single axis has the form

$$\Phi_{\text{vel, acc}} = \begin{bmatrix} \phi_{1,1} & \phi_{1,2} \\ \phi_{2,1} & \phi_{2,2} \end{bmatrix} \tag{10.9}$$

$$\phi_{1,1} = \exp\left(-\Delta t/\tau_{\text{vel}}\right) \tag{10.10}$$

$$\phi_{1,2} = \frac{\tau_{\text{vel}}\tau_{\text{acc}}\left[\exp\left(-\Delta t/\tau_{\text{vel}}\right) - \exp\left(-\Delta t/\tau_{\text{acc}}\right)\right]}{\tau_{\text{vel}} - \tau_{\text{acc}}} \tag{10.11}$$

$$\phi_{2,1} = 0 \tag{10.12}$$

$$\phi_{2,2} = \exp\left(-\Delta t/\tau_{\text{acc}}\right), \tag{10.13}$$

and the corresponding 2×2 submatrix of $\mathbf{Q}$ will be

$$\mathbf{Q}_{\text{vel, acc}} = \begin{bmatrix} 0 & 0 \\ 0 & \sigma_{\text{jerk}}^2\,\Delta t^2 \end{bmatrix}. \tag{10.14}$$

The corresponding steady-state Riccati equation 10.6 can be solved for

$$\sigma_{\text{jerk}}^2 = \sigma_{\text{acc}}^2\left[1 - \exp\left(-2\Delta t/\tau_{\text{acc}}\right)\right]/\Delta t^2, \tag{10.15}$$

the analog of Eq. 10.8.

The steady-state Riccati equation can also be solved for the correlation coefficient

$$\begin{array}{rl} \rho_{\text{vel, acc}} = -\tau_{\text{vel}}\tau_{\text{acc}}\sigma_{\text{acc}} & \exp(-\Delta t/\tau_{\text{acc}})[\exp(-\Delta t/\tau_{\text{vel}}) - \exp(-\Delta t/\tau_{\text{acc}})] \\ \sigma_{\text{vel}}[1- & \exp(-\Delta t/\tau_{\text{acc}})\exp(-\Delta t/\tau_{\text{vel}})](\tau_{\text{vel}} - \tau_{\text{acc}}) \end{array} \tag{10.16}$$

between the velocity and acceleration components. But solving the remaining element $p_{\infty,1,1} = 0$ of the Riccati equation 10.6 for τ_{vel} as variable dependent on the independent variables requires solving a transcendental equation. It is solved numerically in the MATLAB function `Damp2Params.m` on the accompanying CD-ROM.

Figure 10.2 was generated by the MATLAB m-file `Damp2eval.m`, and Fig. 8.21 was generated by the m-file `SchmidtKalmanTest.m` on the accompanying CD-ROM. Both include solutions of the Riccati equation for a DAMP2 GNSS position filter.

10.2.2.6 DAMP3 Tracking Model This type of filter is designed for vehicles with limited but nonzero position variation, such as the altitudes of some surface watercraft (e.g., riverboats) and land vehicles. Ships that remain a sea level are at zero altitude, by definition. They need no vertical navigation, unless they are trying to estimated tides. Flatwater boats and land vehicles in very flat areas can probably do without vertical navigation, as well.

Continuous-Time Solutions It is generally easier to solve the steady-state covariance equation in continuous time

$$0 = \mathbf{F}_3\mathbf{P}_3 + \mathbf{P}_3\mathbf{F}_3^T + \mathbf{Q}_3 \tag{10.17}$$

for the parameters in the steady-state solution

$$\mathbf{P}_3 = \begin{bmatrix} p_{1,1} & p_{1,2} & p_{1,3} \\ p_{1,2} & p_{2,2} & p_{2,3} \\ p_{1,3} & p_{2,3} & p_{3,3} \end{bmatrix}, \tag{10.18}$$

where the other model parameters are

$$\mathbf{F}_3 = \begin{bmatrix} -\tau_{\text{pos}}{}^{-1} & 1 & 0 \\ 0 & -\tau_{\text{vel}}{}^{-1} & 1 \\ 0 & 0 & -\tau_{\text{acc}}{}^{-1} \end{bmatrix}, \tag{10.19}$$

$$\mathbf{Q}_3 = \begin{bmatrix} 0 & 0 & 0 \\ 0 & 0 & 0 \\ 0 & 0 & q_{c,3,3} \end{bmatrix}. \tag{10.20}$$

The six scalar equations equivalent to the symmetric 3×3 matrix equation (10.17) are

$$\left.\begin{array}{rcll} 0 & = & -p_{1,1} + p_{1,2}\,\tau_{\text{pos}} & (\text{Eq}_{1,1}) \\ 0 & = & -p_{1,2}\,\tau_{\text{vel}} + p_{2,2}\,\tau_{\text{pos}}\tau_{\text{vel}} - p_{1,2}\,\tau_{\text{pos}} + p_{1,3}\,\tau_{\text{pos}}\tau_{\text{vel}} & (\text{Eq}_{1,2}) \\ 0 & = & -p_{1,3}\,\tau_{\text{acc}} + p_{2,3}\,\tau_{\text{pos}}\tau_{\text{acc}} - p_{1,3}\,\tau_{\text{pos}} & (\text{Eq}_{1,3}) \\ 0 & = & -p_{2,2} + p_{2,3}\,\tau_{\text{vel}} & (\text{Eq}_{2,2}) \\ 0 & = & -p_{2,3}\,\tau_{\text{acc}} + p_{3,3}\,\tau_{\text{vel}}\tau_{\text{acc}} - p_{2,3}\,\tau_{\text{vel}} & (\text{Eq}_{2,3}) \\ 0 & = & -2\,p_{3,3} + q_{c,3,3}\,\tau_{\text{acc}} & (\text{Eq}_{3,3}) \end{array}\right\}. \tag{10.21}$$

We wish to solve for the steady-state covariance matrix **P**, where the following independent variables are to be specified

τ_{acc}, the acceleration correlation time constant

$p_{1,1}$, the mean-squared position excursion

$p_{2,2}$, the mean-squared velocity variation

$p_{3,3}$, the mean-squared acceleration variation

are to be specified, and the dependent variables are to be determined from Eq. 10.21:

τ_{vel}, the velocity correlation time constant

τ_{pos}, the position correlation time constant

$p_{1,2}$, the cross-covariance of position and velocity

$p_{1,3}$, the cross-covariance of position and acceleration

$p_{2,3}$, the cross-covariance of velocity and acceleration

$q_{c,3,3}$, the continuous-time disturbance noise variance

From the last of these (labeled $\text{Eq}_{3,3}$), we obtain

$$q_{c,3,3} = 2\,\frac{p_{3,3}}{\tau_{\text{acc}}}. \tag{10.22}$$

From $\text{Eq}_{2,2}$ and $\text{Eq}_{2,3}$, we have

$$p_{2,3} = \frac{p_{2,2}}{\tau_{\text{vel}}}, \tag{10.23}$$

$$= \frac{p_{3,3}\,\tau_{\text{vel}}\tau_{\text{acc}}}{\tau_{\text{acc}} + \tau_{\text{vel}}}, \tag{10.24}$$

and from equating the two solutions, we obtain

$$\tau_{\text{vel}} = \frac{p_{2,2} + \sqrt{p_{2,2}{}^2 + 4\,p_{3,3}\,\tau_{\text{acc}}{}^2\,p_{2,2}}}{2\,p_{3,3}\,\tau_{\text{acc}}}. \tag{10.25}$$

Similarly, from $\text{Eq}_{1,1}$ and $\text{Eq}_{1,2}$, we have

$$p_{1,2} = \frac{p_{1,1}}{\tau_{\text{pos}}}, \tag{10.26}$$

$$p_{1,3} = -\frac{p_{2,2}\,\tau_{\text{pos}}{}^2\tau_{\text{vel}} - p_{1,1}\,\tau_{\text{vel}} - p_{1,1}\,\tau_{\text{pos}}}{\tau_{\text{pos}}{}^2\tau_{\text{vel}}}, \tag{10.27}$$

and from $\text{Eq}_{1,3}$, we obtain

$$p_{1,3} = \frac{p_{2,3}\,\tau_{\text{pos}}\tau_{\text{acc}}}{\tau_{\text{acc}} + \tau_{\text{pos}}}. \tag{10.28}$$

By equating the two independent solutions for $p_{1,3}$, we obtain a cubic polynomial in τ_{pos}:

$$0 = c_0 + c_1\,\tau_{\text{pos}} + c_2\,\tau_{\text{pos}}^2 + c_3\,\tau_{\text{pos}}^3 \tag{10.29}$$

with coefficients

$$c_0 = p_{1,1}\,\tau_{\text{vel}}\tau_{\text{acc}}, \tag{10.30}$$

$$c_1 = p_{1,1}\,(\tau_{\text{acc}} + \tau_{\text{vel}}), \tag{10.31}$$

$$c_2 = -p_{2,2}\,\tau_{\text{vel}}\tau_{\text{acc}} + p_{1,1}, \tag{10.32}$$

$$c_3 = -\tau_{\text{vel}}\left(p_{2,3}\,\tau_{\text{acc}} + p_{2,2}\right), \tag{10.33}$$

which can be solved numerically using the MATLAB function `roots`.

The MATLAB solution sequence is then

0 Given: $p_{1,1}$, $p_{2,2}$, $p_{3,3}$ and τ_{acc}.
1 Solve for τ_{vel} using Eq. 10.25.
2 Solve for $p_{2,3}$ using Eq. 10.23.
3 Solve for τ_{pos} using Eq. 10.29 and the MATLAB function `roots`.
4 Solve for $p_{1,2}$ using Eq, 10.26.
5 Solve for $p_{1,3}$ using Eq. 10.28.

This solution is implemented in the MATLAB m-file `DAMP3Params.m` on the CD-ROM.

This leaves the problem of solving for the discrete-time process noise covariance matrix $\mathbf{Q}_{3,\,\text{discrete}}$, which is not the same as the analogous matrix $\mathbf{Q}_3$ in continuous time (solved using Eq. 10.22). There is a solution formula

$$\mathbf{Q}_{3,\,\text{discrete}} = \exp\left(\Delta t\;\mathbf{F}\right)\left[\int_0^{\Delta t} \exp\left(-s\;\mathbf{F}\right)\mathbf{Q}_3 \exp\left(-s\;\mathbf{F}^T\right)\,ds\right]\exp\left(\Delta t\;\mathbf{F}^T\right), \tag{10.34}$$

but—given Φ and $\mathbf{P}_\infty$—it is easier to use the steady-state formula

$$\mathbf{Q}_{3,\,\text{discrete}} = \mathbf{P}_\infty - \Phi P_3 \Phi^T, \tag{10.35}$$

which is how the solution is implemented in `DAMP3Params.m`.

10.2.2.7 Tracking Models for Highly Constrained Trajectories Racecars in some televised races have begun using integrated GPS/INS on each vehicle to determine their positions on the track. The estimated positions are telemetered to the television control system, where they are used in generating television graphics (e.g., an arrow or dagger icon) to designate on the video images where each car is on the track at all times. The integrating filters constrain the cars to be on the 2D track surface, which improves the estimation accuracy considerably.

FIG8 Tracking Model As a simple example of how this works, we will use a one-dimensional "figure-8" track model, with the position on the track completely specified by the down-track distance from a reference point.

The trajectory of a vehicle on the track is specified in terms of a formula,

$$\delta_{\text{pos}} = \begin{bmatrix} \text{Northing} \\ \text{Easting} \\ -\text{Altitude} \end{bmatrix} \tag{10.36}$$

$$= \begin{bmatrix} 3\,S\sin(\omega\,t+\phi) \\ 2\,S\sin(\omega\,t+\phi)\cos(\omega\,t+\phi) \\ -1/2\,h\cos(\omega\,t+\phi) \end{bmatrix} \tag{10.37}$$

where S is a track scaling parameter, $\approx$ [track length (m)]/14.94375529901562, h is half the vertical separation where the track crosses over itself, $\omega = 2\pi \times$ [average speed (m/s)]/[track length (m)], and ϕ is an arbitrary phase angle (rad).

The phase rate $\dot{\phi}$ can be modeled as a random walk or exponentially correlated process, to simulate speed variations. This model is implemented in the MATLAB function `Fig8Mod1D`, which also calculates vehicle velocity, acceleration, attitude, and attitude rates. This m-file is on the accompanying CD-ROM. The resulting trajectory is illustrated in Fig. 10.3.

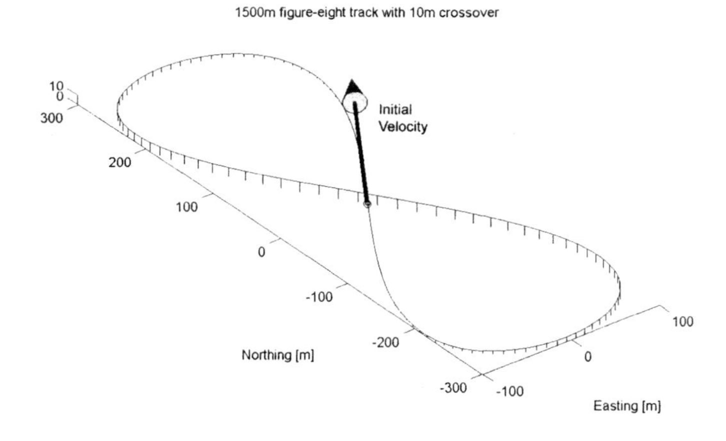

Fig. 10.3 Figure-8 trajectory of length 1500 m.

The resulting Kalman filter is implemented in the MATLAB m-file `GPSTrackingDemo.m` on the CD-ROM. This particular implementation is for a 1.5-km track with vehicle speeds of 90 km/h $\pm 10\%$ RMS random variation. The Kalman filter model in `GPSTrackingDemo.m` uses only two vehicle states: (1) the phase angle ϕ and (2) its derivative $\dot{\phi}$, which is modeled as an exponentially correlated random process with a correlation time constant of 10 s and RMS value equivalent to $\pm 10\%$ variation in speed.

10.2.2.8 Filters for Spacecraft Unpowered vehicles in space do not have sufficiently random dynamics to justify a tracking filter. They may have unknown, quasiconstant orbit parameters, but their trajectories over the short term are essentially defined by a finite set of parameters. GNSS vehicle tracking then becomes an orbit determination problem. The orbital parameters may change during brief orbit changes, but the problem remains an orbit determination problem with increased uncertainty in initial conditions (velocity, in particular).

10.2.2.9 Other Specialized Vehicle Filter Models The list of models in Table 10.1 is by no means exhaustive. It does not include the FIG8 filter described above. Other specialized filters have been designed for vehicles confined to narrow corridors within a limited area, such as race cars on a 2D track or motor vehicles on streets and highways. Specialized filters are also required for trains, which need to know where they are on a 1D track, and possibly which set of parallel rails they are on.

Still, the models listed in Table 10.1 and described above should cover the majority of GNSS applications.

10.2.2.10 Filters for Different Host Vehicle Types Table 10.2 lists some generic host vehicle types, along with names of models in Table 10.1 that might be used for GNSS position tracking on such vehicles.

TABLE 10.2. Filter Models for Unknown Vehicle Dynamics

Host Vehicles	Filter Models	
	Horizontal Directions	Vertical Direction
None (fixed to earth)	Unknown constant	Unknown constant
Parked	Damping harmonic resonator	Damping harmonic resonator
Ships	DAMP1, DAMP2	Unknown constant
Land vehicles	DAMP1, DAMP2	DAMP3
Aircraft and missiles	DAMP1, DAMP2	DAMP2, DAMP3
Spacecraft	*In free fall*, use orbit estimation models *After maneuvers*, increment velocity uncertainty	

TABLE 10.3. Statistical Parameters of Host Vehicle Dynamics

Symbol	Definition
σ^2_{pos}	Mean-squared position excursions[a]
σ^2_{vel}	Mean-squared vehicle velocity ($\mathrm{E}\langle\lvert\mathbf{v}\rvert^2\rangle$)
σ^2_{acc}	Mean-squared vehicle acceleration ($\mathrm{E}\langle\lvert\mathbf{a}\rvert^2\rangle$)
σ^2_{jerk}	Mean-squared jerk ($\mathrm{E}\langle\lvert\dot{\mathbf{a}}\rvert^2\rangle$)
$\rho_{i,j}$	Correlation coefficients between position, velocity, and acceleration variations
τ_{pos}	Position correlation time
τ_{vel}	Velocity correlation time
τ_{acc}	Acceleration correlation time
ω_{resonant}	Suspension resonant frequency
τ_{damping}	Suspension damping time constant

[a] Mean-squared position excursions generally grow without bound, except for altitudes of ships (and possibly land vehicles).

10.2.2.11 Parameters for Vehicle Dynamics Table 10.3 contains descriptions of the tracking filter parameters shown in Table 10.1. These are statistical parameters for characterizing random dynamics of the host vehicle.

10.2.2.12 Empirical Modeling of Vehicle Dynamics The most reliable vehicle dynamic models are those based on data from representative vehicle dynamics. Empirical modeling of the uncertain dynamics of host vehicles requires data (i.e., position and attitude, and their derivatives) recorded under conditions representing the intended mission applications.

The ideal sensor for this purpose is an INS, or at least an inertial sensor assembly (ISA) capable of measuring and recording 3D accelerations and attitude rates (or attitudes) during maneuvers of the host vehicle.

The resulting data are the sum of three types of motion:

1. Internal motions due to vibrating modes of the vehicle, excited by propulsion noise and flow noise in the surrounding medium. The oscillation periods for this noise generally scale with the size of the host vehicle, but are generally in the order of a second or less.
2. Short-term perturbations of the host vehicle that are corrected by steering, such as turbulence acting on aircraft or potholes acting on wheeled vehicles. These also excite the vibrational modes of the vehicle.
3. The intended rigid-body motions of the whole vehicle to follow the planned trajectory. The frequency range of these motions is generally $\ll 10$ Hz and often < 1 Hz.

Only the last of these is of interest in tracking. It can often be separated from the high-frequency noise by lowpass filtering, ignoring the high-frequency end of the power spectral densities and cross-spectral densities of the data. The inverse Fourier transforms of the low-end power spectral data will yield autocovariance functions that are useful for modeling purposes. The statistics of interest in these autocovariance functions are the variances σ^2 (values at zero correlation time) and the approximate exponential decay times τ of the autocovariances.

10.2.3 Vehicle Tracking Filter Comparison

The alternative GNSS receiver tracking filters of the previous section were evaluated using the figure-8 track model described in Section 10.2.2.7. This is a trajectory confined in all three dimensions, and more in some dimensions than others.

10.2.3.1 Simulated Trajectory The simulated trajectory is that of an automobile on a banked figure-8 track, as illustrated in Fig. 10.3. The MATLAB m-file `Fig8TrackDemo.m` on the CD-ROM generates a series of plots and statistics of the simulated trajectory. It calls the MATLAB function `Fig8Mod1D`, which generates the simulated dynamic conditions on the track, and it outputs the following statistics of nominal dynamic conditions:

```
   RMS N-S Position Excursion= 212.9304 meter
   RMS E-W Position Excursion= 70.9768 meter
 RMS Vert. Position Excursion= 3.5361 meter
             RMS N-S Velocity= 22.3017 m/s
             RMS E-W Velocity= 14.8678 m/s
           RMS Vert. Velocity= 0.37024 m/s
         RMS N-S Acceleration= 2.335 m/s/s
         RMS E-W Acceleration= 3.1134 m/s/s
       RMS Vert. Acceleration= 0.038778 m/s/s
     RMS Delta Velocity North= 0.02335 m/s at Δt =0.01 sec.
                             = 2.334 m/s at Δt =1 sec.
      RMS Delta Velocity East= 0.031134 m/s at Δt =0.01 sec.
                             = 3.1077 m/s at Δt =1 sec.
      RMS Delta Velocity Down= 0.00038771 m/s at Δt =0.01 sec.
                             = 0.038754 m/s at Δt =1 sec.
 N. Position Correlation Time= 13.4097 sec.
 E. Position Correlation Time= 7.6696 sec.
 Vertical Position Corr. Time= 9.6786 sec.
 N. Velocity Correlation Time= 9.6786 sec.
 E. Velocity Correlation Time= 21.4921 sec.
 Vertical Velocity Corr. Time= 13.4097 sec.
 N. Acceler. Correlation Time= 13.4097 sec.
 E. Acceler. Correlation Time= 7.6696 sec.
 Vertical Acceler. Corr. Time= 9.6786 sec.
```

TABLE 10.4. Comparison of Alternative GNSS Filters on 1.5-km Figure-8 Track Simulation

GNSS	RMS Position Estimated Errors[a] (m)		
Filter	North	East	Down
TYPE2	42.09	40.71	4.84
DAMP2	22.98	25.00	3.51
DAMP3	7.34	10.52	3.31
FIG8	0.53	0.31	0.01

[a]Clock errors not included.

These statistics are used for "tuning" the filter parameters for each of the alternative vehicle tracking filters—within the capabilities of the tracking filter.

10.2.3.2 Results The MATLAB m-file `GPSTrackingDemo.m` on the accompanying CD-ROM simulates the GPS satellites, the vehicle, and all four types of filters on a common set of pseudorange measurements over a period of 2 h. The position estimation results are summarized in Table 10.4 for one particular simulation.

The m-file `GPSTrackingDemo.m` generates many more plots to demonstrate how the different filters are working, including plots of the simulated and estimated pseudorange errors for each of the 29 satellites, some of which are not in view. Because the simulation uses a pseudo-random-number generator, the results can change from run to run.

10.2.3.3 Model Dimension versus Model Constraints These results indicate that dilution of information is not just a matter of state vector dimension. One might expect that the more variables there are to estimate, the less information will be available for each variable. In Table 10.4, the DAMP3 model has three more state variables than do the TYPE2 or DAMP2 models, yet it produces better results. The other issue at work here is the degree to which the model constrains the solution, and this factor better explains the ordering of estimation accuracy. The degree to which the model constrains the solution increases downward in the table, and simulated performance improves monotonically with the degree of constraint.

10.2.3.4 Role of Model Fidelity These results strongly suggest some performance advantage to be gained by tuning the vehicle tracking filter structure and parameters to the problem at hand. For the simulated trajectory, the accelerations, velocities and position excursions are all constrained, and the model that takes greatest advantage of that is FIG8, the track-specific model.

10.3 LOOSELY COUPLED INTEGRATION

10.3.1 Overall Approach

At an abstract level, loosely coupled implementations represent the two sensor systems in a mathematical model of the sort

$$\mathbf{z}_{\mathrm{GNSS}} = \mathbf{h}_{\mathrm{GNSS}}\left(\mathbf{x}_{\mathrm{hostveh}}\right) + \boldsymbol{\delta}_{\mathrm{GNSS}} \tag{10.38}$$

$$\frac{d}{dt}\boldsymbol{\delta}_{\mathrm{GNSS}} \approx \mathbf{f}_{\mathrm{GNSS}}\left(\boldsymbol{\delta}_{\mathrm{GNSS}},\ \mathbf{x}_{\mathrm{hostveh}}\right) + \mathbf{w}_{\mathrm{GNSS}}(t) \tag{10.39}$$

$$\mathbf{z}_{\mathrm{INS}} = \mathbf{h}_{\mathrm{INS}}\left(\mathbf{x}_{\mathrm{hostveh}}\right) + \boldsymbol{\delta}_{\mathrm{INS}} \tag{10.40}$$

$$\frac{d}{dt}\boldsymbol{\delta}_{\mathrm{INS}} \approx \mathbf{f}_{\mathrm{INS}}\left(\boldsymbol{\delta}_{\mathrm{INS}},\ \mathbf{x}_{\mathrm{hostveh}}\right) + \mathbf{w}_{\mathrm{INS}}(t) \tag{10.41}$$

where $\mathbf{z}_{\mathrm{GNSS}}$ represents the GNSS output; $\mathbf{z}_{\mathrm{INS}}$ represents the INS output; $\mathbf{x}_{\mathrm{hostveh}}$ represents either the "true" navigation state of the host vehicle (the one both GNSS receiver and INS are mounted on), or the best available estimate of the navigation state; $\boldsymbol{\delta}_{\mathrm{GNSS}}$ represents GNSS output error; $\mathbf{f}_{\mathrm{GNSS}}\left(\boldsymbol{\delta}_{\mathrm{GNSS}},\ \mathbf{x}_{\mathrm{hostveh}}\right)$ and the white-noise process $\mathbf{w}_{\mathrm{GNSS}}(t)$ represent the dynamic model for $\boldsymbol{\delta}_{\mathrm{GNSS}}$ assumed in the GNSS/INS integration scheme; $\boldsymbol{\delta}_{\mathrm{INS}}$ represents INS output error; and $\mathbf{f}_{\mathrm{INS}}\left(\boldsymbol{\delta}_{\mathrm{INS}},\ \mathbf{x}_{\mathrm{hostveh}}\right)$ and the white-noise process $\mathbf{w}_{\mathrm{INS}}(t)$ represent the dynamic model for $\boldsymbol{\delta}_{\mathrm{INS}}$ assumed in the GNSS/INS integration scheme.

Different approaches to loosely coupled integration are free to assume quite different mathematical forms for the functions $\mathbf{f}_{\mathrm{GNSS}}$ and $\mathbf{f}_{\mathrm{INS}}$, yet each approach would still be considered loosely coupled.

10.3.2 GNSS Error Models

10.3.2.1 Empirical Modeling The trajectory of the antenna has some influence on receiver positioning errors. However, getting data on the actual GNSS position errors is not that easy unless the receiver antenna is stationary. Otherwise, in order to measure the actual receiver position errors, one needs an independent, more accurate means of measuring antenna position while it is moving—or use simulation of receiver errors during simulation.

We can use the results of the figure-8 track simulation to illustrate how the trajectory influences positioning errors, and how receiver error data for a stationary antenna can be used to develop a model for loosely coupled GNSS/INS integration.

Figure-8 Track Simulation Results Figures 10.4–10.6 show plots of the power spectral densities of position errors from using TYPE2, DAMP2 and DAMP3 filters, respectively, in the simulations summarized in Table 10.4 and generated by the MATLAB m-file `GPSTrackingDemo.m`.

Harmonic Errors These simulations were for one-minute trajectories around a figure-8 track, and they exhibit harmonics at one cycle per minute (≈ 0.01667 Hz)

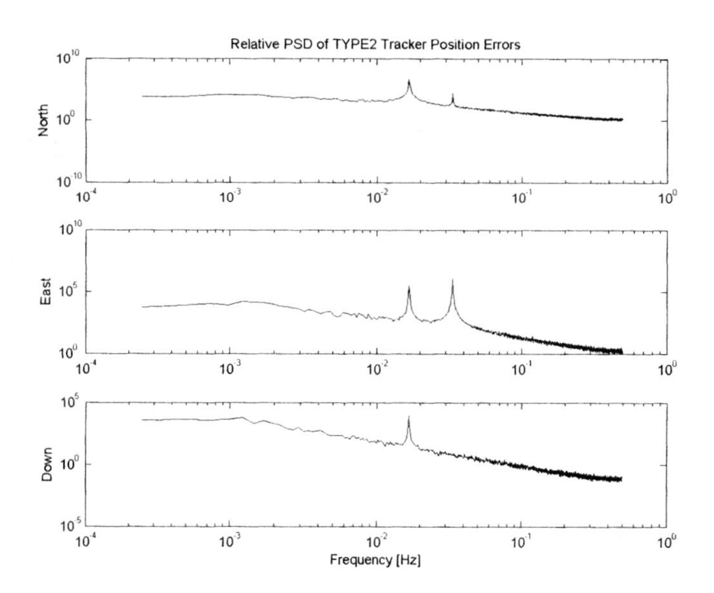

Fig. 10.4 Power spectral densities of TYPE2 GPS filter errors.

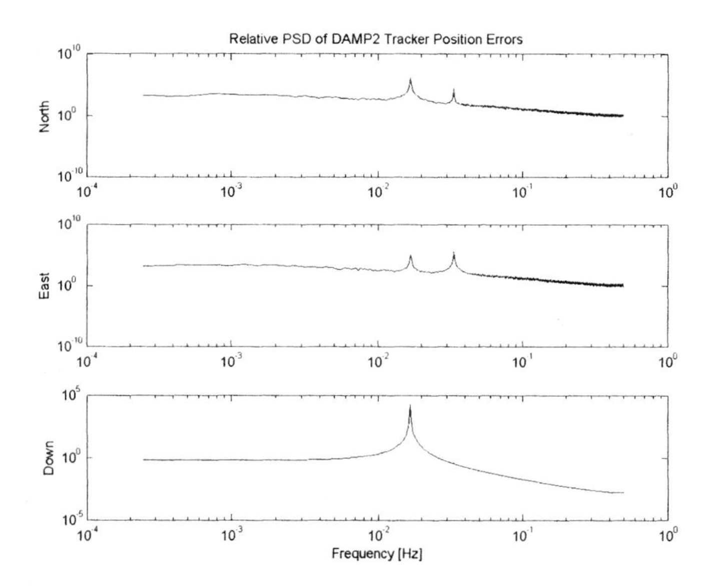

Fig. 10.5 Power spectral densities of DAMP2 GPS filter errors.

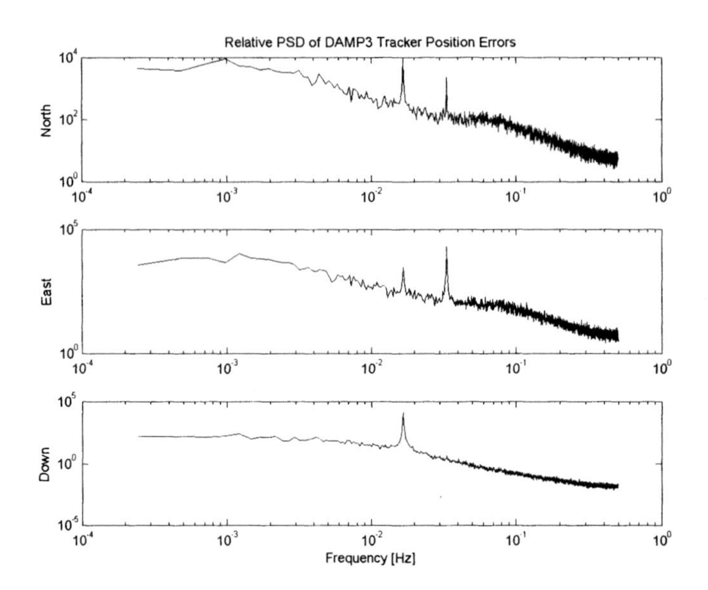

Fig. 10.6 Power spectral densities of DAMP3 GPS filter errors.

and two cycles per minute (≈ 0.0333 Hz), standing 10–30 dB above background noise. The slight broadening of these spectral peaks is due to simulated random vehicle speed variations of $\pm 10\%$ RMS from the mean speed.

These spectral peaks are most likely due to filter lags.[4] The track altitude model has only one harmonic at one cycle per minute, and that is the only harmonic peak evident in the plots of the vertical components of position errors.

PSD for Stationary GPS Antenna Figure 10.7 is a plot of the PSDs analogous to Figs. 10.4–10.6, but with a stationary antenna. The spectra are almost identical to those in Figs. 10.4–10.6, but without the harmonic peaks. This would indicate that the harmonic peaks in Fig. 10.6 are the dominant effects of the figure-8 trajectory, and the rest of the spectrum is due to the other GPS tracking error sources (random ranging errors).

Exponentially Correlated Errors The background noise in the power spectral densities in Fig. 10.7 looks much like the exponentially correlated noise in Fig. 8.9, which flattens out at the low-frequency end but falls off at about -20 dB/decade at frequencies $\gg 1/\tau$, where τ is the correlation time. These correlated errors are most likely due to the correlated errors in the simulated GPS pseudoranges (i.e., signal delay errors exponentially correlated with one minute correlation time).

[4]A diligent engineer would always take the PSD of receiver error data to help in understanding its statistical properties. Seeing these pronounced harmonic peaks, she or he would probably be lead to using an alternative tracking filter such as the FIG8 model.

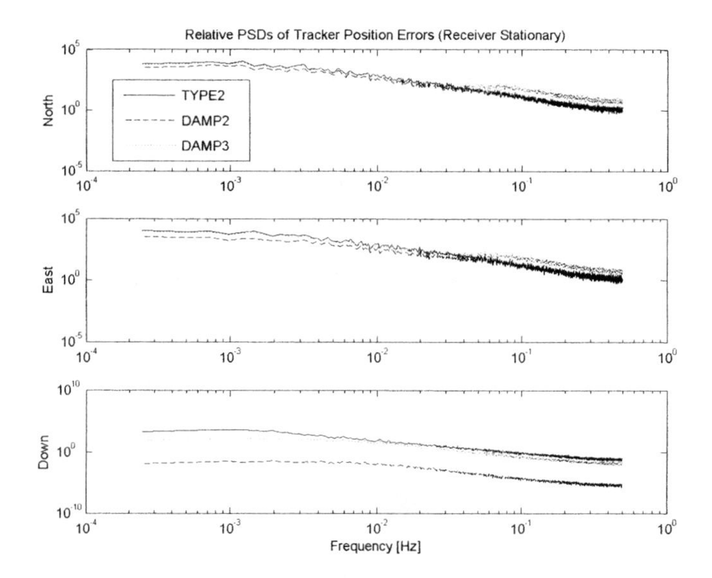

Fig. 10.7 Error power spectral densities without motion.

Indeed, the empirically calculated autocorrelations functions plotted in Fig. 10.8 do appear to be exponential functions. Exponential functions would look like straight lines on these semi-log plots, and that is about what the plots show.

The estimated exponential decay time-constants will equal the lag times at which these straight lines cross the $1/e$ threshold, and these are the values shown in Table 10.5. These vary a bit, depending on the receiver filter used, but the correlation times of the horizontal position components are relatively consistent for a given filter.

10.3.3 Receiver Position Error Model

We will use the DAMP3 receiver filter model results to show how they can be used in deriving an appropriate receiver position error model for loosely-coupled GPS/INS integration. This will be an exponentially damped position error model of the sort

$$\frac{d}{dt}\begin{bmatrix} \delta p_{\text{GNSS N}} \\ \delta p_{\text{GNSS E}} \\ \delta p_{\text{GNSS D}} \end{bmatrix} = \begin{bmatrix} -1/\tau_{\text{hor}} & 0 & 0 \\ 0 & -1/\tau_{\text{hor}} & 0 \\ 0 & 0 & -1/\tau_{\text{vert}} \end{bmatrix} \begin{bmatrix} \delta p_{\text{GNSS N}} \\ \delta p_{\text{GNSS E}} \\ \delta p_{\text{GNSS D}} \end{bmatrix} + \begin{bmatrix} w_{\text{hor}}(t) \\ w_{\text{hor}}(t) \\ w_{\text{vert}}(t) \end{bmatrix}, \tag{10.42}$$

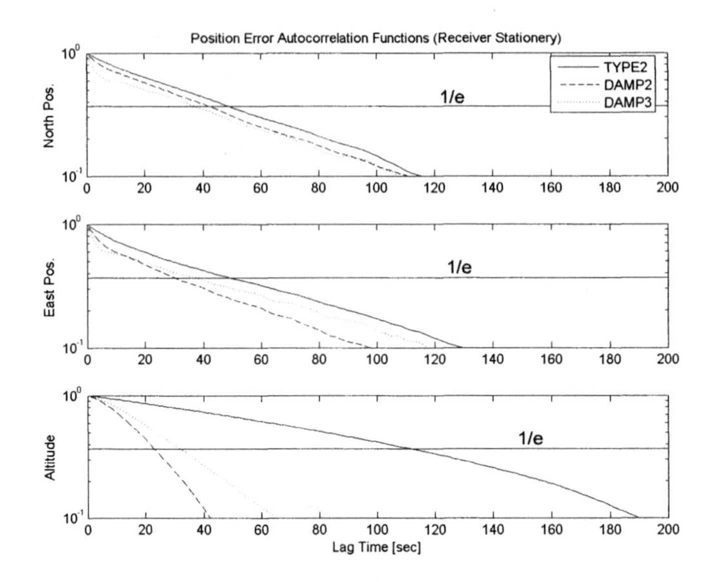

Fig. 10.8 Error autocorrelation functions without motion.

TABLE 10.5. Receiver Position Error Correlation Times

Receiver Filter	Correlation Times (s)		
	North	East	Down
TYPE2	49	49	111
DAMP2	42	30	23
DAMP3	37	38	32

with independent noise on all three channels.

10.3.4 INS Error Models

Error models for GNSS receivers will depend on uncertainty about the internal clock and host vehicle dynamics, and the internal filtering in the receiver for coping with it. The distributions of "upstream" errors in the signal information at the receiver antenna are essentially fixed by the GNSS system error budget, which is constant.

Models for INS errors, on the other hand, depend to some lesser degree on vehicle dynamics, but are dominated by the INS error budget, which depends very much on system design and the quality of sensors used. We will first use a very simple INS model to demonstrate how loosely coupled GNSS/INS integration performance depends on INS performance. This simplified model splits the loosely coupled GNSS/INS integration problem into two problems:

1. An altitude problem, solved by a two-state Kalman filter using GNSS altitude outputs to stabilize the otherwise unstable INS vertical channel.
2. The remaining horizontal navigation problem, which is solved using an independent eight-state Kalman filter.

This splitting ignores the coriolis coupling between the vertical and horizontal channels, but the coriolis coupling is not a dominating error mechanism anyway.

10.3.4.1 Using GNSS Altitude for INS Vertical Channel Stabilization This is quite similar to the barometric altimeter implementation used in Section 9.5.4.1, with a GNSS altitude error model replacing the altimeter error model.

10.3.4.2 Random-Walk Tilt Model This is a generic INS system-level error model designed to demonstrate how INS CEP rate influences integrated GPS/INS performance. It is not a particularly faithful error model for any INS, but it does serve to demonstrate in very general terms how the gross error characteristics of inertial navigators are mitigated by GNSS/INS integration.

Without carouseling, INS errors tend to be dominated by tilt errors, which can result from accelerometer bias errors or gyro errors. A relatively simple model for horizontal error dynamics uses the six state variables:

$$\begin{aligned}
\delta p_N &= \text{north component of position error.}\\
\delta p_E &= \text{east component of position error.}\\
\delta v_N &= \text{north component of velocity error.}\\
\delta v_E &= \text{east component of velocity error.}\\
\rho_N &= \text{tilt error rotation about north axis.}\\
\rho_E &= \text{tilt error rotation about east axis.}
\end{aligned}$$

The corresponding dynamic model for this state vector will be

$$\frac{d}{dt}\begin{bmatrix} \delta p_N \\ \delta p_E \\ \delta v_N \\ \delta v_E \\ \rho_N \\ \rho_E \end{bmatrix} = \underbrace{\begin{bmatrix} 0 & 0 & 1 & 0 & 0 & 0 \\ 0 & 0 & 0 & 1 & 0 & 0 \\ -\Omega^2_{\text{Sch.}} & 0 & 0 & -2s\Omega_{\odot} & -g & 0 \\ 0 & -\Omega^2_{\text{Sch.}} & 2s\Omega_{\odot} & 0 & 0 & g \\ 0 & 0 & 0 & 0 & 0 & 0 \\ 0 & 0 & 0 & 0 & 0 & 0 \end{bmatrix}}_{\mathbf{F}_{\text{hor}}} \begin{bmatrix} \delta p_N \\ \delta p_E \\ \delta v_N \\ \delta v_E \\ \rho_N \\ \rho_E \end{bmatrix}$$

$$+ \begin{bmatrix} 0 \\ 0 \\ 0 \\ 0 \\ w_\rho(t) \\ w_\rho(t) \end{bmatrix}, \tag{10.43}$$

$$s = \sin(\text{Latitude}), \tag{10.44}$$

$$\Omega_{\text{Sch.}} \stackrel{\text{def}}{=} \text{Schuler frequency} \tag{10.45}$$

$$\approx 0.00124 \text{ (rad/s)}, \tag{10.46}$$

$$\Omega_{\odot} \stackrel{\text{def}}{=} \text{earth ratation rate} \tag{10.47}$$

$$\approx 7.3 \times 10^{-5} \text{ (rad/s)}, \tag{10.48}$$

$$g \stackrel{\text{def}}{=} \text{gravitational acceleration} \tag{10.49}$$

$$\approx 9.8 \text{ m s}^{-2}. \tag{10.50}$$

The CEP rate of the modeled INS will depend only on the single parameter

$$q_{\text{continuous}} = \mathop{\mathrm{E}}_{t}\left\langle w_{\rho}^{2}(t)\right\rangle$$

or its equivalent discrete-time counterpart

$$q_{\text{discrete}} = \mathop{\mathrm{E}}_{k}\left\langle w_{k}\right\rangle.$$

The resulting relationship between CEP rate and the discrete-time parameter to q_{discrete} is computed by the MATLAB m-file `HorizINSperfModel.m` on the accompanying CD-ROM, the result of which is plotted in Fig. 10.9. The values of q for some specific CEP rates are listed in Table 10.6.

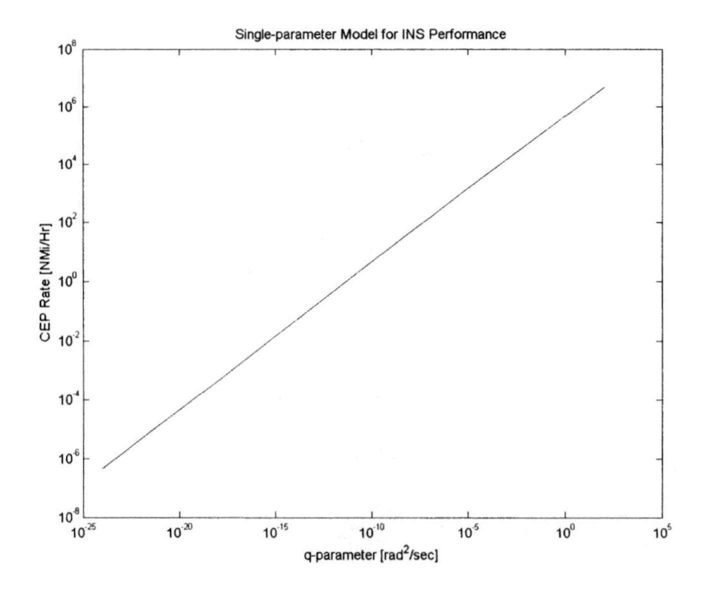

Fig. 10.9 CEP rate versus model parameter q.

TABLE 10.6. *q*-Parameter Values versus CEP Rates

CEP Rate (nmi/h)[a]	q (rad^2/s)
0.01	4.5579×10^{-16}
0.1	4.5579×10^{-14}
1	4.5579×10^{-12}
10	4.5579×10^{-10}
100	4.5579×10^{-8}
1000	4.5579×10^{-6}

[a]Nautical miles per hour.

10.3.4.3 Performance Analysis Performance advantages for GNSS/INS integration include improvements in velocity, acceleration, attitude, and attitude rate estimates, as well as improving position estimates. Nevertheless, the results shown here are for position uncertainty, only.

The full Kalman filtering model used for loosely coupled integration includes 20 state variables:

- Three state variables for time-correlated GPS output position errors.
- The nine state variables of the DAMP3 vehicle dynamics model (three components each of acceleration, velocity, and position).
- Eight state variables for INS output errors (three components each of horizontal velocity and position errors, plus two tilt errors). This part of the model includes one parameter characterizing the CEP rate of the INS navigator, and one parameter (RMS accelerometer noise) characterizing relative vertical channel instability.

The six measurements include three position outputs from the GPS receiver and three position outputs from the INS.

Figures 10.10 and 10.11 were generated by the MATLAB m-file `GPSINSwGPSpos.m` to show the theoretical relative performance of a GPS receiver, an INS, and a loosely coupled integrated GPS/INS system. Different plots show performances for several levels of INS performance, as characterized by CEP rate (for horizontal accuracy) and accelerometer noise (for vertical channel accuracy).

Figure 10.10 shows predicted RMS horizontal position uncertainty as a function of time over 2 h, and INS performances ranging from 0.1 nmi/h (nautical miles per hour) ("high accuracy") to 100 nmi/h (very low accuracy). The plot illustrates how INS-only performance degrades over time, GPS-only performance remains essentially constant over time, and loosely coupled integrated GPS/INS performance is better than either.

Figure 10.11 shows predicted RMS altitude uncertainty as a function of time over 2 h, and accelerometer noise ranging from 0.001 m/s^2 per root hour to 1 m/s^2 per root hour. The plot illustrates how INS-only altitude diverges over

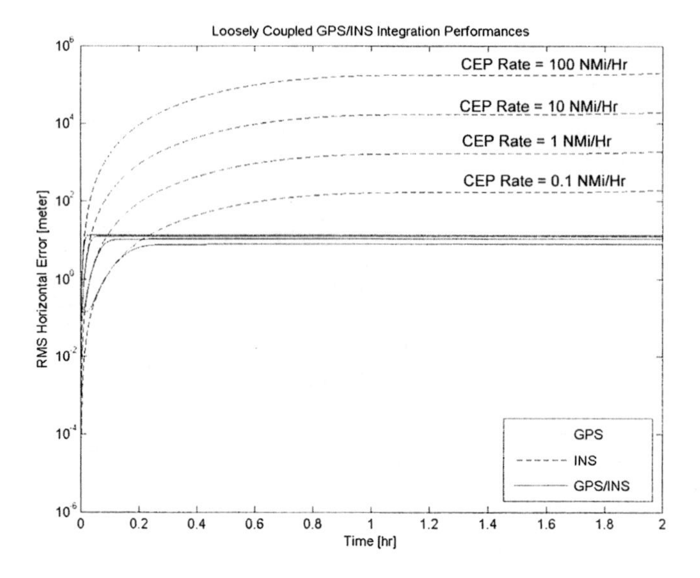

Fig. 10.10 Theoretical horizontal performance of loosely coupled GPS/INS integration.

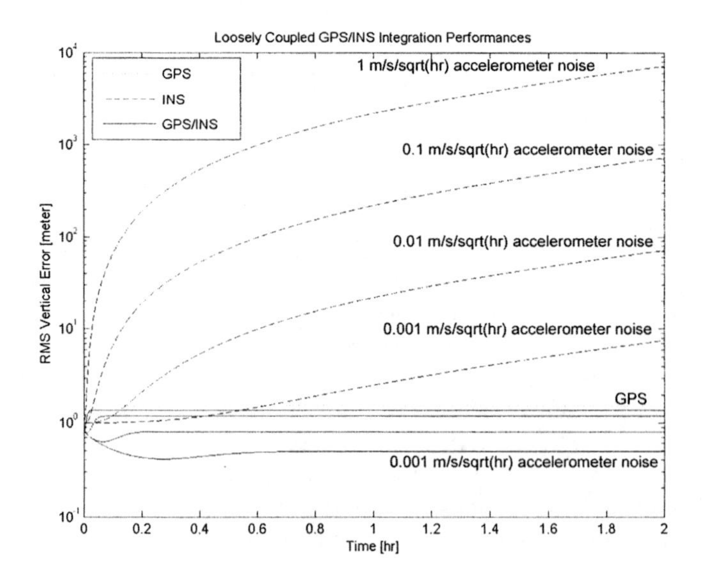

Fig. 10.11 Theoretical vertical performance of loosely coupled GPS/INS integration.

time, GPS-only performance remains essentially constant over time, and loosely coupled integrated GPS/INS performance is better than either. In this example, GPS altitude uncertainty is about 1.5 m RMS, because the DAMP3 filter used for vehicle tracking constrains vehicle altitude to a small dynamic range (based on figure-8 track dynamics).

There are many other ways to implement loosely coupled GNSS/INS integration, but this particular simple example[5] demonstrates how weakly loosely coupled GPS/INS performance depends on standalone INS performance. This can have a profound effect on overall system cost, because INS cost tends to vary inversely as CEP rate. This, in turn, has spurred greater interest in developing lower-cost, lower-performance inertial sensors for integrated systems. The major advantages of GNSS/INS integration come from improvements in the accuracy of velocity, attitude, and other parameters, not just position alone, and these other performance improvements are also driving lower-cost sensor development interests.

10.4 TIGHTLY COUPLED INTEGRATION

There are many GNSS/INS integration approaches requiring changes in the internal software and hardware implementations of GNSS receivers and/or inertial navigation systems. The following subsections describe some that are in use or under development.

10.4.1 Using GNSS for INS Vertical Channel Stabilization

Inertial navigators designed for submarines have used depth (water pressure) sensors for vertical channel stabilization,[6] and early inertial systems designed for aircraft or surface vehicles[7] used a barometric altimeter interface for the same reason. GNSS/INS integration has limited applicability for submarines, whose commanders are reluctant to pop up an antenna while submerged, for fear of being detected. For aircraft and surface vehicles, however, GNSS is more accurate than a barometric altimeter.

This would be considered a less-than-loosely coupled integration, because it requires some INS software changes. However, because vertical channel stabilization has always been part of INS implementations, the changeover to using GNSS may require only relatively minor changes to INS software.

[5]This example would seem to imply that one can effectively stabilize the INS vertical channel without modifying the internal INS software, but this cannot be done in practice. The coriolis effect would couple large vertical velocity errors within the INS implementation into horizontal acceleration errors. INS vertical channel stabilization is best implemented within the INS software.

[6]Submarines can also use electromagnetic waterspeed sensors (EM logs) for velocity error damping.

[7]Except those operating at the sea surface, which can stabilize the vertical channel by fixing altitude to a constant.

10.4.2 Using INS Accelerations to Aid GNSS Signal Tracking

In addition to the effects due to dilution of information, vehicle dynamics have a serious impact on the ability of a GNSS receiver to maintain phase lock on the satellite signals, and to reacquire lock after it is lost. These effects can be mitigated significantly in GNSS/INS integration, by using the acceleration measurements of the INS to predict the Doppler shift changes before they accumulate as phase tracking errors.

If a_{LOS} is the sensed acceleration component along the line of sight toward a particular GNSS satellite, the rate of change of a carrier frequency f_{carrier} from that satellite, due to vehicle acceleration, will be

$$\frac{d}{dt} f_{\text{carrier}} = \frac{a_{\text{LOS}} f_{\text{carrier}}}{c}, \tag{10.51}$$

where $c = 299792458$ m/s is the speed of light. This computed value of frequency shift rate can be applied to the carrier phase-lock loop to head off an otherwise rapidly accumulating carrier phase tracking error.

The phase-lock loops used for maintaining GNSS signal phase lock are primarily PI controllers with control margins designed for relatively benign vehicle dynamics. Their control margins can be extended to more highly maneuverable vehicles by using accelerations in a PID controller, because acceleration is related to the derivative of frequency. This level of GNSS/INS integration generally requires hardware and software changes to the receiver signal interfaces and phase lock control, and may require INS output signal changes, as well.

10.4.3 Using GNSS Pseudoranges

All GNSS navigation receivers use pseudoranges as their fundamental range measurements. The filtering they do to estimate antenna position increases data latency at the receiver outputs. If these outputs are filtered again for GNSS/INS integration, the lags only increase. Time tagging and output of the pseudoranges helps to control and reduce data latency in GNSS/INS integration.

GNSS/INS integration software using pseudorange data may include receiver clock bias and drift estimation, even though this function is already implemented in the receiver software. The reason for wanting to include clock error estimation in the GNSS/INS integration software is that the integration filter uses pseudoranges which are corrupted by residual clock errors. The integrating filter has access to INS measurements, which should allow it to make even better estimates of clock errors than those within the receiver.

Using these improved clock corrections within the receiver may not be necessary, especially if the improvements are rather modest. It simplifies the receiver interface if the clock control loop can be maintained within the receiver itself.

10.4.3.1 Example This is essentially the same model used in Section 10.3.4.3, but with pseudoranges used in place of GPS receiver position estimates. It is implemented in the MATLAB m-file `GPSINSwPRs.m` on the accompanying CD-ROM. As before, the INS system-level error model is a generic model designed to demonstrate how INS CEP rate influences integrated GPS/INS performance. In this case, however, simulated GPS satellite geometry variations will introduce some irregularity into the plots.

The full Kalman filter sensor integration model for this example includes 46 state variables:

- The nine state variables of the DAMP3 vehicle dynamics model (three components each of acceleration, velocity, and position).
- Twenty-nine state variables for time-correlated pseudorange errors. (This had been three state variables for GPS receiver position errors in the loosely coupled example.)
- Eight state variables for INS output errors (three components each of horizontal velocity and position errors, plus two tilt errors). This part of the model includes one parameter characterizing the CEP rate of the INS navigator, and one parameter (RMS accelerometer noise) characterizing relative vertical channel instability.

The measurements include three position outputs from the INS and the pseudoranges from all acquired satellite signals.

Simulation results are plotted in Figures 10.12–10.16. Some of this is "comparing apples and oranges," because the vertical disturbance model for INS errors is driven by vertical accelerometer noise, which is not present in the GPS-only model.

10.4.4 Real-Time INS Recalibration

There are some potential GNSS/INS integration applications for which the host vehicle must navigate accurately through GNSS signal outages lasting a minute or more. These applications include manned or unmanned vehicles operating under leaf cover, through tunnels or in steep terrain (including urban canyons). They also include "standoff" precision weapons for attacking high-value targets protected by GNSS jamming. If the INS can use GNSS to continuously calibrate its sensors while GNSS is available, then the INS can continue to navigate over the short term without GNSS, starting with freshly recalibrated sensors. If and when GNSS signals become available again, the improved position and velocity information from the INS can speed up signal reacquisition.

10.4.4.1 Example This example integration architecture is tightly coupled in two ways:

- It uses pseudoranges directly for a GNSS receiver.
- It feeds back the estimated calibration parameters to the INS software.

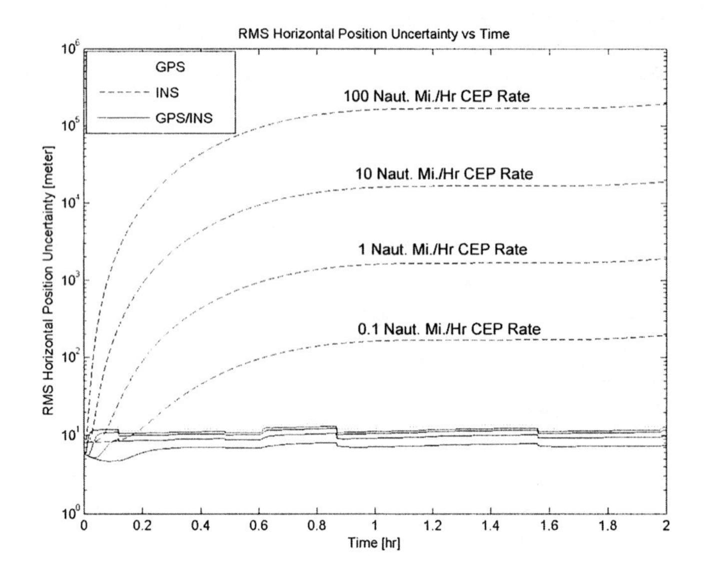

Fig. 10.12 Theoretical RMS horizontal position of tightly coupled GPS/INS integration.

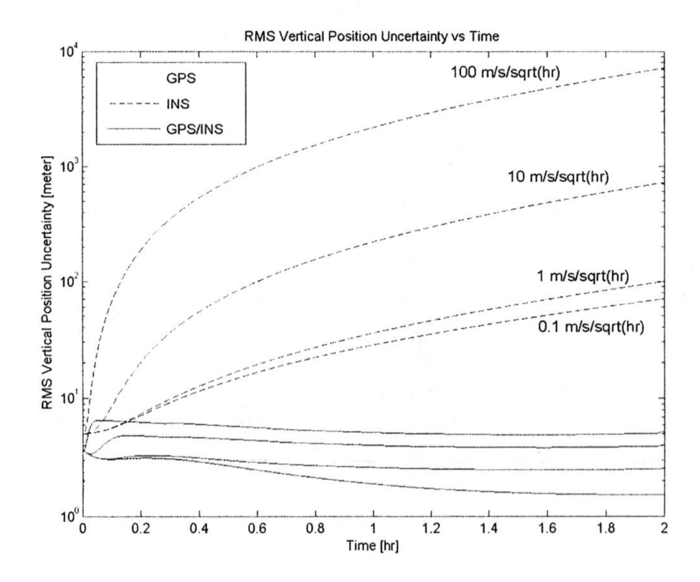

Fig. 10.13 Theoretical RMS altitude of tightly coupled GPS/INS integration.

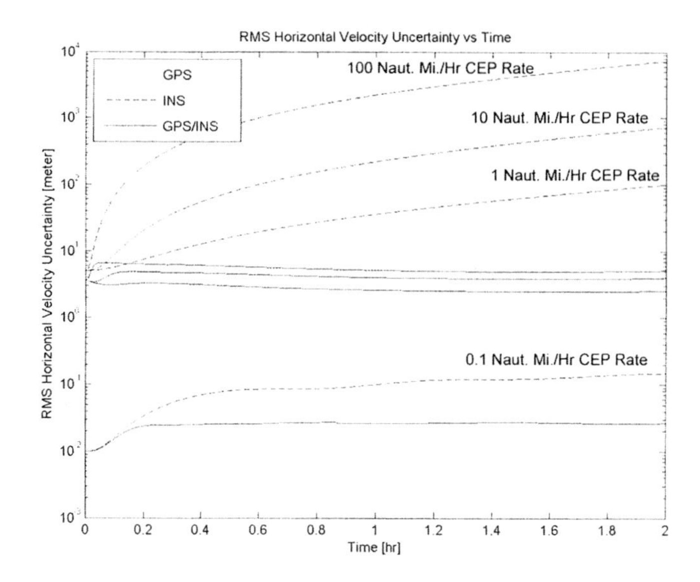

Fig. 10.14 Theoretical RMS horizontal velocity of tightly coupled GPS/INS integration.

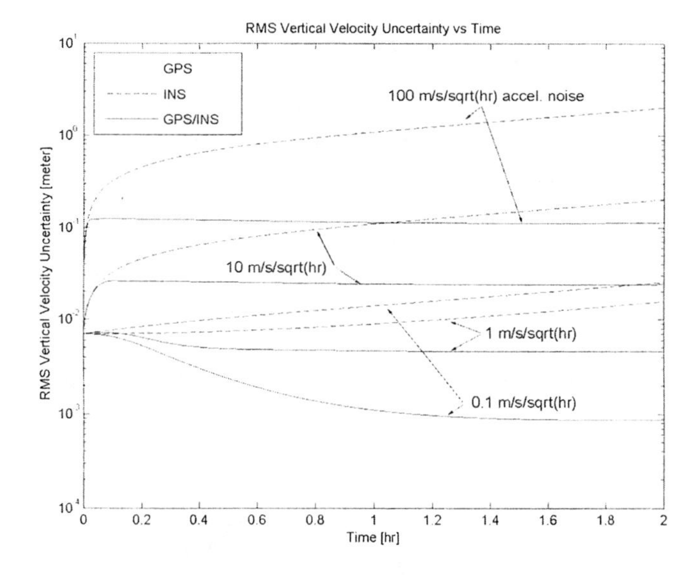

Fig. 10.15 Theoretical RMS vertical velocity of tightly coupled GPS/INS integration.

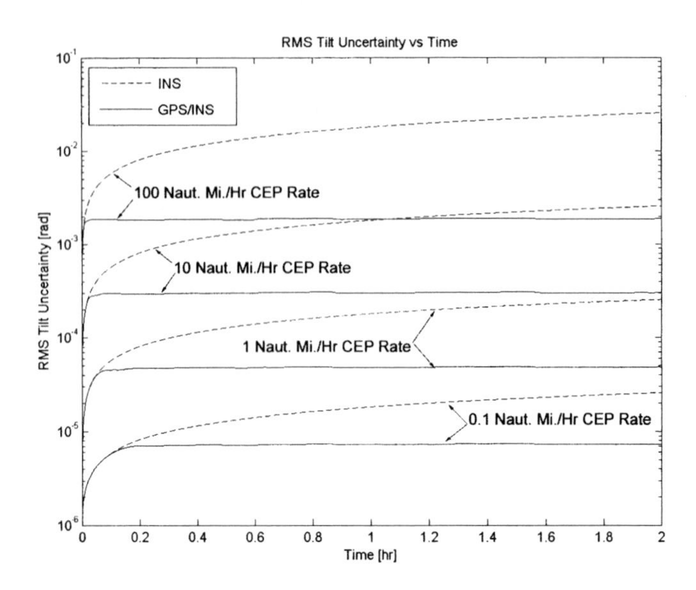

Fig. 10.16 Theoretical RMS tilt uncertainty of tightly coupled GPS/INS integration.

Performance statistics of interest for this example problem would include RMS position, velocity, and attitude uncertainty after periods of a minute or more without GNSS signals.

The software for GNSS/INS integration can run in a relatively slow loop (1 s) compared to the INS dynamic integration loop (1–10 ms, typically), because it is primarily tracking the slowly changing parameters of the inertial sensors. This reduces the added computational requirements considerably.

The example GNSS/INS integration model shown here uses 58 state variables:

15 state variables related to INS navigation errors (position, velocity, acceleration, attitude, and attitude rate) due to sensor calibration errors. The entire INS system-level error model used before had only 8 state variables, but these are insufficient for modeling the effects of calibration errors (below).

12 state variables related to calibration parameter errors (2 state variables to model time-varying calibration parameters of each of the six inertial sensors). This increases the INS error model state vector size to 27 variables, where there had been only 8 before.

2 state variables for the receiver clock (bias and drift). These are included because residual clock errors will corrupt the pseudorange measurements output from the receiver.

29 state variables for pseudorange biases (one for each of the 29 GPS satellites for the March 2006 ephemerides used in simulations)

The last 31 of these are essentially nuisance variables that must be included in the state vector because they corrupt the measurements (pseudoranges, plus INS errors in position, velocity, acceleration, attitude, and attitude rate). Some reductions are possible. The number of state variables devoted to pseudorange biases can be reduced to just those usable at any time, or it can be eliminated altogether by using Schmidt–Kalman filtering. The number of variables used for sensor calibration can be reduced to just those deemed sufficiently unstable to warrant recalibration.

The number of variables used for sensor calibration can also be increased to include drift rates of the calibration parameters.

Integration Filter Models Mathematically, the Kalman filtering model for GNSS/INS integration is specified by

1. The model state vector **x** and two matrix parameters of the state dynamic model:
   (a) The state transition matrix Φ_k or dynamic coefficient matrix $\mathbf{F}(t)$, related by

$$\Phi_k = \exp\left(\int_{t_{k-1}}^{t_k} \mathbf{F}(s)\, ds\right).$$

   (b) The dynamic disturbance noise covariance matrix **Q**.
2. The model measurement vector **z** and two matrix parameters of the measurement model:
   (a) The measurement sensitivity matrix **H**.
   (b) The measurement noise covariance matrix **R**.

Potential State Variables The components of a potential state vector **x** for this example problem include the following:

1	δ_{lat}	=	error in INS latitude (rad),
2	δ_{lon}	=	error in INS longitude (rad),
3	δ_{alt}	=	error in INS altitude (m),
4	δ_{vN}	=	error in north velocity of INS (m/s),
5	δ_{vE}	=	error in east velocity of INS (m/s),
6	δ_{vD}	=	error in downward velocity of INS (m/s),
7	δ_{aN}	=	error in north acceleration of INS(m $s^1 s^{-1}$),
8	δ_{aE}	=	error in east acceleration of INS(m $s^1 s^{-1}$),
9	δ_{aD}	=	error in downward acceleration of ISA(m $s^1 s^{-1}$),
10	$\delta_{\rho N}$	=	north component of INS attitude error rotation vector, representing INS tilt (rad),
11	$\delta_{\rho E}$	=	east component of INS attitude error (rad),
12	$\delta_{\rho \text{D}}$	=	downward component of INS attitude error (rad),
13	$\delta_{\omega N}$	=	north component of vehicle rotation rate error (rad/s),
14	$\delta_{\omega E}$	=	east component of vehicle rotation rate error (rad/s),

15	$\delta_{\omega D}$	=	downward component of vehicle rotation rate error (rad/s),
16	δ_{ab1}	=	first accelerometer bias error(m $s^1 s^{-1}$),
17	δ_{ab2}	=	second accelerometer bias error(m $s^1 s^{-1}$),
18	δ_{ab3}	=	third accelerometer bias error(m $s^1 s^{-1}$),
19	δ_{aS1}	=	relative error in first accelerometer scale factor (unitless),
20	δ_{aS2}	=	relative error in second accelerometer scale factor (unitless),
21	δ_{aS3}	=	relative error in third accelerometer scale factor (unitless),
22	δ_{gb1}	=	first gyroscope bias error (rad/s),
23	δ_{gb2}	=	second gyroscope bias error (rad/s),
24	δ_{gb3}	=	third gyroscope bias error (rad/s),
25	δ_{gS1}	=	relative error in first gyroscope scale factor (unitless),
26	δ_{gS2}	=	relative error in second gyroscope scale factor (unitless),
27	δ_{gS3}	=	relative error in third gyroscope scale factor (unitless),
28	δ_{clockb}	=	residual receiver clock bias error (m),
29	δ_{clockd}	=	residual receiver clock drift rate error (m/s),
30	δ_{PRN1}	=	first satellite pseudorange bias error (m),
31	δ_{PRN2}	=	second satellite pseudorange bias error (m),
		$\vdots$	
58	δ_{PRN29}	=	last satellite pseudorange bias error (m).

The variables $\delta_{ab1}, \ldots, \delta_{aS1}, \ldots, \delta_{gb1}, \ldots, \delta_{gS1}, \ldots$ are sensor calibration coefficient errors. They are numbered 1–3 because the sensor input axis directions are not fixed in RPY coordinates (as they would be for noncarouseled strapdown systems) or NED coordinates (as they would be for noncarouseled gimbaled systems).

A rotation vector $\boldsymbol{\rho}$ representing attitude has been introduced in Chapter 9 for coning error correction. In that case and in this case, the representation should work because the magnitude $|\boldsymbol{\rho}| \ll \pi$, where the rotation vector model behaves badly.

State Vector Dynamics The first-order differential equations defining the state dynamic model are as follows:

$$\frac{d}{dt}\delta_{\text{lat}} = \frac{\delta_{vN}}{R_N(\phi_{\text{lat}}) + h_{\text{lat}}}, \tag{10.52}$$

$$\frac{d}{dt}\delta_{\text{lon}} = \frac{\delta_{vE}}{[R_E(\phi_{\text{lat}}) + h_{\text{alt}}] / \cos\phi_{\text{lat}}}, \tag{10.53}$$

$$\frac{d}{dt}\delta_{\text{alt}} = -\delta_{vD}, \tag{10.54}$$

$$\frac{d}{dt}\delta_{vN} = \delta_{aN} - 2\Omega_{\odot}\sin\phi_{\text{lat}}\delta_{vE} + w_{a1N}(t) + w_{a2N}(t) + w_{a3N}(t), \tag{10.55}$$

$$\frac{d}{dt}\delta_{vE} = \delta_{aE} + 2\Omega_{\odot}\sin\phi_{\text{lat}}\delta_{vN} + 2\Omega_{\odot}\cos\phi_{\text{lat}}\delta_{vD} + w_{a1E}(t) + w_{a2E}(t) + w_{a3E}(t), \tag{10.56}$$

$$\frac{d}{dt}\delta_{vD} = \delta_{aD} - 2\Omega_{\odot}\cos\phi_{\text{lat}}\delta_{vN} + w_{a1D}(t) + w_{a2D}(t) + w_{a3D}(t), \tag{10.57}$$

$$\frac{d}{dt}\delta_{aN} = 0, \tag{10.58}$$

$$\frac{d}{dt}\delta_{aE} = 0, \tag{10.59}$$

$$\frac{d}{dt}\delta_{aD} = 0, \tag{10.60}$$

$$\frac{d}{dt}\delta_{\rho N} = \frac{1}{2}\rho_D\omega_N - \frac{1}{2}\rho_N\omega_D + \left[1 - \frac{1}{2}\frac{\psi\sin(\psi)}{1-\cos(\psi)}\right] \times \left[\frac{\rho_D(-\rho_D\omega_E + \rho_E\omega_D) - \rho_N(-\rho_E\omega_N + \rho_N\omega_E)}{\psi^2}\right], \tag{10.61}$$

$$\frac{d}{dt}\delta_{\rho E} = \frac{1}{2}\rho_D\omega_N - \frac{1}{2}\rho_N\omega_D + \left(1 - \frac{1}{2}\frac{\psi\sin(\psi)}{1-\cos(\psi)}\right) \times \left[\frac{\rho_D(-\rho_D\omega_E + \rho_E\omega_D) - \rho_N(-\rho_E\omega_N + \rho_N\omega_E)}{\psi^2}\right], \tag{10.62}$$

$$\frac{d}{dt}\delta_{\rho D} = -\frac{1}{2}\rho_E\omega_N + \frac{1}{2}\rho_N\omega_E + \left(1 - \frac{1}{2}\frac{\psi\sin(\psi)}{1-\cos(\psi)}\right) \times \left[\frac{-\rho_E(-\rho_D\omega_E + \rho_E\omega_D) + \rho_N(\rho_D\omega_N - \rho_N\omega_D)}{\psi^2}\right], \tag{10.63}$$

$$\psi \stackrel{\text{def}}{=} \sqrt{\rho_{\text{N}}^2 + \rho_{\text{E}}^2 + \rho_{\text{D}}^2}, \tag{10.64}$$

$$\frac{d}{dt}\delta_{\omega N} = w_{\omega 1}(t), \tag{10.65}$$

$$\frac{d}{dt}\delta_{\omega E} = w_{\omega 2}(t), \tag{10.66}$$

$$\frac{d}{dt}\delta_{\omega D} = w_{\omega 3}(t), \tag{10.67}$$

$$\frac{d}{dt}\delta_{ab\,1} = w_{ba\,1}(t), \tag{10.68}$$

$$\frac{d}{dt}\delta_{ab\,2} = w_{ba\,2}(t), \tag{10.69}$$

$$\frac{d}{dt}\delta_{ab\,3} = w_{ba\,3}(t), \tag{10.70}$$

$$\frac{d}{dt}\delta_{aS\,1} = w_{Sa\,1}(t), \tag{10.71}$$

$$\frac{d}{dt}\delta_{aS\,2} = w_{Sa\,2}(t), \tag{10.72}$$

$$\frac{d}{dt}\delta_{aS\,3} = w_{Sa\,3}(t), \tag{10.73}$$

$$\frac{d}{dt}\delta_{gb\,1} = w_{bg\,1}(t), \tag{10.74}$$

$$\frac{d}{dt}\delta_{gb\,2} = w_{bg\,2}(t), \tag{10.75}$$

$$\frac{d}{dt}\delta_{gb\,3} = w_{bg\,3}(t), \tag{10.76}$$

$$\frac{d}{dt}\delta_{gS\,1} = w_{Sg\,1}(t), \tag{10.77}$$

$$\frac{d}{dt}\delta_{gS\,2} = w_{Sg\,2}(t), \tag{10.78}$$

$$\frac{d}{dt}\delta_{gS\,3} = w_{Sg\,3}(t), \tag{10.79}$$

$$\frac{d}{dt}\delta_{\text{clock b}} = w_{b\,\text{clock}}(t), \tag{10.80}$$

$$\frac{d}{dt}\delta_{\text{clock d}} = w_{b\,\text{clock}}(t), \tag{10.81}$$

$$\frac{d}{dt}\delta_{\text{PRN}\,1} = w_{\text{PRN}\,1}(t), \tag{10.82}$$

$$\frac{d}{dt}\delta_{\text{PRN}\,2} = w_{\text{PRN}\,2}(t), \tag{10.83}$$

$$\vdots$$

$$\frac{d}{dt}\delta_{\text{PRN}\,29} = w_{\text{PRN}\,29}(t), \tag{10.84}$$

where $\Omega_{\odot}$ is the earth rotation rate in radians per second, $R_{\text{N}}(\phi_{\text{lat}})$ is the meridional radius of curvature of the geoid model at latitude ϕ_{lat}, and $R_E(\phi_{\text{lat}})$ is the transverse radius of curvature.

Equations 10.61–10.63 are from Eq. 9.40, the Bortz model for rotational dynamics. These can be integrated numerically for propagating the estimated state vector, and the related covariance matrix of estimation uncertainty can be propagated using the first-order approximation

$$\frac{d}{dt}\mathbf{P}_\rho = \mathbf{F}_\rho \mathbf{P}_\rho + \mathbf{P}_\rho \mathbf{F}_\rho^{\mathrm{T}} + \mathbf{Q}, \tag{10.85}$$

$$\mathbf{F}_\rho = \frac{\partial \mathbf{f}_{\mathrm{Bortz}}}{\partial \boldsymbol{\rho}}, \tag{10.86}$$

where $\mathbf{f}_{\mathrm{Bortz}}$ is as defined by Eq. 9.40. The function $\mathbf{f}_{\mathrm{Bortz}}$ (required for numerical integration of $\boldsymbol{\rho}$) is implemented in the MATLAB m-file `fBortz.m`, and the matrix $\mathbf{F}_\rho$ is implement in the MATLAB m-file `BortzF.m`—both on the accompanying CD-ROM.

Measurement Variables Measurements include outputs of the inertial sensors and the pseudoranges output from the GNSS receiver:

1	a_1	=	first accelerometer output,
2	a_2	=	second accelerometer output,
3	a_3	=	third accelerometer output,
4	ω_1	=	first gyroscope output
5	ω_2	=	second gyroscope output,
6	ω_3	=	third gyroscope output,
7	$\rho_{\mathrm{PRN}\ 1}$	=	pseudorange from first GNSS satellite,
8	$\rho_{\mathrm{PRN}\ 2}$	=	pseudorange from second GNSS satellite,
		$\vdots$	
35	$\rho_{\mathrm{PRN}\ 29}$	=	pseudorange from last GNSS satellite,

Only those pseudoranges from acquired satellites need be considered, however, so the actual measurements count may be more like half of 35.

Model Implementation The further derivation, development and evaluation of this integration model is left as an exercise for the reader.

10.5 FUTURE DEVELOPMENTS

GPS and INS were both developed for worldwide navigation capability, and together they have taken that capability to new levels of performance that neither approach could achieve on its own.

The payoff in military costs and capabilities had driven development of GPS by the Department of Defense of the United States of America. However, early pioneers in GPS development had already foreseen many of the markets for GNSS/INS integration, including such applications as automating field equipment operations for farming and grading. These automated control applications require inertial sensors for precise and reliable operation under dynamic conditions, and integration with GPS has brought the costs and capabilities of the resulting systems to very practical levels. The results of integrating inertial systems with GPS has made enormous improvements in achievable operational speed and efficiency.

GNSS systems architectures continue to change with the addition of more systems, more satellites, more signal channels and more aiding systems, and integration-compatible inertial systems are also likely to continue to improve as the market expands, driving hardware costs further downward. It is a part of the "silicon revolution," harnessing the enormous power and low cost of electronic systems to make our lives more enjoyable and efficient. As costs continue downward, the potential applications market continues to expand.

We have only begun to explore applications for GNSS/INS. We are limited only by our ability to imagine them. The opportunities are there for you to shape the future.

APPENDIX A

SOFTWARE

A.1 SOFTWARE SOURCES

The MATLAB m-files on the accompanying CD-ROM are the implementations used to produce many of the examples and figures illustrating GNSS/INS implementation methods and performance evaluation methods. These are intended to demonstrate to the reader how these methods work. This is not "commercial grade" software, and it is not intended to be used as part of any commercial design process or product implementation software. The authors and publisher do not claim that this software meets any standards of mercantibility, and we cannot assume any responsibility for the results if they are used for such purposes.

There is better, more reliable commercial software available for GNSS and INS analysis, implementation and integration. We have used the MATLAB INS and GPS toolboxes from GPSoft to generate some of the figures, and there are other commercial products available for these purposes, as well. Many of the providers of such software maintain internet websites describing their products and services, and the interested user is encouraged to search the internet to shop for suitable sources.

The following sections contain short descriptions of the MATLAB m-files on the accompanying CD-ROM, organized by they chapters in which they are mentioned.

Global Positioning Systems, Inertial Navigation, and Integration, Second Edition, by M. S. Grewal, L. R. Weill, and A. P. Andrews

A.2 SOFTWARE FOR CHAPTER 3

The MATLAB script `ephemeris.m` calculates a GPS satellite position in ECEF coordinates from its ephemeris parameters. The ephemeris parameters comprise a set of Keplerian orbital parameters and describe the satellite orbit during a particular time interval. From these parameters, ECEF coordinates are calculated using the equations from the text. Note that time t is the GPS time at transmission and t_k (`tk`) in the script is the total time different between time t and the epoch time t_{oe} (`toe`). Kepler's equation for eccentric anomaly is nonlinear in E_k (`Ek`) and is solved numerically using the Newton–Raphson method.

The following MATLAB script calculates satellite position for 24 h using almanac data stored on the CD:

`GPS_position(PRN#)` plots satellite position using PRNs one at a time for all satellites.

`GPS_position_3D` plots satellite position for all PRNs in three dimensions. Use rotate option in MATLAB to see the satellite positions from the equator, north pole, south pole, and so on.

`GPS_el_az (PRN#, 33.8825, -117.8833)` plots satellite trajectory for a PRN from Fullerton, California (GPS laboratory located at California State University, Fullerton).

`GPS_el_az_all (33.8825, -117.8833)` plots satellite trajectories for all satellites from Fullerton, California (GPS laboratory located at California State University, Fullerton).

`GPS_el_az_one_time (14.00, 33.8825, -117.8833)` plots the location of satellites at 2 p.m. (14:00 h) for visible satellites from GPS laboratory located at California State University, Fullerton.

A.3 SOFTWARE FOR CHAPTER 5

A.3.1 Ionospheric Delays

The following MATLAB scripts compute and plot ionospheric delays using Klobuchar models:

`Klobuchar_fix` plots the ionospheric delays for GEO stationary satellites for 24 h, such as AOR-W, POR, GEO 3, and GEO 4.

`Klobuchar (PRN#)` plots the ionospheric delays for a satellite specified by the argument PRN, when that satellite is visible.

`Iono_delay (PRN#)` plots the ionospheric delays for a PRN using dual-frequency data, when a satellite is visible. It uses the pseudorange carrier phase data for L_1 and L_2 signals. Plots are overlaid for comparison.

A.4 SOFTWARE FOR CHAPTER 8

osc_ekf.m demonstrates an extended Kalman filter tracking the phase, amplitude, frequency, and damping factor of a harmonic oscillator with randomly time-varying parameters.

GPS_perf.m performs covariance analysis of expected performance of a GPS receiver using a Kalman filter.

init_var initializes parameters and variables for GPS_perf.m.

choose_sat chooses satellite set or use default for GPS_perf.m.

gps_init initializes GPS satellites for GPS_perf.m.

calcH calculates H matrix for GPS_perf.m.

gdop calculates GDOP for chosen constellation for GPS_perf.m.

covar solves Riccati equation for GPS_perf.m.

plot_covar plots results from GPS_perf.m.

SchmidtKalmanTest.m compares Schmidt–Kalman filter and Kalman filter for GPS navigation with time-correlated pseudorange errors.

shootout.m compares performance of several square root covariance filtering methods on an ill conditioned problem from P. Dyer and S. McReynolds, "Extension of Square-Root Filtering to Include Process Noise," *Journal of Optimization Theory and Applications* **3**, 444–458 (1969).

joseph called by shootout.m to implement "Joseph stabilized" Kalman filter.

josephb called by shootout.m to implement "Joseph–Bierman" Kalman filter.

josephdv called by shootout.m to implement "Joseph–DeVries" Kalman filter.

potter called by shootout.m to implement Potter square-root filter.

carlson called by shootout.m to implement Carlson square-root filter.

bierman called by shootout.m to implement Bierman square-root filter.

A.5 SOFTWARE FOR CHAPTER 9

ConingMovie.m generates a MATLAB movie of coning motion, showing how the body-fixed coordinate axes move relative to inertial coordinates. (The built-in MATLAB function movie2avi can convert this to an avi-file.)

VertChanErr.m implements the Riccati equations for the INS vertical channel error covariance with accelerometer noise levels in the order of 10^{-2}, 10^{-1}, 1, 10, and 100 m/s/$\sqrt{\text{h}}$, with and without aiding by a barometric altimeter. Generates the plot shown in Fig. 9.32.

Euler2CTMat converts from Euler angles to coordinate transformation matrices.

CTMat2Euler converts from coordinate transformation matrices to Euler angles.

`RotVec2Quat` converts from rotation vectors to quaternions.

`Quat2RotVec` converts from quaternions to rotation vectors.

`Quat2CTMat` converts from quaternions to coordinate transformation matrices.

`CTMat2Quat` converts from coordinate transformation matrices to quaternions.

`RotVec2CTMat` converts from rotation vectors to coordinate transformation matrices.

`CTMat2RotVec` converts from coordinate transformation matrices to rotation vectors.

`fBortz.m` computes the nonlinear function $\mathbf{f}_{\text{Bortz}}$ for integrating the Bortz "noncommutative" attitude integration formula.

`FBortz(rho,omega)` computes the dynamic coefficient matrix for integrating the Riccati equation for rotation rates.

`AccBiasCarousel.m` simulates the propagation of accelerometer bias error in an inertial navigation system and creates the plot shown in Fig. 9.33.

A.6 SOFTWARE FOR CHAPTER 10

`HSatSim.m` generates measurement sensitivity matrix **H** for GPS satellite simulation.

`Damp2eval.m` evaluates DAMP2 GPS position tracking filters for a range of host vehicle RMS accelerations and acceleration correlation times.

`YUMAdata` loads GPS almanac data from www.navcen.uscg.gov - /ftp/GPS/almanacs/yuma/ for Wednesday, March 08, 2006 10:48 AM, converts to arrays of right ascension and phase angles for 29 satellites. (Used by Damp2eval.m)

`Damp2Params.m` solves transcendental equation for alternative parameters in DAMP2 GPS tracking filter.

`Damp3Params.m` solves transcendental equation for alternative parameters in DAMP3 GPS tracking filter.

`GPSTrackingDemo.m` applies the GPS vehicle tracking filters TYPE2, DAMP2, DAMP3 and FIG8 to the same problem (tracking a vehicle moving on a figure-8 test track).

`Fig8TrackDemo.m` generates a series of plots and statistics of the simulated figure-8 test track trajectory.

`HorizINSperfModel.m` calculates INS error model parameters as a function of CEP rate.

`Fig8Mod1D.m` simulates trajectory of a vehicle going around a figure-8 test track.

`GPSINSwGPSpos.m` simulates GPS/INS loosely-coupled integration, using only standard GPS and INS output position values.

`GPSINSwPRs.m` simulates GPS/INS tightly-coupled integration, using GPS pseudoranges and INS position outputs.

APPENDIX B

VECTORS AND MATRICES

The "S" in "GPS" and in "INS" stands for "system," and *"systems science"* for modeling, analysis, design, and integration of such systems is based largely on linear algebra and matrix theory. Matrices model the ways that components of systems interact dynamically and how overall system performance depends on characteristics of components and subsystems and on the ways they are used within the system.

This appendix presents an overview of matrix theory used for GPS/INS integration and the matrix notation used in this book. The level of presentation is intended for readers who are already somewhat familiar with vectors and matrices. A more thorough treatment can be found in most college-level textbooks on linear algebra and matrix theory.

B.1 SCALARS

Vectors and matrices are arrays composed of scalars, which we will assume to be real numbers. Unless constrained by other conventions, we represent scalars by italic lowercase letters.

In computer implementations, these real numbers will be approximated by floating-point numbers, which are but a finite subset of the rational numbers. The default MATLAB representation for real numbers on 32-bit personal computers is in 64-bit ANSI standard floating point, with a 52-bit mantissa.

Global Positioning Systems, Inertial Navigation, and Integration, Second Edition, by M. S. Grewal, L. R. Weill, and A. P. Andrews

B.2 VECTORS

B.2.1 Vector Notation

Vectors are arrays of scalars, either column vectors,

$$\mathbf{v} = \begin{bmatrix} v_1 \\ v_2 \\ v_3 \\ \vdots \\ v_n \end{bmatrix},$$

or row vectors,

$$\mathbf{y} = [y_1, y_2, y_3, \ldots, y_m].$$

Unless specified otherwise, vectors can be assumed to be column vectors.

The scalars v_k or y_k are called the *components* of $\mathbf{v}$ or $\mathbf{y}$, respectively. The number of components of a vector (rows in a column vector or columns in a row vector) is called its *dimension*. The dimension of $\mathbf{v}$ shown above is the integer n and the dimension of $\mathbf{y}$ is m. An n-dimensional vector is also called an n-vector.

Vectors are represented by boldface lowercase letters, and the corresponding italic lowercase letters with subscripts represent the scalar components of the associated vector.

B.2.2 Unit Vectors

A *unit vector* (i.e., a vector with magnitude equal to 1) is represented by the symbol $\mathbf{1}$.

B.2.3 Subvectors

Vectors can be partitioned and represented in *block form* as a vector of subvectors:

$$\mathbf{x} = \begin{bmatrix} \mathbf{x}_1 \\ \mathbf{x}_2 \\ \mathbf{x}_3 \\ \vdots \\ \mathbf{x}_\ell \end{bmatrix},$$

where each subvector $\mathbf{x}_k$ is also a vector, as indicated by boldfacing.

B.2.4 Transpose of a Vector

Vector *transposition,*, represented by the post-superscript T transforms row vectors to column vectors, and *vice versa*:

$$\mathbf{v}^{\mathrm{T}} = [v_1, v_2, v_3, \ldots, v_n], \qquad \mathbf{y}^T = \begin{bmatrix} y_1 \\ y_2 \\ y_3 \\ \vdots \\ y_m \end{bmatrix}.$$

In MATLAB, the transpose of vector $\mathbf{v}$ is written as $\mathbf{v}'$.

B.2.5 Vector Inner Product

The inner product or dot product of two m-vectors is the sum of the products of their corresponding components:

$$\mathbf{x}^{\mathrm{T}}\mathbf{y} \quad \text{or} \quad \mathbf{x} \cdot \mathbf{y} \stackrel{\text{def}}{=} \sum_{k=1}^{m} x_k y_k.$$

B.2.6 Orthogonal Vectors

Vectors $\mathbf{x}$ and $\mathbf{y}$ are called *orthogonal* or *normal* if their inner product is zero.

B.2.7 Magnitude of a Vector

The *magnitude* of a vector is the root-sum-squared of its components, denoted by $|\cdot|$ and defined as

$$\begin{aligned} |\mathbf{v}| &\stackrel{\text{def}}{=} \sqrt{\mathbf{v}\mathbf{v}^{\mathrm{T}}} \quad \text{(row vector)} \\ &= \sqrt{\sum_{k=1}^{n} v_k^2}, \\ |\mathbf{y}| &\stackrel{\text{def}}{=} \sqrt{\mathbf{y}^{\mathrm{T}}\mathbf{y}} \quad \text{(column vector)} \\ &= \sqrt{\sum_{k=1}^{m} y_k^2}. \end{aligned}$$

B.2.8 Unit Vectors and Orthonormal Vectors

A unit vector has magnitude equal to 1, and a pair or set of mutually orthogonal unit vectors is called *orthonormal*.

B.2.9 Vector Norms

The magnitude of a column n-vector $\mathbf{x}$ is also called its *Euclidean norm*. This is but one of a class of norms called "*Hölder norms*,"[1] "l_p *norms*," or simply "*p-norms*":

$$\|x\|_p \stackrel{\text{def}}{=} \left[\sum_{i=1}^{n} |x_i|^p\right]^{1/p},$$

and in the limit (as $p \to \infty$) as the *sup*[2] norm, or ∞ norm:

$$\|x\|_\infty \stackrel{\text{def}}{=} \max_i |x_i|.$$

These norms satisfy the *Hölder inequality*:

$$|x^{\mathrm{T}}y| \leq \|x\|_p \|y\|_q \qquad \text{for } \frac{1}{p} + \frac{1}{q} = 1.$$

They are also related by inequalities such as

$$\|x\|_\infty \leq \|x\|_E \leq \|x\|_1 \leq n\|x\|_\infty.$$

The Euclidean norm (Hölder 2-norm) is the default norm for vectors. When no other norm is specified, the implied norm is the Euclidean norm.

B.2.10 Vector Cross-product

Vector cross-products are only defined for vectors with three components (i.e., 3-vectors). For any two 3-vectors $\mathbf{x}$ and $\mathbf{y}$, their vector cross-products are defined as

$$\mathbf{x} \otimes \mathbf{y} \stackrel{\text{def}}{=} \begin{bmatrix} x_2 y_3 - x_3 y_2 \\ x_3 y_1 - x_1 y_3 \\ x_1 y_2 - x_2 y_1 \end{bmatrix},$$

which has the properties

$$\mathbf{x} \otimes \mathbf{y} = -\mathbf{y} \otimes \mathbf{x},$$
$$\mathbf{x} \otimes \mathbf{x} = 0,$$
$$|\mathbf{x} \otimes \mathbf{y}| = \sin(\theta)|\mathbf{x}||\mathbf{y}|,$$

where θ is the angle between the vectors $\mathbf{x}$ and $\mathbf{y}$.

[1]Named for the German mathematician Otto Ludwig Hölder (1859–1937).
[2]"Sup" (sounds like "soup") stands for *supremum*, a mathematical term for the *least upper bound* of a set of real numbers. The maximum (max) is the supremum over a finite set.

B.2.11 Right-Handed Coordinate Systems

A Cartesian coordinate system in three dimensions is considered "right handed" if its three coordinate axes are numbered consecutively such that the unit vectors $\mathbf{1}_k$ along its respective coordinate axes satisfy the cross-product rules

$$\mathbf{1}_1 \otimes \mathbf{1}_2 = \mathbf{1}_3, \tag{B.1}$$

$$\mathbf{1}_2 \otimes \mathbf{1}_3 = \mathbf{1}_1, \tag{B.2}$$

$$\mathbf{1}_3 \otimes \mathbf{1}_1 = \mathbf{1}_2. \tag{B.3}$$

B.2.12 Vector Outer Product

The vector outer product of two column vectors

$$\mathbf{x} = \begin{bmatrix} x_1 \\ x_2 \\ x_3 \\ \vdots \\ x_n \end{bmatrix}, \qquad \mathbf{y} = \begin{bmatrix} y_1 \\ y_2 \\ y_3 \\ \vdots \\ y_m \end{bmatrix}$$

is defined as the $n \times m$ array

$$\mathbf{x}\mathbf{y}^{\mathrm{T}} \stackrel{\text{def}}{=} \begin{bmatrix} x_1 y_1 & x_1 y_2 & x_1 y_2 & \dots & x_1 y_m \\ x_2 y_1 & x_2 y_2 & x_2 y_2 & \dots & x_2 y_m \\ x_3 y_1 & x_3 y_2 & x_3 y_2 & \dots & x_3 y_m \\ \vdots & \vdots & \vdots & \ddots & \vdots \\ x_n y_1 & x_n y_2 & x_n y_2 & \dots & x_n y_m \end{bmatrix},$$

a matrix.

B.3 MATRICES

B.3.1 Matrix Notation

For positive integers m and n, an *m-by-n* real *matrix* **A** is a two-dimensional rectangular array of scalars, designated by the subscript notation a_{ij}, and usually displayed in the following format:

$$\mathbf{A} = \begin{bmatrix} a_{11} & a_{12} & a_{13} & \dots & a_{1n} \\ a_{21} & a_{22} & a_{23} & \dots & a_{2n} \\ a_{31} & a_{32} & a_{33} & \dots & a_{3n} \\ \vdots & \vdots & \vdots & \ddots & \vdots \\ a_{m1} & a_{m2} & a_{m3} & \dots & a_{mn} \end{bmatrix}.$$

The scalars a_{ij} are called the *elements* of **A**. Uppercase *bolded* letters are used for matrices, with the corresponding lowercase letter denoting scalar elements of the associated matrices.

Row and Column Subscripts The first subscript (i) on the element a_{ij} refers to the *row* in which the element occurs, and the second subscript (j) refers to the *column* in which a_{ij} occurs in this format. The integers i and j in this notation are also called *indices* of the elements. The first index is called the *row index*, and the second index is called the *column index* of the element. The term "(*ij*)th *position*" in the matrix **A** refers to the position of a_{ij}, and a_{ij} is called the "(*ij*)th *element*" of **A**:

←		columns		→		rows
1st	2nd	3rd	...	nth		
↓	↓	↓		↓		
a_{11}	a_{12}	a_{13}	...	a_{1n}	←	1st
a_{21}	a_{22}	a_{23}	...	a_{2n}	←	2nd
a_{31}	a_{32}	a_{33}	...	a_{3n}	←	3rd
$\vdots$	$\vdots$	$\vdots$	$\ddots$	$\vdots$	$\vdots$	$\vdots$
a_{m1}	a_{m2}	a_{m3}	...	a_{mn}	←	mth

If juxtaposition of subscripts leads to confusion, they may be separated by commas. The element in the eleventh row and first column of the matrix **A** would then be denoted by $a_{11,1}$, not a_{111}.

Dimensions The positive integers m and n are called the *dimensions* of a matrix **A**: m is called the *row dimension* of **A** and n is called the *column dimension* of **A**. The dimensions of **A** may also be represented as "$m \times n$," which is to be read as "m by n." The symbol "$\times$" in this notation does not indicate multiplication. (The number of elements in the matrix **A** equals the product mn, however, and this is important for determining memory requirements for data structures to hold **A**.)

B.3.2 Special Matrix Forms

Square Matrices A matrix is called *square* if it has the same row and column dimensions. The *main diagonal* of a square matrix **A** is the set of elements a_{ij} for which $i = j$. The other elements are called *off-diagonal*. If all the off-diagonal elements of a square matrix **A** are zero, **A** is called a *diagonal* matrix. This and other special forms of square matrices are illustrated in Fig. B.1.

Sparse and Dense Matrices A matrix with a "significant fraction" (typically, half or more) of zero elements is called *sparse*. Matrices that are decidedly not sparse are called *dense*, although both sparsity and density are matters of degree. All the forms except symmetric shown in Fig. B.1 are sparse, although

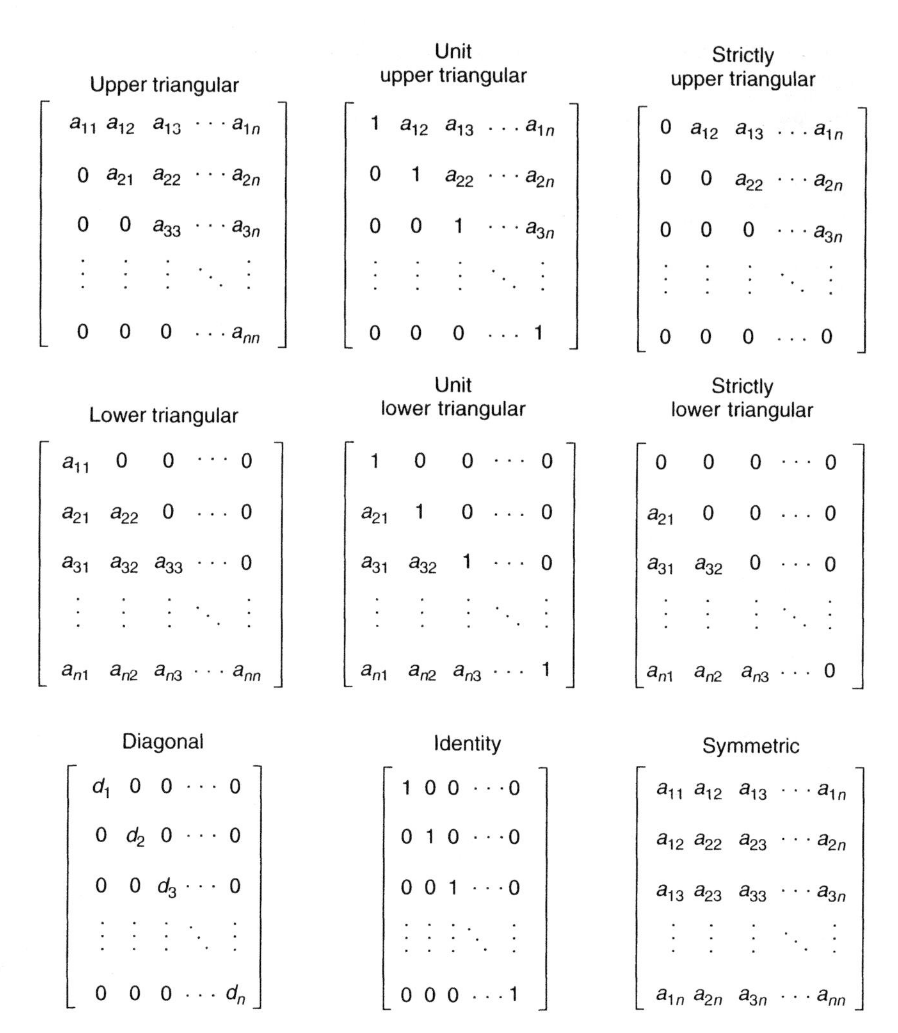

Fig. B.1 Special forms of square matrices.

sparse matrices do not have to be square. Sparsity is an important characteristic for implementation of matrix methods, because it can be exploited to reduce computer memory and computational requirements.

Zero Matrices The ultimate sparse matrix is a matrix in which *all* elements are 0 (zero). It is called a *zero matrix*, and it is represented by the symbol "0" (zero). The equation $A = 0$ indicates that **A** is a zero matrix. Whenever it is necessary to specify the dimensions of a zero matrix, they may be indicated by subscripting: $0_{m \times n}$ will indicate an $m \times n$ zero matrix. If the matrix is square, only one subscript will be used: 0_n will mean an $n \times n$ zero matrix.

Identity Matrices The identity matrix will be represented by the symbol **I**. If it is necessary to denote the dimension of **I** explicitly, it will be indicated by subscripting the symbol: $\mathbf{I}_n$ denotes the $n \times n$ identity matrix.

B.4 MATRIX OPERATIONS

B.4.1 Matrix Transposition

The *transpose* of **A** is the matrix $\mathbf{A}^{\mathrm{T}}$ (with the superscript "T" denoting the transpose operation), obtained from **A** by interchanging rows and columns:

$$\begin{bmatrix} a_{11} & a_{12} & a_{13} & \dots & a_{1n} \\ a_{21} & a_{22} & a_{23} & \dots & a_{2n} \\ a_{31} & a_{32} & a_{33} & \dots & a_{3n} \\ \vdots & \vdots & \vdots & \ddots & \vdots \\ a_{m1} & a_{m2} & a_{m3} & \dots & a_{mn} \end{bmatrix}^{\mathrm{T}} = \begin{bmatrix} a_{11} & a_{21} & a_{31} & \dots & a_{m1} \\ a_{12} & a_{22} & a_{32} & \dots & a_{m2} \\ a_{13} & a_{23} & a_{33} & \dots & a_{m3} \\ \dots & \dots & \dots & \ddots & \vdots \\ a_{1n} & a_{2n} & a_{3n} & \dots & a_{mn} \end{bmatrix}.$$

The transpose of an $m \times n$ matrix is an $n \times m$ matrix.

The transpose of the matrix **M** in MATLAB is written as M'.

Symmetric Matrices A matrix **A** is called *symmetric* if $\mathbf{A}^{\mathrm{T}} = \mathbf{A}$ and *skew symmetric* (or *anti-symmetric*) if $\mathbf{A}^{\mathrm{T}} = -\mathbf{A}$. Only square matrices can be symmetric or skew symmetric. Therefore, whenever a matrix is said to be symmetric or skew-symmetric, it is implied that it is a square matrix. Any square matrix **A** can be expressed as a sum of its symmetric and antisymmetric parts:

$$\mathbf{A} = \underbrace{\tfrac{1}{2}(\mathbf{A} + \mathbf{A}^{\mathrm{T}})}_{\text{symmetric}} + \underbrace{\tfrac{1}{2}(\mathbf{A} - \mathbf{A}^{\mathrm{T}})}_{\text{antisymmetric}}.$$

Cross-Product Matrices The vector cross-product $\boldsymbol{\rho} \otimes \boldsymbol{\alpha}$ can also be expressed in matrix form as

$$\boldsymbol{\rho} \otimes \boldsymbol{\rho} = \begin{bmatrix} \rho_1 \\ \rho_2 \\ \rho_3 \end{bmatrix} \otimes \begin{bmatrix} \alpha_1 \\ \alpha_2 \\ \alpha_3 \end{bmatrix} \tag{B.4}$$

$$= \begin{bmatrix} \rho_2\alpha_3 - \rho_3\alpha_2 \\ \rho_3\alpha_1 - \rho_1\alpha_3 \\ \rho_1\alpha_2 - \rho_2\alpha_1 \end{bmatrix} \tag{B.5}$$

$$= [\boldsymbol{\rho}\otimes]\boldsymbol{\alpha} \tag{B.6}$$

$$= \begin{bmatrix} 0 & -\rho_3 & \rho_2 \\ \rho_3 & 0 & -\rho_1 \\ -\rho_2 & \rho_1 & 0 \end{bmatrix} \begin{bmatrix} \alpha_1 \\ \alpha_2 \\ \alpha_3 \end{bmatrix}, \tag{B.7}$$

where the "cross-product matrix"

$$[\boldsymbol{\rho}\otimes] \stackrel{\text{def}}{=} \begin{bmatrix} 0 & -\rho_3 & \rho_2 \\ \rho_3 & 0 & -\rho_1 \\ -\rho_2 & \rho_1 & 0 \end{bmatrix} \tag{B.8}$$

is skew-symmetric.

B.4.2 Subscripted Matrix Expressions

Subscripts represent an operation on a matrix that extracts the designated matrix element. Subscripts may also be applied to matrix expressions. The element in the (ij)th position of a matrix expression can be indicated by subscripting the expression, as in

$$\{\mathbf{A}^T\}_{ij} = a_{ij}.$$

Here, we have used braces {} to indicate the scope of the expression to which the subscripting applies. This is a handy device for defining matrix operations.

B.4.3 Multiplication of Matrices by Scalars

Multiplication of a matrix $\mathbf{A}$ by a scalar s is equivalent to multiplying every element of $\mathbf{A}$ by s:

$$\{\mathbf{A}s\}_{ij} = \{s\mathbf{A}\}_{ij} = sa_{ij}.$$

B.4.4 Addition and Multiplication of Matrices

Addition of Matrices Is Associative and Commutative Matrices can be added together if and only if they share the same dimensions. If $\mathbf{A}$ and $\mathbf{B}$ have the same dimensions, then addition is defined by adding corresponding elements:

$$\{\mathbf{A} + \mathbf{B}\}_{ij} = a_{ij} + b_{ij}.$$

Addition of matrices is *commutative* and *associative*. That is, $\mathbf{A} + \mathbf{B} = \mathbf{B} + \mathbf{A}$ and $\mathbf{A} + (\mathbf{B} + \mathbf{C}) = (\mathbf{A} + \mathbf{B}) + \mathbf{C}$.

Additive Inverse of a Matrix The product of a matrix $\mathbf{A}$ by the scalar -1 yields its *additive inverse* $-\mathbf{A}$:

$$(-1)\mathbf{A} = -\mathbf{A}, \qquad \mathbf{A} + (-\mathbf{A}) = \mathbf{A} - \mathbf{A} = 0.$$

Here, we have followed the not uncommon practice of using the symbol "$-$" both as a unary (additive inverse) and binary (subtraction) operator. *Subtraction* of a matrix $\mathbf{A}$ from a matrix $\mathbf{B}$ is equivalent to adding the additive inverse of $\mathbf{A}$ to $\mathbf{B}$:

$$\mathbf{B} - \mathbf{A} = \mathbf{B} + (-\mathbf{A}).$$

Multiplication of Matrices is Associative but Not Commutative Multiplication of an $m \times n$ matrix $\mathbf{A}$ by a matrix $\mathbf{B}$ on the right-hand side of $\mathbf{A}$, as in the matrix product $\mathbf{AB}$, is defined only if *the row dimension of* $\mathbf{B}$ *equals the column dimension of* $\mathbf{A}$. That is, we can multiply an $m \times n$ matrix $\mathbf{A}$ by a $p \times q$ matrix $\mathbf{B}$ in this order only if $n = p$. In that case, the matrices $\mathbf{A}$ and $\mathbf{B}$ are said to be *conformable* for multiplication in that order, and the matrix product is defined element by element by

$$\{\mathbf{AB}\}_{ij} \stackrel{\text{def}}{=} \sum_{k=1}^{n} a_{ik} b_{kj},$$

the result of which is an $m \times q$ matrix. Whenever matrices appear as a product in an expression, it is implied that they are conformable for multiplication.

Products with Identity Matrices Multiplication of any $m \times n$ matrix $\mathbf{A}$ by a conformable *identity matrix* yields the original matrix $\mathbf{A}$ as the product:

$$\mathbf{AI}_n = \mathbf{A}, \qquad \mathbf{I}_m\mathbf{A} = \mathbf{A}.$$

B.4.5 Powers of Square Matrices

Square matrices can always be multiplied by themselves, and the resulting matrix products are again conformable for multiplication. Consequently, one can define the pth power of a square matrix $\mathbf{A}$ as

$$\mathbf{A}^p = \underbrace{\mathbf{A} \times \mathbf{A} \times \mathbf{A} \times \cdots \times \mathbf{A}}_{p \text{ elements}}.$$

B.4.6 Matrix Inversion

If $\mathbf{A}$ and $\mathbf{B}$ are square matrices of the same dimension, and such that their product

$$\mathbf{AB} = \mathbf{I},$$

then $\mathbf{B}$ is the *matrix inverse* of $\mathbf{A}$ and $\mathbf{A}$ is the matrix inverse of $\mathbf{B}$. (It turns out that $\mathbf{BA} = \mathbf{AB} = \mathbf{I}$ in this case.) The inverse of a matrix $\mathbf{A}$ is unique, if it exists, and is denoted by $\mathbf{A}^{-1}$. Not all matrices have inverses. Matrix *inversion* is the process of finding a matrix inverse, if it exists. If the inverse of a matrix $\mathbf{A}$ does not exist, $\mathbf{A}$ is called *singular*. Otherwise, it is called *non-singular*.

B.4.7 Generalized Matrix Inversion

Even nonsquare and/or singular matrices can have *generalized inverses*. The *Moore-Penrose generalized inverse* of an $m \times n$ matrix $\mathbf{A}$ is the $n \times m$ matrix $\mathbf{A}^+$ such that

$$\begin{aligned} \mathbf{AA^+A} &= \mathbf{A}, \\ \mathbf{A^+AA^+} &= \mathbf{A^+}, \\ (\mathbf{AA^+})^{\mathrm{T}} &= \mathbf{AA^+}, \\ (\mathbf{A^+A})^{\mathrm{T}} &= \mathbf{A^+A}. \end{aligned}$$

B.4.8 Orthogonal Matrices

A square matrix $\mathbf{A}$ is called *orthogonal* if $\mathbf{A}^{\mathrm{T}} = \mathbf{A}^{-1}$. Orthogonal matrices have several useful properties:

- Orthogonality of a matrix $\mathbf{A}$ implies that the row vectors of $\mathbf{A}$ are jointly orthonormal vectors, and the column vectors of $\mathbf{A}$ are also jointly orthonormal vectors.
- The dot products of vectors are invariant under multiplication by a conformable orthogonal matrix. That is, if $\mathbf{A}$ is orthogonal, then $\mathbf{x}^T\mathbf{y} = (\mathbf{Ax})^T(\mathbf{Ay})$ for all conformable $\mathbf{x}$ and $\mathbf{y}$.
- Products and inverses of orthogonal matrices are orthogonal.

As a rule, multiplications by orthogonal matrices tend to be numerically well conditioned, compared to general matrix multiplications. (The inversion of orthogonal matrices is obviously extremely well conditioned.)

B.5 BLOCK MATRIX FORMULAS

B.5.1 Submatrices, Partitioned Matrices, and Blocks

For any $m \times n$ matrix $\mathbf{A}$ and any subset $S_{\text{rows}} \subseteq \{1, 2, 3, \ldots, m\}$ of the row indices and subset $S_{\text{cols}} \subseteq \{1, 2, 3, \ldots, n\}$ of the column indices, the subset of elements

$$\mathbf{A}' = \{a_{ij} | i \in S_{\text{rows}}, j \in S_{\text{cols}}\}$$

is called a *submatrix* of $\mathbf{A}$.

A *partitioning* of an integer n is an exhaustive collection of contiguous subsets S_k of the form

$$\overbrace{1, 2, 3, \ldots, \ell_1}^{S_1}, \overbrace{(\ell_1 + 1), \ldots, \ell_2}^{S_2}, \ldots, \overbrace{(\ell_{p-1} + 1), \ldots, n}^{S_p}.$$

The collection of submatrices formed by partitionings of the row and column dimensions of a matrix is called a *partitioning* of the matrix, and the matrix is said to be *partitioned* by that partitioning. Each submatrix of a partitioned matrix

A is called a *partitioned submatrix, partition*, submatrix *block, subblock*, or *block* of **A**. Each block of a partitioned matrix **A** can be represented by a conformable matrix expression, and **A** can be displayed as a *block matrix*:

$$\mathrm{A} = \begin{bmatrix} \mathbf{B} & \mathbf{C} & \mathbf{D} & \dots & \mathbf{F} \\ \mathbf{G} & \mathbf{H} & \mathbf{J} & \dots & \mathbf{L} \\ \mathbf{M} & \mathbf{N} & \mathbf{P} & \dots & \mathbf{R} \\ \vdots & \vdots & \vdots & \ddots & \vdots \\ \mathbf{V} & \mathbf{W} & \mathbf{X} & \dots & \mathbf{Z} \end{bmatrix}$$

where **B, C, D,** ... stand for matrix expressions. Whenever a matrix is displayed as a block matrix, it is implied that all block submatrices in the same row have the same row dimension and that all block submatrices in the same column have the same column dimension.

A block matrix of the form

$$\begin{bmatrix} \mathbf{A} & 0 & 0 & \dots & 0 \\ 0 & \mathbf{B} & 0 & \dots & 0 \\ 0 & 0 & \mathbf{C} & \dots & 0 \\ \vdots & \vdots & \vdots & \ddots & \vdots \\ 0 & 0 & 0 & \dots & \mathbf{M} \end{bmatrix},$$

in which the off-diagonal block submatrices are zero matrices, is called a *block diagonal matrix*, and a block matrix in which the block submatrices on one side of the diagonal are zero matrices is called a *block triangular matrix*.

Columns and Rows as Blocks There are two special partitionings of matrices in which the block submatrices are vectors. The *column vectors* of an $m \times n$ matrix **A** are the block submatrices of the partitioning of **A** for which all column dimensions are 1 and all row dimensions are m. The *row vectors* of **A** are the block submatrices of the partitioning for which all row dimensions are 1 and all column dimensions are n. All column vectors of an $m \times n$ matrix are m-vectors, and all row vectors are n-vectors.

B.5.2 Rank and Linear Dependence

A *linear combination* of a finite set of n-vectors $\{\mathbf{v}_i\}$ is a summation of the sort $\sum_i a_i v_i$ for some set of scalars $\{a_i\}$. If some linear combination $\sum a_i \mathbf{v}_i = 0$ and at least one coefficient $a_i \neq 0$, the set of vectors $\{\mathbf{v}_i\}$ is called *linearly dependent*. Conversely, if the only linear combination for which $\sum a_i \mathbf{v}_i = 0$ is the one for which all the $a_i = 0$, then the set of vectors $\{v_i\}$ is called *linearly independent*.

The *rank* of a $n \times m$ matrix **A** equals the size of the *largest* collection of its column vectors that is linearly independent. Note that any such linear combination can be expressed in the form **Aa**, where the nonzero elements of the column m-vector **A** are the associated scalars of the linear combination, and the number of nonzero components of $\mathbf{A}$ is the size of the collection of column vectors in the linear combination. The same value for the rank of a matrix is obtained if the test is applied to its row vectors, where any linear combination of row vectors can be expressed in the form $\mathbf{a}^{\mathrm{T}}\mathbf{A}$ for some column n-vector $\mathbf{A}$.

An $n \times n$ matrix is nonsingular if and only if its rank equals its dimension n.

B.5.3 Conformable Block Operations

Block matrices with conformable partitionings may be transposed, added, subtracted, and multiplied in block format. For example,

$$\begin{bmatrix} \mathbf{A} & \mathbf{B} \\ \mathbf{C} & \mathbf{D} \end{bmatrix}^{\mathrm{T}} = \begin{bmatrix} \mathbf{A}^{\mathrm{T}} & \mathbf{C}^{\mathrm{T}} \\ \mathbf{B}^{\mathrm{T}} & \mathbf{D}^{\mathrm{T}} \end{bmatrix},$$

$$\begin{bmatrix} \mathbf{A} & \mathbf{B} \\ \mathbf{C} & \mathbf{D} \end{bmatrix} + \begin{bmatrix} \mathbf{E} & \mathbf{F} \\ \mathbf{G} & \mathbf{H} \end{bmatrix} = \begin{bmatrix} \mathbf{A}+\mathbf{E} & \mathbf{B}+\mathbf{F} \\ \mathbf{C}+\mathbf{G} & \mathbf{D}+\mathbf{H} \end{bmatrix},$$

$$\begin{bmatrix} \mathbf{A} & \mathbf{B} \\ \mathbf{C} & \mathbf{D} \end{bmatrix} \times \begin{bmatrix} \mathbf{E} & \mathbf{F} \\ \mathbf{G} & \mathbf{H} \end{bmatrix} = \begin{bmatrix} \mathbf{AE}+\mathbf{BG} & \mathbf{AF}+\mathbf{BH} \\ \mathbf{CE}+\mathbf{DG} & \mathbf{CF}+\mathbf{DH} \end{bmatrix}.$$

B.5.4 Block Matrix Inversion Formula

The inverse of a partitioned matrix with square diagonal blocks may be represented in block form as [53]

$$\begin{bmatrix} \mathbf{A} & \mathbf{B} \\ \mathbf{C} & \mathbf{D} \end{bmatrix}^{-1} = \begin{bmatrix} \mathbf{E} & \mathbf{F} \\ \mathbf{G} & \mathbf{H} \end{bmatrix},$$

where

$$\mathbf{E} = \mathbf{A}^{-1} + \mathbf{A}^{-1}\mathbf{BHCA}^{-1},$$

$$\mathbf{F} = -\mathbf{A}^{-1}\mathbf{BH},$$

$$\mathbf{G} = -\mathbf{HCA}^{-1},$$

$$\mathbf{H} = [\mathbf{D} - \mathbf{CA}^{-1}\mathbf{B}]^{-1}.$$

This formula can be proved by multiplying the original matrix times its alleged inverse and verifying that the result is the identity matrix.

B.5.5 Inversion Formulas for Matrix Expressions

Sherman—Morrison Formula A "rank 1" modification of a square matrix $\mathbf{A}$ is a sum of the form $\mathbf{A} + \mathbf{bc}^{\mathrm{T}}$, where $\mathbf{b}$ and $\mathbf{c}$ are conformable column vectors. Its inverse is given by the formula

$$[\mathbf{A} + \mathbf{bc}^{\mathrm{T}}]^{-1} = \mathbf{A}^{-1} - \frac{\mathbf{A}^{-1}\mathbf{bc}^{\mathrm{T}}\mathbf{A}^{-1}}{1 + \mathbf{c}^{\mathrm{T}}\mathbf{A}^{-1}\mathbf{b}}.$$

Sherman—Morrison—Woodbury Formula This is the generalization of the above formula for conformable matrices in place of vectors:

$$[\mathbf{A} + \mathbf{B}\mathbf{C}^{\mathrm{T}}]^{-1} = \mathbf{A}^{-1} - \mathbf{A}^{-1}\mathbf{B}[\mathbf{I} + \mathbf{C}^{\mathrm{T}}\mathbf{A}^{-1}\mathbf{B}]^{-1}\mathbf{C}^{\mathrm{T}}\mathbf{A}^{-1}.$$

Hemes Inversion Formula A further generalization of this formula (used in the derivation of the Kalman filter equations) includes an additional conformable square matrix factor in the modification:

$$[\mathbf{A} + \mathbf{B}\mathbf{C}^{-1}\mathbf{D}^{\mathrm{T}}]^{-1} = \mathbf{A}^{-1} - \mathbf{A}^{-1}\mathbf{B}[\mathbf{C} + \mathbf{D}^{\mathrm{T}}\mathbf{A}^{-1}\mathbf{B}]^{-1}\mathbf{D}^{\mathrm{T}}\mathbf{A}^{-1}. \tag{B.9}$$

B.6 FUNCTIONS OF SQUARE MATRICES

B.6.1 Determinants and Characteristic Values

Elementary Permutation Matrices An *elementary permutation matrix* is formed by interchanging rows or columns of an identity matrix $\mathbf{I}_n$:

$$\mathbf{P}_{[ij]} = \begin{array}{c} \\ \\ i \\ \\ j \\ \\ \\ \end{array} \begin{pmatrix} 1 & \dots & 0 & \dots & 0 & \dots & 0 \\ \vdots & \ddots & \vdots & & \vdots & & \vdots \\ 0 & \dots & 0 & \dots & 1 & \dots & 0 \\ \vdots & & \vdots & \ddots & \vdots & & \vdots \\ 0 & \dots & 1 & \dots & 0 & \dots & 0 \\ \vdots & & \vdots & & \vdots & \ddots & \vdots \\ 0 & \dots & 0 & \dots & 0 & \dots & 1 \end{pmatrix}.$$

(columns i and j marked above the third and fifth columns)

Multiplication of a vector $\mathbf{x}$ by $\mathbf{P}_{[ij]}$ permutes the ith and jth elements of $\mathbf{x}$. Note that $\mathbf{P}_{[ij]}$ is an *orthogonal* matrix and that $\mathbf{P}_{[ii]} = \mathbf{I}_n$, the identity matrix.

Determinants of Elementary Permutation Matrices The *determinant* of an elementary permutation matrix $\mathbf{P}_{[ij]}$ is defined to be -1, unless $i = j$ (i.e., $\mathbf{P}_{[ij]} = \mathbf{I}_n$):

$$\det(\mathbf{P}_{[ij]}) \stackrel{\text{def}}{=} \begin{cases} -1, & i \neq j, \\ +1, & i = j. \end{cases}$$

Permutation Matrices A *permutation matrix* is any product of elementary permutation matrices. These are also orthogonal matrices. Let $\mathcal{P}_n$ denote the set of all distinct $n \times n$ permutation matrices. There are $n! = 1 \times 2 \times 3 \times \cdots \times n$ of them, corresponding to the $n!$ permutations of n indices.

Determinants of Permutation Matrices The determinant of a permutation matrix can be defined by the rule that the determinant of a product of matrices is the product of the determinants:

$$\det(\mathbf{AB}) = \det(\mathbf{A})\,\det(\mathbf{B}).$$

Therefore, the determinant of a permutation matrix will be either $+1$ or -1. A permutation matrix is called "even" if its determinant is $+1$ and "odd" if its determinant equals -1.

Determinants of Square Matrices The determinant of any $n \times n$ matrix $\mathbf{A}$ can be defined as follows:

$$\det(\mathbf{A}) \stackrel{\text{def}}{=} \sum_{\mathbf{P}\in\mathcal{P}_n} \det(\mathbf{P}) \prod_{i=1}^{n} \{\mathbf{AP}\}_{ii}.$$

This formula has $\mathcal{O}(n \times n!)$ computational complexity (for a sum over $n!$ products of n elements each).

Characteristic Values of Square Matrices For a free variable λ, the polynomial

$$p_A(\lambda) \stackrel{\text{def}}{=} \det[\mathbf{A} - \lambda\mathbf{I}] = \sum_{i=0}^{n} a_i \lambda^i$$

is called the *characteristic polynomial* of $\mathbf{A}$. The roots of $p_A(\lambda)$ are called the *characteristic values* (or *eigenvalues*) of $\mathbf{A}$. The determinant of $\mathbf{A}$ equals the product of its characteristic values, with each characteristic value occurring as many times in the product as the multiplicity of the associated root of the characteristic polynomial.

Definiteness of Symmetric Matrices If $\mathbf{A}$ is symmetric, all its characteristic values are real numbers, which implies that they can be ordered. They are usually expressed in descending order:

$$\lambda_1(\mathbf{A}) \ge \lambda_2(\mathbf{A}) \ge \lambda_3(\mathbf{A}) \ge \cdots \ge \lambda_n(\mathbf{A}).$$

A real square symmetric matrix $\mathbf{A}$ is called

$$\begin{aligned} \textit{positive definite} \quad & \text{if} \quad \lambda_n(\mathbf{A}) > 0, \\ \textit{non-negative definite} \quad & \text{if} \quad \lambda_n(\mathbf{A}) \ge 0, \\ \textit{indefinite} \quad & \text{if} \quad \lambda_1(\mathbf{A}) > 0 \text{ and } \lambda_n(\mathbf{A}) < 0, \\ \textit{non-positive definite} \quad & \text{if} \quad \lambda_1(\mathbf{A}) \le 0, \text{ and} \\ \textit{negative definite} \quad & \text{if} \quad \lambda_1(\mathbf{A}) < 0. \end{aligned}$$

Non-negative definite matrices are also called *positive semidefinite*, and non-positive definite matrices are also called *negative semidefinite*.

Characteristic Vectors For each real characteristic value $\lambda_i(\mathbf{A})$ of a real symmetric $\mathbf{A}$, there is a corresponding *characteristic vector* (or *eigenvector*) $\mathbf{e}_i(\mathbf{A})$ such that $\mathbf{e}_i(\mathbf{A}) \neq 0$ and $\mathbf{A}\mathbf{e}_i(\mathbf{A}) = \lambda_i(\mathbf{A})\mathbf{e}_i(\mathbf{A})$. The characteristic vectors corresponding to distinct characteristic values are mutually orthogonal.

B.6.2 B.6.2 The Matrix Trace

The *trace* of a square matrix is the sum of its diagonal elements. It also equals the sum of the characteristic values and has the property that the trace of the product of conformable matrices is independent of the order of multiplication-a very useful attribute:

$$\text{trace}(\mathbf{AB}) = \sum_i \{\mathbf{AB}\}_{ii} \tag{B.10}$$

$$= \sum_i \sum_j \mathbf{A}_{ij}\mathbf{B}_{ji} \tag{B.11}$$

$$= \sum_j \sum_i \mathbf{B}_{ji}\mathbf{A}_{ij} \tag{B.12}$$

$$= \text{trace}(\mathbf{BA}). \tag{B.13}$$

Note the product $\mathbf{AB}$ is conformable for the trace function only if it is a square matrix, which requires that $\mathbf{A}$ and $\mathbf{B}^{\mathrm{T}}$ have the same dimensions. If they are $m \times n$ (or $n \times m$), then the computation of the trace of their product requires mn multiplications, whereas the product itself would require m^2n (or mn^2) multiplications.

B.6.3 Algebraic Functions of Matrices

An algebraic function may be defined by an expression in which the independent variable (a matrix) is a free variable, such as the truncated power series

$$f(\mathbf{A}) = \sum_{k=-n}^{n} \mathbf{B}_k \mathbf{A}^k,$$

where the negative power $\mathbf{A}^{-p} = \{\mathbf{A}^{-1}\}^p = \{\mathbf{A}^p\}^{-1}$. In this representation, the matrix $\mathbf{A}$ is the independent (free) variable and the other matrix parameters ($\mathbf{B}_k$) are assumed to be known and fixed.

B.6.4 Analytic Functions of Matrices

An analytic function is defined in terms of a convergent power series. It is necessary that the power series converge to a limit, and the matrix norms defined in Section B.1.7 must be used to define and prove convergence of a power series. This level of rigor is beyond the scope of this book, but we do need to use one particular analytic function, the exponential function.

Matrix Exponential Function The power series

$$e^{\mathbf{A}} \stackrel{\text{def}}{=} \sum_{k=0}^{\infty} \frac{1}{k!}\mathbf{A}^k, \tag{B.14}$$

$$k! \stackrel{\text{def}}{=} 1 \times 2 \times 3 \cdots \times k, \tag{B.15}$$

does converge[3] for all square matrices **A**. It defines the exponential function of the matrix **A**. This definition is sufficient to prove some elementary properties of the exponential function for matrices, such as

- $e^{0_n} = \mathbf{I}_n$ for 0_n, the $n \times n$ zero matrix.
- $e^{\mathbf{I}_n} = e\mathbf{I}_n$ for $\mathbf{I}_n$, the $n \times n$ identity matrix.
- $e^{\mathbf{A}^{\mathrm{T}}} = \{e^{\mathbf{A}}\}^{\mathrm{T}}$.
- $(d/dt)e^{\mathbf{A}t} = \mathbf{A}e^{\mathbf{A}t} = e^{\mathbf{A}t}\mathbf{A}$.
- The exponential of a skew-symmetric matrix is an orthogonal matrix.
- The characteristic vectors of **A** are also the characteristic vectors of $e^{\mathbf{A}}$.
- If λ is a characteristic value of **A**, then e^{λ} is a characteristic value of $e^{\mathbf{A}}$.

Powers and Exponentials of Cross-product Matrices The fact that exponential functions of skew-symmetric matrices are orthogonal matrices will have important consequences for coordinate transformations (Appendix C), because the matrices transforming vectors from one right-handed coordinate system (defined in Section B.1.2.11) to another can can be represented as the exponentials of cross-product matrices (defined in Eq. B.9). We show here how to represent the exponential of a cross-product matrix

$$[\boldsymbol{\rho}\otimes] = \begin{bmatrix} 0 & -\rho & \rho_2 \\ \rho_3 & 0 & -\rho_1 \\ -\rho_2 & \rho_1 & 0 \end{bmatrix}$$

in closed form. The first few powers can be calculated by hand, as

$$\begin{aligned}
[\boldsymbol{\rho}\otimes]^0 &= \mathbf{I}_3, \\
[\boldsymbol{\rho}]^1 &= [\boldsymbol{\rho}\otimes], \\
[\boldsymbol{\rho}\otimes]^2 &= \begin{bmatrix} -\rho_3^2 - \rho_2^2 & \rho_2\rho_1 & \rho_3\rho_1 \\ \rho_2\rho_1 & -\rho_3^2 - \rho_1^2 & \rho_3\rho_2 \\ \rho_3\rho_1 & \rho_3\rho_2 & -\rho_2^2 - \rho_1^2 \end{bmatrix} \\
&= \boldsymbol{\rho}\boldsymbol{\rho}^{\mathrm{T}} - |\boldsymbol{\rho}|^2\mathbf{I}_3,
\end{aligned}$$

[3]However, convergence is not fast enough to make this a reasonable general-purpose formula for approximating the exponential of **A**. More reliable and efficient methods can be found, e.g., in [41].

$$
\begin{aligned}
[\boldsymbol{\rho}\otimes]^3 &= [\boldsymbol{\rho}\otimes][\boldsymbol{\rho}\otimes]^2 \\
&= [\boldsymbol{\rho}\otimes][\boldsymbol{\rho}\boldsymbol{\rho}^{\mathrm{T}} - |\boldsymbol{\rho}|^2\mathbf{I}_3]^2 \\
&= -|\boldsymbol{\rho}|^2[\boldsymbol{\rho}\otimes], \\
[\boldsymbol{\rho}\otimes]^4 &= -|\boldsymbol{\rho}|^2[\boldsymbol{\rho}\otimes]^2, \\
&\vdots \\
[\boldsymbol{\rho}\otimes]^{2k+1} &= (-1)^k|\boldsymbol{\rho}|^{2k}[\boldsymbol{\rho}\otimes], \\
[\boldsymbol{\rho}\otimes]^{2k+2} &= (-1)^k|\boldsymbol{\rho}|^{2k}[\boldsymbol{\rho}\otimes]^2,
\end{aligned}
$$

so that the exponential expansion

$$
\begin{aligned}
\exp([\boldsymbol{\rho}\otimes]) &= \sum_{l=1}^{+\infty} \frac{1}{\ell!}[\boldsymbol{\rho}\otimes]^l \\
&= [\boldsymbol{\rho}\otimes]^0 + \frac{1}{|\boldsymbol{\rho}|}\left\{\sum_{k=0}^{+\infty} \frac{(-1)^k|\boldsymbol{\rho}|^{2k+1}}{(2k+1)!}\right\}[\boldsymbol{\rho}\otimes] \\
&\quad + \frac{1}{|\boldsymbol{\rho}|^2}\left\{\sum_{k=0}^{+\infty} \frac{(-1)^k|\boldsymbol{\rho}|^{2k+2}}{(2k+2)!}\right\}[\boldsymbol{\rho}\otimes]^2 \\
&= \cos(|\boldsymbol{\rho}|)\mathbf{I}_2 + \frac{1-\cos(|\boldsymbol{\rho}|)}{|\boldsymbol{\rho}|^2}\boldsymbol{\rho}\boldsymbol{\rho}^{\mathrm{T}} + \frac{\sin(|\boldsymbol{\rho}|)}{|\boldsymbol{\rho}|}\begin{bmatrix} 0 & -\rho & \rho_2 \\ \rho_3 & 0 & -\rho_1 \\ -\rho_2 & \rho_1 & 0 \end{bmatrix},
\end{aligned}
\tag{B.16}
$$

where ! denots the factorial function (defined in Eq. B.16).

B.6.5 Similarity Transformations and Analytic Functions

For any $n \times n$ nonsingular matrix $\mathbf{A}$, the transform $\boldsymbol{X} \rightarrow \mathbf{A}^{-1}\boldsymbol{X}\mathbf{A}$ is called a *similarity transformation* of the $n \times n$ matrix $\boldsymbol{X}$. It is a useful transformation for analytic functions of matrices

$$f(\boldsymbol{X}) = \sum_{k=0}^{\infty} a_k \boldsymbol{X}^k,$$

because

$$f(\mathbf{A}^{-1}\boldsymbol{X}\mathbf{A}) = \sum_{k=0}^{\infty} a_k (\mathbf{A}^{-1}\boldsymbol{X}\mathbf{A})^k$$

$$= \mathbf{A}^{-1}(\sum_{k=0}^{\infty} a_k X^k)\mathbf{A}$$
$$= \mathbf{A}^{-1} f(X)\mathbf{A}.$$

If the characteristic values of X are distinct, then the similarity transform performed with the characteristic vectors of X as the column vectors of $\mathbf{A}$ will diagonalize X with its characteristic values along the main diagonal:

$$\mathbf{A}^{-1}X\mathbf{A} = \text{diag}_\ell\{\lambda_\ell\},$$
$$f(\mathbf{A}^{-1}X\mathbf{A}) = \text{diag}_\ell\{\mathbf{F}(\lambda_l\ell)\},$$
$$f(X) = \mathbf{A}\text{diag}_\ell\{\mathbf{F}(\lambda_\ell)\}\mathbf{A}^{-1}.$$

(Although this is a useful analytical approach for demonstrating functional dependencies, it is not considered a robust numerical method.)

B.7 NORMS

B.7.1 Normed Linear Spaces

Vectors and matrices can be considered as elements of *linear spaces*, in that they can be added and multiplied by scalars. A *norm* is *any* nonnegative real-valued function $\|\cdot\|$ defined on a linear space such that, for any scalar s and elements $\mathbf{x}$ and $\mathbf{y}$ of the linear space (vectors *or* matrices),

$$\begin{aligned} \|x\| = 0 \quad &\text{iff} \quad x = 0, \\ \|x\| > 0 \quad &\text{iff} \quad x \neq 0, \\ \|sx\| \quad &= \quad |s|\|x\|, \\ \|x + y\| \quad &\leq \quad \|x\| + \|y\|, \end{aligned}$$

where iff stands for "if and only if." These constraints are rather loose, and many possible norms can be defined for a particular linear space. A linear space with a specified norm is called a *normed linear space*. The norm induces a *topology* on the linear space, which is used to define continuity and convergence. Norms are also used in numerical analysis for establishing error bounds and in sensitivity analysis for bounding sensitivities. The multiplicity of norms is useful in these applications, because the user is free to pick the one that works best for her or his particular problem.

We define here many of the more popular norms, some of which are known by more than one name.

B.7.2 Matrix Norms

Many norms have been defined for matrices. Two general types are presented here. Both are derived from vector norms, but by different means.

Generalized Vector Norms Vector norms can be generalized to matrices by treating the matrix like a doubly-subscripted vector. For example, the Hölder norms for vectors can be generalized to matrices as

$$\|\mathbf{A}\|_{(p)} = \left\{ \sum_{i=1}^{m} \sum_{j=1}^{n} |a_{i,j}|^p \right\}^{1/p}.$$

The matrix (2)-norm defined in this way is also called the *Euclidean norm, Schur norm*, or *Frobenius norm.* We will use the notation $\|\cdot\|_F$ in place of $\|\cdot\|_{(2)}$ for the Frobenius norm.

The reason for putting the parentheses around the subscript p in the above definition is that there is another way that the vector p-norms are used to define matrix norms, and it is with this alternative definition that they are usually allowed to wear an unadorned p subscript. These alternative norms also have the following desirable properties.

Desirable Multiplicative Properties of Matrix Norms Because matrices can be multiplied, one could also apply the additional constraint that

$$\|\mathbf{AB}\|_M \leq \|\mathbf{A}\|_M \|\mathbf{B}\|_M$$

for conformable matrices **A** and **B** and a matrix norm $\|\cdot\|_M$. This is a good property to have for some applications. One might also insist on a similar property with respect to multiplication by vector **x**, for which a norm $\|\cdot\|_{V_1}$ may already be defined:

$$|\mathbf{A}x\|_{V_2} \leq \|\mathbf{A}\|_M \|x\|_{V_1}.$$

This property is called *compatibility* between the matrix norm $||\cdot||_M$ and the vector norms $||\cdot||_{V_1}$ and $||\cdot||_{V_2}$. (Note that there can be two distinct vector norms associated with a matrix norm: one in the normed linear space containing **x** and one in the space containing $\mathbf{A}x$.)

Matrix Norms Subordinate to Vector Hölder Norms There is a family of alternative matrix "p-norms" [but not (p)-norms] defined by the formula

$$||\mathbf{A}||_p \stackrel{\text{def}}{=} \sup_{||x|| \neq 0} \frac{||\mathbf{A}x||_p}{||x||_p},$$

where the norms on the right-hand side are the vector Hölder norms and the induced matrix norms on the left are called *subordinate* to the corresponding Hölder norms. The 2-norm defined in this way is also called the *spectral norm* of **A**. It has the properties:

$$||\text{diag}_i\{\lambda_i\}||_2 = \max_i |\lambda_i| \quad \text{and} \quad ||\mathbf{A}x||_2 \leq ||\mathbf{A}||_2 ||x||_2.$$

The first of these properties implies that $||\mathbf{I}||_2 = 1$. The second property is compatibility between the spectral norm and the vector Euclidean norm. (Subordinate matrix norms are guaranteed to be compatible with the vector norms used to define them.) All matrix norms subordinate to vector norms also have the property that $||\mathbf{I}|| = 1$.

Computation of Matrix Hölder Norms The following formulas may be used in computing 1-norms and ∞-norms of $m \times n$ matrices $\mathbf{A}$:

$$||\mathbf{A}||_1 = \max_{i \leq j \leq n} \left\{ \sum_{i=1}^{m} |a_{ij}| \right\},$$

$$||\mathbf{A}||_\infty = \max_{1 \leq i \leq m} \left\{ \sum_{j=1}^{n} |a_{ij}| \right\}.$$

The norm $||\mathbf{A}||_2$ can be computed as the square root of the largest characteristic value of $\mathbf{A}^{\mathrm{T}}\mathbf{A}$, which takes considerably more effort.

Default Matrix Norm When the type of norm applied to a matrix is not specified (by an appropriate subscript), the default will be the spectral norm (Hölder matrix 2-norm). It satisfies the following bounds with respect to the Frobenius norm and the other matrix Hölder norms for $m \times n$ matrices $\mathbf{A}$:

$$\begin{aligned}
||\mathbf{A}||_2 &\leq ||\mathbf{A}||_F \leq \sqrt{n}||\mathbf{A}||_2, \\
\frac{1}{\sqrt{m}}||\mathbf{A}||_1 &\leq ||\mathbf{A}||_2 \leq \sqrt{n}||\mathbf{A}||_1, \\
\frac{1}{\sqrt{n}}||\mathbf{A}||_\infty &\leq ||\mathbf{A}||_2 \leq \sqrt{m}||\mathbf{A}||_\infty, \\
\max_{\substack{1 \leq i \leq m \\ 1 \leq j \leq n}} |a_{ij}| &\leq ||\mathbf{A}||_F \leq \sqrt{mn} \max_{\substack{1 \leq i \leq m \\ 1 \leq j \leq n}} |a_{ij}|.
\end{aligned}$$

B.8 FACTORIZATIONS AND DECOMPOSITIONS

Decompositions are also called *factorizations* of matrices. These are generally represented by algorithms or formulas for representing a matrix as a product of matrix factors with useful properties. The two factorization algorithms described here have either triangular or diagonal factors in addition to orthogonal factors.

Decomposition methods are algorithms for computing the factors, given the matrix to be "decomposed."

B.8.1 Cholesky Decomposition

This decomposition is named after André Louis Cholesky [9], who was perhaps not the first discoverer of the method for factoring a symmetric, positive-definite matrix $\mathbf{P}$ as a product of triangular factors.

Cholesky Factors A Cholesky factor of a symmetric positive-definite matrix **P** is a matrix **C** such that

$$\mathbf{C}\mathbf{C}^{\mathrm{T}} = \mathbf{P}. \tag{B.17}$$

Note that it does not matter whether we write this equation in the alternative form $\mathbf{F}^{\mathrm{T}}\mathbf{F} = \mathbf{P}$, because the two solutions are related by $\mathbf{F} = \mathbf{C}^{\mathrm{T}}$.

Cholesky factors are not unique, however. If **C** is a Cholesky factor of **P**, then for any conformable orthogonal matrix **M**, the matrix

$$\mathbf{A} \stackrel{\mathrm{def}}{=} \mathbf{C}\mathbf{M}$$

satisfies the equation

$$\begin{aligned} \mathbf{A}\mathbf{A}^{\mathrm{T}} &= \mathbf{C}\mathbf{M}(\mathbf{C}\mathbf{M})^{\mathrm{T}} \\ &= \mathbf{C}\mathbf{M}\mathbf{M}^{\mathrm{T}}\mathbf{C}^{\mathrm{T}} \\ &= \mathbf{C}\mathbf{C}^{\mathrm{T}} \\ &= \mathbf{P}. \end{aligned} \tag{B.18}$$

That is, **A** is also a legitimate Cholesky factor. The ability to transform one Cholesky factor into another using orthogonal matrices will turn out to be very important in square-root filtering (in Section 8.1.6).

Cholesky Factoring Algorithms There are two possible forms of the Cholesky factorization algorithm, corresponding to two possible forms of the defining equation:

$$\mathbf{P} = \mathbf{L}_1\mathbf{L}_1^{\mathrm{T}} = \mathbf{U}_1^{\mathrm{T}}\mathbf{U}_1 \tag{B.19}$$

$$= \mathbf{U}_2\mathbf{U}_2^{\mathrm{T}} = \mathbf{L}_2^{\mathrm{T}}\mathbf{L}_2, \tag{B.20}$$

where the Cholesky factors $\mathbf{U}_1, \mathbf{U}_2$ are upper triangular and their respective transposes $\mathbf{L}_1, \mathbf{L}_2$ are lower triangular.

The first of these is implemented by the built-in MATLAB function `chol (P)`, with argument **P** a symmetric positive-definite matrix. The call `chol (P)` returns an upper triangular matrix $\mathbf{U}_1$ satisfying Eq. B.20. The MATLAB m-file `chol2.m` on the accompanying diskette implements the solution to Eq. B.21. The call `chol2 (P)` returns an upper triangular matrix $\mathbf{U}_2$ satisfying Eq. B.21.

Modified Cholesky Factorization The algorithm for Cholesky factorization of a matrix requires taking square roots, which can be avoided by using a *modified Cholesky factorization* in the form

$$\mathbf{P} = \mathbf{U}\mathbf{D}\mathbf{U}^{\mathrm{T}}, \tag{B.21}$$

where **D** is a diagonal matrix with positive diagonal elements and **U** is a *unit triangular matrix* (i.e., **U** has 1's along its main diagonal). This algorithm is implemented in the m-file `modchol.m` on the accompanying diskette.

B.8.2 QR Decomposition (Triangularization)

The **QR** *decomposition* of a matrix **A** is a representation in the form

$$\mathbf{A} = \mathbf{QR},$$

where **Q** is an orthogonal matrix and **R** is a triangular matrix. Numerical methods for **QR** decomposition are also called "triangularization" methods. Some of these methods are an integral part of square-root Kalman filtering and are presented in Section 8.1.6.3.

B.8.3 Singular-Value Decomposition

The *singular-value decomposition* of an $m \times n$ matrix **A** is a representation in the form $\mathbf{A} = \mathbf{T}_m\mathbf{D}\mathbf{T}_n$, where $\mathbf{T}_m$ and $\mathbf{T}_n$ are orthogonal matrices (with square dimensions as specified by their subscripts) and **D** is an $m \times n$ matrix filled with zeros everywhere except along the main diagonal of its maximal upper left square submatrix. This decomposition will have either of three forms:

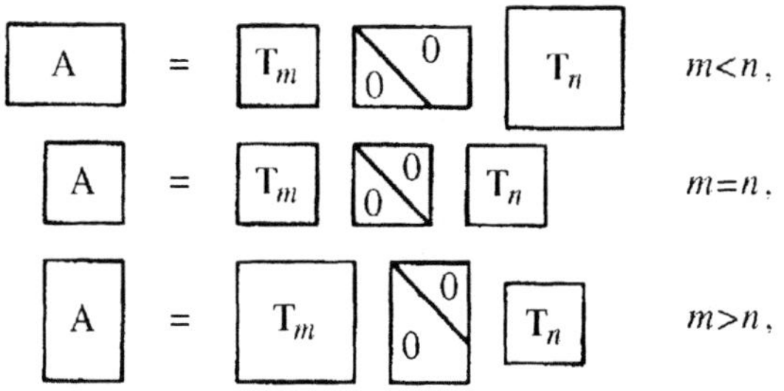

A = T_m [0 \ 0] T_n $m<n$,

A = T_m [0 \ 0] T_n $m=n$,

A = T_m [0 \ 0] T_n $m>n$,

depending on the relative values of m and n. The middle matrix **D** has the block form

$$\mathbf{D} = \begin{cases} [\text{diag}_i\{\sigma_i\}|0_{m\times(n-m)}] & \text{if } m < n, \\ \text{diag}_i\{\sigma_i\} & \text{if } m = n, \\ \begin{bmatrix} \text{diag}_i\{\sigma_i\} \\ 0_{(m-n)\times n} \end{bmatrix} & \text{if } m > n, \end{cases}$$

$$\sigma_1 \geq \sigma_2 \geq \sigma_3 \geq \cdots \geq \sigma_p \geq 0,$$

$$p = \min(m, n).$$

That is, the diagonal nonzero elements of **D** are in *descending order*, and nonnegative. These are called the *singular values* of **A**. For a proof that this decomposition exists, and an algorithm for computing it, see the book by Golub and Van Loan [41].

The singular values of a matrix characterize many useful matrix properties, such as

$$||\mathbf{A}||_2 = \sigma_1(\mathbf{A}),$$

rank $(\mathbf{A}) = r$ such that $\sigma_r > 0$ and either $\sigma_{r+1} = 0$ or $r = p$ (the rank of a matrix is defined in Section B.1.5.2), and

the *condition number* of $\mathbf{A}$ equals σ_1/σ_p.

The condition number of the matrix $\mathbf{A}$ in the linear equation $\mathbf{A}x = b$ bounds the sensitivity of the solution $\mathbf{x}$ to variations in b and the sensitivity of the solution to roundoff errors in determining it. The singular-value decomposition may also be used to define the "pseudorank" of $\mathbf{A}$ as the smallest singular value σ_i such that $\sigma_i > \varepsilon\sigma_1$, where ε is a processor- and precision-dependent constant such that $0 < \varepsilon \ll 1$ and $1 + \varepsilon \equiv 1$ in machine precision.

These relationships are useful for the analysis of state transition matrices Φ of Kalman filters, which can be singular or close enough to being singular that numerical roundoff can cause the product $\Phi P \Phi^{\mathrm{T}}$ to be essentially singular.

B.8.4 Eigenvalue–Eigenvector Decompositions of Symmetric Matrices

Symmetric QR Decomposition The so-called "symmetric QR" decomposition of an $n \times n$ symmetric real matrix $\mathbf{A}$ has the special form $\mathbf{A} = \boldsymbol{TDT}^T$, where the right orthogonal matrix is the transposed left orthogonal matrix and the diagonal matrix

$$\mathbf{D} = \operatorname{diag}_i\{\lambda_i\}.$$

That is, the diagonal elements are the characteristic values of the symmetric matrix. Furthermore, the column vectors of the orthogonal matrix T are the associated characteristic vectors $\mathbf{e}_i$ of $\mathbf{A}$:

$$\begin{aligned}
\mathbf{A} &= \boldsymbol{TDT}^{\mathrm{T}} \\
&= \sum_{i=1}^{n} \lambda_i \mathbf{e}_i \mathbf{e}_i^{\mathrm{T}}, \\
\boldsymbol{T} &= [\mathbf{e}_1 \quad \mathbf{e}_2 \quad \mathbf{e}_3 \quad \dots \quad \mathbf{e}_n].
\end{aligned}$$

These relationships are useful for the analysis of covariance matrices, which are constrained to have nonnegative characteristic values, although their numerical values may stray enough in practice (due to computer roundoff errors) to develop negative characteristic values.

B.9 QUADRATIC FORMS

Bilinear and Quadratic Forms For a matrix $\mathbf{A}$ and all conformable column vectors $\mathbf{x}$ and $\mathbf{y}$, the functional mapping $(x, y) \to x^{\mathrm{T}}\mathbf{A}y$ is called a *bilinear form.* As a function of $\mathbf{x}$ and $\mathbf{y}$, it is linear in both $\mathbf{x}$ and $\mathbf{y}$ and hence *bilinear.* In the case that $x = y$, the functional mapping $x \to x^{\mathrm{T}}\mathbf{A}x$ is called a *quadratic form.* The matrix $\mathbf{A}$ of a quadratic form is always a square matrix.

B.9.1 Symmetric Decomposition of Quadratic Forms

Any square matrix $\mathbf{A}$ can be represented uniquely as the sum of a symmetric matrix and a skew-symmetric matrix:

$$\mathbf{A} = \tfrac{1}{2}(\mathbf{A} + \mathbf{A}^{\mathrm{T}}) + \tfrac{1}{2}(\mathbf{A} - \mathbf{A}^{\mathrm{T}}),$$

where $\frac{1}{2}(\mathbf{A} + \mathbf{A}^{\mathrm{T}})$ is called the *symmetric part* of $\mathbf{A}$ and $\frac{1}{2}(\mathbf{A} - \mathbf{A}^{\mathrm{T}})$ is called the *skew-symmetric part* of $\mathbf{A}$. The quadratic form $x^{\mathrm{T}}\mathbf{A}x$ depends only on the symmetric part of $\mathbf{A}$:

$$x^{\mathrm{T}}\mathbf{A}x = x^{\mathrm{T}}\{\tfrac{1}{2}(\mathbf{A} + \mathbf{A}^{\mathrm{T}})\}x.$$

Therefore, one can always assume that the matrix of a quadratic form is symmetric, and one can express the quadratic form in summation form as

$$\begin{aligned} x^{\mathrm{T}}\mathbf{A}x &= \sum_{i=1}^{n}\sum_{j=1}^{n} a_{ij},x_i x_j = \sum_{i=j} a_{ij}x_i x_j + \sum_{i\neq j} a_{ij}x_i x_j \\ &= \sum_{i=1}^{n} a_{ii}x_i^2 + 2\sum_{i<j} a_{ij}x_i x_j \end{aligned}$$

for symmetric $\mathbf{A}$.

Ranges of Quadratic Forms The domain of a quadratic form for an $n \times n$ matrix is n-dimensional Euclidean space, and the range is in $(-\infty, +\infty)$, the real line. In the case that $x \neq 0$,

$$\begin{aligned}
&\text{if } \mathbf{A} \text{ is } \textit{positive definite}, && \text{the range of } x \to x^{\mathrm{T}}\mathbf{A}x \text{ is} && (0, +\infty);\\
&\text{if } \mathbf{A} \text{ is } \textit{non-negative definite}, && \text{the range of } x \to x^{\mathrm{T}}\mathbf{A}x \text{ is} && [0, +\infty);\\
&\text{if } \mathbf{A} \text{ is } \textit{indefinite}, && \text{the range of } x \to x^{\mathrm{T}}\mathbf{A}x \text{ is} && (-\infty, +\infty);\\
&\text{if } \mathbf{A} \text{ is } \textit{non-positive definite}, && \text{the range of } x \to x^{\mathrm{T}}\mathbf{A}x \text{ is} && (-\infty, 0];\\
&\text{if } \mathbf{A} \text{ is } \textit{negative definite}, && \text{the range of } x \to x^{\mathrm{T}}\mathbf{A}x \text{ is} && (-\infty, 0).
\end{aligned}$$

If $x^{\mathrm{T}}x = 1$, then $\lambda_n(\mathbf{A}) \leq x^{\mathrm{T}}\mathbf{A}x \leq \lambda_1(\mathbf{A})$. That is, the quadratic form maps the unit n-sphere onto the closed interval $[\lambda_n(\mathbf{A}), \lambda_1(\mathbf{A})]$.

B.10 DERIVATIVES OF MATRICES

B.10.1 Derivatives of Matrix-Valued Functions

The derivative of a matrix with respect to a scalar is the matrix of derivatives of its elements:

$$\mathbf{F}(t) = \begin{bmatrix} f_{11}(t) & f_{12}(t) & f_{13}(t) & \cdots & f_{1n}(t) \\ f_{21}(t) & f_{22}(t) & f_{23}(t) & \cdots & f_{2n}(t) \\ f_{31}(t) & f_{32}(t) & f_{33}(t) & \cdots & f_{3n}(t) \\ \vdots & \vdots & \vdots & & \vdots \\ f_{m1}(t) & f_{m2}(t) & f_{m3}(t) & \cdots f_{mn}(t) & \end{bmatrix},$$

$$d/dt\mathbf{F}(t) = \begin{bmatrix} d/dtf_{11}(t) & d/dtf_{12}(t) & d/dtf_{13}(t) & \cdots & d/dtf_{1n}(t) \\ d/dtf_{21}(t) & d/dtf_{22}(t) & d/dtf_{23}(t) & \cdots & d/dtf_{2n}(t) \\ d/dtf_{31}(t) & d/dtf_{32}(t) & d/dtf_{33}(t) & \cdots & d/dtf_{3n}(t) \\ \vdots & \vdots & \vdots & & \vdots \\ d/dtf_{m1}(t) & d/dtf_{m2}(t) & d/dtf_{m3}(t) & \ldots & d/dtf_{mn}(t) \end{bmatrix}.$$

The rule for the derivative of a product applies also to matrix products:

$$d/dt[\mathbf{A}(t)\mathbf{B}(t)] = [d/dt\mathbf{A}(t)]\mathbf{B}(t) + \mathbf{A}(t)[d/dt\mathbf{B}(t)],$$

provided that the order of the factors is preserved.

Derivative of Matrix Inverse If $\mathbf{F}(t)$ is square and nonsingular, then $\mathbf{F}(t)\mathbf{F}^{-1}(t) = \mathbf{I}$, a constant. As a consequence, its derivative will be zero. This fact can be used to derive the formula for the derivative of a matrix inverse:

$$\begin{aligned} 0 &= d/dt\mathbf{I} \\ &= d/dt[\mathbf{F}(t)\mathbf{F}^{-1}(t)] \\ &= [d/dt\mathbf{F}(t)]\mathbf{F}^{-1}(t) + \mathbf{F}(t)[d/dt\mathbf{F}^{-1}(t)], \\ d/dt\mathbf{F}^{-1}(t) &= -\mathbf{F}^{-1}[d/dt\mathbf{F}(t)]\mathbf{F}^{-1}. \end{aligned} \tag{B.22}$$

Derivative of Orthogonal Matrix If the $\mathbf{F}(t)$ is orthogonal, its inverse $\mathbf{F}^{-1}(t) = \mathbf{F}^{\mathrm{T}}(t)$, its transpose, and because

$$d/dt\mathbf{F}^{\mathrm{T}}(t) = [d/dt\mathbf{F}(t)]^{\mathrm{T}} = \dot{\mathbf{F}}^{\mathrm{T}},$$

one can show that orthogonal matrices satisfy matrix differential equations with antisymmetric dynamic coefficient matrices:

$$\begin{aligned} 0 &= \frac{d}{dt}\mathbf{I} & 0 &= \frac{d}{dt}\mathbf{I} \\ &= \frac{d}{dt}[\mathbf{F}(t)\mathbf{F}^{\mathrm{T}}(t)] & &= \frac{d}{dt}[\mathbf{F}^{\mathrm{T}}(t)\mathbf{F}(t)] \\ &= \dot{\mathbf{F}}(t)\mathbf{F}^{\mathrm{T}}(t) + \mathbf{F}(t)\dot{\mathbf{F}}^{\mathrm{T}}(t), & &= \dot{\mathbf{F}}^{\mathrm{T}}(t)\mathbf{F}(t) + \mathbf{F}^{\mathrm{T}}(t)\dot{\mathbf{F}}(t), \end{aligned}$$

$$\begin{aligned}
\dot{\mathbf{F}}(t)\mathbf{F}^{\mathrm{T}}(t) &= -[\mathbf{F}(t)\cdot\dot{\mathbf{F}}^{\mathrm{T}}(t)] \\
&= -[\dot{\mathbf{F}}(t)\mathbf{F}^{\mathrm{T}}(t)]^{\mathrm{T}} \\
&= \text{antisymmetric matrix} \\
&= \mathbf{\Omega}_{\text{left}}, \\
\dot{\mathbf{F}}(t) &= \mathbf{\Omega}_{left}\mathbf{F}(t),
\end{aligned}
\qquad
\begin{aligned}
\mathbf{F}^{\mathrm{T}}(t)\dot{\mathbf{F}}(t) &= -[\dot{\mathbf{F}}(t)\mathbf{F}(t)\mathbf{f}(t)] \\
&= -[\mathbf{F}^{\mathrm{T}}(t)\dot{\mathbf{F}}(t)]^{\mathrm{T}} \\
&= \text{antisymmetric matrix} \\
&= \mathbf{\Omega}_{\text{right}}, \\
\dot{\mathbf{F}}(t) &= \mathbf{F}(t)\mathbf{\Omega}_{\text{right}}.
\end{aligned}$$

That is, all time-differentiable orthogonal matrices $\mathbf{F}(t)$ satisfy dynamic equations with antisymmetric coefficient matrices, which can be either left- or right-side coefficient matrices.

B.10.2 Gradients of Quadratic Forms

If $f(x)$ is a differentiable scalar-valued function of an n-vector $\mathbf{x}$, then the vector

$$\frac{\partial f}{\partial x} = \left[\frac{\partial f}{\partial x_1}, \frac{\partial f}{\partial x_2}, \frac{\partial f}{\partial x_3}, \ldots, \frac{\partial f}{\partial x_n}\right]^{\mathrm{T}}$$

is called the *gradient* of f with respect to $\mathbf{x}$. In the case that f is a quadratic form with symmetric matrix $\mathbf{A}$, then the ith component of its gradient will be

$$\begin{aligned}
\left[\frac{\partial}{\partial x}(x^{\mathrm{T}}\mathbf{A}x)\right]_i &= \frac{\partial f}{\partial x_i}\left(\sum_j a_{jj}x_j^2 + 2\sum_{j<k} a_{jk}x_jx_k\right) \\
&= \left(2a_{ii}x_i + 2\sum_{i<k} a_{ik}x_k + 2\sum_{j<i} a_{ji}x_j\right) \\
&= \left(2a_{ii}x_i + 2\sum_{i\neq k} a_{ik}x_k\right) \\
&= 2\sum_{k=1}^{n} a_{ik}x_k \\
&= [2\mathbf{A}x]_i.
\end{aligned}$$

That is, the gradient vector can be expressed as

$$\frac{\partial}{\partial x}(x^T\mathbf{A}x) = 2\mathbf{A}x.$$

APPENDIX C

COORDINATE TRANSFORMATIONS

C.1 NOTATION

We use the notation $\mathbf{C}_{\text{to}}^{\text{from}}$ to denote a coordinate transformation matrix from one coordinate frame (designated by "from") to another coordinated frame (designated by "to"). For example,

$\mathbf{C}_{\text{ENU}}^{\text{ECI}}$ denotes the coordinate transformation matrix from earth-centered inertial (ECI) coordinates to earth-fixed east-north-up (ENU) local coordinates and

$\mathbf{C}_{\text{NED}}^{\text{RPY}}$ denotes the coordinate transformation matrix from vehicle body-fixed roll-pitch-yaw (RPY) coordinates to earth-fixed north-east-down (NED) coordinates.

Coordinate transformation matrices satisfy the composition rule

$$\mathbf{C}_C^B \mathbf{C}_B^A = \mathbf{C}_C^A,$$

where A, B, and C represent different coordinate frames.

What we mean by a coordinate transformation matrix is that if a vector $\mathbf{v}$ has the representation

$$\mathbf{v} = \begin{bmatrix} v_x \\ v_y \\ v_z \end{bmatrix} \tag{C.1}$$

Global Positioning Systems, Inertial Navigation, and Integration, Second Edition, by M. S. Grewal, L. R. Weill, and A. P. Andrews

in XYZ coordinates and the same vector $\mathbf{v}$ has the alternative representation

$$\mathbf{v} = \begin{bmatrix} v_u \\ v_v \\ v_w \end{bmatrix} \tag{C.2}$$

in UVW coordinates, then

$$\begin{bmatrix} v_x \\ v_y \\ v_z \end{bmatrix} = \mathbf{C}_{XYW}^{UVW} \begin{bmatrix} v_u \\ v_v \\ v_w \end{bmatrix}, \tag{C.3}$$

where "XYZ" and "UVW" stand for any two Cartesian coordinate systems in three-dimensional space.

The components of a vector in either coordinate system can be expressed in terms of the vector components along unit vectors parallel to the respective coordinate axes. For example, if one set of coordinate axes is labeled X, Y and Z, and the other set of coordinate axes are labeled U, V, and W, then the same vector $\mathbf{v}$ can be expressed in either coordinate frame as

$$\mathbf{v} = v_x \mathbf{1}_x + v_y \mathbf{1}_y + v_z \mathbf{1}_z \tag{C.4}$$

$$= v_u \mathbf{1}_u + v_v \mathbf{1}_v + v_w \mathbf{1}_w, \tag{C.5}$$

where

- the unit vectors $\mathbf{1}_x$, $\mathbf{1}_y$, and $\mathbf{1}_z$ are along the XYZ axes;
- the scalars v_x, v_y, and v_z are the respective components of $\mathbf{v}$ along the XYZ axes;
- the unit vectors $\mathbf{1}_u$, $\mathbf{1}_v$, and $\mathbf{1}_w$ are along the UVW axes; and
- the scalars v_u, v_v, and v_w are the respective components of $\mathbf{v}$ along the UVW axes.

The respective components can also be represented in terms of dot products of $\mathbf{v}$ with the various unit vectors,

$$v_x = \mathbf{1}_x^{\mathrm{T}} \mathbf{v} = v_u \mathbf{1}_x^{\mathrm{T}} \mathbf{1}_u + v_v \mathbf{1}_x^{\mathrm{T}} \mathbf{1}_v + v_w \mathbf{1}_x^{\mathrm{T}} \mathbf{1}_w, \tag{C.6}$$

$$v_y = \mathbf{1}_y^{\mathrm{T}} \mathbf{v} = v_u \mathbf{1}_y^{\mathrm{T}} \mathbf{1}_u + v_v \mathbf{1}_y^{\mathrm{T}} \mathbf{1}_v + v_w \mathbf{1}_y^{\mathrm{T}} \mathbf{1}_w, \tag{C.7}$$

$$v_z = \mathbf{1}_z^{\mathrm{T}} \mathbf{v} = v_u \mathbf{1}_z^{\mathrm{T}} \mathbf{1}_u + v_v \mathbf{1}_z^{\mathrm{T}} \mathbf{1}_v + v_w \mathbf{1}_z^{\mathrm{T}} \mathbf{1}_w, \tag{C.8}$$

which can be represented in matrix form as

$$\begin{bmatrix} v_x \\ v_y \\ v_z \end{bmatrix} = \begin{bmatrix} \mathbf{1}_x^{\mathrm{T}}\mathbf{1}_u & \mathbf{1}_x^{\mathrm{T}}\mathbf{1}_v & \mathbf{1}_x^{\mathrm{T}}\mathbf{1}_w \\ \mathbf{1}_y^{\mathrm{T}}\mathbf{1}_u & \mathbf{1}_y^{\mathrm{T}}\mathbf{1}_v & \mathbf{1}_y^{\mathrm{T}}\mathbf{1}_w \\ \mathbf{1}_z^{\mathrm{T}}\mathbf{1}_u & \mathbf{1}_z^{\mathrm{T}}\mathbf{1}_v & \mathbf{1}_z^{\mathrm{T}}\mathbf{1}_w \end{bmatrix} \begin{bmatrix} v_u \\ v_v \\ v_w \end{bmatrix} \tag{C.9}$$

$$\stackrel{\text{def}}{=} \mathbf{C}_{XYZ}^{UVW} \begin{bmatrix} v_u \\ v_v \\ v_w \end{bmatrix}, \tag{C.10}$$

which defines the coordinate transformation matrix $\mathbf{C}_{XYZ}^{UVW}$ from UVW to XYZ coordinates in terms of the dot products of unit vectors. However, dot products of unit vectors also satisfy the cosine rule (defined in Section B.1.2.5)

$$\mathbf{1}_a^T \mathbf{1}_b = \cos(\theta_{ab}), \tag{C.11}$$

where θ_{ab} is the angle between the unit vectors $\mathbf{1}_a$ and $\mathbf{1}_b$. As a consequence, the coordinate transformation matrix can also be written in the form

$$C_{XYZ}^{UVW} = \begin{bmatrix} \cos(\theta_{xu}) & \cos(\theta_{xv}) & \cos(\theta_{xw}) \\ \cos(\theta_{yu} & \cos(\theta_{yv}) & \cos(\theta_{yw}) \\ \cos(\theta_{zu}) & \cos(\theta_{zv}) & \cos(\theta_{zw}) \end{bmatrix}, \tag{C.12}$$

which is why coordinate transformation matrices are also called "direction cosines matrices."

Navigation makes use of coordinates that are natural to the problem at hand: inertial coordinates for inertial navigation, orbital coordinates for GPS navigation, and earth-fixed coordinates for representing locations on the earth.

The principal coordinate systems used in navigation, and the transformations between these different coordinate systems, are summarized in this appendix. These are primarily Cartesian (orthogonal) coordinates, and the transformations between them can be represented by orthogonal matrices. However, the coordinate transformations can also be represented by rotation vectors or quaternions, and all representations are used in the derivations and implementation of GPS/INS integration.

C.2 INERTIAL REFERENCE DIRECTIONS

C.2.1 Vernal Equinox

The equinoxes are those times of year when the length of the day equals the length of the night (the meaning of "equinox"), which only happens when the sun is over the equator. This happens twice a year: when the sun is passing from the Southern Hemisphere to the Northern Hemisphere (vernal equinox) and again when it is passing from the Northern Hemisphere to the Southern Hemisphere (autumnal equinox). The time of the vernal equinox defines the beginning of spring (the meaning of "vernal") in the Northern Hemisphere, which usually occurs around March 21–23.

The direction from the earth to the sun at the instant of the vernal equinox is used as a "quasi-inertial" direction in some navigation coordinates. This direction is defined by the intersection of the equatorial plane of the earth with the ecliptic (earth-sun plane). These two planes are inclined at about 23.45°, as illustrated in Fig. C.1. The inertial direction of the vernal equinox is changing ever so slowly, on the order of 5 arc seconds per year, but the departure from truly inertial directions is neglible over the time periods of most navigation problems. The vernal equinox was in the constellation Pisces in the year 2000. It was in the constellation Aries at the time of Hipparchus (190-120 BCE) and is sometimes still called "the first point of Aries."

C.2.2 Polar Axis of Earth

The one inertial reference direction that remains invariant in earth-fixed coordinates as the earth rotates is its polar axis, and that direction is used as a reference direction in inertial coordinates. Because the polar axis is (by definition) orthogonal to the earth's equatorial plane and the vernal equinox is (by definition) in the earth's equatorial plane, the earth's polar axis will always be orthogonal to the vernal equinox.

A third orthogonal axis can then be defined (by their cross-product) such that the three axes define a right-handed (defined in Section B.2.11) orthogonal coordinate system.

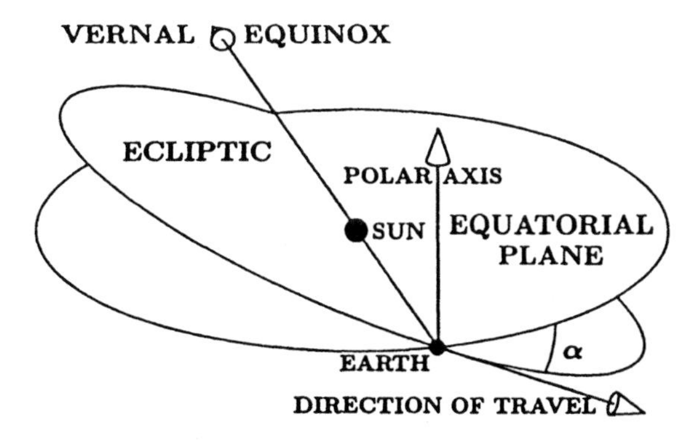

Fig. C.1 Direction of vernal equinox.

C.3 COORDINATE SYSTEMS

Although we are concerned exclusively with coordinate systems in the three dimensions of the observable world, there are many ways of representing a location in that world by a set of coordinates. The coordinates presented here are those used in navigation with GPS and/or INS.

C.3.1 Cartesian and Polar Coordinates

René Descartes (1596–1650) introduced the idea of representing points in three-dimensional space by a triplet of coordinates, called "Cartesian coordinates" in his honor. They are also called "Euclidean coordinates," but not because Euclid discovered them first. The Cartesian coordinates (x, y, z) and polar coordinates (θ, ϕ, r) of a common reference point, as illustrated in Fig. C.2, are related by the equations

$$x = r\cos(\theta)\cos(\phi), \tag{C.13}$$

$$y = r\sin(\theta)\cos(\phi), \tag{C.14}$$

$$z = r\sin(\phi), \tag{C.15}$$

$$r = \sqrt{x^2 + y^2 + z^2}, \tag{C.16}$$

$$\phi = \arcsin\left(\frac{z}{r}\right) \qquad \left(-\frac{1}{2}\pi \le \phi \le +\frac{1}{2}\pi\right), \tag{C.17}$$

$$\theta = \arctan\left(\frac{y}{x}\right) \qquad (-\pi < \theta \le +\pi), \tag{C.18}$$

with the angle θ (in radians) undefined if $\phi = \pm\frac{1}{2}\pi$.

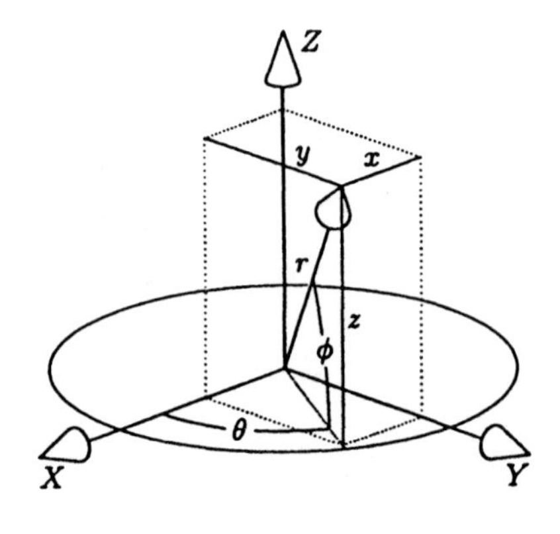

Fig. C.2 Cartesian and polar coordinates.

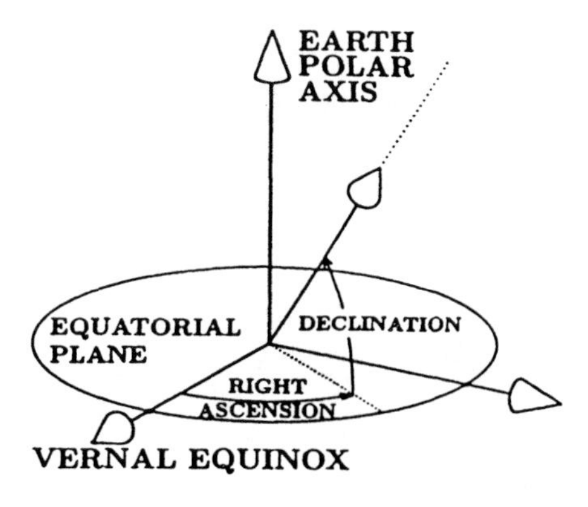

Fig. C.3 Celestial coordinates.

C.3.2 Celestial Coordinates

The "celestial sphere" is a system for inertial directions referenced to the polar axis of the earth and the vernal equinox. The polar axis of these celestial coordinates is parallel to the polar axis of the earth and its prime meridian is fixed to the vernal equinox. Polar celestial coordinates are *right ascension* (the celestial analog of longitude, measured eastward from the vernal equinox) and *declination* (the celestial analog of latitude), as illustrated in Fig. C.3. Because the celestial sphere is used primarily as a reference for direction, no origin need be specified.

Right ascension is zero at the vernal equinox and increases eastward (in the direction the earth turns). The units of right ascension (RA) can be radians, degrees, or hours (with 15 deg/h as the conversion factor).

By convention, declination is zero in the equatorial plane and increases toward the north pole, with the result that celestial objects in the Northern Hemisphere have positive declinations. Its units can be degrees or radians.

C.3.3 Satellite Orbit Coordinates

Johannes Kepler (1571–1630) discovered the geometric shapes of the orbits of planets and the minimum number of parameters necessary to specify an orbit (called "Keplerian" parameters). Keplerian parameters used to specify GPS satellite orbits in terms of their orientations relative to the equatorial plane and the vernal equinox (defined in Section C.2.1 and illustrated in Fig. C.1) include the following:

- Right ascension of the ascending node and orbit inclination, specifying the orientation of the orbital plane with respect to the vernal equinox and equatorial plane, is illustrated in Fig. C.4.
  (a) Right ascension is defined in the previous section and is shown in Fig. C.3.

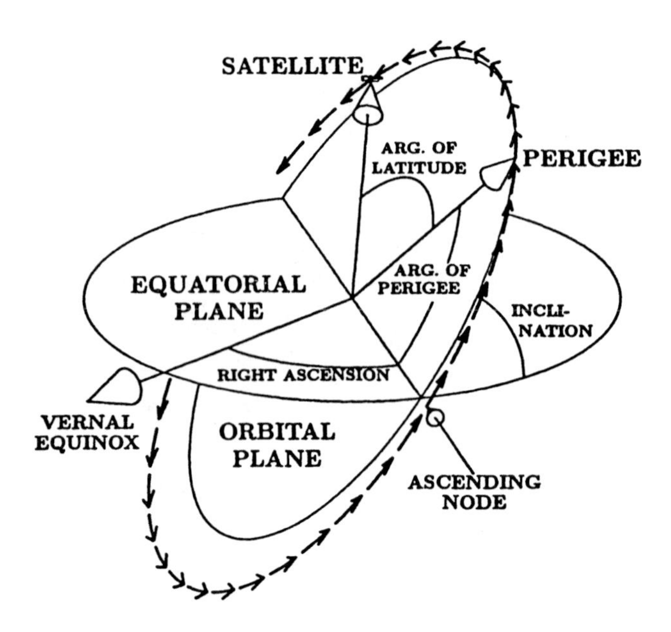

Fig. C.4 Keplerian parameters for satellite orbit.

(b) The intersection of the orbital plane of a satellite with the equatorial plane is called its "line of nodes," where the "nodes" are the two intersections of the satellite orbit with this line. The two nodes are dubbed "ascending"[1] (i.e., ascending from the Southern Hemisphere to the Northern Hemisphere) and "descending". The right ascension of the ascending node (RAAN) is the angle in the equatorial plane from the vernal equinox to the ascending node, measured counterclockwise as seen looking down from the north pole direction.

(c) Orbital inclination is the dihedral angle between the orbital plane and the equatorial plane. It ranges from zero (orbit in equatorial plane) to 90° (polar orbit).

- Semimajor axis a and semiminor axis b (defined in Section C.3.5.2 and illustrated in Fig. C.6) specify the size and shape of the elliptical orbit within the orbital plane.
- Orientation of the ellipse within its orbital plane, specified in terms of the "argument of perigee," the angle between the ascending node and the perigee of the orbit (closest approach to earth), is illustrated in Fig. C.4.
- Position of the satellite relative to perigee of the elliptical orbit, specified in terms of the angle from perigee, called the "argument of latitude" or "true anomaly," is illustrated in Fig. C.4.

[1]The astronomical symbol for the ascending node is ☊, often read as "earphones."

For computer simulation demonstrations, GPS satellite orbits can usually be assumed to be circular with radius $a = b = R = 26{,}560$ km and inclined at 55° to the equatorial plane. This eliminates the need to specify the orientation of the elliptical orbit within the orbital plane. (The argument of perigee becomes overly sensitive to orbit perturbations when eccentricity is close to zero.)

C.3.4 ECI Coordinates

Earth-centered inertial (ECI) coordinates are the favored inertial coordinates in the near-earth environment. The origin of ECI coordinates is at the center of gravity of the earth, with (Fig. C.5)

1. axis in the direction of the vernal equinox,
2. axis direction parallel to the rotation axis (north polar axis) of the earth, and
3. an additional axis to make this a right-handed orthogonal coordinate system, with the polar axis as the third axis (hence the numbering).

The equatorial plane of the earth is also the equatorial plane of ECI coordinates, but the earth itself is rotating relative to the vernal equinox at its sidereal rotation rate of about $7{,}292{,}115{,}167 \times 10^{-14}$ rad/s, or about 15.04109 deg/h, as illustrated in Fig. C.5.

C.3.5 ECEF Coordinates

Earth-centered, earth-fixed (ECEF) coordinates have the same origin (earth center) and third (polar) axis as ECI coordinates but rotate with the earth, as shown

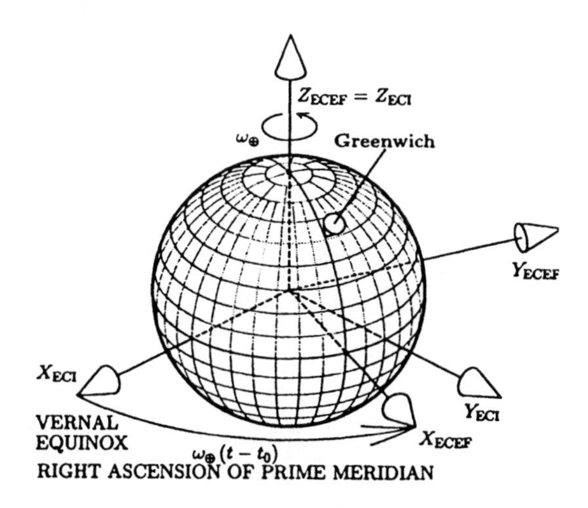

Fig. C.5 ECI and ECEF Coordinates.

in Fig. C.5. As a consequence, ECI and ECEF longitudes differ only by a linear function of time.

Longitude in ECEF coordinates is measured east (+) and west (−) from the prime meridian passing through the principal transit instrument at the observatory at Greenwich, UK, a convention adopted by 41 representatives of 25 nations at the International Meridian Conference, held in Washington, DC, in October of 1884.

Latitudes are measured with respect to the equatorial plane, but there is more than one kind of "latitude." *Geocentric latitude* would be measured as the angle between the equatorial plane and a line from the reference point to the center of the earth, but this angle could not be determined accurately (before GPS) without running a transit survey over vast distances. The angle between the pole star and the local vertical direction could be measured more readily, and that angle is more closely approximated as *geodetic latitude*. There is yet a third latitude (parametric latitude) that is useful in analysis. The latter two latitudes are defined in the following subsections.

C.3.5.1 Ellipsoidal Earth Models *Geodesy* is the study of the size and shape of the earth and the establishment of physical control points defining the origin and orientation of coordinate systems for mapping the earth. Earth shape models are very important for navigation using either GPS or INS, or both. INS alignment is with respect to the local vertical, which does not generally pass through the center of the earth. That is because the earth is not spherical.

At different times in history, the earth has been regarded as being flat (first-order approximation), spherical (second-order), and ellipsoidal (third-order). The third-order model is an ellipsoid of revolution, with its shorter radius at the poles and its longer radius at the equator.

C.3.5.2 Parametric Latitude For geoids based on ellipsoids of revolution, every meridian is an ellipse with equatorial radius a (also called "semimajor axis") and polar radius b (also called "semiminor axis"). If we let z be the Cartesian coordinate in the polar direction and $x_{\text{meridional}}$ be the equatorial coordinate in the meridional plane, as illustrated in Fig. C.6, then the equation for this ellipse will be

$$\frac{x_{\text{meridional}}^2}{a^2} + \frac{z^2}{b^2} = 1 \tag{C.19}$$

$$= \cos^2(\phi_{\text{parametric}}) + \sin^2(\phi_{\text{parametric}}) \tag{C.20}$$

$$= \frac{a^2\cos^2(\phi_{\text{parametric}})}{a^2} + \frac{b^2\sin^2(\phi_{\text{parametric}})}{b^2} \tag{C.21}$$

$$= \frac{[a\cos(\phi_{\text{parametric}})]^2}{a^2} + \frac{[b\sin(\phi_{\text{parametric}})]^2}{b^2}. \tag{C.22}$$

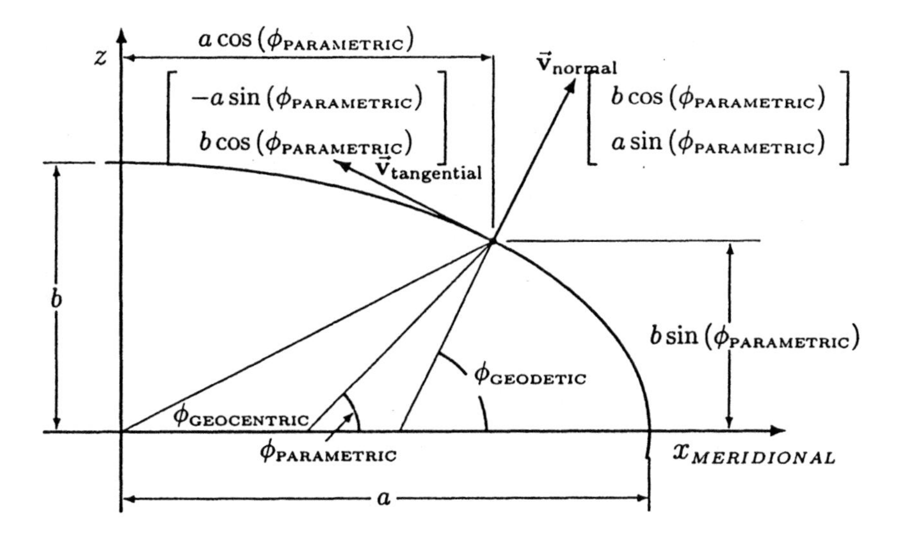

Fig. C.6 Geocentric, parametric, and geodetic latitudes in meridional plane.

That is, a parametric solution for the ellipse is

$$x_{\text{meridional}} = a\cos(\phi_{\text{parametric}}), \tag{C.23}$$

$$z = b\sin(\phi_{\text{parametric}}), \tag{C.24}$$

as illustrated in Fig. C.6. Although the parametric latitude $\phi_{\text{parametric}}$ has no physical significance, it is quite useful for relating geocentric and geodetic latitude, which do have physical significance.

C.3.5.3 Geodetic Latitude Geodetic latitude is defined as the elevation angle above (+) or below (−) the equatorial plane of the normal to the ellipsoidal surface. This direction can be defined in terms of the parametric latitude, because it is orthogonal to the meridional tangential direction.

The vector tangential to the meridian will be in the direction of the derivative to the elliptical equation solution with respect to parametric latitude:

$$\mathbf{v}_{\text{tangential}} \propto \frac{\partial}{\partial\phi_{\text{parametric}}}\begin{bmatrix} a\cos(\phi_{\text{parametric}}) \\ b\sin(\phi_{\text{parametric}}) \end{bmatrix} \tag{C.25}$$

$$= \begin{bmatrix} -a\sin(\phi_{\text{parametric}}) \\ b\cos(\phi_{\text{parametric}}) \end{bmatrix}, \tag{C.26}$$

and the meridional normal direction will be orthogonal to it, or

$$\mathbf{v}_{\text{normal}} \propto \begin{bmatrix} b\cos(\phi_{\text{parametric}}) \\ a\sin(\phi_{\text{parametric}}) \end{bmatrix}, \tag{C.27}$$

as illustrated in Fig. C.6.

The tangent of geodetic latitude is then the ratio of the z- and x-components of the surface normal vector, or

$$\tan(\phi_{\text{geodetic}}) = \frac{a\sin(\phi_{\text{parametric}})}{b\cos(\phi_{\text{parametric}})} \tag{C.28}$$

$$= \frac{a}{b}\tan(\phi_{\text{parametric}}), \tag{C.29}$$

from which, using some standard trigonometric identities,

$$\sin(\phi_{\text{geodetic}}) = \frac{\tan(\phi_{\text{geodetic}})}{\sqrt{1+\tan^2(\phi_{\text{geodetic}})}} \tag{C.30}$$

$$= \frac{a\sin(\phi_{\text{parametric}})}{\sqrt{a^2\sin^2(\phi_{\text{parametric}}) + b^2\cos^2(\phi_{\text{parametric}})}}, \tag{C.31}$$

$$\cos(\phi_{\text{geodetic}}) = \frac{1}{\sqrt{1+\tan^2(\phi_{\text{geodetic}})}} \tag{C.32}$$

$$= \frac{b\cos(\phi_{\text{parametric}})}{\sqrt{a^2\sin^2(\phi_{\text{parametric}}) + b^2\cos^2(\phi_{\text{parametric}})}}. \tag{C.33}$$

The inverse relationship is

$$\tan(\phi_{\text{parametric}}) = \frac{b}{a}\tan(\phi_{\text{geodetic}}), \tag{C.34}$$

from which, using the same trigonometric identities as before,

$$\sin(\phi_{\text{parametric}}) = \frac{\tan(\phi_{\text{parametric}})}{\sqrt{1+\tan^2(\phi_{\text{parametric}})}} \tag{C.35}$$

$$= \frac{b\sin(\phi_{\text{geodetic}})}{\sqrt{a^2\cos^2(\phi_{\text{geodetic}}) + b^2\sin^2(\phi_{\text{geodetic}})}}, \tag{C.36}$$

$$\cos(\phi_{\text{parametric}}) = \frac{1}{\sqrt{1+\tan^2(\phi_{\text{parametric}})}} \tag{C.37}$$

$$= \frac{a\cos(\phi_{\text{geodetic}})}{\sqrt{a^2\cos^2(\phi_{\text{geodetic}}) + b^2\sin^2(\phi_{\text{geodetic}})}}, \tag{C.38}$$

and the two-dimensional X-Z Cartesian coordinates in the meridional plane of a point on the geoid surface will

$$x_{\text{meridional}} = a\cos(\phi_{\text{parametric}}) \tag{C.39}$$

$$= \frac{a^2\cos(\phi_{\text{geodetic}})}{\sqrt{a^2\cos^2(\phi_{\text{geodetic}}) + b^2\sin^2(\phi_{\text{geodetic}})}}, \tag{C.40}$$

$$z = b\sin(\phi_{\text{parametric}}) \tag{C.41}$$

$$= \frac{b^2\sin(\phi_{\text{geodetic}})}{\sqrt{a^2\cos^2(\phi_{\text{geodetic}}) + b^2\sin^2(\phi_{\text{geodetic}})}} \tag{C.42}$$

in terms of geodetic latitude.

Equations C.40 and C.42 apply only to points on the geoid surface. Orthometric height h above (+) or below (−) the geoid surface is measured along the surface normal, so that the X-Z coordinates for a point with altitude h will be

$$x_{\text{meridional}} = \cos(\phi_{\text{geodetic}}) \times \left(h + \frac{a^2}{\sqrt{a^2\cos^2(\phi_{\text{geodetic}}) + b^2\sin^2(\phi_{\text{geodetic}})}}\right), \tag{C.43}$$

$$z = \sin(\phi_{\text{geodetic}}) \times \left(h + \frac{b^2}{\sqrt{a^2\cos^2(\phi_{\text{geodetic}}) + b^2\sin^2(\phi_{\text{geodetic}})}}\right). \tag{C.44}$$

In three-dimensional ECEF coordinates, with the X-axis passing through the equator at the prime meridian (at which longitude $\theta = 0$),

$$x_{\text{ECEF}} = \cos(\theta)x_{\text{meridional}} \tag{C.45}$$

$$= \cos(\theta)\cos(\phi_{\text{geodetic}}) \times \left(h + \frac{a^2}{\sqrt{a^2\cos^2(\phi_{\text{geodetic}}) + b^2\sin^2(\phi_{\text{geodetic}})}}\right), \tag{C.46}$$

$$y_{\text{ECEF}} = \sin(\theta)x_{\text{meridional}} \tag{C.47}$$

$$= \sin(\theta)\cos(\phi_{\text{geodetic}}) \times \left(h + \frac{a^2}{\sqrt{a^2\cos^2(\phi_{\text{geodetic}}) + b^2\sin^2(\phi_{\text{geodetic}})}} \right), \tag{C.48}$$

$$z_{\text{ECEF}} = \sin(\phi_{\text{geodetic}}) \times \left(h + \frac{b^2}{\sqrt{a^2\cos^2(\phi_{\text{geodetic}}) + b^2\sin^2(\phi_{\text{geodetic}})}} \right), \tag{C.49}$$

in terms of geodetic latitude ϕ_{geodetic}, longitude θ, and orthometric altitude h with respect to the reference geoid.

The inverse transformation, from ECEF *XYZ* to geodetic longitude–latitude–altitude coordinates, is

$$\theta = \text{atan2}(y_{\text{ECEF}}, x_{\text{ECEF}}), \tag{C.50}$$

$$\phi_{\text{geodetic}} = \text{atan2}\left(z_{\text{ECEF}} + \frac{e^2 a^2 \sin^3(\zeta)}{b}, \xi - e^2 a \cos^3(\zeta) \right), \tag{C.51}$$

$$h = \frac{\xi}{\cos(\phi)} - r_T, \tag{C.52}$$

where atan2 is the four-quadrant arctangent function in MATLAB and

$$\zeta = \text{atan2}(a z_{\text{ECEF}}, b\xi), \tag{C.53}$$

$$\xi = \sqrt{x_{\text{ECEF}}^2 + y_{\text{ECEF}}^2}, \tag{C.54}$$

$$r_T = \frac{a}{\sqrt{1 - e^2\sin(\phi)}}, \tag{C.55}$$

where r_T is the transverse radius of curvature on the ellipsoid, a is the equatorial radius, b is the polar radius, and e is elliptical eccentricity.

C.3.5.4 Geocentric Latitude For points on the geoid surface, the tangent of geocentric latitude is the ratio of distance above (+) or below (−) the equator $[z = b\sin(\phi_{\text{parametric}})]$ to the distance from the polar axis $[(x_{\text{meridional}} = a\cos(\phi_{\text{parametric}})]$, or

$$\tan(\phi_{\text{GEOCENTRIC}}) = \frac{b\sin(\phi_{\text{parametric}})}{a\cos(\phi_{\text{parametric}})} \tag{C.56}$$

$$= \frac{b}{a}\tan(\phi_{\text{parametric}}) \tag{C.57}$$

$$= \frac{b^2}{a^2}\tan(\phi_{\text{geodetic}}), \tag{C.58}$$

from which, using the same trigonometric identities as were used for geodetic latitude,

$$\sin(\phi_{\text{geocentric}}) = \frac{\tan(\phi_{\text{geocentric}})}{\sqrt{1 + \tan^2(\phi_{\text{geocentric}})}} \tag{C.59}$$

$$= \frac{b \sin(\phi_{\text{parametric}})}{\sqrt{a^2 \cos^2(\phi_{\text{parametric}}) + b^2 \sin^2(\phi_{\text{parametric}})}} \tag{C.60}$$

$$= \frac{b^2 \sin(\phi_{\text{geodetic}})}{\sqrt{a^4 \cos^2(\phi_{\text{geodetric}}) + b^4 \sin^2(\phi_{\text{geodetic}})}}, \tag{C.61}$$

$$\cos(\phi_{\text{geocentric}}) = \frac{1}{\sqrt{1 + \tan^2(\phi_{\text{geocentric}})}} \tag{C.62}$$

$$= \frac{a \cos(\phi_{\text{parametric}})}{\sqrt{a^2 \cos^2(\phi_{\text{parametric}}) + b^2 \sin^2(\phi_{\text{parametric}})}} \tag{C.63}$$

$$= \frac{a^2 \cos(\phi_{\text{geodetic}})}{\sqrt{a^4 \cos^2(\phi_{\text{geodetic}}) + b^4 \sin^2(\phi_{\text{geodetic}})}}. \tag{C.64}$$

The inverse relationships are

$$\tan(\phi_{\text{parametric}}) = \frac{a}{b} \tan(\phi_{\text{geocentric}}), \tag{C.65}$$

$$\tan(\phi_{\text{geodetic}}) = \frac{a^2}{b^2} \tan(\phi_{\text{geocentric}}), \tag{C.66}$$

from which, using the same trigonometric identities again,

$$\sin(\phi_{\text{parametric}}) = \frac{\tan(\phi_{\text{parametric}})}{\sqrt{1 + \tan^2(\phi_{\text{parametric}})}} \tag{C.67}$$

$$= \frac{a \sin(\phi_{\text{geocentric}})}{\sqrt{a^2 \sin^2(\phi_{\text{geocentric}}) + b^2 \cos^2(\phi_{\text{geocentric}})}}, \tag{C.68}$$

$$\sin(\phi_{\text{geodetic}}) = \frac{a^2 \sin(\phi_{\text{geocentric}})}{\sqrt{a^4 \sin^2(\phi_{\text{geocentric}}) + b^4 \cos^2(\phi_{\text{geocentric}})}}, \tag{C.69}$$

$$\cos(\phi_{\text{parametric}}) = \frac{1}{\sqrt{1 + \tan^2(\phi_{\text{parametric}})}} \tag{C.70}$$

$$= \frac{b\cos(\phi_{\text{geocentric}})}{\sqrt{a^2\sin^2(\phi_{\text{geocentric}}) + b^2\cos^2(\phi_{\text{geocentric}})}}, \tag{C.71}$$

$$\cos(\phi_{\text{geodetic}}) = \frac{b^2\cos(\phi_{\text{geocentric}})}{\sqrt{a^4\sin^2(\phi_{\text{geocentric}}) + b^4\cos^2(\phi_{\text{geocentric}})}}. \tag{C.72}$$

C.3.6 LTP Coordinates

Local tangent plane (LTP) coordinates, also called "locally level coordinates," are a return to the first-order model of the earth as being flat, where they serve as local reference directions for representing vehicle attitude and velocity for operation on or near the surface of the earth. A common orientation for LTP coordinates has one horizontal axis (the north axis) in the direction of increasing latitude and the other horizontal axis (the east axis) in the direction of increasing longitude, as illustrated in Fig. C.7. Horizontal location components in this local coordinate frame are called "relative northing" and "relative easting."

C.3.6.1 Alpha Wander Coordinates Maintaining east–north orientation was a problem for some INSs at the poles, where north and east directions change by 180°. Early gimbaled inertial systems could not slew the platform axes fast enough for near-polar operation. This problem was solved by letting the platform axes "wander" from north but keeping track of the angle α between north and a reference platform axis, as shown in Fig. C.8. This LTP orientation came to be called "alpha wander."'

C.3.6.2 ENU/NED Coordinates East–north–up (ENU) and north–east–down (NED) are two common right-handed LTP coordinate systems. ENU coordinates may be preferred to NED coordinates because altitude increases in the upward direction. But NED coordinates may also be preferred over ENU coordinates because the direction of a right (clockwise) turn is in the positive direction

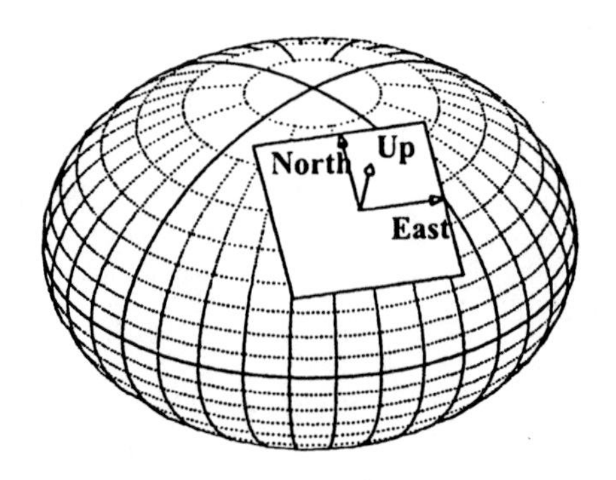

Fig. C.7 ENU coordinates.

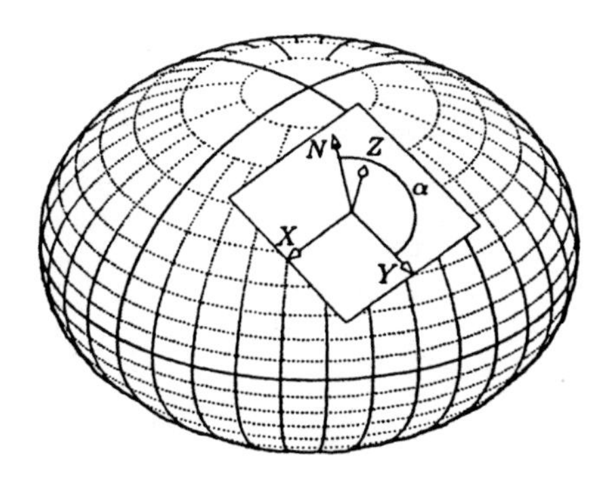

Fig. C.8 Alpha wander.

with respect to a downward axis, and NED coordinate axes coincide with vehicle-fixed roll–pitch–yaw (RPY) coordinates (Section C.3.7) when the vehicle is level and headed north.

The coordinate transformation matrix $\mathbf{C}_{\text{NED}}^{\text{ENU}}$ from ENU to NED coordinates and the transformation matrix $\mathbf{C}_{\text{ENU}}^{\text{NED}}$ from NED to ENU coordinates are one and the same:

$$\mathbf{C}_{\text{NED}}^{\text{ENU}} = \mathbf{C}_{\text{ENU}}^{\text{NED}} = \begin{bmatrix} 0 & 1 & 0 \\ 1 & 0 & 0 \\ 0 & 0 & -1 \end{bmatrix}. \tag{C.73}$$

C.3.6.3 ENU/ECEF Coordinates The unit vectors in local *east*, *north*, and *up* directions, as expressed in ECEF Cartesian coordinates, will be

$$\mathbf{1}_E = \begin{bmatrix} -\sin(\theta) \\ \cos(\theta) \\ 0 \end{bmatrix}, \tag{C.74}$$

$$\mathbf{1}_N = \begin{bmatrix} -\cos(\theta)\sin(\phi_{\text{geodetic}}) \\ -\sin(\theta)\sin(\phi_{\text{geodetic}}) \\ \cos(\phi_{\text{geodetic}}) \end{bmatrix}, \tag{C.75}$$

$$\mathbf{1}_U = \begin{bmatrix} \cos(\theta)\cos(\phi_{\text{geodetic}}) \\ \sin(\theta)\cos(\phi_{\text{geodetic}}) \\ \sin(\phi_{\text{geodetic}}) \end{bmatrix}, \tag{C.76}$$

and the unit vectors in the ECEF X, Y, and Z directions, as expressed in ENU coordinates, will be

$$\mathbf{1}_X = \begin{bmatrix} -\sin(\theta) \\ -\cos(\theta)\sin(\phi_{\text{geodetic}}) \\ \cos(\theta)\cos(\phi_{\text{geodetic}}) \end{bmatrix}, \tag{C.77}$$

$$\mathbf{1}_Y = \begin{bmatrix} \cos(\theta) \\ -\sin(\theta)\sin(\phi_{\text{geodetic}}) \\ \sin(\theta)\cos(\phi_{\text{geodetic}}) \end{bmatrix}, \tag{C.78}$$

$$\mathbf{1}_Z = \begin{bmatrix} 0 \\ \cos(\phi_{\text{geodetic}}) \\ \sin(\phi_{\text{geodetic}}) \end{bmatrix}. \tag{C.79}$$

C.3.6.4 NED/ECEF Coordinates It is more natural in some applications to use NED directions for locally level coordinates. This coordinate system coincides with vehicle-body-fixed RPY coordinates (shown in Fig. C.9) when the vehicle is level headed north. The unit vectors in local *north*, *east*, and *down* directions, as expressed in ECEF Cartesian coordinates, will be

$$\mathbf{1}_N = \begin{bmatrix} -\cos(\theta)\sin(\phi_{\text{geodetic}}) \\ -\sin(\theta)\sin(\phi_{\text{geodetic}}) \\ \cos(\phi_{\text{geodetic}}) \end{bmatrix}, \tag{C.80}$$

$$\mathbf{1}_E = \begin{bmatrix} -\sin(\theta) \\ \cos(\theta) \\ 0 \end{bmatrix}, \tag{C.81}$$

$$\mathbf{1}_D = \begin{bmatrix} -\cos(\theta)\cos(\phi_{\text{geodetic}}) \\ -\sin(\theta)\cos(\phi_{\text{geodetic}}) \\ -\sin(\phi_{\text{geodetic}}) \end{bmatrix}, \tag{C.82}$$

and the unit vectors in the ECEF X, Y, and Z directions, as expressed in NED coordinates, will be

$$\mathbf{1}_X = \begin{bmatrix} -\cos(\theta)\sin(\phi_{\text{geodetic}}) \\ -\sin(\theta) \\ -\cos(\theta)\cos(\phi_{\text{geodetic}}) \end{bmatrix}, \tag{C.83}$$

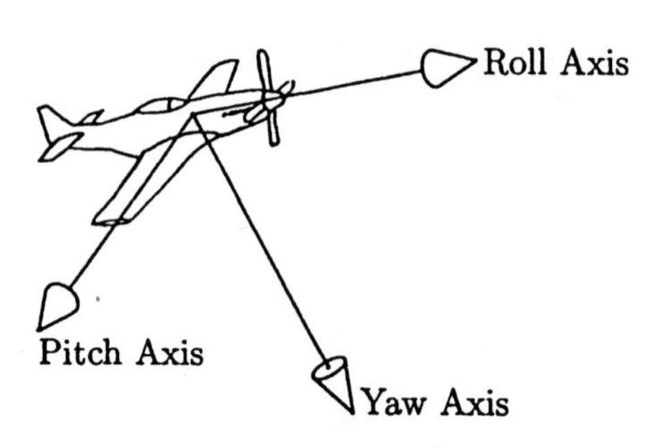

Fig. C.9 Roll-pitch-yaw axes.

$$\mathbf{1}_Y = \begin{bmatrix} -\sin(\theta)\sin(\phi_{\text{geodetic}}) \\ \cos(\theta) \\ -\sin(\theta)\cos(\phi_{\text{geodetic}}) \end{bmatrix}, \tag{C.84}$$

$$\mathbf{1}_Z = \begin{bmatrix} \cos(\phi_{\text{geodetic}}) \\ 0 \\ -\sin(\phi_{\text{geodetic}}) \end{bmatrix}, \tag{C.85}$$

C.3.7 RPY Coordinates

The RPY coordinates are vehicle fixed, with the roll axis in the nominal direction of motion of the vehicle, the pitch axis out the right-hand side, and the yaw axis such that turning to the right is positive, as illustrated in Fig.C.9. The same orientations of vehicle-fixed coordinates are used for surface ships and ground vehicles. They are also called "SAE coordinates," because they are the standard body-fixed coordinates used by the Society of Automotive Engineers.

For rocket boosters with their roll axes vertical at lift-off, the pitch axis is typically defined to be orthogonal to the plane of the boost trajectory (also called the "pitch plane" or "ascent plane").

C.3.8 Vehicle Attitude Euler Angles

The attitude of the vehicle body with respect to local coordinates can be specified in terms of rotations about the vehicle roll, pitch, and yaw axes, starting with these axes aligned with NED coordinates. The angles of rotation about each of these axes are called *Euler angles*, named for the Swiss mathematician Leonard Euler (1707–1783). It is always necessary to specify the order of rotations when specifying Euler (pronounced "oiler") angles.

A fairly common convention for vehicle attitude Euler angles is illustrated in Fig. C.10, where, starting with the vehicle level with roll axis pointed north:

1. *Yaw/Heading*. Rotate through the yaw angle (Y) about the vehicle yaw axis to the intended azimuth (heading) of the vehicle roll axis. Azimuth is measured clockwise (east) from north.

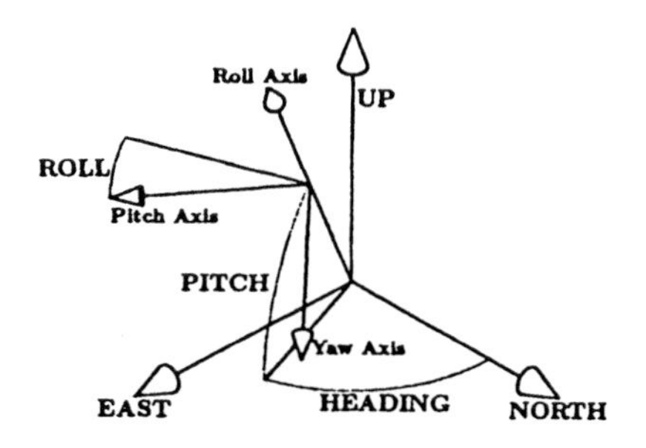

Fig. C.10 Vehicle Euler angles.

2. *Pitch.* Rotate through the pitch angle (P) about the vehicle pitch axis to bring the vehicle roll axis to its intended elevation. Elevation is measured positive upward from the local horizontal plane.
3. *Roll.* Rotate through the roll angle (R) about the vehicle roll axis to bring the vehicle attitude to the specified orientation.

Euler angles are redundant for vehicle attitudes with 90° pitch, in which case the roll axis is vertical. In that attitude, heading changes also rotate the vehicle about the roll axis. This is the attitude of most rocket boosters at lift-off. Some boosters can be seen making a roll maneuver immediately after lift-off to align their yaw axes with the launch azimuth in the ascent plane. This maneuver may be required to correct for launch delays on missions for which launch azimuth is a function of launch time.

C.3.8.1 RPY/ENU Coordinates With vehicle attitude specified by yaw angle (Y), pitch angle (P), and roll angle (R) as specified above, the resulting unit vectors of the roll, pitch, and yaw axes in ENU coordinates will be

$$\mathbf{1}_R = \begin{bmatrix} \sin(Y)\cos(P) \\ \cos(Y)\cos(P) \\ \sin(P) \end{bmatrix}, \tag{C.86}$$

$$\mathbf{1}_P = \begin{bmatrix} \cos(R)\cos(Y) + \sin(R)\sin(Y)\sin(P) \\ -\cos(R)\sin(Y) + \sin(R)\cos(Y)\sin(P) \\ -\sin(R)\cos(P) \end{bmatrix}, \tag{C.87}$$

$$\mathbf{1}_Y = \begin{bmatrix} -\sin(R)\cos(Y) + \cos(R)\sin(Y)\sin(P) \\ \sin(R)\sin(Y) + \cos(R)\cos(Y)\sin(P) \\ -\cos(R)\cos(P) \end{bmatrix}; \tag{C.88}$$

the unit vectors of the east, north, and up axes in RPY coordinates will be

$$\mathbf{1}_E = \begin{bmatrix} \sin(Y)\cos(P) \\ \cos(R)\cos(Y) + \sin(R)\sin(Y)\sin(P) \\ -\sin(R)\cos(Y) + \cos(R)\sin(Y)\sin(P) \end{bmatrix}, \tag{C.89}$$

$$\mathbf{1}_N = \begin{bmatrix} \cos(Y)\cos(P) \\ -\cos(R)\sin(Y) + \sin(R)\cos(Y)\sin(P) \\ \sin(R)\sin(Y) + \cos(R)\cos(Y)\sin(P) \end{bmatrix}, \tag{C.90}$$

$$\mathbf{1}_U = \begin{bmatrix} \sin(P) \\ -\sin(R)\cos(P) \\ -\cos(R)\cos(P) \end{bmatrix}; \tag{C.91}$$

and the coordinate transformation matrix from RPY coordinates to ENU coordinates will be

$$C_{\mathrm{ENU}}^{\mathrm{RPY}} = [\mathbf{1}_R \quad \mathbf{1}_P \quad \mathbf{1}_Y] = \begin{bmatrix} \mathbf{1}_E^{\mathrm{T}} \\ \mathbf{1}_N^{\mathrm{T}} \\ \mathbf{1}_U^{\mathrm{T}} \end{bmatrix} \tag{C.92}$$

$$= \begin{bmatrix} S_Y C_P & C_R C_Y + S_R S_Y S_P & -S_R C_Y + C_R S_Y S_P \\ C_Y C_P & -C_R S_Y + S_R C_Y S_P & S_R S_Y + C_R C_Y S_P \\ S_P & -S_R C_P & -C_R C_P \end{bmatrix}, \tag{C.93}$$

where

$$S_R = \sin(R), \tag{C.94}$$

$$C_R = \cos(R), \tag{C.95}$$

$$S_P = \sin(P), \tag{C.96}$$

$$C_P = \cos(P), \tag{C.97}$$

$$S_Y = \sin(Y), \tag{C.98}$$

$$C_Y = \cos(Y). \tag{C.99}$$

C.3.9 GPS Coordinates

The parameter Ω in Fig. C.12 is the RAAN, which is the ECI longitude where the orbital plane intersects the equatorial plane as the satellite crosses from the Southern Hemisphere to the Northern Hemisphere. The orbital plane is specified by Ω and α, the inclination of the orbit plane with respect to the equatorial plane ($\alpha \approx 55^\circ$ for GPS satellite orbits). The θ parameter represents the location of the satellite within the orbit plane, as the angular phase in the circular orbit with respect to ascending node.

For GPS satellite orbits, the angle θ changes at a nearly constant rate of about 1.4584×10^{-4} rad/s and a period of about 43,082s (half a day).

The nominal satellite position in ECEF coordinates is then given as

$$x = R[\cos\theta \cos\Omega - \sin\theta \sin\Omega \cos\alpha], \tag{C.100}$$

$$y = R[\cos\theta \cos\Omega + \sin\theta \sin\Omega \cos\alpha], \tag{C.101}$$

$$z = R \sin\theta \sin\alpha, \tag{C.102}$$

$$\theta = \theta_0 + (t - t_0)\frac{360}{43,082}\mathrm{deg}, \tag{C.103}$$

$$\Omega = \Omega_0 - (t - t_0)\frac{360}{86,164}\mathrm{deg}, \tag{C.104}$$

$$R = 26{,}560{,}000 \text{ m}. \tag{C.105}$$

GPS satellite positions in the transmitted navigation message are specified in the ECEF coordinate system of WGS 84. A locally level x^1, y^1, z^1 reference coordinate system (described in Section C.3.6) is used by an observer location on the earth, where the $x^1 - y^1$ plane is tangential to the surface of the earth, x^1 pointing east, y^1 pointing north, and z^1 normal to the plane. See Fig. C.11. Here,

$$X_{\mathrm{ENU}} = \mathbf{C}_{\mathrm{ENU}}^{\mathrm{ECEF}} X_{\mathrm{ECEF}} + \mathbf{S},$$

$$\mathbf{C}_{\mathrm{ENU}}^{\mathrm{ECEF}} = \text{coordinate transformation matrix from ECEF to ENU},$$

$$\mathbf{S} = \text{coordinate origin shift vector from ECEF to local reference},$$

$$\mathbf{C}_{\mathrm{ENU}}^{\mathrm{ECEF}} = \begin{bmatrix} -\sin\theta & \cos\theta & 0 \\ -\sin\phi\cos\theta & -\sin\phi\sin\theta & \cos\phi \\ \cos\phi\cos\theta & \cos\phi\sin\theta & \sin\phi \end{bmatrix},$$

$$S = \begin{bmatrix} X_U \sin\theta - Y_U \cos\theta \\ X_U \sin\phi\cos\theta + Y_U \sin\phi\cos\theta - Z_U \cos\phi \\ -X_U \cos\phi\cos\theta - Y_U \cos\phi\sin\theta - Z_U \sin\phi \end{bmatrix},$$

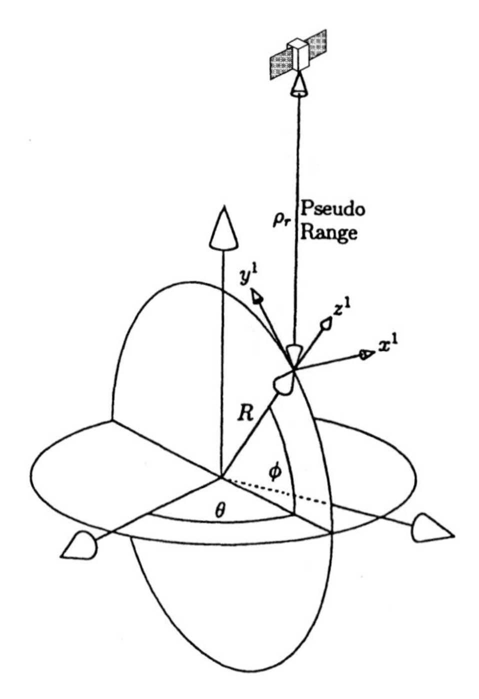

Fig. C.11 Pseudorange.

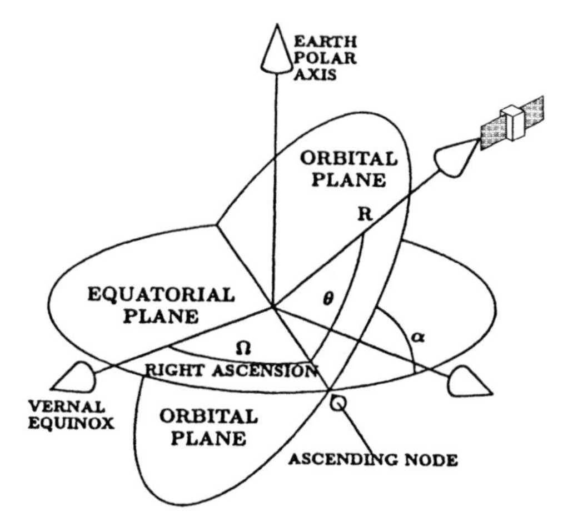

Fig. C.12 Satellite coordinates.

$$X_U, Y_U, Z_U = \text{user's position},$$
$$\theta = \text{local reference longitude},$$
$$\phi = \text{local geometric latitude}.$$

C.4 COORDINATE TRANSFORMATION MODELS

Coordinate transformations are methods for transforming a vector represented in one coordinate system into the appropriate representation in another coordinate system. These coordinate transformations can be represented in a number of different ways, each with its advantages and disadvantages.

These transformations generally involve translations (for coordinate systems with different origins) and rotations (for Cartesian coordinate systems with different axis directions) or transcendental transformations (between Cartesian and polar or geodetic coordinates). The transformations between Cartesian and polar coordinates have already been discussed in Section C.3.1 and translations are rather obvious, so we will concentrate on the rotations.

C.4.1 Euler Angles

Euler angles were used for defining vehicle attitude in Section C.3.8, and vehicle attitude representation is a common use of Euler angles in navigation.

Euler angles are used to define a coordinate transformation in terms of a set of three angular rotations, performed in a specified sequence about three specified orthogonal axes, to bring one coordinate frame to coincide with another.

The coordinate transformation from RPY coordinates to NED coordinates, for example, can be composed from three Euler rotation matrices:

$$C_{\text{NED}}^{\text{RPY}} = \overbrace{\begin{bmatrix} C_Y & -S_Y & 0 \\ S_Y & C_Y & 0 \\ 0 & 0 & 1 \end{bmatrix}}^{\text{Yaw}} \overbrace{\begin{bmatrix} C_P & 0 & S_P \\ 0 & 1 & 0 \\ -S_P & 0 & C_P \end{bmatrix}}^{\text{Pitch}} \overbrace{\begin{bmatrix} 1 & 0 & 0 \\ 0 & C_R & -S_R \\ 0 & S_R & C_R \end{bmatrix}}^{\text{Roll}} \tag{C.106}$$

$$= \underbrace{\begin{bmatrix} C_Y P_P & -S_Y C_R + C_Y S_P S_R & S_Y S_R + C_Y S_P C_R \\ S_Y C_P & C_Y C_R + S_Y S_P S_R & -C_Y S_R + S_Y S_P C_R \\ -S_P & C_P S_R & C_P C_R \\ (\text{rollaxis}) & (\text{pitchaxis}) & (\text{yawaxis}) \end{bmatrix}}_{\text{in NED coordinates}}, \tag{C.107}$$

where the matrix elements are defined in Eqs. C.94–C.99. This matrix also rotates the NED coordinate axes to coincide with RPY coordinate axes. (Compare this with the transformation from RPY to ENU coordinates in Eq. C.93.)

For example, the coordinate transformation for nominal booster rocket launch attitude (roll axis straight up) would be given by Eq. with pitch angle $P = \frac{1}{2}\pi$ ($C_P = 0$, $S_P = 1$), which becomes

$$\mathbf{C}_{\text{NED}}^{\text{RPY}} = \begin{bmatrix} 0 & \sin(R - Y) & \cos(R - Y) \\ 0 & \cos(R - Y) & -\sin(R - Y) \\ 1 & 10 & 0 \end{bmatrix}.$$

That is, the coordinate transformation in this attitude depends only on the difference between roll angle (R) and yaw angle (Y). Euler angles are a concise representation for vehicle attitude. They are handy for driving cockpit displays such as compass cards (using Y) and artificial horizon indicators (using R and P), but they are not particularly handy for representing vehicle attitude dynamics. The reasons for the latter include the following:

- Euler angles have discontinuities analogous to "gimbal lock" (Section 6.4.1.2) when the vehicle roll axis is pointed upward, as it is for launch of many rockets. In that orientation, tiny changes in vehicle pitch or yaw cause $\pm 180^\circ$ changes in heading angle. For aircraft, this creates a slewing rate problem for electromechanical compass card displays.
- The relationships between sensed body rates and Euler angle rates are mathematically complicated.

C.4.2 Rotation Vectors

All right-handed orthogonal coordinate systems with the same origins in three dimensions can be transformed one onto another by single rotations about fixed axes. The corresponding *rotation vectors* relating two coordinate systems are defined by the direction (rotation axis) and magnitude (rotation angle) of that transformation.

For example, the rotation vector for rotating ENU coordinates to NED coordinates (and vice versa) is

$$\boldsymbol{\rho}_{\text{NED}}^{\text{ENU}} = \begin{bmatrix} \pi/\sqrt{2} \\ \pi/\sqrt{2} \\ 0 \end{bmatrix}, \tag{C.108}$$

which has magnitude $|\boldsymbol{\rho}_{\text{NED}}^{\text{ENU}}| = \pi(180^\circ)$ and direction north—east, as illustrated in Fig. C.13. (For illustrative purposes only, NED coordinates are shown as being translated from ENU coordinates in Fig. C.13. In practice, rotation vectors represent pure rotations, without any translation.)

The rotation vector is another minimal representation of a coordinate transformation, along with Euler angles. Like Euler angles, rotation vectors are concise but also have some drawbacks:

1. It is not a unique representation, in that adding multiples of $\pm 2\pi$ to the magnitude of a rotation vector has no effect on the transformation it represents.
2. It is a nonlinear and rather complicated representation, in that the result of one rotation followed by another is a third rotation, the rotation vector for which is a fairly complicated function of the first two rotation vectors.

But, unlike Euler angles, rotation vector models do not exhibit "gimbal lock."

C.4.2.1 Rotation Vector to Matrix The rotation represented by a rotation vector

$$\boldsymbol{\rho} = \begin{bmatrix} \rho_1 \\ \rho_2 \\ \rho_3 \end{bmatrix} \tag{C.109}$$

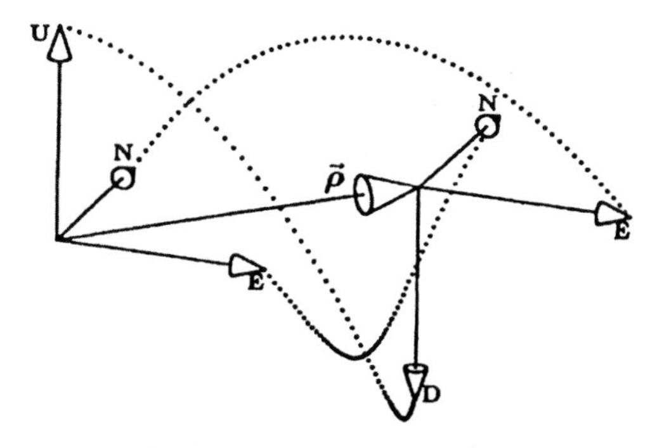

Fig. C.13 Rotation from ENU to NED coordinates.

can be implemented as multiplication by the matrix

$$\mathbf{C}(\boldsymbol{\rho}) \stackrel{\text{def}}{=} \exp(\boldsymbol{\rho}\otimes) \tag{C.110}$$

$$\stackrel{\text{def}}{=} \exp\left(\begin{bmatrix} 0 & -\rho_3 & \rho_2 \\ \rho_3 & 0 & -\rho_1 \\ -\rho_2 & \rho_1 & 0 \end{bmatrix}\right) \tag{C.111}$$

$$= \cos(|\boldsymbol{\rho}|)\mathbf{I}_3 + \frac{1-\cos(|\boldsymbol{\rho}|)}{|\boldsymbol{\rho}|^2}\boldsymbol{\rho}\boldsymbol{\rho}^{\mathrm{T}} + \frac{\sin(|\boldsymbol{\rho}|)}{|\boldsymbol{\rho}|}\begin{bmatrix} 0 & -\rho_3 & \rho_2 \\ \rho_3 & 0 & -\rho_1 \\ -\rho_2 & \rho_1 & 0 \end{bmatrix} \tag{C.112}$$

$$= \cos(\theta)\mathbf{I}_3 + (1-\cos(\theta))\mathbf{1}_\rho\mathbf{1}_\rho^{\mathrm{T}} + \sin(\theta)\begin{bmatrix} 0 & -u_3 & u_2 \\ u_3 & 0 & -u_1 \\ -u_2 & u_1 & 0 \end{bmatrix}, \tag{C.113}$$

$$\theta \stackrel{\text{def}}{=} |\boldsymbol{\rho}|, \tag{C.114}$$

$$\mathbf{1}_\rho \stackrel{\text{def}}{=} \frac{\boldsymbol{\rho}}{\boldsymbol{\rho}|}, \tag{C.115}$$

which was derived in Eq. B.17. That is, for any three-rowed column vector $\mathbf{v}$, $\mathbf{C}(\boldsymbol{\rho})\mathbf{v}$ rotates it through an angle of $|\boldsymbol{\rho}|$ radians about the vector $\boldsymbol{\rho}$.

The form of the matrix in Eq. C.113[2] is better suited for computation when $\theta \approx 0$, but the form of the matrix in Eq. C.112 is useful for computing sensitivities using partial derivatives (used in Chapter 8).

For example, the rotation vector $\rho_{\mathrm{NED}}^{\mathrm{ENU}}$ in Eq. C.108 transforming between ENU and NED has magnitude and direction

$$\theta = \pi \qquad [\sin(\theta) = 0, \cos(\theta) = -1],$$

$$\mathbf{1}_\rho = \begin{bmatrix} 1/\sqrt{2} \\ 1/\sqrt{2} \\ 0 \end{bmatrix},$$

respectively, and the corresponding rotation matrix

$$\mathbf{C}_{\mathrm{NED}}^{\mathrm{ENU}} = \cos(\pi)\mathbf{I}_3 + [1-\cos(\pi)]\mathbf{1}_\rho\mathbf{1}_\rho + \sin(\pi)\begin{bmatrix} 0 & -u_3 & u_2 \\ u_3 & 0 & -u_1 \\ -u_2 & u_1 & 0 \end{bmatrix}$$

$$= -\mathbf{I}_3 + 2\mathbf{1}_\rho\mathbf{1}_\rho^{\mathrm{T}} + 0$$

[2]Linear combinations of the sort $a_1\mathbf{I}_{3\times 3} + a_2[\mathbf{1}_\rho\otimes] + a_3\mathbf{1}_\rho\mathbf{1}_\rho^{\mathrm{T}}$, where $\mathbf{1}$ is a unit vector, form a subalgebra of 3×3 matrices with relatively simple rules for multiplication, inversion, etc.

$$= \begin{bmatrix} -1 & 0 & 0 \\ 0 & -1 & 0 \\ 0 & 0 & -1 \end{bmatrix} + \begin{bmatrix} 1 & 1 & 0 \\ 1 & 1 & 0 \\ 0 & 0 & 0 \end{bmatrix}$$

$$= \begin{bmatrix} 0 & 1 & 0 \\ 1 & 0 & 0 \\ 0 & 0 & -1 \end{bmatrix}$$

transforms from ENU to NED coordinates. (Compare this result to Eq. C.73.) Because coordinate transformation matrices are orthogonal matrices and the matrix $\mathbf{C}_{\text{NED}}^{\text{ENU}}$ is also symmetric, $\mathbf{C}_{\text{NED}}^{\text{ENU}}$ is its own inverse. That is,

$$\mathbf{C}_{\text{NED}}^{\text{ENU}} = \mathbf{C}_{\text{ENU}}^{\text{NED}}. \tag{C.116}$$

C.4.2.2 ***Matrix to Rotation Vector*** Although there is a unique coordinate transformation matrix for each rotation vector, the converse is not true. Adding multiples of 2π to the magnitude of a rotation vector has no effect on the resulting coordinate transformation matrix. The following approach yields a unique rotation vector with magnitude $|\boldsymbol{\rho}| \leq \pi$.

The trace tr($\mathbf{C}$) of a square matrix $\mathbf{M}$ is the sum of its diagonal values. For the coordinate transformation matrix of Eq. C.112,

$$\text{tr}[\mathbf{C}(\boldsymbol{\rho})] = 1 + 2\cos(\theta), \tag{C.117}$$

from which the rotation angle

$$|\boldsymbol{\rho}| = \theta \tag{C.118}$$

$$= \arccos\left(\frac{\text{tr}[\mathbf{C}(\boldsymbol{\rho})] - 1}{2}\right), \tag{C.119}$$

a formula that will yield a result in the range $0 < \theta < \pi$, but with poor fidelity near where the derivative of the cosine equals zero at $\theta = 0$ and $\theta = \pi$.

The values of θ near $\theta = 0$ and $\theta = \pi$ can be better estimated using the sine of θ, which can be recovered using the antisymmetric part of $\mathbf{C}(\boldsymbol{\rho})$,

$$\mathbf{A} = \begin{bmatrix} 0 & -a_{21} & a_{13} \\ a_{21} & 0 & -a_{32} \\ -a_{13} & a_{32} & 0 \end{bmatrix} \tag{C.120}$$

$$\stackrel{\text{def}}{=} \tfrac{1}{2}[\mathbf{C}(\boldsymbol{\rho}) - \mathbf{C}^{\text{T}}(\boldsymbol{\rho})] \tag{C.121}$$

$$= \frac{\sin(\theta)}{\theta} \begin{bmatrix} 0 & -\rho_3 & \rho_2 \\ \rho_3 & 0 & -\rho_1 \\ -\rho_2 & \rho_1 & 0 \end{bmatrix}, \tag{C.122}$$

from which the vector

$$\begin{bmatrix} a_{32} \\ a_{13} \\ a_{21} \end{bmatrix} = \sin(\theta)\frac{1}{|\boldsymbol{\rho}|}\boldsymbol{\rho} \tag{C.123}$$

will have magnitude

$$\sqrt{a_{32}^2 + a_{13}^2 + a_{21}^2} = \sin(\theta) \tag{C.124}$$

and the same direction as $\boldsymbol{\rho}$. As a consequence, one can recover the magnitude θ of $\boldsymbol{\rho}$ from

$$\theta = \text{atan2}\left(\sqrt{a_{32}^2 + a_{13}^2 + a_{21}^2}, \frac{\text{tr}[\mathbf{C}(\boldsymbol{\rho})] - 1}{2}\right) \tag{C.125}$$

using the MATLAB function atan2, and then the rotation vector $\boldsymbol{\rho}$ as

$$\boldsymbol{\rho} = \frac{\theta}{\sin(\theta)}\begin{bmatrix} a_{32} \\ a_{13} \\ a_{21} \end{bmatrix} \tag{C.126}$$

when $0 < \theta < \pi$.

C.4.2.3 *Special Cases for* $\sin(\theta) \approx 0$ For $\theta \approx 0$, $\boldsymbol{\rho} \approx 0$, although Eq. C.126 may still work adequately for $\theta > 10^{-6}$, say.

For $\theta \approx \pi$, the symmetric part of $\mathbf{C}(\boldsymbol{\rho})$,

$$\mathbf{S} = \begin{bmatrix} s_{11} & s_{12} & s_{13} \\ s_{12} & s_{22} & s_{23} \\ s_{13} & s_{23} & s_{33} \end{bmatrix} \tag{C.127}$$

$$\stackrel{\text{def}}{=} \tfrac{1}{2}[\mathbf{C}(\boldsymbol{\rho}) + \mathbf{C}^{\mathrm{T}}(\boldsymbol{\rho})] \tag{C.128}$$

$$= \cos(\theta)\mathbf{I}_3 + \frac{1 - \cos(\theta)}{\theta_2}\boldsymbol{\rho}\boldsymbol{\rho}^{\mathrm{T}} \tag{C.129}$$

$$\approx -\mathbf{I}_3 + \frac{2}{\theta^2}\boldsymbol{\rho}\boldsymbol{\rho}^{\mathrm{T}} \tag{C.130}$$

and the unit vector

$$\mathbf{1}_\rho \stackrel{\text{def}}{=} \frac{1}{\theta}\boldsymbol{\rho} \tag{C.131}$$

satisfies

$$\mathbf{S} \approx \begin{bmatrix} 2u_1^2 - 1 & 2u_1u_2 & 2u_1u_3 \\ 2u_1u_2 & 2u_2^2 - 1 & 2u_2u_3 \\ 2u_1u_3 & 2u_2u_3 & 2u_3^2 - 1 \end{bmatrix}, \tag{C.132}$$

which can be solved for a unique $\mathbf{u}$ by assigning $u_k > 0$ for

$$k = \operatorname{argmax}\left(\begin{bmatrix} s_{11} \\ s_{22} \\ s_{33} \end{bmatrix}\right), \tag{C.133}$$

$$u_k = \sqrt{\tfrac{1}{2}(s_{kk} + 1)} \tag{C.134}$$

then, depending on whether $k = 1$, $k = 2$, or $k = 3$,

$$\left.\begin{array}{cccc} & k = 1 & k = 2 & k = 3 \\ u_1 \approx & \sqrt{\dfrac{s_{11}+1}{2}} & \dfrac{s_{12}}{2u_2} & \dfrac{s_{13}}{2u_3} \\ u_2 \approx & \dfrac{s_{12}}{2u_1} & \sqrt{\dfrac{s_{22}+1}{2}} & \dfrac{s_{23}}{2u_2} \\ u_3 \approx & \dfrac{s_{13}}{2u_1} & \dfrac{s_{23}}{2u_2} & \sqrt{\dfrac{s_{11}+1}{2}} \end{array}\right\} \tag{C.135}$$

and

$$\boldsymbol{\rho} = \theta \begin{bmatrix} u_1 \\ u_2 \\ u_3 \end{bmatrix}. \tag{C.136}$$

C.4.2.4 Time Derivatives of Rotation Vectors The mathematical relationships between rotation rates ω_k and the time derivatives of the corresponding rotation vector $\boldsymbol{\rho}$ are fairly complicated, but they can be derived from Eq. C.221 for the dynamics of coordinate transformation matrices.

Let $\boldsymbol{\rho}_{\text{ENU}}$ be the rotation vector represented in earth-fixed ENU coordinates that rotates earth-fixed ENU coordinate axes into vehicle body-fixed RPY axes, and let $\mathbf{C}(\boldsymbol{\rho})$ be the corresponding rotation matrix, so that, in ENU coordinates,

$$\mathbf{1}_E = [\ 1 \quad 0 \quad 0\ ]^{\mathrm{T}}, \quad \mathbf{1}_N = [\ 0 \quad 1 \quad 0\ ]^{\mathrm{T}}, \quad \mathbf{1}_U = [\ 0 \quad 0 \quad 1\ ]^{\mathrm{T}},$$

$$\mathbf{C}(\boldsymbol{\rho}_{\text{ENU}})\mathbf{1}_E = \mathbf{1}_R, \qquad \mathbf{C}(\boldsymbol{\rho}_{\text{ENU}})\mathbf{I}_N = \mathbf{1}_P, \qquad \mathbf{C}(\boldsymbol{\rho}_{\text{ENU}})\mathbf{1}_U = \mathbf{1}_Y,$$

$$\begin{aligned}
\mathbf{C}_{\text{ENU}}^{\text{RPY}} &= [\ \mathbf{1}_R \quad \mathbf{1}_P \quad \mathbf{1}_Y\ ], \\
&= [\mathbf{C}(\boldsymbol{\rho}_{\text{ENU}})\mathbf{1}_E \quad \mathbf{C}(\boldsymbol{\rho}_{\text{ENU}})\mathbf{1}_N \quad \mathbf{C}(\boldsymbol{\rho}_{\text{ENU}})\mathbf{1}_U] \\
&= \mathbf{C}(\boldsymbol{\rho}_{\text{ENU}})[\ \mathbf{1}_E \quad \mathbf{1}_N \quad \mathbf{1}_U\ ] \qquad (C.137) \\
&= \mathbf{C}(\boldsymbol{\rho}_{\text{ENU}}) \begin{bmatrix} 1 & 0 & 0 \\ 0 & 1 & 0 \\ 0 & 0 & 1 \end{bmatrix}
\end{aligned}$$

$$\mathbf{C}_{\text{ENU}}^{\text{RPY}} = \mathbf{C}(\boldsymbol{\rho}_{\text{ENU}}). \qquad (C.138)$$

That is, $\mathbf{C}(\boldsymbol{\rho}_{\text{ENU}})$ is the coordinate transformation matrix from RPY coordinates to ENU coordinates. As a consequence, from Eq. C.221,

$$\frac{d}{dt}\mathbf{C}(\boldsymbol{\rho}_{\text{ENU}}) = \frac{d}{dt}\mathbf{C}_{\text{ENU}}^{\text{RPY}} \qquad (C.139)$$

$$= \begin{bmatrix} 0 & \omega_U & -\omega_N \\ -\omega_U & 0 & \omega_E \\ \omega_N & -\omega_E & 0 \end{bmatrix} \mathbf{C}_{\text{ENU}}^{\text{RPY}} + \mathbf{C}_{\text{ENU}}^{\text{RPY}} \begin{bmatrix} 0 & -\omega_Y & \omega_P \\ \omega_Y & 0 & -\omega_R \\ -\omega_P & \omega_R & 0 \end{bmatrix}, \qquad (C.140)$$

$$\frac{d}{dt}\mathbf{C}(\boldsymbol{\rho}_{\text{ENU}}) = \begin{bmatrix} 0 & \omega_U & -\omega_N \\ -\omega_U & 0 & \omega_E \\ \omega_N & -\omega_E & 0 \end{bmatrix} \mathbf{C}(\boldsymbol{\rho}_{\text{ENU}}) + \mathbf{C}(\boldsymbol{\rho}_{\text{ENU}}) \begin{bmatrix} 0 & -\omega_Y & \omega_P \\ \omega_Y & 0 & -\omega_R \\ -\omega_P & \omega_R & 0 \end{bmatrix}, \qquad (C.141)$$

where

$$\boldsymbol{\omega}_{\text{RPY}} = \begin{bmatrix} \omega_R \\ \omega_P \\ \omega_Y \end{bmatrix} \qquad (C.142)$$

is the vector of inertial rotation rates of the vehicle body, expressed in RPY coordinates, and

$$\boldsymbol{\omega}_{\text{ENU}} = \begin{bmatrix} \omega_E \\ \omega_N \\ \omega_U \end{bmatrix} \qquad (C.143)$$

is the vector of inertial rotation rates of the ENU coordinate frame, expressed in ENU coordinates.

The 3×3 matrix equation C.141 is equivalent to nine scalar equations:

$$\frac{\partial c_{11}}{\partial \rho_E}\dot{\rho}_E + \frac{\partial c_{11}}{\partial \rho_N}\dot{\rho}_N + \frac{\partial c_{11}}{\partial \rho_U}\dot{\rho}_U = -c_{1,3}\omega_P + c_{1,2}\omega_Y - c_{3,1}\omega_N + c_{2,1}\omega_U,$$

$$\frac{\partial c_{12}}{\partial \rho_E}\dot{\rho}_E + \frac{\partial c_{12}}{\partial \rho_N}\dot{\rho}_N + \frac{\partial c_{12}}{\partial \rho_U}\dot{\rho}_U = c_{1,3}\omega_R - c_{1,1}\omega_Y - c_{3,2}\omega_N + c_{2,2}\omega_U,$$

$$\frac{\partial c_{13}}{\partial \rho_E}\dot{\rho}_E + \frac{\partial c_{13}}{\partial \rho_N}\dot{\rho}_N + \frac{\partial c_{13}}{\partial \rho_U}\dot{\rho}_U = -c_{1,2}\omega_R + c_{1,1}\omega_P - c_{3,3}\omega_N + c_{2,3}\omega_U,$$

$$\frac{\partial c_{21}}{\partial \rho_E}\dot{\rho}_E + \frac{\partial c_{21}}{\partial \rho_N}\dot{\rho}_N + \frac{\partial c_{21}}{\partial \rho_U}\dot{\rho}_U = -c_{2,3}\omega_P + c_{2,2}\omega_Y + c_{3,1}\omega_E - c_{1,1}\omega_U,$$

$$\frac{\partial c_{22}}{\partial \rho_E}\dot{\rho}_E + \frac{\partial c_{22}}{\partial \rho_N}\dot{\rho}_N + \frac{\partial c_{22}}{\partial \rho_U}\dot{\rho}_U = c_{2,3}\omega_R - c_{2,1}\omega_Y + c_{3,2}\omega_E - c_{1,2}\omega_U,$$

$$\frac{\partial c_{23}}{\partial \rho_E}\dot{\rho}_E + \frac{\partial c_{23}}{\partial \rho_N}\dot{\rho}_N + \frac{\partial c_{23}}{\partial \rho_U}\dot{\rho}_U = -c_{2,2}\omega_R + c_{2,1}\omega_P + c_{3,3}\omega_E - c_{1,3}\omega_U,$$

$$\frac{\partial c_{31}}{\partial \rho_E}\dot{\rho}_E + \frac{\partial c_{31}}{\partial \rho_N}\dot{\rho}_N + \frac{\partial c_{31}}{\partial \rho_U}\dot{\rho}_U = -c_{3,3}\omega_P + c_{3,2}\omega_Y - c_{2,1}\omega_E + c_{1,1}\omega_N,$$

$$\frac{\partial c_{32}}{\partial \rho_E}\dot{\rho}_E + \frac{\partial c_{32}}{\partial \rho_N}\dot{\rho}_N + \frac{\partial c_{32}}{\partial \rho_U}\dot{\rho}_U = c_{3,3}\omega_R - c_{3,1}\omega_Y - c_{2,2}\omega_E + c_{1,2}\omega_N,$$

$$\frac{\partial c_{33}}{\partial \rho_E}\dot{\rho}_E + \frac{\partial c_{33}}{\partial \rho_N}\dot{\rho}_N + \frac{\partial c_{33}}{\partial \rho_U}\dot{\rho}_U = -c_{3,2}\omega_R + c_{3,1}\omega_P - c_{2,3}\omega_E + c_{1,3}\omega_N,$$

where

$$\begin{bmatrix} c_{11} & c_{12} & c_{13} \\ c_{21} & c_{22} & c_{23} \\ c_{31} & c_{32} & c_{33} \end{bmatrix} \stackrel{\text{def}}{=} \mathbf{C}(\boldsymbol{\rho}_{\text{ENU}})$$

and the partial derivatives

$$\frac{\partial c_{11}}{\partial \rho_E} = \frac{u_E(1-u_E^2)\{2[1-\cos(\theta)] - \theta\sin(\theta)\}}{\theta},$$

$$\frac{\partial c_{11}}{\partial \rho_N} = \frac{u_N\{-2u_E^2[1-\cos(\theta)] - \theta\sin(\theta)(1-u_E^2)\}}{\theta},$$

$$\frac{\partial c_{11}}{\partial \rho_U} = \frac{u_U\{-2u_E^2[1-\cos(\theta)] - \theta\sin(\theta)(1-u_E^2)\}}{\theta},$$

$$\frac{\partial c_{12}}{\partial \rho_E} = \frac{u_N(1-2u_E^2)[1-\cos(\theta)] + u_E u_U\sin(\theta) - \theta u_E u_U\cos(\theta) + \theta u_N u_E^2\sin(\theta)}{\theta},$$

$$\frac{\partial c_{12}}{\partial \rho_N} = \frac{u_E(1-2u_N^2)[1-\cos(\theta)] + u_U u_N\sin(\theta) - \theta u_N u_U\cos(\theta) + \theta u_E u_N^2\sin(\theta)}{\theta},$$

$$\frac{\partial c_{12}}{\partial \rho_U} = \frac{-2u_E u_N u_U[1-\cos(\theta)] - (1-u_U^2)\sin(\theta) - \theta u_U^2 \cos(\theta) + \theta u_U u_N u_E \sin(\theta)}{\theta},$$

$$\frac{\partial c_{13}}{\partial \rho_E} = \frac{u_U(1-2u_E^2)[1-\cos(\theta)] - u_E u_N \sin(\theta) + \theta u_E u_N \cos(\theta) + \theta u_U u_E^2 \sin(\theta)}{\theta},$$

$$\frac{\partial c_{13}}{\partial \rho_N} = \frac{-2u_E u_N u_U[1-\cos(\theta)] + (1-u_N^2)\sin(\theta) + \theta u_N^2 \cos(\theta) + \theta u_U u_N u_E \sin(\theta)}{\theta},$$

$$\frac{\partial c_{13}}{\partial \rho_U} = \frac{u_E(1-2u_U^2)[1-\cos(\theta)] - u_U u_N \sin(\theta) + \theta u_N u_U \cos(\theta) + \theta u_E u_U^2 \sin(\theta)}{\theta},$$

$$\frac{\partial c_{21}}{\partial \rho_E} = \frac{u_N(1-2u_E^2)[1-\cos(\theta)] - u_E u_U \sin(\theta) + \theta u_E u_U \cos(\theta) + \theta u_N u_E^2 \sin(\theta)}{\theta},$$

$$\frac{\partial c_{21}}{\partial \rho_N} = \frac{u_E(1-2u_N^2)[1-\cos(\theta)] - u_U u_N \sin(\theta) + \theta u_N u_U \cos(\theta) + \theta u_E u_N^2 \sin(\theta)}{\theta},$$

$$\frac{\partial c_{21}}{\partial \rho_U} = \frac{-2u_E u_N u_U[1-\cos(\theta)] + \sin(\theta)(1-u_U^2) + \theta u_U^2 \cos(\theta) + \theta u_U u_N u_E \sin(\theta)}{\theta},$$

$$\frac{\partial c_{22}}{\partial \rho_E} = \frac{u_E\{-2u_N^2[1-\cos(\theta)] - \theta(1-u_N^2)\sin(\theta)\}}{\theta},$$

$$\frac{\partial c_{22}}{\partial \rho_N} = \frac{u_N(1-u_N^2)\{2[1-\cos(\theta)] - \theta\sin(\theta)\}}{\theta},$$

$$\frac{\partial c_{22}}{\partial \rho_U} = \frac{u_U\{-2u_N^2[1-\cos(\theta)] - \theta(1-u_N^2)\sin(\theta)\}}{\theta},$$

$$\frac{\partial c_{23}}{\partial \rho_E} = \frac{-2u_E u_N u_U[1-\cos(\theta)] - (1-u_E^2)\sin(\theta) - \theta u_E^2 \cos(\theta) + \theta u_E u_N u_U \sin(\theta)}{\theta},$$

$$\frac{\partial c_{23}}{\partial \rho_N} = \frac{u_U(1-2u_N^2)[1-\cos(\theta)] + u_E u_N \sin(\theta) - \theta u_E u_N \cos(\theta) + \theta u_N^2 u_U \sin(\theta)}{\theta},$$

$$\frac{\partial c_{23}}{\partial \rho_U} = \frac{u_N(1-2u_U^2)[1-\cos(\theta)] + u_E u_U \sin(\theta) - \theta u_E u_U \cos(\theta) + \theta u_U^2 u_N \sin(\theta)}{\theta},$$

$$\frac{\partial c_{31}}{\partial \rho_E} = \frac{u_U(1-2u_E^2)[1-\cos(\theta)] + u_E u_N \sin(\theta) - \theta u_E u_N \cos(\theta) + \theta u_U u_E^2 \sin(\theta)}{\theta},$$

$$\frac{\partial c_{31}}{\partial \rho_N} = \frac{-2u_E u_N u_U[1-\cos(\theta)] - (1-u_N^2)\sin(\theta) - \theta u_N^2 \cos(\theta) + \theta u_U u_N u_E \sin(\theta)}{\theta},$$

$$\frac{\partial c_{31}}{\partial \rho_U} = \frac{u_E(1-2u_U^2)[1-\cos(\theta)] + u_U u_N \sin(\theta) - \theta u_N u_U \cos(\theta) + \theta u_E u_U^2 \sin(\theta)}{\theta},$$

$$\frac{\partial c_{32}}{\partial \rho_E} = \frac{-2u_E u_N u_U[1-\cos(\theta)] + (1-u_E^2)\sin(\theta) + \theta u_E^2 \cos(\theta) + \theta u_U u_N u_E \sin(\theta)}{\theta},$$

$$\frac{\partial c_{32}}{\partial \rho_N} = \frac{u_U(1-2u_N^2)[1-\cos(\theta)] - u_E u_N \sin(\theta) + \theta u_E u_N \cos(\theta) + \theta u_N^2 u_U \sin(\theta)}{\theta},$$

$$\frac{\partial c_{32}}{\partial \rho_U} = \frac{u_N(1-2u_U^2)[1-\cos(\theta)] - u_E u_U \sin(\theta) + \theta u_E u_U \cos(\theta) + \theta u_U^2 u_N \sin(\theta)}{\theta},$$

$$\frac{\partial c_{33}}{\partial \rho_E} = \frac{u_E\{-2u_U^2[1-\cos(\theta)] - \theta \sin(\theta)(1+u_U^2)\}}{\theta},$$

$$\frac{\partial c_{33}}{\partial \rho_N} = \frac{u_N\{-2u_U^2[1-\cos(\theta)] - \theta \sin(\theta)(1+u_U^2)\}}{\theta},$$

$$\frac{\partial c_{33}}{\partial \rho_U} = \frac{u_U(1-u_U^2)\{2[1-\cos(\theta)] - \theta \sin(\theta)\}}{\theta}$$

for

$$\theta \stackrel{\text{def}}{=} |\boldsymbol{\rho}_{\text{ENU}}|,$$

$$u_E \stackrel{\text{def}}{=} \frac{\rho_E}{\theta}, \qquad u_N \stackrel{\text{def}}{=} \frac{\rho_N}{\theta}, \qquad u_u \stackrel{\text{def}}{=} \frac{\rho_U}{\theta}.$$

These nine scalar linear equations can be put into matrix form and solved in least squares fashion as

$$\mathbf{L} \begin{bmatrix} \dot{\rho}_E \\ \dot{\rho}_N \\ \dot{\rho}_U \end{bmatrix} = \mathbf{R} \begin{bmatrix} \omega_R \\ \omega_P \\ \omega_Y \\ \omega_E \\ \omega_N \\ \omega_U \end{bmatrix}, \tag{C.144}$$

$$\begin{bmatrix} \dot{\rho}_E \\ \dot{\rho}_N \\ \dot{\rho}_U \end{bmatrix} = \underbrace{[\mathbf{L}^{\mathrm{T}}\mathbf{L}] \backslash [\mathbf{L}^{\mathrm{T}}\mathbf{R}]}_{\partial \dot{\boldsymbol{\rho}}/\partial \boldsymbol{\omega}} \begin{bmatrix} \boldsymbol{\omega}_{\text{RPY}} \\ \boldsymbol{\omega}_{\text{ENU}} \end{bmatrix}. \tag{C.145}$$

The matrix product $\mathbf{L}^T\mathbf{L}$ will always be invertible because its determinant

$$\det[\mathbf{L}^{\mathrm{T}}\mathbf{L}] = 32\frac{[1-\cos(\theta)]^2}{\theta^4}, \tag{C.146}$$

$$\lim_{\theta \to 0} \det[\mathbf{L}^{\mathrm{T}}\mathbf{L}] = 8, \tag{C.147}$$

and the resulting equation for $\boldsymbol{\rho}_{\text{ENU}}$ can be put into the form

$$\dot{\boldsymbol{\rho}}_{\text{ENU}} = \left[\frac{\partial \dot{\boldsymbol{\rho}}}{\partial \boldsymbol{\omega}}\right] \begin{bmatrix} \boldsymbol{\omega}_{\text{RPY}} \\ \boldsymbol{\omega}_{\text{ENU}} \end{bmatrix}. \tag{C.148}$$

The 3×6 matrix $\partial \boldsymbol{\rho}/\partial \boldsymbol{\omega}$ can be partitioned as

$$\left[\frac{\partial \dot{\boldsymbol{\rho}}}{\partial \boldsymbol{\omega}}\right] = \left[\frac{\partial \dot{\boldsymbol{\rho}}}{\partial \boldsymbol{\omega}_{\text{RPY}}} \middle| \frac{\partial \dot{\boldsymbol{\rho}}}{\partial \boldsymbol{\omega}_{\text{ENU}}}\right] \tag{C.149}$$

with 3×3 submatrices

$$\frac{\partial \boldsymbol{\rho}}{\partial \boldsymbol{\omega}_{\mathrm{RPY}}} = \left[\frac{1}{|\boldsymbol{\rho}|^2} - \frac{\sin(|\boldsymbol{\rho}|)}{2|\boldsymbol{\rho}|[1-\cos(|\boldsymbol{\rho}|)]}\right]\boldsymbol{\rho}\boldsymbol{\rho}^{\mathrm{T}} + \frac{|\boldsymbol{\rho}|\sin(|\boldsymbol{\rho}|)}{2[1-\cos(|\boldsymbol{\rho}|)]}\mathbf{I} + \frac{1}{2}[\boldsymbol{\rho}\otimes] \tag{C.150}$$

$$= \mathbf{1}_\rho \mathbf{1}_\rho^{\mathrm{T}} + \frac{\theta \sin(\theta)}{2[1-\cos(\theta)]}[\mathbf{I} - \mathbf{1}_\rho \mathbf{1}\rho^{\mathrm{T}}] + \frac{\theta}{2}[\mathbf{1}_\rho\otimes], \tag{C.151}$$

$$\lim_{P\to 0} |\boldsymbol{\rho} \frac{P\partial\dot{\boldsymbol{\rho}}}{\partial \boldsymbol{\omega}_{\mathrm{RPY}}} = \mathbf{I}, \tag{C.152}$$

$$\frac{\partial \dot{\boldsymbol{\rho}}}{\partial \boldsymbol{\omega}_{\mathrm{ENU}}} = -\left[\frac{1}{|\boldsymbol{\rho}|^2} - \frac{\sin(|\boldsymbol{\rho}|)}{2|\boldsymbol{\rho}|[1-\cos(|\boldsymbol{\rho}|)]}\right]\boldsymbol{\rho}\boldsymbol{\rho}^{\mathrm{T}} - \frac{|\boldsymbol{\rho}|\sin(|\boldsymbol{\rho}|)}{2[1-\cos(|\boldsymbol{\rho}|)]}\mathbf{I} + \frac{1}{2}[\boldsymbol{\rho}\otimes]4 \tag{C.153}$$

$$= -\mathbf{1}_\rho \mathbf{1}_\rho^{\mathrm{T}} - \frac{\theta \sin(\theta)}{2[1-\cos(\theta)]}[\mathbf{I} - \mathbf{1}_\rho \mathbf{1}_\rho^{\mathrm{T}}] + \frac{\theta}{\mathbf{2}}[\mathbf{1}_\rho\otimes]. \tag{C.154}$$

$$\lim_{|\boldsymbol{\rho}|\to 0} \frac{\partial \dot{\boldsymbol{\rho}}}{\partial \boldsymbol{\omega}_{\mathrm{ENU}}} = -\mathbf{I}. \tag{C.155}$$

For locally leveled gimbaled systems, $\boldsymbol{\omega}_{\mathrm{RPH}} = \mathbf{0}$. That is, the gimbals normally keep the accelerometer axes aligned to the ENU or NED coordinate axes, a process modeled by $\boldsymbol{\omega}_{\mathrm{ENU}}$ alone.

C.4.2.5 Time Derivatives of Matrix Expressions The Kalman filter implementation for integrating GPS with a strapdown INS in Chapter 8 will require derivatives with respect to time of the matrices

$$\frac{\partial \dot{\boldsymbol{\rho}}_{\mathrm{ENU}}}{\partial \boldsymbol{\omega}_{\mathrm{RPH}}} \quad \text{(Eq.C.150)} \quad \text{and} \quad \frac{\partial \dot{\boldsymbol{\rho}}_{\mathrm{ENU}}}{\partial \boldsymbol{\omega}_{\mathrm{ENU}}} \quad \text{(Eq. C.153)}.$$

We derive here a general-purpose formula for taking such derivatives and then apply it to these two cases.

General Formulas There is a general-purpose formula for taking the time derivatives $(d/dt)\mathbf{M}(\boldsymbol{\rho})$ of matrix expressions of the sort

$$\mathbf{M}(\boldsymbol{\rho}) = \mathbf{M}(s_1(\boldsymbol{\rho}), s_2(\boldsymbol{\rho}), s_3(\boldsymbol{\rho})) \tag{C.156}$$

$$= s_1(\boldsymbol{\rho})\mathbf{I}_3 + s_2(\boldsymbol{\rho})[\boldsymbol{\rho}\otimes] + s_3(\boldsymbol{\rho})\boldsymbol{\rho}\boldsymbol{\rho}^{\mathrm{T}}, \tag{C.157}$$

that is, as linear combinations of $\mathbf{I}_3$, $\boldsymbol{\rho}\otimes$, and $\boldsymbol{\rho}\boldsymbol{\rho}^{\mathrm{T}}$ with scalar functions of $\boldsymbol{\rho}$ as the coefficients.

The derivation uses the time derivatives of the basis matrices,

$$\frac{d}{dt}\mathbf{I}_3 = \mathbf{0}_3, \tag{C.158}$$

$$\frac{d}{dt}[\boldsymbol{\rho}\otimes] = [\dot{\boldsymbol{\rho}}\otimes], \tag{C.159}$$

$$\frac{d}{dt}\boldsymbol{\rho}\boldsymbol{\rho}^{\mathrm{T}} = \dot{\boldsymbol{\rho}}\boldsymbol{\rho}^{\mathrm{T}} + \boldsymbol{\rho}\dot{\boldsymbol{\rho}}^{\mathrm{T}}, \tag{C.160}$$

where the vector

$$\dot{\boldsymbol{\rho}} = \frac{d}{dt}\boldsymbol{\rho}, \tag{C.161}$$

and then uses the chain rule for differentiation to obtain the general formula

$$\begin{aligned}\frac{d}{dt}\mathbf{M}(\boldsymbol{\rho}) = {} & \frac{\partial s_1(\boldsymbol{\rho})}{\partial \boldsymbol{\rho}}\dot{\boldsymbol{\rho}}\mathbf{I}_3 + \frac{\partial s_2(\boldsymbol{\rho})}{\partial \boldsymbol{\rho}}\dot{\boldsymbol{\rho}}[\boldsymbol{\rho}\otimes] + s_2(\boldsymbol{\rho})[\dot{\boldsymbol{\rho}}\otimes], \\ & + \frac{\partial s_3(\boldsymbol{\rho})}{\partial \boldsymbol{\rho}}\dot{\boldsymbol{\rho}}[\boldsymbol{\rho}\boldsymbol{\rho}^{\mathrm{T}}] + s_3(\boldsymbol{\rho})[\dot{\boldsymbol{\rho}}\boldsymbol{\rho}^{\mathrm{T}} + \boldsymbol{\rho}\dot{\boldsymbol{\rho}}^{\mathrm{T}}],\end{aligned} \tag{C.162}$$

where the gradients $\partial s_i(\boldsymbol{\rho})/\partial \boldsymbol{\rho}$ are to be computed as row vectors and the inner products $[\partial s_i(\boldsymbol{\rho})/\partial \boldsymbol{\rho}]\dot{\boldsymbol{\rho}}$ will be scalars.

Equation C.162 is the general-purpose formula for the matrix forms of interest, which differ only in their scalar functions $s_i(\boldsymbol{\rho})$. These scalar functions $s_i(\boldsymbol{\rho})$ are generally rational functions of the following scalar functions (shown in terms of their gradients):

$$\frac{\partial}{\partial \boldsymbol{\rho}}|\boldsymbol{\rho}|^p = p|\boldsymbol{\rho}|^{p-2}\boldsymbol{\rho}^{\mathrm{T}}, \tag{C.163}$$

$$\frac{\partial}{\partial \boldsymbol{\rho}}\sin(|\boldsymbol{\rho}|) = \cos(|\boldsymbol{\rho}|)|\boldsymbol{\rho}|^{-1}\boldsymbol{\rho}^{\mathrm{T}}, \tag{C.164}$$

$$\frac{\partial}{\partial \boldsymbol{\rho}}\cos(|\boldsymbol{\rho}|) = -\sin(|\boldsymbol{\rho}|)|\boldsymbol{\rho}|^{-1}\boldsymbol{\rho}^{\mathrm{T}} \tag{C.165}$$

Time Derivative of $\partial \dot{\boldsymbol{\rho}}_{\mathrm{ENU}}/\partial \boldsymbol{\omega}_{\mathrm{RPY}}$ In this case (Eq. C.150).

$$s_1(\boldsymbol{\rho}) = \frac{|\boldsymbol{\rho}|\sin(|\boldsymbol{\rho}|)}{2[1-\cos(|\boldsymbol{\rho}|)]}, \tag{C.166}$$

$$\frac{\partial s_1(\boldsymbol{\rho})}{\partial \boldsymbol{\rho}} = -\frac{1-|\boldsymbol{\rho}|^{-1}\sin(|\boldsymbol{\rho}|)}{2[1-\cos(|\boldsymbol{\rho}|)]}\boldsymbol{\rho}^{\mathrm{T}}, \tag{C.167}$$

$$s_2(\boldsymbol{\rho}) = \tfrac{1}{2}, \tag{C.168}$$

$$\frac{\partial s_2}{\partial \boldsymbol{\rho}} = \mathbf{0}_{1\times 3}, \tag{C.169}$$

$$s_3(\boldsymbol{\rho}) = \left[\frac{1}{|\boldsymbol{\rho}|^2} - \frac{\sin(|\boldsymbol{\rho}|)}{2|\boldsymbol{\rho}|[1-\cos(|\boldsymbol{\rho}|)]}\right], \tag{C.170}$$

$$\frac{\partial s_3(\boldsymbol{\rho})}{\partial \boldsymbol{\rho}} = \frac{1 + |\boldsymbol{\rho}|^{-1}\sin(|\boldsymbol{\rho}|) - 4|\boldsymbol{\rho}|^{-2}[1-\cos(|\boldsymbol{\rho}|)]}{2|\boldsymbol{\rho}|^2[1-\cos(|\boldsymbol{\rho}|)]}\boldsymbol{\rho}^{\mathrm{T}}, \tag{C.171}$$

$$\begin{aligned}\frac{d}{dt}\frac{\partial \dot{\boldsymbol{\rho}}_{\mathrm{ENU}}}{\partial \boldsymbol{\omega}_{\mathrm{RPY}}} &= \frac{\partial s_1(\boldsymbol{\rho})}{\partial \boldsymbol{\rho}}\dot{\boldsymbol{\rho}}\mathbf{I}_3 + \frac{\partial s_2(\boldsymbol{\rho})}{\partial \boldsymbol{\rho}}\dot{\boldsymbol{\rho}}[\boldsymbol{\rho}\otimes] + s_2(\boldsymbol{\rho})[\dot{\boldsymbol{\rho}}\otimes] \\ &\quad + \frac{\partial s_3(\boldsymbol{\rho})}{\partial \boldsymbol{\rho}}\dot{\boldsymbol{\rho}}[\boldsymbol{\rho}\boldsymbol{\rho}^{\mathrm{T}}] + s_3(\boldsymbol{\rho})[\dot{\boldsymbol{\rho}}\boldsymbol{\rho}^{\mathrm{T}} + \boldsymbol{\rho}\dot{\boldsymbol{\rho}}^{\mathrm{T}}],\end{aligned} \tag{C.172}$$

$$\begin{aligned} &= -\left(\frac{1 - |\boldsymbol{\rho}|^{-1}\sin(|\boldsymbol{\rho}|)}{2[1-\cos(|\boldsymbol{\rho}|)]}\right)(\boldsymbol{\rho}^{\mathrm{T}}\dot{\boldsymbol{\rho}})\mathbf{I}_3 + \frac{1}{2}[\dot{\boldsymbol{\rho}}\otimes], \\ &\quad + \left(\frac{1 + |\boldsymbol{\rho}|^{-1}\sin(|\boldsymbol{\rho}|) - 4|\boldsymbol{\rho}|^{-2}[1-\cos(|\boldsymbol{\rho}|)]}{2|\boldsymbol{\rho}|^2[1-\cos(|\boldsymbol{\rho}|)]}\right) \times (\boldsymbol{\rho}^{\mathrm{T}}\dot{\boldsymbol{\rho}})[\boldsymbol{\rho}\boldsymbol{\rho}^{\mathrm{T}}], \\ &\quad + \left(\frac{1}{|\boldsymbol{\rho}|^2} - \frac{\sin(|\boldsymbol{\rho}|)}{2|\boldsymbol{\rho}|[1-\cos(|\boldsymbol{\rho}|)]}\right)[\dot{\boldsymbol{\rho}}\boldsymbol{\rho}^{\mathrm{T}} + \boldsymbol{\rho}\dot{\boldsymbol{\rho}}^{\mathrm{T}}].\end{aligned} \tag{C.173}$$

Time Derivative of $\partial \dot{\boldsymbol{\rho}}_{\mathrm{ENU}}/\partial \boldsymbol{\omega}_{\mathrm{ENU}}$ *In this case (Eq. C.153),*

$$s_1(\boldsymbol{\rho}) = -\frac{|\boldsymbol{\rho}|\sin(|\boldsymbol{\rho}|)}{2[1-\cos(|\boldsymbol{\rho}|)]}, \tag{C.174}$$

$$\frac{\partial s_1(\boldsymbol{\rho})}{\partial \boldsymbol{\rho}} = \frac{1 - |\boldsymbol{\rho}|^{-1}\sin(|\boldsymbol{\rho}|)}{2[1-\cos(|\boldsymbol{\rho}|)]}\boldsymbol{\rho}^{\mathrm{T}}, \tag{C.175}$$

$$s_2(\boldsymbol{\rho}) = \tfrac{1}{2}, \tag{C.176}$$

$$\frac{\partial s_2}{\partial \boldsymbol{\rho}} = \mathbf{0}_{1\times 3}, \tag{C.177}$$

$$s_3(\boldsymbol{\rho}) = -\left[\frac{1}{|\boldsymbol{\rho}|^2} - \frac{\sin(|\boldsymbol{\rho}|)}{2|\boldsymbol{\rho}|[1-\cos(|\boldsymbol{\rho}|)]}\right], \tag{C.178}$$

$$\frac{\partial s_3(\boldsymbol{\rho})}{\partial \boldsymbol{\rho}} = -\frac{1 + |\boldsymbol{\rho}|^{-1}\sin(|\boldsymbol{\rho}|) - 4|\boldsymbol{\rho}|^{-2}[1-\cos(|\boldsymbol{\rho}|)]}{2|\boldsymbol{\rho}|^2[1-\cos(|\boldsymbol{\rho}|)]}\boldsymbol{\rho}^{\mathrm{T}}, \tag{C.179}$$

$$\begin{aligned}\frac{d}{dt}\frac{\partial \dot{\boldsymbol{\rho}}_{\mathrm{ENU}}}{\partial \boldsymbol{\omega}_{\mathrm{ENU}}} &= \frac{\partial s_1(\boldsymbol{\rho})}{\partial \boldsymbol{\rho}}\dot{\boldsymbol{\rho}}\mathbf{I}_3 + \frac{\partial s_2(\boldsymbol{\rho})}{\partial \boldsymbol{\rho}}\dot{\boldsymbol{\rho}}[\boldsymbol{\rho}\otimes] + s_2(\boldsymbol{\rho})[\dot{\boldsymbol{\rho}}\otimes], \\ &\quad + \frac{\partial s_3(\boldsymbol{\rho})}{\partial \boldsymbol{\rho}}\dot{\boldsymbol{\rho}}[\boldsymbol{\rho}\boldsymbol{\rho}^{\mathrm{T}}] + s_3(\boldsymbol{\rho})[\dot{\boldsymbol{\rho}}\boldsymbol{\rho}^{\mathrm{T}} + \boldsymbol{\rho}\dot{\boldsymbol{\rho}}^{\mathrm{T}}]\end{aligned} \tag{C.180}$$

$$= \left(\frac{1 - |\boldsymbol{\rho}|^{-1}\sin(|\boldsymbol{\rho}|)}{2[1-\cos(|\boldsymbol{\rho}|)]} \right)(\boldsymbol{\rho}^{\mathrm{T}}\dot{\boldsymbol{\rho}})\mathbf{I}_3 + \frac{1}{2}(\boldsymbol{\rho})[\dot{\boldsymbol{\rho}}\otimes],$$

$$- \left(\frac{1 + |\boldsymbol{\rho}|^{-1}\sin(|\boldsymbol{\rho}|) - 4|\boldsymbol{\rho}|^{-2}[1-\cos(|\boldsymbol{\rho}|)]}{2|\boldsymbol{\rho}|^2[1-\cos(|\boldsymbol{\rho}|)]} \right) \times (\boldsymbol{\rho}^{\mathrm{T}}\dot{\boldsymbol{\rho}})[\boldsymbol{\rho}\boldsymbol{\rho}^{\mathrm{T}}],$$

$$- \left(\frac{1}{|\boldsymbol{\rho}|^2} - \frac{\sin(|\boldsymbol{\rho}|)}{2|\boldsymbol{\rho}|[1-\cos(|\boldsymbol{\rho}|)]} \right)[\dot{\boldsymbol{\rho}}\boldsymbol{\rho}^{\mathrm{T}} + \boldsymbol{\rho}\dot{\boldsymbol{\rho}}^{\mathrm{T}}]. \tag{C.181}$$

C.4.2.6 Partial Derivatives with Respect to Rotation Vectors Calculation of the dynamic coefficient matrices **F** and measurement sensitivity matrices **H** in linearized or extended Kalman filtering with rotation vectors $\boldsymbol{\rho}_{\mathrm{ENU}}$ as part of the system model state vector requires taking derivatives with respect to $\boldsymbol{\rho}_{\mathrm{ENU}}$ of associated vector-valued **f**- or **h**-functions, as

$$\mathbf{F} = \frac{\partial \mathbf{f}(\boldsymbol{\rho}_{\mathrm{ENU}}, \mathbf{v})}{\partial \boldsymbol{\rho}_{\mathrm{ENU}}}, \tag{C.182}$$

$$\mathbf{H} = \frac{\partial \mathbf{h}(\boldsymbol{\rho}_{\mathrm{ENU}}, \mathbf{v})}{\partial \boldsymbol{\rho}_{\mathrm{ENU}}}, \tag{C.183}$$

where the vector-valued functions will have the general form

$$\mathbf{f}(\boldsymbol{\rho}_{\mathrm{ENU}}, \mathbf{v}) \text{ or } \mathbf{h}(\boldsymbol{\rho}_{\mathrm{ENU}}, \mathbf{v})$$
$$= \{s_0(\boldsymbol{\rho}_{\mathrm{ENU}})\mathbf{I}_3 + s_1(\boldsymbol{\rho}_{\mathrm{ENU}})[\boldsymbol{\rho}_{\mathrm{ENU}}\otimes] + s_2(\boldsymbol{\rho}_{\mathrm{ENU}})\boldsymbol{\rho}_{\mathrm{ENU}}\boldsymbol{\rho}_{\mathrm{ENU}}^{\mathrm{T}}\}\mathbf{v}, \tag{C.184}$$

and s_0, s_1, s_2 are scalar-valued functions of $\boldsymbol{\rho}_{\mathrm{ENU}}$ and $\mathbf{v}$ is a vector that does not depend on $\boldsymbol{\rho}_{\mathrm{ENU}}$. We will derive here the general formulas that can be used for taking the partial derivatives $\partial \mathbf{f}(\boldsymbol{\rho}_{\mathrm{ENU}}, \mathbf{v})/\partial \boldsymbol{\rho}_{\mathrm{ENU}}$ or $\partial \mathbf{h}(\boldsymbol{\rho}_{\mathrm{ENU}}, \mathbf{v})/\partial \boldsymbol{\rho}_{\mathrm{ENU}}$. These formulas can all be derived by calculating the derivatives of the different factors in the functional forms and then using the chain rule for differentiation to obtain the final result.

Derivatives of Scalars The derivatives of the scalar factors s_0, s_1, s_2 are

$$\frac{\partial}{\partial \boldsymbol{\rho}_{\mathrm{ENU}}} s_i(\boldsymbol{\rho}_{\mathrm{ENU}}) = \left[\frac{\partial s_i(\boldsymbol{\rho}_{\mathrm{ENU}})}{\partial \rho_E} \; \frac{\partial s_i(\boldsymbol{\rho}_{\mathrm{ENU}})}{\partial \rho_N} \; \frac{\partial s_i(\boldsymbol{\rho}_{\mathrm{ENU}})}{\partial \rho_U} \right], \tag{C.185}$$

a row vector. Consequently, for any vector-valued function $\mathbf{g}(\boldsymbol{\rho}_{\mathrm{ENU}})$ by the chain rule, the derivatives of the vector-valued product $s_i(\boldsymbol{\rho}_{\mathrm{ENU}})\mathbf{g}(\boldsymbol{\rho}_{\mathrm{ENU}})$ are

$$\frac{\partial \{s_i(\boldsymbol{\rho}_{\mathrm{ENU}})\mathbf{g}(\boldsymbol{\rho}_{\mathrm{ENU}})\}}{\partial \boldsymbol{\rho}_{\mathrm{ENU}}} = \underbrace{\mathbf{g}(\boldsymbol{\rho}_{\mathrm{ENU}})\frac{\partial s_i(\boldsymbol{\rho}_{\mathrm{ENU}})}{\partial \boldsymbol{\rho}_{\mathrm{ENU}}}}_{3\times 3\text{matrix}} + s_i(\boldsymbol{\rho}_{\mathrm{ENU}}) \underbrace{\frac{\partial \mathbf{g}(\boldsymbol{\rho}_{\mathrm{ENU}})}{\partial \boldsymbol{\rho}_{\mathrm{ENU}}}}_{3\times 3\text{matrix}}, \tag{C.186}$$

the result of which will be the 3×3 Jacobian matrix of that subexpression in **f** or **h**.

Derivatives of Vectors The three potential forms of the vector-valued function $\mathbf{g}$ in Eq. C.186 are

$$\mathbf{g}(\boldsymbol{\rho}_{\text{ENU}}) = \begin{cases} \mathbf{I}\mathbf{v} = \mathbf{v}, \\ \boldsymbol{\rho}_{\text{ENU}} \otimes \mathbf{v}, \\ \boldsymbol{\rho}_{\text{ENU}} \boldsymbol{\rho}_{\text{ENU}}^{T} \mathbf{v}, \end{cases} \tag{C.187}$$

each of which is considered independently:

$$\frac{\partial \mathbf{v}}{\partial \boldsymbol{\rho}_{\text{ENU}}} = 0_{3\times 3}, \tag{C.188}$$

$$\frac{\partial \boldsymbol{\rho}_{\text{ENU}} \otimes \mathbf{v}}{\partial \boldsymbol{\rho}_{\text{ENU}}} = \frac{\partial[-\mathbf{v} \otimes \boldsymbol{\rho}_{\text{ENU}}]}{\partial \boldsymbol{\rho}_{\text{ENU}}}, \tag{C.189}$$

$$= -[\mathbf{v}\otimes], \tag{C.190}$$

$$= -\begin{bmatrix} 0 & -v_3 & v_2 \\ v_3 & 0 & -v_1 \\ -v_2 & v_1 & 0 \end{bmatrix}, \tag{C.191}$$

$$\frac{\partial \boldsymbol{\rho}_{\text{ENU}} \boldsymbol{\rho}_{\text{ENU}}^{\text{T}} \mathbf{v}}{\partial \boldsymbol{\rho}_{\text{ENU}}} = (\boldsymbol{\rho}_{\text{ENU}}^{\text{T}} \mathbf{v}) \frac{\partial \boldsymbol{\rho}_{\text{ENU}}}{\partial \boldsymbol{\rho}_{\text{ENU}}} + \boldsymbol{\rho}_{\text{ENU}} \frac{\partial \boldsymbol{\rho}_{\text{ENU}}^{\text{T}} \mathbf{v}}{\partial \boldsymbol{\rho}_{\text{ENU}}}, \tag{C.192}$$

$$= (\boldsymbol{\rho}_{\text{ENU}}^{\text{T}} \mathbf{v}) \mathbf{I}_{3\times 3} + \boldsymbol{\rho}_{\text{ENU}} \mathbf{v}^{\text{T}}. \tag{C.193}$$

General Formula Combining the above formulas for the different parts, one can obtain the following general-purpose formula:

$$\begin{aligned}
&\frac{\partial}{\partial \boldsymbol{\rho}_{\text{ENU}}} \{s_0(\boldsymbol{\rho}_{\text{ENU}})\mathbf{I}_3 + s_1(\boldsymbol{\rho}_{\text{ENU}})[\boldsymbol{\rho}_{\text{ENU}}\otimes] + s_2(\boldsymbol{\rho}_{\text{ENU}})\boldsymbol{\rho}_{\text{ENU}}\boldsymbol{\rho}_{\text{ENU}}^{T}\}\mathbf{v} \\
&\quad = \mathbf{v} \left[\frac{\partial s_0(\boldsymbol{\rho}_{\text{ENU}})}{\partial \rho_E} \; \frac{\partial s_0(\boldsymbol{\rho}_{\text{ENU}})}{\partial \rho_N} \; \frac{\partial s_0(\boldsymbol{\rho}_{\text{ENU}})}{\partial \rho_U} \right] \\
&\quad + [\boldsymbol{\rho}_{\text{ENU}} \otimes \mathbf{v}] \left[ccc \frac{\partial s_1(\boldsymbol{\rho}_{\text{ENU}})}{\partial \rho_E} \; \frac{\partial s_1(\boldsymbol{\rho}_{\text{ENU}})}{\partial \rho_N} \; \frac{\partial s_1(\boldsymbol{\rho}_{\text{ENU}})}{\partial \rho_U} \right] \\
&\quad - s_1(\boldsymbol{\rho}_{\text{ENU}})[\mathbf{v}\otimes] \\
&\quad + (\boldsymbol{\rho}_{\text{ENU}}^{\text{T}} \mathbf{v}) \boldsymbol{\rho}_{\text{ENU}} \left[\frac{\partial s_2(\boldsymbol{\rho}_{\text{ENU}})}{\partial \rho_E} \; \frac{\partial s_2(\boldsymbol{\rho}_{\text{ENU}})}{\partial \rho_N} \; \frac{\partial s_2(\boldsymbol{\rho}_{\text{ENU}})}{\partial \rho_U} \right] \\
&\quad + s_2(\boldsymbol{\rho}_{\text{ENU}})[(\boldsymbol{\rho}_{\text{ENU}}^{\text{T}} \mathbf{v}) \mathbf{I}_{3\times 3} + \boldsymbol{\rho}_{\text{ENU}} \mathbf{v}^{\text{T}}],
\end{aligned} \tag{C.194}$$

applicable for any differentiable scalar functions s_0, s_1, s_2.

C.4.3 Direction Cosines Matrix

We have demonstrated in Eq.C.12 that the coordinate transformation matrix between one orthogonal coordinate system and another is a matrix of direction cosines between the unit axis vectors of the two coordinate systems,

$$\mathbf{C}_{XYZ}^{UVW} = \begin{bmatrix} \cos(\theta_{XU}) & \cos(\theta_{XV}) & \cos(\theta_{XW}) \\ \cos(\theta_{YU}) & \cos(\theta_{YV}) & \cos(\theta_{YW}) \\ \cos(\theta_{ZU}) & \cos(\theta_{ZV}) & \cos(\theta_{ZW}) \end{bmatrix}. \tag{C.195}$$

Because the angles do not depend on the order of the direction vectors (i.e., $\theta_{ab} = \theta_{ba}$), the inverse transformation matrix

$$\mathbf{C}_{UVW}^{XYZ} = \begin{bmatrix} \cos(\theta_{UX}) & \cos(\theta_{UY}) & \cos(\theta_{UZ}) \\ \cos(\theta_{VX}) & \cos(\theta_{VY}) & \cos(\theta_{VZ}) \\ \cos(\theta_{WX}) & \cos(\theta_{WY}) & \cos(\theta_{WX}) \end{bmatrix}, \tag{C.196}$$

$$= \begin{bmatrix} \cos(\theta_{XU}) & \cos(\theta_{XV}) & \cos(\theta_{XW}) \\ \cos(\theta_{YU}) & \cos(\theta_{YV}) & \cos(\theta_{YW}) \\ \cos(\theta_{ZU}) & \cos(\theta_{ZV}) & \cos(\theta_{ZW}) \end{bmatrix}^{\mathrm{T}}, \tag{C.197}$$

$$= (\mathbf{C}_{XYZ}^{UVW})^{\mathrm{T}}. \tag{C.198}$$

That is, the inverse coordinate transformation matrix is the transpose of the forward coordinate transformation matrix. This implies that the coordinate transformation matrices are orthogonal matrices.

C.4.3.1 Rotating Coordinates Let "rot" denote a set of rotating coordinates, with axes X_{rot}, Y_{rot}, Z_{rot}, and let "non" represent a set of non-rotating (i.e., inertial) coordinates, with axes X_{non}, Y_{non}, Z_{non}, as illustrated in Fig. C.14.

Any vector $\mathbf{v}_{\text{rot}}$ in rotating coordinates can be represented in terms of its nonrotating components and unit vectors parallel to the nonrotating axes, as

$$\mathbf{v}_{\text{rot}} = v_{x,\text{non}}\mathbf{1}_{x,\text{non}} + v_{y,\text{non}}\mathbf{1}_{y,\text{non}} + v_{z,\text{non}}\mathbf{1}_{z,\text{non}} \tag{C.199}$$

$$= \begin{bmatrix} \mathbf{1}_{x,\text{non}} & \mathbf{1}_{y,\text{non}} & \mathbf{1}_{z,\text{non}} \end{bmatrix} \begin{bmatrix} v_{x,\text{non}} \\ v_{y,\text{non}} \\ v_{z,\text{non}} \end{bmatrix} \tag{C.200}$$

$$= \mathbf{C}_{\text{rot}}^{\text{non}}\mathbf{v}_{\text{non}}, \tag{C.201}$$

where $v_{x,\text{non}}$, $v_{y,\text{non}}$, $v_{z,\text{non}}$ are nonrotating components of the vector, $\mathbf{1}_{x,\text{non}}$, $\mathbf{1}_{y,\text{non}}$, $\mathbf{1}_{z,\text{non}}$ = unit vectors along X_{non}, Y_{non}, Z_{non} axes, as expressed in rotating coordinates

$\mathbf{v}_{\text{rot}}$ = vector $\mathbf{v}$ expressed in RPY coordinates

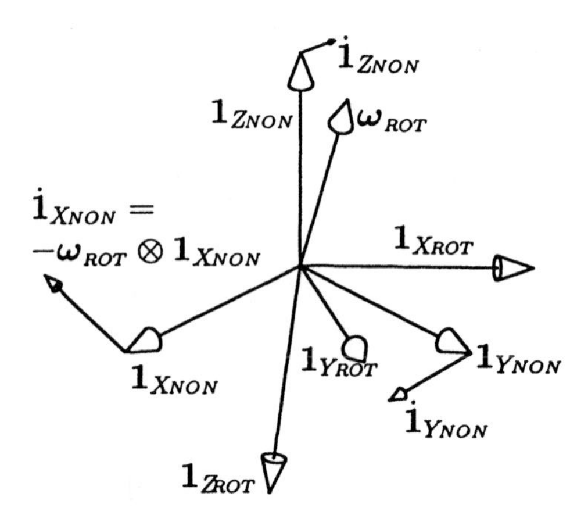

Fig. C.14 Rotating coordinates.

$\mathbf{v}_{\text{non}}$ = vector $\mathbf{v}$ expressed in ECI coordinates,

$\mathbf{C}_{\text{rot}}^{\text{non}}$ = coordinate transformation matrix from nonrotating coordinates to rotating coordinates

and

$$\mathbf{C}_{\text{rot}}^{\text{non}} = [\ \mathbf{1}_{x,\text{non}} \quad \mathbf{1}_{y,\text{non}} \quad \mathbf{1}_{z,\text{non}}\ ]. \tag{C.202}$$

The time derivative of $\mathbf{C}_{\text{rot}}^{\text{non}}$, as viewed from the non-rotating coordinate frame, can be derived in terms of the dynamics of the unit vectors $\mathbf{1}_{x,\text{non}}$, $\mathbf{1}_{y,\text{non}}$ and $\mathbf{1}_{z,\text{non}}$ in rotating coordinates.

As seen by an observer fixed with respect to the nonrotating coordinates, the nonrotating coordinate directions will appear to remain fixed, but the external inertial reference directions will appear to be changing, as illustrated in Fig. C.14. Gyroscopes fixed in the rotating coordinates would measure three components of the inertial rotation rate vector

$$\boldsymbol{\omega}_{\text{rot}} = \begin{bmatrix} \omega_{x,\text{rot}} \\ \omega_{y,\text{rot}} \\ \omega_{z,\text{rot}} \end{bmatrix} \tag{C.203}$$

in rotating coordinates, but the non-rotating unit vectors, as viewed in rotating coordinates, appear to be changing in the opposite sense, as

$$\frac{d}{dt}\mathbf{1}_{x,\text{non}} = -\boldsymbol{\omega}_{\text{rot}} \otimes \mathbf{1}_{x,\text{non}}, \tag{C.204}$$

$$\frac{d}{dt}\mathbf{1}_{y,\text{non}} = -\boldsymbol{\omega}_{\text{rot}} \otimes \mathbf{1}_{y,\text{non}}, \tag{C.205}$$

$$\frac{d}{dt}\mathbf{1}_{z,\text{non}} = -\boldsymbol{\omega}_{\text{rot}} \otimes \mathbf{1}_{z,\text{non}}, \tag{C.206}$$

as illustrated in Fig. C.14. The time-derivative of the coordinate transformation represented in Eq. C.202 will then be

$$\frac{d}{dt}\mathbf{C}_{\mathrm{rot}}^{\mathrm{non}} = \left[\frac{d}{dt}\mathbf{1}_{x,\mathrm{non}} \quad \frac{d}{dt}\mathbf{1}_{y,\mathrm{non}} \quad \frac{d}{dt}\mathbf{1}_{z,\mathrm{non}}\right] \tag{C.207}$$

$$= [-\boldsymbol{\omega}_{\mathrm{rot}} \otimes \mathbf{1}_{x,\mathrm{non}} \quad -\boldsymbol{\omega}_{\mathrm{rot}} \otimes \mathbf{1}_{y,\mathrm{non}} \quad -\boldsymbol{\omega}_{\mathrm{rot}} \otimes \mathbf{1}_{z,\mathrm{non}}]$$

$$= -[\boldsymbol{\omega}_{\mathrm{rot}}\otimes][\mathbf{1}_{x,\mathrm{non}} \quad \mathbf{1}_{y,\mathrm{non}} \quad \mathbf{1}_{z,\mathrm{non}}]$$

$$= -[\boldsymbol{\omega}_{\mathrm{rot}}\otimes]\mathbf{C}_{\mathrm{rot}}^{\mathrm{non}}, \tag{C.208}$$

$$[\boldsymbol{\omega}_{\mathrm{rot}}\otimes] \stackrel{\mathrm{def}}{=} \begin{bmatrix} 0 & -\omega_{z,\mathrm{rot}} & \omega_{y,\mathrm{rot}} \\ \omega_{z,\mathrm{rot}} & 0 & -\omega_{x,\mathrm{rot}} \\ -\omega_{y,\mathrm{rot}} & \omega_{x,\mathrm{rot}} & 0 \end{bmatrix}. \tag{C.209}$$

The inverse coordinate transformation

$$\mathbf{C}_{\mathrm{non}}^{\mathrm{rot}} = (\mathbf{C}_{\mathrm{rot}}^{\mathrm{non}})^{-1} \tag{C.210}$$

$$= (\mathbf{C}_{\mathrm{rot}}^{\mathrm{non}})^{\mathrm{T}}, \tag{C.211}$$

the transpose of $\mathbf{C}_{\mathrm{rot}}^{\mathrm{non}}$, and its derivative

$$\frac{d}{dt}\mathbf{C}_{\mathrm{non}}^{\mathrm{rot}} = \frac{d}{dt}(\mathbf{C}_{\mathrm{rot}}^{\mathrm{non}})^{\mathrm{T}} \tag{C.212}$$

$$= \left(\frac{d}{dt}\mathbf{C}_{\mathrm{rot}}^{\mathrm{non}}\right)^{\mathrm{T}} \tag{C.213}$$

$$= (-[\boldsymbol{\omega}_{\mathrm{rot}}\otimes]\mathbf{C}_{\mathrm{rot}}^{\mathrm{non}})^{\mathrm{T}} \tag{C.214}$$

$$= -(\mathbf{C}_{\mathrm{rot}}^{\mathrm{non}})^{\mathrm{T}}[\boldsymbol{\omega}_{\mathrm{rot}}\otimes]^{\mathrm{T}}, \tag{C.215}$$

$$= \mathbf{C}_{\mathrm{non}}^{\mathrm{rot}}[\boldsymbol{\omega}_{\mathrm{rot}}\otimes]. \tag{C.216}$$

In the case that "rot" is "RPY" (roll-pitch-yaw coordinates) and "non" is "ECI" (earth centered inertial coordinates), Eq. C.216 becomes

$$\frac{d}{dt}\mathbf{C}_{\mathrm{ECI}}^{\mathrm{RPY}} = \mathbf{C}_{\mathrm{ECI}}^{\mathrm{RPY}}[\boldsymbol{\omega}_{\mathrm{RPY}}\otimes], \tag{C.217}$$

and in the case that "rot" is "ENU" (east-north-up coordinates) and "non" is "ECI" (earth centered inertial coordinates), Eq. C.208 becomes

$$\frac{d}{dt}\mathbf{C}_{\mathrm{ENU}}^{\mathrm{ECI}} = -[\boldsymbol{\omega}_{\mathrm{ENU}}\otimes]\mathbf{C}_{\mathrm{ENU}}^{\mathrm{ECI}}, \tag{C.218}$$

and the derivative of their product

$$\mathbf{C}_{\text{ENU}}^{\text{RPY}} = \mathbf{C}_{\text{ENU}}^{\text{ECI}} \mathbf{C}_{\text{ECI}}^{\text{RPY}}, \tag{C.219}$$

$$\begin{aligned}
\frac{d}{dt}\mathbf{C}_{\text{ENU}}^{\text{RPY}} &= \left[\frac{d}{dt}\mathbf{C}_{\text{ENU}}^{\text{ECI}}\right]\mathbf{C}_{\text{ECI}}^{\text{RPY}} + \mathbf{C}_{\text{ENU}}^{\text{ECI}}\left[\frac{d}{dt}\mathbf{C}_{\text{ECI}}^{\text{RPY}}\right] \qquad \text{(C.220)} \\
&= [[-\boldsymbol{\omega}_{\text{ENU}}\otimes]\mathbf{C}_{\text{ENU}}^{\text{ECI}}]\mathbf{C}_{\text{ECI}}^{\text{RPY}} + \mathbf{C}_{\text{ENU}}^{\text{ECI}}[\mathbf{C}_{\text{ECI}}^{\text{RPY}}[\boldsymbol{\omega}_{\text{RPY}}\otimes]] \\
&= [-\boldsymbol{\omega}_{\text{ENU}}\otimes]\underbrace{\mathbf{C}_{\text{ENU}}^{\text{ECI}}\mathbf{C}_{\text{ECI}}^{\text{RPY}}}_{\mathbf{C}_{\text{ENU}}^{\text{RPY}}} + \underbrace{\mathbf{C}_{\text{ENU}}^{\text{ECI}}\mathbf{C}_{\text{ECI}}^{\text{RPY}}}_{\mathbf{C}_{\text{ENU}}^{\text{RPY}}}[\boldsymbol{\omega}_{\text{RPY}}\otimes],
\end{aligned}$$

$$\frac{d}{dt}\mathbf{C}_{\text{ENU}}^{\text{RPY}} = -[\boldsymbol{\omega}_{\text{ENU}}\otimes]\mathbf{C}_{\text{ENU}}^{\text{RPY}} + \mathbf{C}_{\text{ENU}}^{\text{RPY}}[\boldsymbol{\omega}_{\text{RPY}}\otimes]. \tag{C.221}$$

Equation C.221 was originally used for maintaining vehicle attitude information in strapdown INS implementations, where the variables

$$\boldsymbol{\omega}_{\text{RPY}} = \text{vector of inertial rates measured by the gyroscopes}, \tag{C.222}$$

$$\boldsymbol{\omega}_{\text{ENU}} = \boldsymbol{\omega}_{\text{earthrate}} + \boldsymbol{\omega}_{v_E} + \boldsymbol{\omega}_{v_N},$$

$$\boldsymbol{\omega}_{\oplus} = \omega_{\oplus}\begin{bmatrix} 0 \\ \cos(\phi_{\text{geodetic}}) \\ \sin(\phi_{\text{geodetic}}) \end{bmatrix}, \tag{C.223}$$

$$\boldsymbol{\omega}_{v_E} = \frac{v_E}{r_T + h}\begin{bmatrix} 0 \\ 1 \\ 0 \end{bmatrix}, \tag{C.224}$$

$$\boldsymbol{\omega}_{v_N} = \frac{v_N}{r_M + h}\begin{bmatrix} -1 \\ 0 \\ 0 \end{bmatrix}, \tag{C.225}$$

and
where

$\omega_{\oplus}$ = earth rotation rate
ϕ_{geodetic} = geodetic latitude
v_E = the east component of velocity with respect to the surface of the earth
r_T = transverse radius of curvature of the ellipsoid (Eq. 6.41)
v_N = north component of velocity with respect to the surface of the earth
r_M = meridional radius of curvature of the ellipsoid (Eq. 6.38)
h = altitude above (+) or below (−) the reference ellipsoid surface (≈mean sea level)

Unfortunately, Eq. C.221 was found to be not particularly well suited for accurate integration in finite-precision arithmetic. This integration problem was eventually solved using quaternions.

C.4.4 Quaternions

The term *quaternions* is used in several contexts to refer to sets of four. In mathematics, it refers to an algebra in four dimensions discovered by the Irish physicist and mathematician Sir William Rowan Hamilton (1805–1865). The utility of quaternions for representing rotations (as points on a sphere in four dimensions) was known before strapdown systems, they soon became the standard representation of coordinate transforms in strapdown systems, and they have since been applied to computer animation.

C.4.4.1 Quaternion Matrices For people already familiar with matrix algebra, the algebra of quaternions can be defined by using an isomorphism between 4×1 *quaternion vectors* $\mathbf{q}$ and real 4×4 *quaternion matrices* $\mathbf{Q}$:

$$\mathbf{q} = \begin{bmatrix} q_1 \\ q_2 \\ q_3 \\ q_4 \end{bmatrix} \leftrightarrow \mathbf{Q} = \begin{bmatrix} q_1 & -q_2 & -q_3 & -q_4, \\ q_2 & q_1 & -q_4 & q_3, \\ q_3 & q_4 & q_1 & -q_2, \\ q_4 & -q_3 & q_2 & q_1 \end{bmatrix} \tag{C.226}$$

$$= q_1 \mathcal{Q}_1 + q_2 \mathcal{Q}_2 + q_3 \mathcal{Q}_3 + q_4 \mathcal{Q}_4, \tag{C.227}$$

$$\mathcal{Q}_1 \stackrel{\text{def}}{=} \begin{bmatrix} 1 & 0 & 0 & 0, \\ 0 & 1 & 0 & 0, \\ 0 & 0 & 1 & 0, \\ 0 & 0 & 0 & 1 \end{bmatrix}, \tag{C.228}$$

$$\mathcal{Q}_2 \stackrel{\text{def}}{=} \begin{bmatrix} 0 & -1 & 0 & 0 \\ 1 & 0 & 0 & 0 \\ 0 & 0 & 0 & -1 \\ 0 & 0 & 1 & 0 \end{bmatrix}, \tag{C.229}$$

$$\mathcal{Q}_3 \stackrel{\text{def}}{=} \begin{bmatrix} 0 & 0 & -1 & 0 \\ 0 & 0 & 0 & 1 \\ 1 & 0 & 0 & 0 \\ 0 & -1 & 0 & 0 \end{bmatrix}, \tag{C.230}$$

$$\mathcal{Q}_4 \stackrel{\text{def}}{=} \begin{bmatrix} 0 & 0 & 0 & -1 \\ 0 & 0 & -1 & 0 \\ 0 & 1 & 0 & 0 \\ 1 & 0 & 0 & 0 \end{bmatrix}, \tag{C.231}$$

in terms of four 4×4 *quaternion basis matrices*, $\mathcal{Q}_1, \mathcal{Q}_2, \mathcal{Q}_3, \mathcal{Q}_4$, the first of which is an identity matrix and the rest of which are antisymmetric.

C.4.4.2 Addition and Multiplication Addition of quaternion vectors is the same as that for ordinary vectors. Multiplication is defined by the usual rules for matrix multiplication applied to the four quaternion basis matrices, the multiplication table for which is given in Table C.1. Note that, like matrix multiplication,

TABLE C.1. Multiplication of Quaternion Basis Matrices

First Factor	Second Factor			
	$\mathcal{Q}_1$	$\mathcal{Q}_2$	$\mathcal{Q}_3$	$\mathcal{Q}_4$
$\mathcal{Q}_1$	$\mathcal{Q}_1$	$\mathcal{Q}_2$	$\mathcal{Q}_3$	$\mathcal{Q}_4$
$\mathcal{Q}_2$	$\mathcal{Q}_2$	$-\mathcal{Q}_1$	$\mathcal{Q}_4$	$-\mathcal{Q}_3$
$\mathcal{Q}_3$	$\mathcal{Q}_3$	$-\mathcal{Q}_4$	$-\mathcal{Q}_1$	$\mathcal{Q}_2$
$\mathcal{Q}_4$	$\mathcal{Q}_4$	$\mathcal{Q}_3$	$-\mathcal{Q}_2$	$-\mathcal{Q}_1$

quaternion multiplication is noncommutative. That is, the result depends on the order of multiplication.

Using the quaternion basis matrix multiplication Table (C.1), the ordered product **AB** of two quaternion matrices

$$\mathbf{A} = a_1\mathcal{Q}_1 + a_2\mathcal{Q}_2 + a_3\mathcal{Q}_3 + a_4\mathcal{Q}_4, \tag{C.232}$$

$$\mathbf{B} = b_1\mathcal{Q}_1 + b_2\mathcal{Q}_2 + b_3\mathcal{Q}_3 + b_4\mathcal{Q}_4 \tag{C.233}$$

can be shown to be

$$\begin{aligned} \mathbf{AB} = {} & (a_1b_1 - a_2b_2 - a_3b_3 - a_4b_4)\mathcal{Q}_1 \\ & +(a_2b_1 + a_1b_2 - a_4b_3 + a_3b_4)\mathcal{Q}_2 \\ & +(a_3b_1 + a_4b_2 + a_1b_3 - a_2b_4)\mathcal{Q}_3 \\ & +(a_4b_1 - a_3b_2 + a_2b_3 + a_1b_4)\mathcal{Q}_4 \end{aligned} \tag{C.234}$$

in terms of the coefficients a_k, b_k and the quaternion basis matrices.

C.4.4.3 Conjugation Conjugation of quaternions is a unary operation analogous to conjugation of complex numbers, in that the real part (the first component of a quaternion) is unchanged and the other parts change sign. For quaternions, this is equivalent to transposition of the associated quaternion matrix

$$\mathbf{Q} = q_1\mathcal{Q}_1 + q_2\mathcal{Q}_2 + q_3\mathcal{Q}_3 + q_4\mathcal{Q}_4, \tag{C.235}$$

so that

$$\mathbf{Q}^{\mathrm{T}} = q_1\mathcal{Q}_1 - q_2\mathcal{Q}_2 - q_3\mathcal{Q}_3 - q_4\mathcal{Q}_4 \tag{C.236}$$

$$\leftrightarrow \mathbf{q}^*, \tag{C.237}$$

$$\mathbf{Q}^{\mathrm{T}}\mathbf{Q} = (q_1^2 + q_2^2 + q_3^2 + q_4^2)\mathcal{Q}_1 \tag{C.238}$$

$$\leftrightarrow \mathbf{q}^*\mathbf{q} = |\mathbf{q}|^2. \tag{C.239}$$

C.4.4.4 Representing Rotations The problem with rotation vectors as representations for rotations is that the rotation vector representing successive rotations $\boldsymbol{\rho}_1, \boldsymbol{\rho}_2, \boldsymbol{\rho}_3, \ldots, \boldsymbol{\rho}_n$ is not a simple function of the respective rotation vectors.

This representation problem is solved rather elegantly using quaternions, such that the quaternion representation of the successive rotations is represented by the quaternion product $\mathbf{q}_n \times \mathbf{q}_{n-1} \times \cdots \times \mathbf{q}_3 \times \mathbf{q}_2 \times \mathbf{q}_1$. That is, each successive rotation can be implemented by a single quaternion product.

The quaternion equivalent of the rotation vector $\boldsymbol{\rho}$ with $|\boldsymbol{\rho}| = \theta$,

$$\boldsymbol{\rho} \stackrel{\text{def}}{=} \begin{bmatrix} \rho_1 \\ \rho_2 \\ \rho_3 \end{bmatrix} \stackrel{\text{def}}{=} \theta \begin{bmatrix} u_1 \\ u_2 \\ u_3 \end{bmatrix} \tag{C.240}$$

(i.e., where $\mathbf{u}$ is a unit vector), is

$$\mathbf{q}(\boldsymbol{\rho}) \stackrel{\text{def}}{=} \begin{bmatrix} \cos\left(\dfrac{\theta}{2}\right) \\ \dfrac{\rho_1 \sin(\theta/2)}{\theta} \\ \dfrac{\rho_2 \sin(\theta/2)}{\theta} \\ \dfrac{\rho_3 \sin(\theta/2)}{\theta} \end{bmatrix} = \begin{bmatrix} \cos\left(\dfrac{\theta}{2}\right) \\ u_1 \sin\left(\dfrac{\theta}{2}\right) \\ u_2 \sin\left(\dfrac{\theta}{2}\right) \\ u_3 \sin\left(\dfrac{\theta}{2}\right) \end{bmatrix}, \tag{C.241}$$

and the vector $\mathbf{w}$ resulting from the rotation of any three-dimensional vector

$$\mathbf{v} \stackrel{\text{def}}{=} \begin{bmatrix} v_1 \\ v_2 \\ v_3 \end{bmatrix}$$

through the angle θ about the unit vector $\mathbf{u}$ is implemented by the quaternion product

$$\mathbf{q}(\mathbf{w}) \stackrel{\text{def}}{=} \mathbf{q}(\boldsymbol{\rho})\mathbf{q}(\mathbf{v})\mathbf{q}^*(\boldsymbol{\rho}) \tag{C.242}$$

$$\stackrel{\text{def}}{=} \begin{bmatrix} \cos\left(\dfrac{\theta}{2}\right) \\ u_1 \sin\left(\dfrac{\theta}{2}\right) \\ u_2 \sin\left(\dfrac{\theta}{2}\right) \\ u_3 \sin\left(\dfrac{\theta}{2}\right) \end{bmatrix} \times \begin{bmatrix} 0 \\ v_1 \\ v_2 \\ v_3 \end{bmatrix} \times \begin{bmatrix} \cos\left(\dfrac{\theta}{2}\right) \\ -u_1 \sin\left(\dfrac{\theta}{2}\right) \\ -u_2 \sin\left(\dfrac{\theta}{2}\right) \\ -u_3 \sin\left(\dfrac{\theta}{2}\right) \end{bmatrix} \tag{C.243}$$

$$= \begin{bmatrix} 0 \\ w_1 \\ w_2 \\ w_3 \end{bmatrix}, \tag{C.244}$$

$$\begin{aligned} w_1 &= \cos(\theta)v_1 + [1 - \cos(\theta)][u_1(u_1v_1 + u_2v_2 + u_3v_3)] \\ &\quad + \sin(\theta)(u_2v_3 - u_3v_2), \end{aligned} \tag{C.245}$$

$$\begin{aligned} w_2 &= \cos(\theta)v_2 + [1 - \cos(\theta)][u_2(u_1v_1 + u_2v_2 + u_3v_3)] \\ &\quad + \sin(\theta)(u_3v_1 - u_1v_3), \end{aligned} \tag{C.246}$$

$$\begin{aligned} w_3 &= \cos(\theta)v_3 + [1 - \cos(\theta)][u_3(u_1v_1 + u_2v_2 + u_3v_3)] \\ &\quad + \sin(\theta)(u_1v_2 - u_2v_1), \end{aligned} \tag{C.247}$$

or

$$\begin{bmatrix} w_1 \\ w_2 \\ w_3 \end{bmatrix} = \mathbf{C}(\boldsymbol{\rho}) \begin{bmatrix} v_1 \\ v_2 \\ v_3 \end{bmatrix}, \tag{C.248}$$

where the rotation matrix $\mathbf{C}(\boldsymbol{\rho})$ is defined in Eq. C.113 and Eq. C.242 implements the same rotation of $\mathbf{v}$ as the matrix product $\mathbf{C}(\boldsymbol{\rho})\mathbf{v}$. Moreover, if

$$\mathbf{q}(\mathbf{w}_k) \stackrel{\text{def}}{=} \mathbf{v} \tag{C.249}$$

and

$$\mathbf{q}(\mathbf{w}_k) \stackrel{\text{def}}{=} \mathbf{q}(\boldsymbol{\rho}_k)\mathbf{q}(\mathbf{w}_{k-1})\mathbf{q}^*(\boldsymbol{\rho}_k) \tag{C.250}$$

for $k = 1, 2, 3, \ldots, n$, then the nested quaternion product

$$\mathbf{q}(\mathbf{w}_n) = \mathbf{q}(\boldsymbol{\rho}_n) \cdots \mathbf{q}(\boldsymbol{\rho}_2)\mathbf{q}(\boldsymbol{\rho}_1)\mathbf{q}(\mathbf{v})\mathbf{q}^*(\boldsymbol{\rho}_1)\mathbf{q}^*(\boldsymbol{\rho}_2) \cdots \mathbf{q}^*(\boldsymbol{\rho}_n) \tag{C.251}$$

implements the succession of rotations represented by the rotation vectors $\boldsymbol{\rho}_1$, $\boldsymbol{\rho}_2$, $\boldsymbol{\rho}_3, \ldots, \boldsymbol{\rho}_n$, and the single quaternion

$$\mathbf{q}_{[n]} \stackrel{\text{def}}{=} \mathbf{q}(\boldsymbol{\rho}_n)\mathbf{q}(\boldsymbol{\rho}_{n-1}) \cdots \mathbf{q}(\boldsymbol{\rho}_3)\mathbf{q}(\boldsymbol{\rho}_2)\mathbf{q}(\boldsymbol{\rho}_1) \tag{C.252}$$

$$= \mathbf{q}(\boldsymbol{\rho}_n)\mathbf{q}_{[n-1]} \tag{C.253}$$

then represents the net effect of the successive rotations as

$$\mathbf{q}(\mathbf{w}_n) = \mathbf{q}_{[n]}\mathbf{q}(\mathbf{w}_0)\mathbf{q}^*_{[n]}. \tag{C.254}$$

The initial value $\mathbf{q}_{[0]}$ for the rotation quaternion will depend upon the inital orientation of the two coordinate systems. The initial value

$$\mathbf{q}_{[0]} \stackrel{\text{def}}{=} \begin{bmatrix} 1 \\ 0 \\ 0 \\ 0 \end{bmatrix} \tag{C.255}$$

applies to the case that the two coordinate systems are aligned. In strapdown system applications, the initial value $\mathbf{q}_{[0]}$ is determined during the INS alignment procedure.

Equation C.252 is the much-used quaternion representation for successive rotations, and Eq. C.254 is how it is used to perform coordinate transformations of any vector $\mathbf{w}_0$.

This representation uses the four components of a unit quaternion to maintain the transformation from one coordinate frame to another through a succession of rotations. In practice, computer roundoff may tend to alter the magnitude of the alegedly unit quaternion, but it can easily be rescaled to a unit quaternion by dividing by its magnitude.

REFERENCES

[1] R. Ahmadi, G. S. Becker, S. R. Peck, F. Choquette, T. F. Gerard, A. J. Mannucci, B. A. Iijima, and A. W. Moore, "Validation Analysis of the WAAS GIVE and UIVE Algorithms," *Proceedings of the Institute of Navigation, ION '98* (Santa Monica, CA), ION, Alexandria, VA, Jan. 1998.

[2] D. W. Allan, *The Measurement of Frequency and Frequency Stability of Precision Oscillators*, NBS Technical Note 669, pp. 1–27, 1975.

[3] D. W. Allan, "Fine-Tuning Time in the Space Age," *IEEE Spectrum* **35**(3), 43–51 (1998).

[4] D. W. Allan, "Time-Domain Spectrum of GPS SA," *Proceedings of the 6th International Technical Meeting of the Satellite Division of the Institute of Navigation (ION) GPS-93* (Salt Lake City, UT), Sept. 22–24, 1993, ION, Alexandria, VA, 1993, pp. 129–136.

[5] D. W. Allan, N. Ashby, and C. C. Hodge, *The Science of Timekeeping*, Hewlett-Packard Application Note 1289, Palo Alto, CA, 1997.

[6] B. D. O. Anderson and J. B. Moore, *Optimal Filtering*, MIT Press, Cambridge, MA, 2006 (reprint of 1979 edition published by Prentice-Hall, Englewood Cliffs, NJ).

[7] A. Andrews, "Calibrating the Drift Rates of Strapdown Electrostatic Gyroscopes," *IEEE National Aerospace Electronics Conference Proceedings*, 1973.

[8] M. Barbour, "Inertial Components—Past, Present and Future," *Proceedings of 2001 Guidance, Navigation and Control Conference*, Montreal, American Institute of Aeronautics and Astronautics, 2001.

Global Positioning Systems, Inertial Navigation, and Integration, Second Edition, by M. S. Grewal, L. R. Weill, and A. P. Andrews

[9] B. C. Barker et al., "Details of the GPS M Code Signal," *Proceedings of the ION 2000 National Technical Meeting*, Jan. 2000.

[10] T. Barnes, "Selective Availability via the Levinson Predictor," *Proceedings of the Institute of Navigation (ION), GPS '95* (Palm Springs, CA), Sept. 12–15, ION, Alexandria, VA, 1995.

[11] R. R. Bate, D. D. Mueller, and J. E. White, *Fundamentals of Astrodynamics*, Dover, New York, 1971.

[12] R. H. Battin, *Astronautical Guidance*, McGraw-Hill, New York, 1964.

[13] Commandant Benoit, "Sur une méthode de résolution des équations normales provenant de l' application de la méthode des moindes carrés a un système d'équations linéaires en numbre inférieur a celui des inconnues—application de la méthode a la resolution d'un système defini d'équations linéaires (Procédé du Commandant Cholesky)," *Bulletin Géodesique et Géophysique Internationale*, Toulouse, pp. 67–77, 1924.

[14] J. Bernstein, "An overview of MEMS Inertial Sensing Technology," *Sensor Technology and Design* **20**(2) (Feb. 2003).

[15] W. I. Bertiger et al., "A Real-Time Wide Area Differential GPS System," *Institute of Navigation "Redbook," Global Positioning System: Selected Papers on Satellite Based Augmentation Systems (SBASs)*, Vol. VI, Alexandria, VA, 1999.

[16] J. W. Betz, "Binary Offset Carrier Modulations for Radionavigation," *Navigation* (Institute of Navigation) **48**(4), 227–246 (Winter 2001–2002).

[17] G. L. Bierman, *Factorization Methods for Discrete Sequential Estimation, Mathematics in Science and Engineering*, Vol. 128, Academic Press, New York, 1977.

[18] D. J. Biezad, *Integrated Navigation and Guidance Systems*, American Institute of Aeronautics and Astronautics, New York, 1999.

[19] Å. Björck, "Solving Least Squares Problems by Orthogonalization," *BIT* **7**, 1–21 (1967).

[20] H. S. Black, "Stabilized Feedback Amplifiers," *Bell System Technical Journal* **13**, 1–18 (Jan. 1934).

[21] F. R. Bletzacker, D. H. Eller, T. M. Gorgette, G. L. Seibert, J. L. Vavrus, and M. D. Wade, "Kalman Filter Design for Integration of Phase III GPS with an Inertial Navigation System," *Proceedings of the Institute of Navigation* (Santa Barbara, CA), Jan. 26–29, 1988, ION, Alexandria, VA, 1988, pp. 113–129.

[22] G. Blewitt, "An Automatic Editing Algorithm for GPS Data," *Geophysical Research Letters* **17**(3), 199–202 (1990).

[23] J. E. Bortz, "A New Mathematical Formulation for Strapdown Inertial Navigation," *IEEE Transactions on Aerospace and Electronic Systems* **AES-6**, 61–66 (1971).

[24] M. S. Braasch, "Improved Modeling of GPS Selective Availability," *Proceedings of the Institute of Navigation (ION) Annual Technical Meeting*, ION, Alexandria, VA, 1993.

[25] M. S. Braasch, "A Signal Model for GPS," *Navigation: Journal of the Institute of Navigation* **37**(4), 363–379 (1990).

[26] E. A. Bretz, "X Marks the Spot, Maybe," *IEEE Spectrum* **37**(4), 26–36 (2000).

[27] K. R. Britting, *Inertial Navigation Systems Analysis*, Wiley, New York, 1971.

[28] F. P. Brooks, Jr., *The Mythical Man-month*, Addison-Wesley, Reading, MA, 1982.

[29] R. G. Brown and P. Y. C. Hwhang, *Introduction to Random Signals and Applied Kalman Filtering: With Matlab Exercises and Solutions*, 3rd ed., Wiley, New York, 1997.

[30] R. S. Bucy and P. D. Joseph, *Filtering for Stochastic Processes with Applications to Guidance*, Chelsea, New York, 1968 (republished by the American Mathematical Society).

[31] N. A. Carlson, "Fast Triangular Formulation of the Square Root Filter," *AIAA Journal* **11**(9), 1259–1265 (1973).

[32] H. Carvalho, P. Del Moral, A. Monin, and G. Salut, "Optimal Nonlinear Filtering in GPS/INS integration," *IEEE Transactions on Aerospace and Electronic Systems* **33**(3), 835–850 (1997).

[33] E. N. Carolipio, N. Pandya, and M. S. Grewal, "GEO Orbit Determination via Covariance Analysis with a Known Clock Error," *Navigation: Journal of the Institute of Navigation* **48**(4), 255–260 (2002).

[34] Chairman of Joint Chiefs of Staff, U.S. Department of Defense, *2003 CJCS Master Positioning, Navigation and Timing Plan*, Report CJCSI 6130.01C, March 2003.

[35] Y. Chao and B. W. Parkinson, "The Statistics of Selective Availability and Its Effects on Differential GPS," *Proceedings of the 6th International Technical Meeting of the Satellite Division of (ION) GPS-93* (Salt Lake City, UT), Sept. 22–24, 1993, ION, Alexandria, VA, 1993, pp. 1509–1516.

[36] A. B. Chatfield, *Fundamentals of High Accuracy Inertial Navigation*, American Institute of Aeronautics and Astronautics, New York, 1997.

[37] S. Cooper and H. Durrant-Whyte, "A Kalman Filter Model for GPS Navigation of Land Vehicles," *Intelligent Robots and Systems*, IEEE, Piscataway, NJ, 1994.

[38] E. Copros, J. Spiller, T. Underwood, and C. Vialet, "An Improved Space Segment for the End-State WAAS and EGNOS Final Operational Capability," *Proceedings of the Institute of Navigation*, ION GPS-96 (Kansas City, MO), ION, Alexandria, VA, Sept. 1996, pp. 1119–1125 .

[39] G. G. de Coriolis, *Sur les équations du mouvement relatif des systèmes de corps*, Ecole Polytechnique, Paris, 1835.

[40] J. P. Costas, "Synchronous Communications," *Proceedings of the IRE* **45**, 1713–1718 (1956).

[41] C. C. Counselman, III, "Multipath-Rejecting GPS Antennas," *Proceedings of the IEEE* **87**(1), 86–91 (1999).

[42] P. Daum, J. Beyer, and T. F. W. Köhler, "Aided Inertial LAnd NAvigation System (ILANA) with a Minimum Set of Inertial Sensors," *Proceedings of IEEE Position, Locations and Navigation Conference*, Las Vegas, 1994.

[43] K. Davies, *Ionospheric Radio*, Peter Peregrinus, London, 1990.

[44] A. J. Van Dierendonck, "Understanding GPS Receiver Terminology: A Tutorial," *GPS WORLD* (Advanstar Communications, Eugene, OR) pp. 34–44 (Jan. 1995).

[45] A. J. Van Dierendonck and C. Hegarty, "The New L5 Civil GPS Signal," *GPS World* (Advanstar Communications, Eugene, OR) pp. 64–71 (Sep. 2000).

[46] M. Djodat, *Comparison of Various Differential Global Positioning Systems*, Master's thesis, California State University, Fullerton, 1996.

[47] C. S. Draper, "Origins of Inertial Navigation," *Journal of Guidance and Control* (American Institute of Aeronautics and Astronautics, New York) **81**, 449–463 (1981).

[48] R. M. du Plessis, *Poor Man's Explanation of Kalman Filtering, or How I Stopped Worrying and Learned to Love Matrix Inversion*, Autonetics Technical Note, Anaheim, CA, 1967, republished by Taygeta Scientific Incorporated, Monterey, CA, 1996.

[49] P. Dyer and S. McReynolds, "Extension of Square-Root Filtering to Include Process Noise," *Journal of Optimization Theory and Applications* **3**, 444–458 (1969).

[50] M. B. El-Arini, Robert S. Conker, Thomas W. Albertson, James K. Reagan, John A. Klobuchar, and Patricia H. Doherty, "Comparison of Real-Time Ionospheric Algorithms for a GPS Wide-Area Augmentation System," *Navigation: Journal of the Institute of Navigation* **41**(4), 393–413 (Winter 1994/1995).

[51] J. A. Farrell and M. Barth, *The Global Positioning System & Inertial Navigation*, McGraw-Hill, New York, 1998.

[52] Federal Aviation Administration (USA), *FAA* Specification WAAS FAA-E-2892 B, Oct. 1997.

[53] C. M. Feit, "GPS Range Updates in an Automatic Flight Inspection System: Simulation, Static and Flight Test Results," *Proceedings of the Institute of Navigation (ION) GPS-92*, ION, Alexandria, VA, Sept. 1992, pp. 75–86 .

[54] W. A. Feess and S. G. Stephens, "Evaluation of GPS Ionospheric Time Delay Algorithm for Single Frequency Users," *Proceedings of the IEEE Position, Location, and Navigation Symposium (PLANS '86)* (Las Vegas, NV), Nov. 4–7, 1986, New York, 1986, pp. 206–213. .

[55] W. A. Feess and S. G. Stephens, "Evaluation of GPS Ionospheric Time Delay Model," *IEEE Transactions on Aerospace and Electronic Systems* **AES-23**(3) (May 1987).

[56] R. D. Fontana et al., "The Modernized L2 Civil Signal," *GPS World* (Advanstar Communications, Eugene, OR) pp. 28–32 (Sept. 2001).

[57] M. E. Frerking, "Fifty Years of Progress in Quartz Crystal Frequency Standards," *Proceedings of the 1996 IEEE International Frequency Control Symposium*, IEEE, New York, 1996, pp. 33–46.

[58] L. Garin, F. van Diggelen, and J. Rousseau, "Strobe and Edge Correlator Multipath Mitigation for Code," *Proceedings of ION GPS-96, the 9th International Technical Meeting of the Satellite Division of the Institute of Navigation* (Kansas City, MO), ION, Alexandria, VA, 1996, pp. 657–664.

[59] A. Gelb (Editor), *Applied Optimal Estimation*, MIT Press, Cambridge, MA, 1974.

[60] G. H. Golub and C. F. Van Loan, *Matrix Computations*, 2nd ed., Johns Hopkins University Press, Baltimore, MD, 1989.

[61] GPS Interface Control Document ICD-GPS-200, Rockwell International Corporation, Satellite Systems Division, Revision B, July 3, 1991.

[62] *GPSoft Inertial Navigation System Toolbox for Matlab*, GPSoft, Athens, OH, 1998.

[63] R. L. Greenspan, "Inertial Navigation Technology from 1970–1995," *Navigation: Journal of the Institute of Navigation* **42**(1), 165–186 (1995).

[64] M. S. Grewal, "GEO Uplink Subsystem (GUS) Clock Steering Algorithms Performance and Validation Results," *Proceedings of 1999 ION Conference, Vision 2010: Present & Future, National Technical Meeting* (San Diego, CA), Jan. 25–27, ION, Alexandria, VA, 1999, pp. 853–859 .

[65] M. S. Grewal and A. P. Andrews, *Application of Kalman Filtering to GPS, INS, & Navigation*, Short Course Notes, Kalman Filtering Consulting Associates, Anaheim, CA, June 2000.

[66] M. S. Grewal and A. P. Andrews, *Kalman Filtering: Theory and Practice Using MATLAB*, 2nd ed., Wiley, New York, 2000.

[67] M. S. Grewal, W. Brown, S. Evans, P. Hsu, and R. Lucy, "Ionospheric Delay Validation Using Dual Frequency Signal from GPS at GEO Uplink Subsystem (GUS) Locations," *Proceedings of ION GPS '99, Satellite Division of the Institute of Navigation 12th International Technical Meeting*, Session C4, Atmospheric Effects, (Nashville, TN), Sept. 14–17, ION, Alexandria, VA, 1999.

[68] M. S. Grewal, W. Brown, and R. Lucy, "Test Results of Geostationary Satellite (GEO) Uplink Sub-System (GUS) Using GEO Navigation Payloads," *Monographs of the Global Positioning System: Papers Published in Navigation ("Redbook")*, Vol. VI, Institute of Navigation, ION, Alexandria, VA, 1999, pp. 339–348.

[69] M. S. Grewal, W. Brown, P. Hsu, and R. Lucy, "GEO Uplink Subsystem (GUS) Clock Steering Algorithms Performance, Validation and Test Results," *Proceedings of 31st Annual Precise Time and Time Interval (PTTI) Systems and Applications Meeting* (Dana Point, CA), Dec. 7–9, Time Services Dept., U.S. Naval Observatory, Washington, DC, 1999.

[70] M. S. Grewal, H. Habereder, and T. R. Schempp, "Overview of the SBAS Integrity Design," Proceedings of ION Conference, Graz, Austria, 2003.

[71] M. S. Grewal, N. Pandya, J. Wu, and E. Carolipio, "Dependence of User Differential Ranging Error (UDRE) on Augmentation Systems—Ground Station Geometries," *Proceedings of the Institute of Navigation's (ION) 2000 National Technical Meeting* (Anaheim, CA), Jan. 26–28, 2000, ION Alexandria, VA, 2000, pp. 80–91.

[72] M. S. Grewal, P. Hsu, and T. W. Plummer, "A New Algorithm for SBAS GEO Uplink Subsystem (GUS) Clock Steering," *ION GPS/GNSS Proceedings*, Sept. 2003, pp. 2712–2719.

[73] M. S. Grewal, M. S., W. Brown, R. Lucy, and P. Hsu, "GEO Uplink Subsystem (GUS) Clock Steering Algorithms Performance and Validation Results," *Proceedings of Institute of Navigation Vision 2010*, Jan. 25–27, 1999, pp. 853–859.

[74] M. S. Grewal, M. S., W. Brown, R. Lucy, and P. Hsu, "GEO Uplink Subsystem (GUS) Clock Steering Algorithms Performance, Validation, and Results," *Proceedings of 31st Annual Precise Time and Time Interval (PTTI) Systems and Applications Meeting*, U.S. Naval Observatory, Washington, DC, Dec. 1999, pp. 173–180.

[75] M. S. Grewal and E. Carolipio, "Comparison of GEO and GPS Orbit Determination," *Proceedings of ION/GPS 2002*, Portland, OR, Sept. 24–27, 2002, pp. 790–800.

[76] M. S. Grewal, L. R. Weill, and A. P. Andrews, *Global Positioning Systems, Inertial Navigation, and Integration*, Wiley, New York, 2001, pp. 71–76.

[77] *Gyros, Platforms, Accelerometers: Technical Information for the Engineer*, No. 2 (7th ed.), Kearfott Systems Division, Nov. 1967.

[78] L. Hagerman, *Effects of Multipath on Coherent and Noncoherent PRN Ranging Receiver*, Aerospace Report TOR-0073(3020-03)-3, Aerospace Corporation, Development Planning Division, El Segundo, CA, May 15, 1973.

[79] C. M. Harris and A. G. Piersol, *Harris' Shock and Vibration Handbook*, 2nd ed., McGraw-Hill, New York, 2002.

[80] B. Hassibi, A. H. Sayed, and T. Kailath, *Indefinite Quadratic Estimation and Control: A Unified Approach to* H^2 *and* H^∞ *Theories*, SIAM, Philadelphia, PA, 1998.

[81] H. V. Henderson and S. R. Searle, "On Deriving the Inverse of a Sum of Matrices," *SIAM Review* **23**, 53–60 (1981).

[82] T. A. Herring, "The Global Positioning System," *Scientific American* 44–50 (Feb. 1996).

[83] D. Herskovitz, "A Sampling of Global Positioning System Receivers," *Journal of Electronic Defense* 61–66 (May 1994).

[84] B. Hofmann-Wellenhof, H. Lichtenegger, and J. Collins, *GPS: Theory and Practice*, Springer-Verlag, Vienna, 1997.

[85] K. W. Hudnut and B. Tims, *GPS L1 Signal Modernization (L1C)*, The Interagency GPS Executive Board, Stewardship Project #204, sponsored by the Department of the Interior, U.S. Geological Survey, and the Navstar Global Positioning System Space and Missile Systems Center, July 30, 2004.

[86] P. Y. C. Hwang, "Recommendations for Enhancement of RTCM-104 Differential Standard and Its Derivatives," *Proceedings of the 6th International Technical Meeting of the Satellite Division of the Institute of Navigation (ION) GPS-93* (Salt Lake City, UT), ION, Alexandria, VA, Sept. 1993, pp. 1501–1508.

[87] *IEEE Standard for Inertial Sensor Terminology*, IEEE Standard 528-2001, Institute of Electrical and Electronics Engineers, New York, 2001.

[88] *IEEE Standard for Inertial System Terminology*, Preliminary IEEE Standard P1559/D29 (April 2006), Institute of Electrical and Electronics Engineers, New York (currently in preparation: expected publication: 2007).

[89] Institute of Navigation, *Monographs of the Global Positioning System: Papers Published in Navigation ("Redbook")*, Vol. I, ION, Alexandria, VA, 1980.

[90] Institute of Navigation, *Monographs of the Global Positioning System: Papers Published in Navigation ("Redbook")*, Vol. II, ION, Alexandria, VA, 1984.

[91] Institute of Navigation, *Monographs of the Global Positioning System: Papers Published in Navigation ("Redbook")*, with Overview by R. Kalafus, Vol. III, ION, Alexandria, VA, 1986.

[92] Institute of Navigation, *Monographs of the Global Positioning System: Papers Published in Navigation ("Redbook")*, with Overview by R. Hatch, Institute of Navigation, Vol. IV, ION, Alexandria, VA, 1993.

[93] Institute of Navigation, *Monographs of the Global Positioning System: Papers Published in Navigation ("Redbook")*, Vol. V, ION, Alexandria, VA, 1998.

[94] Institute of Navigation, *Global Positioning System, Selected Papers on Satellite Based Augmentation Systems (SBASs) ("Redbook")*, Vol. VI, ION Alexandria, VA, 1999.

[95] K. Ito, K. Hoshinoo, and M. Ito, "Differential Positioning Experiment Using Two Geostationary Satellites," *IEEE Transactions on Aerospace and Electronic Systems* **35**(3), 866–878 (1999).

[96] H. W. Janes, R. B. Langley, and S. P. Newby, "Analysis of Tropospheric Delay Prediction Models: Comparisons with Ray-Tracing and Implications for GPS Relative Positioning," *Bulletin Geodisique* **65**(3), 151–161 (1991).

[97] J. M. Janky, *Clandestine Location Reporting by a Missing Vehicle*, U.S. Patent 5,629,693, May 13, 1997.

[98] Javad Navigation Systems, *GPS Electronics Innovations and Features*, section titled "Multipath Reduction with Signal Processing," on Internet Website www.javad.com.

[99] A. H. Jazwinski, *Stochastic Processes and Filtering Theory*, Academic Press, San Diego, CA, 1970.

[100] G. E. Johnson, "Constructions of Particular Random Processes," *Proceedings of the IEEE* **82**(2), (1994).

[101] T. Kailath, A. H. Sayed, and B. Hassibi, "Kalman Filtering Techniques," in *Wiley Encyclopedia of Electrical and Electronics Engineering*, Wiley, New York, 1999.

[102] R. M. Kalafus, *Receiver Autonomous Integrity Monitoring of GPS*, Project Memorandum DOT-TSC-FAA-FA-736-1, U.S. DOT Transportation Systems Center, Cambridge, MA, 1987.

[103] R. M. Kalafus, A. J. Van Dierendonck, and N. Pealer, "Special Committee 104 Recommendations for Differential GPS Service," in *Global Positioning System*, Vol. III, Institute of Navigation, ION, Alexandria, VA, 1986, pp. 101–116.

[104] R. E. Kalman, "A New Approach to Linear Filtering and Prediction Problems," *ASME Transactions, Series D: Journal of Basic Engineering* **82**, 35–45 (1960).

[105] R. E. Kalman and R. S. Bucy, "New Results in Linear Filtering and Prediction Theory," *ASME Transactions, Series D: Journal of Basic Engineering* **83**, 95–108 (1961).

[106] P. G. Kaminski, *Square Root Filtering and Smoothing for Discrete Processes*, Ph.D. thesis, Stanford University, Stanford, CA, 1971.

[107] E. D. Kaplan, *Understanding GPS Principles and Applications*, Artech House, Boston, 1996.

[108] M. Kayton and W. L. Fried, *Avionics Navigation Systems*, 2nd ed., Wiley, New York, 1997.

[109] A. Kelley, *Modern Inertial and Navigation Satellite Systems*, Carnegie Mellon University Robotics Institute, Report CMU-RI-TR-94-15, Pittsburgh, 1994.

[110] A. D. King, "Inertial Navigation —Forty Years of Evolution," *GEC Review* **13**(3), 140–149 (1998).

[111] J. A. Klobuchar, "Ionospheric Time Delay Corrections for Advanced Satellite Ranging Systems," NATO AGARD Conference Proceedings 209, in *Propagation Limitations of Navigation and Positioning Systems*, NATO AGARD, Paris, 1976.

[112] D. T. Knight, "Demonstration of a New, Tightly-Coupled GPS/INS," *Proceedings of the 6th International Technical Meeting, Institute of Navigation, ION GPS-93* (Salt Lake City, UT), Sept. 22–24, 1993, ION, Alexandria, VA, 1993.

[113] D. T. Knight, "Rapid Development of Tightly-Coupled GPS/INS Systems," *IEEE AES Systems Magazine* 14–18 (Feb. 1997).

[114] D. Kügler, "Integration of GPS and Loran-C/Chayka: A European Perspective, *Navigation: Journal of the Institute of Navigation* **46**(1), 1–13 (1999).

[115] K. Lambeck, *The Earth's Variable Rotation,* Cambridge University Press, 1980.

[116] P. Langevin, "Sur la théorie du muovement brownien," *Comptes Rendus de l'Académie des Sciences* (Paris) **146**, 530–533 (1908).

[117] R. B. Langley, "A GPS Glossary," *GPS World* 61–63 (Oct. 1995).

[118] R. B. Langley, "The GPS Observables," *GPS World* 54–59 (April 1993).

[119] J. H. Laning, Jr., *The Vector Analysis of Finite Rotations and Angles*, MIT/IL Special Report 6398-S-3, Massachusetts Institute of Technology, Cambridge, MA, 1949.

[120] A. Lawrence, *Modern Inertial Technology: Navigation, Guidance, and Control,* 2nd ed., Springer-Verlag, New York, 1993.

[121] Y. C. Lee, "Analysis of Range and Position Comparison Methods as a Means to Provide GPS Integrity in the User Receiver," *Proceedings of the Annual Meeting of the Institute of Navigation* (Seattle, WA), June 24–26, 1986, pp. 1–4.

[122] H. Lefevre, *The Fiber-Optic Gyroscope,* Artech House, Norwood, MA, 1993.

[123] A. Leick, "Appendix G," *GPS: Satellite Surveying,* 2nd ed., Wiley, New York, 1995, pp. 534–537.

[124] Litton Aero Products Training Support Services, *Fundamentals of Inertial Navigation,* Litton Aero Products, Woodland Hills, CA, 1975.

[125] T. Logsdon, *The NAVSTAR Global Positioning System,* Van Nostrand Reinhold, New York, 1992, pp. 1–90.

[126] M. A. Lombardi, L. M. Nelson, A. N. Novick, and V. S. Zhang, "Time and Frequency Measurements Using the Global Positioning System," *Measurements Science Conference,* Anaheim, CA, Jan. 19–20, 2001, pp. 26–33.

[127] *Loran-C User Handbook,* Department of Transportation, U.S. Coast Guard, Commandant Instruction M16562.3, Washington, DC, May 1990.

[128] P. F. MacDoran (inventor), *Method and Apparatus for Calibrating the Ionosphere and Application to Surveillance of Geophysical Events*, U.S. Patent 4,463,357, July 31, 1984.

[129] D. Mackenzie, *Inventing Accuracy: An Historical Sociology of Nuclear Missile Guidance,* MIT Press, Cambridge, MA, 1990.

[130] A. Mannucci, B. Wilson, and C. D. Edwards, "A New Method for Monitoring the Earth's Total Electron Content Using the GPS Global Network," *Proceedings of ION GPS-93,* (Salt Lake City, UT), Sept. 1993, pp. 22–24.

[131] A. Mannucci, B. Wilson, and C. D. Edwards, "An Improved Ionospheric Correction Method for Wide-Area Augmentation Systems," *Proceedings of ION GPS-95,* Sept. 4, 1995.

[132] A. Mannucci and B. Wilson, "Instrumental Biases in Ionospheric Measurements Derived from GPS Data," *Proceedings of ION GPS-93,* Sept. 4, 1995.

[133] J. O. Mar and J.-H. Leu, "Simulations of the Positioning Accuracy of Integrated Vehicular Navigation Systems," *IEE Proceedings—Radar Sonar Navigation* **143**(2), 121–128 (1996).

[134] M. B. May, "Inertial Navigation and GPS," *GPS World* 56–65 (Sept. 1993).

[135] P. S. Maybeck, *Stochastic Models, Estimation and Control,* Vols. I and II, Navtech Seminars, Arlington, VA, 1994.

[136] P. S. Maybeck, *Stochastic Models, Estimation and Control* (3 vols.), Academic Press, New York, 1979.

[137] L. A. McGee and S. F. Schmidt, *Discovery of the Kalman Filter as a Practical Tool for Aerospace and Industry*, National Aeronautics and Space Administration Report NASA-TM-86847, 1985.

[138] G. McGraw and M. Braasch, "GNSS Multipath Mitigation Using Gated and High Resolution Correlator Concepts," *Proceedings of the 1999 National Technical Meeting and 19th Biennial Guidance Test Symposium,* Institute of Navigation, San Diego, CA, 1999, pp. 333–342.

[139] J. McGreevy, *Fundamentals of Strapdown Inertial Navigation,* Litton Training Manual TT110 (Revision C), Litton Aero Products, Moorpark (now Northrop Grumman in Woodland Hills), CA, 1989.

[140] J. C. McMillan and D. A. G. Arden, "Sensor Integration Options for Low Cost Position and Attitude Determination," *Proceedings of IEEE Position, Location and Navigation Conference (PLANS '94)* (Las Vegas, NV), IEEE, New York, 1994.

[141] P. Misra and P. Enge, *Global Positioning System: Signals, Measurements, and Performance,* Ganga-Jamuna Press, Lincoln, MA, 2001.

[142] R. Moreno and N. Suard, "Ionospheric Delay Using Only L1: Validation and Application to GPS Receiver Calibration and to Inter-Frequency Bias Estimation," *Proceedings of The Institute of Navigation (ION),* Jan. 25–27, 1999, ION, Alexandria, VA, 1999, pp. 119–129.

[143] M. Morf and T. Kailath, "Square Root Algorithms for Least Squares Estimation," *IEEE Transactions on Automatic Control* **AC-20**, 487–497 (1975).

[144] F. K. Mueller, "A History of Inertial Navigation," *Journal of the British Interplanetary Society* **38**, 180–192 (1985).

[145] NAVSYS Corporation, *WAAS Dual Frequency Ranging and Timing Analysis Design Report,* NAVSYS Corporation Report, Nov. 11, 1996, Colorado Springs, CO.

[146] Novatel, Inc., advertising insert in *GPS World Showcase,* Advanstar Communications, Eugene, OR, Aug. 2001, pp. 28–29.

[147] N. Pandya, "Dependence of GEO UDRE on Ground Station Geometries," in *WAAS Engineering Notebook*, Raytheon Systems Company, Fullerton, CA, Dec. 1, 1999.

[148] N. Pandya, J. S. Wu, E. Carolipio, and M. S. Grewal, "Dependence of GEO UDRE on Ground Station Geometries," *Proceedings of ION National Technical Meeting* (Anaheim, CA), Jan. 2000, pp. 80–90.

[149] B. C. Parker et al., "Overview of the GPS M Code Signal," *Proceedings of the ION 2000 National Technical Meeting* (Anaheim, CA), Jan. 2000, pp. 542–549.

[150] B. W. Parkinson and P. Axelrad, "Autonomous GPS Integrity Monitoring Using the Pseudorange Residual," *Navigation* (Institute of Navigation) **35**(2), 255–274 (1988).

[151] B. W. Parkinson and J. J. Spilker, Jr. (Eds.), *Global Positioning System: Theory and Applications*, Vol. 1, Progress in Astronautics and Aeronautics (series), American Institute of Aeronautics and Astronautics, Washington, DC, 1996.

[152] B. W. Parkinson and J. J. Spilker, Jr. (Eds.), *Global Positioning System: Theory and Applications*, Vol. 2, Progress in Astronautics and Aeronautics (series), American Institute of Aeronautics and Astronautics, Washington, DC, 1996.

[153] B. W. Parkinson, M. L. O'Connor and K. T. Fitzgibbon, "Aircraft Automatic Approach and Landing Using GPS," Chapter 14 in *Global Positioning System: Theory & Applications*, Vol. II, B. W. Parkinson and J. J. Spilker, Jr. (eds.), Progress in Astronautics and Aeronautics (series), Vol. 164, Paul Zarchan editor-in-chief, American Institute of Aeronautics and Astronautics, Washington, DC, 1995, pp. 397–425.

[154] S. Peck, C. Griffith, V. Reinhardt, W. Bertiger, B. Haines, and G. M. R. Winkler, "WAAS Network Time Performance and Validation Results," *Proceedings of the Institute of Navigation* (Santa Monica, CA), ION, Alexandria VA, Jan. 1998.

[155] J. E. Potter and R. G. Stern, "Statistical Filtering of Space Navigation Measurements, *Proceedings of the 1963 AIAA Guidance and Control Conference,* American Institute of Aeronautics and Astronautics, Washington, DC, 1963.

[156] K. D. Rao and L. Narayana, "An Approach for a Faster GPS Tracking Extended Kalman Filter," *Navigation: Journal of the Institute of Navigation,* **42**(4), 619–630 (1995/1996).

[157] H. E. Rauch, F. Tung, and C. T. Streibel, "Maximum Likelihood Estimates of Linear Dynamic Systems," *AIAA Journal* **3**, 1445–1450 (1965).

[158] Raytheon Systems Company, *GEO Uplink System (GUS) Clock Steering Algorithms*, ENB, Fullerton, CA. April 13, 1998.

[159] Raytheon Company, *Software Requirements Specification for the Corrections and Verifications Computer Software Configuration Item for the Wide Area Augmentation System,* CDRL Sequence No. A043-001D, Nov. 1, 1998.

[160] K. C. Redmond and T. M. Smith, *From Whirlwind to MITRE: The R&D Story of the SAGE Air Defense Computer* MIT Press, Cambridge, MA, 2000.

[161] I. Rhee, M. F. Abdel-Hafez, and J. L. Speyer, "Observability of an Integrated GPS/INS during maneuvers," *IEEE Transactions on Aerospace and Electronic Systems* **40**(2), 526–535 (2004).

[162] J. F. Riccati, "Animadversationnes in aequationes differentiales secundi gradus," *Acta Eruditorum Quae Lipside Publicantur Supplimenta* **8**, 66–73 (1724).

[163] D. Roddy, *Satellite Communications,* 2nd ed., McGraw-Hill, New York, 1989.

[164] P. Ross and R. Lawrence, "Averaging Measurements to Minimize SA Errors," Internet communication on AOL, Sept. 18, 1995.

[165] RTCA, *Minimum Operational Performance Standards (MOPS) for Global Positioning System/Wide Area Augmentation System Airborne Equipment*, RTCA/DO-229, Jan. 16, 1996, and subsequent changes, Appendix A, "WAAS System Signal Specification," RTCA, Washington, DC.

[166] RTCA, *Minimum Operational Performance Standards for Airborne Supplemental Navigation Equipment Using Global Positioning System (GPS),* Document RTCA/DO-208, Radio Technical Commission for Aeronautics, Washington, DC, July 1991.

[167] RTCA, *Minimum Operating Performance Standards for Global Positioning System, Wide Area Augmentation System*, Document RTCA/D0-229, Radio Technical Commission for Aeronautics, Washington, DC, 1999.

[168] O. Salychev, *Inertial Systems in Navigation and Geophysics,* Bauman Moscow State Technical University Press, Moscow, Russia, 1998.

[169] P. G. Savage, *Introduction to Strapdown Inertial Navigation Systems,* Vols. 1 and 2, Strapdown Associates, Maple Plain, MN, 1996.

[170] P. G. Savage, *Strapdown Analytics,* Strapdown Associates, Maple Plain, MN, 2000.

[171] T. Schempp et al., "WAAS Algorithm Contribution to Hazardously Misleading Information (HMI)," 14th International Technical Meeting of the Satellite Division of the Institute of Navigation, Salt Lake City, Utah, Sept. 11–14, 2001

[172] T. Schempp et al., "An Application of Gaussian Overbounding for the WAAS Fault Free Error Analysis," *Proceedings of ION GPS,* Portland, OR, Sept. 2002, pp. 766–772.

[173] S. F. Schmidt, "Application of State Space Methods to Navigation Problems," in *Advances in Control Systems,* Vol. 3, C. T. Leondes (Ed.), Academic Press, New York, 1966.

[174] M. Schuler, "Die Storung von Pendul-und Kreiselapparaten durch die Beschleunigung der Fahrzeuges," *Physicalische Zeitschrift* **B**, 24 (1923).

[175] K. P. Schwartz, M. Wei, and M. Van Gelderen, "Aided Versus Embedded: A Comparison of Two Approaches to GPS/INS Integration," *Proceedings of IEEE Position, Locations and Navigation Conference,* (Las Vegas, NV), IEEE, New York, 1994.

[176] J. Sciegienny, R. Nurse, J. Wexler, and P. Kampton, *Inertial Navigation System Standard Software Development*, Vol. 2, *Survey and Analytical Development,* Report R-977, C. S. Draper Laboratory, Cambridge, MA, June 1976.

[177] G. Seeber, *Satellite Geodesy: Foundation, Methods, and Applications*, Walter de Gruyter, Berlin, 1993.

[178] J. Sennott, I.-S. Ahn, and D. Pietraszewski, "Marine Applications" (U.S. government), Chapter 11 in *Global Positioning Systems: Theory & Applications*, Vol. II (see Ref. 98), pp. 303–325.

[179] G. M. Siouris, *Aerospace Avionics Systems,* Academic Press, London, 1993.

[180] T. Stansell, Jr., "RTCM SC-104 Recommended Pseudolite Signal Specification," in *Global Positioning System,* Vol. III, Institute of Navigation, 1986, pp. 117–134.

[181] S. H. Stovall, *Basic Inertial Navigation,* Technical Memorandum NAWCWPNS TM 8128, Naval Air Warfare Center Weapons Division, China Lake, CA, Sept. 1997.

[182] M. A. Sturza, "Navigation System Integrity Monitoring Using Redundant Measurements," *Navigation* **35**(4), 483–501 (1988/89).

[183] M. A. Sturza and A. K. Brown, "Comparison of Fixed and Variable Threshold RAIM Algorithms," *Proceedings of the 3rd International Technical Meeting of the Institute of Navigation,* Satellite Division, ION GPS-90 (Colorado Springs, CO), Sept. 19–21, 1990, pp. 437–443.

[184] X. Sun, C. Xu, and Y. Wang, "Unmanned Space Vehicle Navigation by GPS," *IEEE AES Systems Magazine* **11**, 31–33 (July 1996).

[185] P. Swerling, "First Order Error Propagation in a Stagewise Smoothing Procedure for Satellite Observations," *Journal of the Astronomical Society* **6**, 46–52 (1959).

[186] S. Thomas, *The Last Navigator: A Young Man, An Ancient Mariner, the Secrets of the Sea,* McGraw-Hill, New York, 1997.

[187] C. L. Thornton, *Triangular Covariance Factorizations for Kalman Filtering*, Ph.D. thesis, University of California at Los Angeles, School of Engineering, 1976.

[188] C. L. Thornton and G. J. Bierman, "Gram-Schmidt Algorithms for Covariance Propagation," *International Journal of Control* **25**(2), 243–260 (1977).

[189] D. H. Titterton and J. L. Weston, *Strapdown Inertial Navigation Technology,* Peter Peregrinus, Stevenage, United Kingdom, 1997.

[190] B. Townsend and P. Fenton, "A Practical Approach to the Reduction of Pseudorange Multipath Errors in a L1 GPS Receiver," *Proceedings of ION GPS-94, the 7th International Technical Meeting of the Satellite Division of the Institute of Navigation* (Salt Lake City, UT), Alexandria, VA, 1994, pp. 143–148.

[191] B. Townsend, D. J. R. Van Nee, P. Fenton, and K. Van Dierendonck, "Performance Evaluation of the Multipath Estimating Delay Lock Loop," *Proceedings of the National Technical Meeting,* Institute of Navigation, Anaheim, CA, 1995, pp. 277–283.

[192] P. Tran and J. DiLellio, "Impacts of GEOs as Ranging Sources on Precision Approach Category I Availability," *Proceedings of ION Annual Meeting,* Institute of Navigation, Alexandria, VA, June 2000.

[193] J. B.-Y. Tsui, *Fundamentals of Global Positioning System Receivers,* 2nd ed., Wiley, New York, 2005.

[194] E. Udd, H. C. Lefevre, and K. Hotake, (Eds.), *Fiber Optic Gyros: 20th Anniversary Conference,* SPIE, Bellingham, WA, 1996.

[195] A. J. Van Dierendonck, P. Fenton, and T. Ford, "Theory and Performance of Narrow Correlator Spacing in a GPS Receiver," *Proceedings of the National Technical Meeting,* Institute of Navigation, San Diego, CA, 1992, pp. 115–124.

[196] A. J. Van Dierendonck, S. S. Russell, E. R. Kopitzke, and M. Birnbaum, "The GPS Navigation Message," in *Global Positioning System,* Vol. I, ION, Alexandria, VA, 1980.

[197] H. L. Van Trees, *Detection, Estimation and Modulation Theory,* Part 1, Wiley, New York, 1968.

[198] M. Verhaegen and P. Van Dooren, "Numerical Aspects of Different Kalman Filter Implementations," *IEEE Transactions on Automatic Control* **AC-31**, 907–917 (1986).

[199] A. Wald, *Sequential Analysis*, Wiley, New York, 1947.

[200] G. Watt, T. Walter, B. DeCleene et al., "Lessons Learned in the Certification of Integrity for a Satellite-Based Navigation System," *Proceedings of the ION NTM 2003*, Anaheim, CA, Jan. 2003.

[201] L. Weill, "C/A Code Pseudoranging Accuracy—How Good Can it Get?" *Proceedings of ION GPS-94, the 7th International Technical Meeting of the Satellite Division of the Institute of Navigation* (Salt Lake City, UT), 1994, ION, Alexandria, VA, pp. 133–141.

[202] L. Weill, "GPS Multipath Mitigation by Means of Correlator Reference Waveform Design," *Proceedings of the National Technical Meeting*, Institute of Navigation (Santa Monica, CA), Jan. 1997, ION, Alexandria, VA, pp. 197–206.

[203] L. Weill, "Application of Superresolution Concepts to the GPS Multipath Mitigation Problem," in *Proceedings of the National Technical Meeting*, Institute of Navigation (Long Beach, CA), 1998, ION, Alexandria, VA, pp. 673–682.

[204] L. Weill, "Achieving Theoretical Accuracy Limits for Pseudoranging in the Presence of Multipath," *Proceedings of ION GPS-95, 8th International Technical Meeting of the Satellite Division of the Institute of Navigation* (Palm Springs, CA), 1995, ION, Alexandria, VA, pp. 1521–1530.

[205] L. R. Weill, "Fisher Information Matrix for Delay and Phase" and "Normalized P.S.D. for GPS Signals," unpublished notes, Aug. 28, 2002.

[206] L. R. Weill, "Conquering Multipath: The GPS Accuracy Battle," *GPS World* (Advanstar Communications, Eugene, OR), pp. 59–66 (April 1997).

[207] L. R. Weill, "The Next Generation of a Super Sensitive GPS System," *Proceedings of ION GNSS 2004, 17th International Technical Meeting of the Satellite Division of the Institute of Navigation* (Long Beach, CA), Sept. 21–24, 2004, pp. 1924–1935.

[208] L. Weill and B. Fisher, *Method for Mitigating Multipath Effects in Radio Ranging Systems*, U.S. Patent 6,031,881, Feb. 29, 2000.

[209] J. D. Weiss and D. S. Kee, "A Direct Performance Comparison Between Loosely Coupled and Tightly Coupled GPS/INS Integration Techniques," *Proceedings of 51st Annual Meeting, Institute of Navigation* (Colorado Springs, CO), June 5–7, 1995, ION, Alexandria, VA, 1995, pp. 537–544.

[210] J. D. Weiss, "Analysis of an Upgraded GPS Internal Kalman Filter," *IEEE AES Systems Magazine* 23–26 (Jan. 1996).

[211] *WGS 84 Implementation Manual*, Version 2.4, European Organization for the Safety of Air Navigation (EUROCONTROL), Brussels, Feb. 1998.

[212] C. E. Wheatley III, C. G. Mosley, and E. V. Hunt, *Controlled Oscillator Having Random Variable Frequency*, U.S. Patent 4,646,032, Feb. 24, 1987.

[213] W. S. Widnall and P. A. Grundy, *Inertial Navigation System Error Model*, Report TR-03-73, Intermetrics Inc., 1973.

[214] J. E. D. Williams, *From Sails to Satellites: The Origin and Development of Navigational Science*, Oxford University Press, 1992.

[215] W. Wrigley, "History of Inertial Navigation," *Navigation: Journal of the Institute of Navigation* **24** 1–6 (1977).

[216] C. Yinger, W. A. Feess, R. DiEsposti, A. Chasko, B. Cosentino, D. Syse,, B. Wilson, and B. Wheaton, "GPS Satellite Interfrequency Biases," *Proceedings of The Institute of Navigation* (Cambridge, MA), June 28–30, 1999, ION, Alexandria, VA, pp. 347–355.

[217] T. Yunk et al., *A Robust and Efficient New Approach to Real Time Wide Area Differential GPS Navigation for Civil Aviation*, NASA/JPL Internal Report JPL D-12584, 1995, Pasadena, CA.

[218] P. Zarchan and H. Musoff, *Fundamentals of Kalman Filtering: A Practical Approach*, Vol. 190, Progress in Aeronautics and Astronautics (series), American Institute of Aeronautics and Astronautics, New York, 2000.

INDEX

Global Positioning Systems, Inertial Navigation, and Integration, Second Edition, by M. S. Grewal, L. R. Weill, and A. P. Andrews